CLIO MEETS SCIENCE

EDITED BY

*Robert E. Kohler and
Kathryn M. Olesko*

O S I R I S | 27

A Research Journal Devoted to the
History of Science and Its Cultural Influences

Osiris

Series editor, 2002–2012

KATHRYN OLESKO, *Georgetown University*

Volumes 17 to 27 in this series are designed to dissolve boundaries between history and the history of science. They cast science in the framework of larger issues prominent in the historical discipline but infrequently treated in the history of science, such as the development of civil society, urbanization, and the evolution of international affairs. They aim to open up new categories of analysis, to stimulate fresh areas of investigation, and to explore novel ways of synthesizing major historical problems that demand consideration of the role science has played in them. They are written not only for historians of science, but also for historians and other scholars who wish to integrate issues concerning science into courses on broader themes, as well as for readers interested in viewing science from a general historical perspective. Special attention is paid to the international dimensions of each volume's topic.

Cover Illustration:

Historia, as depicted by Cesare Ripa, *Iconologia* (Padua, 1625), 304. Historia is represented as a woman in the fashion of Victory with wings. She looks behind her, to the past, as she writes in a book resting on Saturn, or Father Time, who carries a scythe symbolizing the harvest of everyone and everything to the dustbin of time.

OSIRIS   2012   SECOND SERIES   VOLUME 27

Introduction:
Clio Meets Science

by Robert E. Kohler* and Kathryn M. Olesko†

ABSTRACT

The introduction discusses the essays in the volume in terms of ongoing changes in the history of science. It addresses the challenges of going beyond the microstudy and the historiographical issues that need to be addressed when doing so: constructing appropriate categories of historical analysis, defining periodization, and casting subjects in relevant historical contexts.

Historians of science today work in a world of subjects and methodologies that is unprecedented in its abundance and diversity. The range of what we take to be our proper objects of study has vastly expanded in recent decades. Sciences once thought to be less scientific than the older canonic ones (and thus less worthy of historical inquiry) are now at the center of our interest. Connections between science and technology and medicine are now so dense as almost to erase the categorical boundaries. History of science has at the same time gone global and comparative across widely different cultures and historical periods. And the range of legitimate participants in science has expanded to include virtually anyone and everyone as historians examine vernacular culture, artisanal and citizen science, and the practices of everyday life. The range of our conceptual tools has likewise vastly diversified, partly from internal elaboration, partly by borrowing from the various branches of history, sociology and anthropology, philosophy (as formerly), and the arts. All this is well known and indicates a discipline enjoying an intellectual and social maturity.

Along with this diverse abundance, however, has come an undercurrent of malaise about how it is all to be managed and what it means for our communal future—a malaise sometimes expressed in public but more often in informal conversation and shoptalk. Abundance and diversity, while enriching our individual pursuits as scholars and teachers, have made it harder to define the shared purposes and communal identity that position the history of science in the wider world of scholarship and public life. It is not just the overload of information in an age of instant access

*Department of History and Sociology of Science, 303 Claudia Cohen Hall, University of Pennsylvania, Philadelphia, PA 19104; rkohler@sas.upenn.edu.
†Department of History, Georgetown University, Washington, DC 20057; oleskok@georgetown.edu.
We would like to thank the authors of the volume and the exceptional editors with whom we have worked—Rachel Kamins and Jennifer Paxton—for the intellectually stimulating and exciting experiences we have had in assembling, compiling, and writing this volume, the last in the series under Olesko's general editorship.

to everything. Scholars in every age have complained of overload, yet have managed to cope, usually by taking on ever-smaller units of specialized study or by devising technical methods of organizing ever-larger amounts of information—paper tools at first, nowadays machines. Information we can handle without too much trouble. The malaise—*crisis* would be too strong a word—is, rather, the side effect of specialization and microstudy: the increased difficulty of taking a larger, communal, view of the meaning of our particular concerns.

If we feel the lack of a common enterprise (and not everyone does), it is in part because historians of science have until relatively recently always had some large vanguard undertaking in which all (in principle) could partake: great chains of big ideas; progress in rationality and scientific method; science and democracy; paradigms and revolutions; the triumph of infrastructure; social construction of knowledge. Not every individual scholar bought into these vanguard intellectual enterprises, obviously, or even approved. And most in their turn—social construction is the exception—have been pecked to death by critics or gone out of fashion, together with the idea of "grand narratives" in general. Yet some general, communal frame was there for those who wanted to connect and contribute. Now, for better or worse, there is no such default frame of reference: we have many, or none, as scholars divide their work into ever-smaller, bounded bits.

For the better part of a generation the default practice of history of science has been the intensive, local microstudy. Arguably it has always been the common mode (peruse any volume of the *Isis* annual bibliography to dispel any doubts about that). But microstudy has not always been the esteemed analytic and literary form that it has been in our own time: not just a convenient practice but the one that best confers credibility and exhibits expertise. Histories on the grand scale are still being written, to be sure, and always will be. Still, our devotion to microstudy has made it harder than ever to engage with our muse, Clio, on a grander scale and in broadly interpretive ways. Abundance has liberated us from grand narratives—yet we cannot help wondering whether there should be a Next Big Thing, and if so, what it might be. Perhaps a cultural studies of science (or is that moment already past?) or (reflecting the current job market) a closer partnership with general history? A revived alliance with science and technology studies or, alternatively, with philosophy of science? Or possibly a new dalliance with historical geography?[1] Moves in all these directions have been made.

We are not forced by empirical abundance to overspecialize; that is just how we choose to deal with abundance—and with reason. For one thing, the social situation of professional academics nowadays strongly favors it, in particular our discipline's recent strong demographic growth. David Kaiser has compiled the statistical profile of this population boom: from 21 new PhDs a year in 1974 to about 50 in 1988 (an annual increase of about 2), to a peak of 189 new additions in 1996 (an annual increase of 14.5), followed by a slight deceleration but still an annual addition of about 160 credentialed colleagues per year between 1994 and 2003—and into a job market

[1] E.g., Peter Dear, "Cultural History of Science: An Overview with Reflections," *Sci. Tech. Hum. Val.* 20 (1995): 150–70; Dear and Sheila Jasanoff, "Dismantling Boundaries in Science and Technology Studies," *Isis* 101 (2010): 759–74; Lorraine Daston, "Science Studies and the History of Science," *Crit. Inq.* 35 (2009): 798–813; Peter Galison, "Ten Problems in History and Philosophy of Science," *Isis* 99 (2008): 111–24; Simon Naylor, ed., "Historical Geographies of Science," *Brit. J. Hist. Sci.* 38, no. 1 (2005).

that was growing slowly and absorbing only a fraction of these.[2] As Kaiser observes, there were two ways of dealing with this demographic squeeze: expand the range of subjects we can claim as our own or divide up older fields into ever-smaller special subfields. Historians of science did both, and the result was a field of disjointed interest groups and specialists who experienced few incentives to see big pictures or cultivate a generalist vision.[3] All those graduate students needing projects doable in ever more stringent time frames; all those junior faculty needing presentations and publications for promotions and tenure; escalating standards of empirical justification and theoretical sophistication plus more aggressive gatekeeping; and the need in a crowded and competitive world to be seen and heard, to stand out in a crowd. Life in a corporatized, professionalized (arguably overprofessionalized), competitive academic marketplace almost inevitably favors intensive microstudy. It is no accident that a period of demographic boom and bust was also one of abundance and diversity, and of scholarship on a small scale. It is how we live now.

Another reason for this pattern lies in the unusual life cycle of social construction—our most recent vanguard project. This project of "naturalizing" science (to use Steven Shapin's apt term) has in the space of a generation or two shown beyond question that science is a product of the society that creates and harbors it: not transcendent of contexts of production and use but always and deeply shaped by them.[4] This understanding was achieved not by abstract reasoning (though social construction was highly theorized) but empirically, by detailed study of local sites of knowledge making. These showed concretely how scientific findings were the products of particular local situations and communal practices with all their historical and social contingencies. Intensive case studies made the concept of naturalized science hard to doubt, or at least hard to argue with. And as the concept became common knowledge, so did its method of microstudy become common practice. It was accessible and applicable to any subject, and was easily taught and learned. It was an ideal practice for a period of demographic growth and pressure for predictable dissertations and quick and steady publication. The empirical method thus acquired a value in itself, independent of its original, epistemic, purpose. Microstudy became an all-purpose practice, whose aim was not to demonstrate that all knowledge is someone's, somewhere—that claim needs no further demonstration—but to pursue particular topics in depth for their own sake. And in that close-up view, large-scale patterns—the stuff of big-picture histories—tend to fade and disappear, or at least are harder to define with certainty in a blur of events and contingencies.

The unintended consequences of microstudy did not go unnoticed. In the early 1990s, at the very height of social construction, there began to be heard public reflections on the need to revive "big-picture" history of science or a "generalists' vision." Perhaps the first were Caspar Hakfoort's (1991) and the little group of exemplary essays organized by James Secord for the British Society for the History of Science (1993). Others followed, among them Kathryn Olesko's editorial introduction to the current series of *Osiris* on the intersection of history of science with themes from

---

[2] Kaiser, "Training and the Generalist's Vision in the History of Science," *Isis* 96 (2005): 244–51.
[3] Ibid., 248.
[4] Shapin, *Never Pure: Historical Studies of Science as if It Was Produced by People with Bodies, Situated in Time, Space, Culture, and Society, and Struggling for Credibility and Authority* (Baltimore, 2010). A good entry point into a vast literature is Jan Golinski's *Making Natural Knowledge: Constructivism and the History of Science* (New York, 1998).

general history (2002), Bernard Lightman's editorial introduction to the new "Focus" section of *Isis* (2004), Robert Kohler's "Focus" section on "the generalist vision" (2005), and Josep Simon and Néstor Herran's project of helping beginning and junior scholars to frame their work in big-picture ways (2008).[5] These and others call for more synthetic and comparative works, review essays, and up-to-date textbooks of history of science. Still others have called upon historians of science to write popular books—despite the pitfalls—in the hope of garnering some of the broad public recognition accorded to science journalists and popular science writers.[6] Works like Jan Golinski's constructivist historiography and John Pickstone's historical taxonomy of ways of knowing evince a desire for comprehensive, universal schemas of one sort or another.[7] And Nicholas Jardine calls on historians to transcend the straitened horizons of microsites and actors' categories by returning, with new tools and in our own terms, to the old quest for causal explanations of the major transformations in the sciences through history.[8]

THINKING BROADLY

Osiris 27 follows in this tradition of calls to transcend microstudy. However, our approach here is somewhat different from any of its predecessors. Our aim in this collection of articles is not to promote big-picture genres or to advocate some all-encompassing conceptual framework. It is, rather, to encourage big thinking about the small subjects of our everyday historical practice. Big thinking is not quite the same as big-picture thinking. It is not necessarily synthetic or comparative (though it can be); nor is it necessarily far ranging in time, space, or subject. The idea here is to develop the generalities that grow out of and remain rooted in specialized research on particular subjects, yet are also recognizably of use to scholars of other special subjects. Our brief to contributors to *Osiris* 27 was to avoid the usual big-picture genres, overtly programmatic schemes, literature reviews, projections of new subject areas, and historiography in the grand manner. Rather, we invited them to begin with their own special research interests and to expatiate and speculate on their larger meanings and connections.

The articles in this volume differ widely in subject matter and methodology—by intent—but share the common purpose of freeing historical imagination and bring-

[5] Hakfoort, "The Missing Syntheses in the Historiography of Science," *Hist. Sci.* 29 (1991): 207–16; Secord, ed., "The Big Picture," *Brit. J. Hist. Sci.* 26, no. 4 (1993); Olesko, "History and the History of Science *Redux*: A Preface," *Osiris* 17 (2002): vii–x; Lightman, editorial, *Isis* 95 (2004): 357–8; Kohler, ed., "Focus: The Generalist Vision in the History of Science," *Isis* 96 (2005): 224–51; Simon and Herran, "Introduction," in *Beyond Borders: Fresh Perspectives in History of Science*, ed. Simon and Herran (Newcastle, 2008), 1–26.

[6] David P. Miller, "The 'Sobel Effect': The Amazing Tale of How Multitudes of Popular Writers Pinched All the Best Stories in the History of Science and Became Rich and Famous while Historians Languished in Accustomed Poverty and Obscurity, and How This Transformed the World," *Metascience* 11 (2002): 185–200; John Gascoigne, "'Getting a Fix': The *Longitude* Phenomenon," *Isis* 98 (2007): 769–78.

[7] Golinski, *Making Natural Knowledge* (cit. n. 4); Pickstone, *Ways of Knowing: A New History of Science, Technology, and Medicine* (Chicago, 2001).

[8] Jardine, "World before Word," *Times Literary Supplement*, September 14, 2007, 10–1; Jardine, "Progress, Power, Cooperation and Topography: Stages towards Understanding How Science Happened," *Times Literary Supplement*, December 16, 2011, 3–4. See also Jardine, "Uses and Abuses of Anachronism in the History of the Sciences," *Hist. Sci.* 38 (2000): 251–70.

ing to light and to life the large aspects of special topics. They exemplify a kind of historical presentation that is less *big* picture—much less grand narrative—than it is *mid*-picture. It is history that begins in, and to an extent stays in, case studies, but that gives them a larger intellectual purview.

A premise of *Osiris* 27 is that scholars must make the most of the social world of scholarship in which they live. Specialized local studies will likely remain what most historians of science do most of the time. The practice is too deeply embedded in our history and too strongly adapted to the present sociology of careers to change any time soon. Nor is the distrust of all grand narratives—a legacy of the era of cultural pluralism—likely to yield to affection. There may be no next big centralizing thing but rather, as at present, a diverse abundance of middling, decentered things. Synthetic and popular histories and textbooks will be produced—by those who have the requisite abilities and opportunities. But for most historians of science such genres are occasional episodes in lives whose days are devoted to deep empirical study of some special topic. And thinking big about special subjects, unlike big-picture genres, is something that anyone and everyone can do, anytime. That, we submit, is the great merit of big-thinking history—if this is not too grandiose a term for something that is, after all, rather commonsensical. It requires no commitment to any grand system or next big thing. Nor does it require that we give up specialized case studies; only that we give them a larger intellectual space. A second premise of this volume is that the practices we most value should be those that are accessible without impediments to all practitioners. The articles here exemplify this kind of history: open to all through the particular interests of each, grassroots, participatory.

In fact, a trend to this kind of midscale history of science is already under way. *Isis* "Focus" sections and thematic volumes of *Osiris* have proved popular and influential. Thematic issues of journals and sessions of society meetings are now common—if often filled out with microstudies. Midlevel concepts of contemporary interest have been put forward—communication and circulation, place and space, publishing and print culture, and lay meanings and uses of science, among others—to take the place of big pictures past.[9] Broad-scope textbooks and books for general as well as expert readers, some very good, now appear with some regularity.[10] And studies of the categories, roles, and practices foundational to the pursuit of science, especially those issuing from Berlin's Max Planck Institute, begin to look like the elements of a

[9] On communication and circulation, see James A. Secord, "Knowledge in Transit," *Isis* 95 (2004): 654–72; Simon Schaffer, Lissa Roberts, Kapil Raj, and James Delbourgo, eds., *The Brokered World: Go-Betweens and Global Intelligence, 1770–1820* (Sagamore Beach, Mass., 2009). On space and boundaries, see Thomas F. Gieryn, *Cultural Boundaries of Science: Credibility on the Line* (Chicago, 1999); David N. Livingstone, *Putting Science in Its Place: Geographies of Natural Knowledge* (Chicago, 2003); Diarmid A. Finnegan, "The Spatial Turn: Geographical Approaches in the History of Science," *J. Hist. Biol.* 41 (2008): 369–88. On print culture, see Secord, *Victorian Sensation: The Extraordinary Publication, Reception, and Secret Authorship of "Vestiges of the Natural History of Creation"* (Chicago, 2000). On lay meanings, see Bernadette Bensaude-Vincent, "A Historical Perspective on Science and Its Others," *Isis* 100 (2009): 359–68; Jim Endersby, "Too Much of a Good Thing?" *Hist. Sci.* 47 (2009): 475–84; Jeremy Vetter, "Introduction: Lay Participation in the History of Scientific Observation," *Sci. Context* 24 (2011): 127–41.

[10] An exemplary midscale textbook is Peter Dear's *Revolutionizing the Sciences: European Knowledge and Its Ambitions, 1500–1700* (Princeton, N.J., 2009). Model books for both general and expert readers are Jim Endersby's *A Guinea Pig's History of Biology: The Plants and Animals Who Taught Us the Facts of Life* (London, 2007) and David Kaiser's *How the Hippies Saved Physics: Science, Counterculture, and the Quantum Revival* (New York, 2011).

midlevel historiographic synthesis taking shape.[11] Overall, the worried tone of previous calls for big-picture history is giving way to an air of confidence.

The problem of generalist history—that is, of bringing it back to life in some form or other—may thus prove less of a problem than it has seemed, because generalist and microhistory are not mutually exclusive but quite compatible. At least, they can be made so—as Lightman has done in his compact yet clear and comprehensive synthesis of the evolving changes in modern science in one of its epicenters, London, over the nineteenth century. The idea, as Lightman puts it, is to discover the connections between specialized studies, then lay out the larger patterns that emerge.[12] Here is the ideal of big-thinking history in practice. Keith Thomas's recent suggestion that the proper business of the humanities today is less *research* than *scholarship*—less digging out novelties than reinvigorating well-dug subjects with novel meanings—is evidence of a wider trend.[13] We will also, as this trend plays out, have to broaden and redefine what counts as "new."

So, although this volume of articles follows in the tradition of calls for regaining larger perspectives, its intent is not to lament what has been lost but to endorse and describe what is already being regained or, more precisely, reinvented. It is less consciousness raising than consolidation. *Osiris* 27 thus marks a shift in tone and attitude, and perhaps in the basic practice of our discipline. It is beginning to look as if the recurring laments about the demise of big-picture history may mark a transitional passage in the history of our field, from a regimen of communal vanguard projects to one of diverse, midlevel generalizing. Microstudy, though it seemed for a time to signal the demise of general or big-picture history, may in fact prove to be the foundation of a new kind of midpicture history. Lynn Nyhart, in her article in this volume, shows how zoologists in the late nineteenth century reconfigured their concepts of special and general, combining an older practice of specializing in a particular animal taxon with a newer one that dealt with structures and functions general to all taxa. Together these became the warp and weft, as Nyhart puts it, of a zoology that was seamlessly special and general; and she speculates that historians of science are at an analogous point in their own history, of rethinking inherited concepts of general and special and devising a kind of history that values both the particularity of phenomena and the general patterns that each displays. Time will tell.

Meanwhile, it is not jeremiad and exhortation that are needed, but exemplars of a historical practice that reconnects special and general in ways that suit an age of abundance, diversity, and informatics. That is the premise, and the promise, of *Osiris* 27.

---

[11] E.g., Lorraine Daston and Peter Galison, *Objectivity* (New York, 2007); Daston and Otto Sibum, eds., "Scientific Personae and Their Histories," *Sci. Context* 16, nos. 1–2 (2003); Daston and Elizabeth Lunbeck, eds., *Histories of Scientific Observation* (Chicago, 2011).

[12] Bernard Lightman, "Refashioning the Spaces of London Science: Elite Epistemes in the Nineteenth Century," in *Geographies of Nineteenth-Century Science*, ed. David N. Livingstone and Charles W. J. Withers (Chicago, 2011), 25–50; Lightman, e-mail message to Kohler, 2 December 2011.

[13] Thomas, "What Are Universities For?" *Times Literary Supplement*, May 7, 2010, 13–5. In the same spirit Adam Gopnik has observed that historians' premier task is to both "see small and think big." "History helps us to understand reality by disassembling the big nouns into the small acts that make them up," he writes. "But if history ignores its responsibilities to the big nouns it isn't doing its job." It becomes "all nuance and no news." Gopnik, "Inquiring Minds: The Spanish Inquisition Revisited," *New Yorker*, January 16, 2012, 70–5, on 73. In fact, big nouns do seem to be making a comeback: e.g., Barbara Weinstein, "History without a Cause? Grand Narratives, World History, and the Postcolonial Dilemma," *Int. Rev. Soc. Hist.* 50 (2005): 71–93. That is news.

The articles in this volume display a range of historical texture from general historiography to case studies of some empirical density. Those closer to particular case studies will be discussed now; the general ones follow in the second, historiographical part of the introduction. The first four articles in the case-study group deal with the external relations of science: subjects include historical practices *within* certain sciences (Lorraine Daston); expertise and the dual logics of its public authority (Thomas Broman); the "thinning" of science for use in the public sphere, and its consequences (Theodore Porter); and the concepts of citizenship that guide scientists' varied relations with state powers (Fa-ti Fan). The next three focus on internal process and organization: how concepts of general and special knowledge shape status relations and change (Lynn Nyhart); how importing field-science practices into lab science reshapes that important social boundary (Bruno Strasser); and how demographic bubbles affect scientists' choices of what to do and how to do it (David Kaiser). Accepted categories are reconfigured. Familiar historiographic topics are given new meanings. Issues on the fringe of historical consciousness become centers of attention.

Since the mid-nineteenth century a categorical boundary has separated humanistic disciplines—which rework old texts in libraries by reading and comment—from natural sciences, which produce new facts in labs by observation and measurement. Yet early modern practices of reading and observing have been shown to be remarkably alike. And, Daston proposes, historical and bookish practices are characteristic of modern science as well, especially of a group that she designates "sciences of the archive." This group is defined less by historical subjects (e.g., evolution) than by empirical practices that depend on abundant data collected by many hands over long periods of time (as in astronomy, taxonomy, or demography) and use methods of analysis that are as much those of history and archives as those of science and labs. With this composite, functional category it is a short step from the particulars of reading history to a wide-open rethinking of disciplinary identities and interconnections that previous categories made (and were meant to make) unthinkable.

Expertise is another such category of concept and practice. Broman opens his re-examination of expertise with the observation that while public belief in science is firm, trust in scientists has become shaky. This paradox arises, he argues, because expertise is justified in two distinct ways: as occupational skilled labor, whose worth can be measured and is bought and sold; and as the ideology of those who possess general principles, whose worth cannot be measured and can only be taken on trust. Taken separately each rationale is familiar and unproblematic; historically, however, they are not separate but variously confused or combined—often simply conflated—with untoward consequences for both science and public life. These shifting categorical entanglements, Broman argues, are the key to understanding the power and the fragility of expertise in modern mass society. Comparing distant yet surprisingly similar experiments in mass education, one from the French Revolution, the other from Cold War America, he shows how a general history of expertise might be conceived.

Porter begins his meditation on the uses and abuses of scientific reasoning and justification with a similar paradox: that while science in practice is thickly descriptive (local, skilled, culturally situated), in the world of business and state administration it is stripped of context, simplified, and reduced to what can be measured and used without being deeply understood. Science is powerful because it travels fast and crosses boundaries easily, and it travels best when it travels light—or thin. And thinned, it thins the world and science itself, to ends both good and ill. As Broman sets us the task of refiguring the intellectual and social dynamics of learned and occupational expertise, so Porter sets us to rethinking the transformations of expert roles and expertise at the public interfaces of science, where knowledge and know-how acquire new uses and new powers. Porter maps the social topography in which the games of expertise are played and the consequences of its thinning revealed.

Fan's fresh look at the relations of scientists to states draws more on comparative and political history. His particular subject is the mass science of Mao's China as exemplified in the mobilization of rural Chinese to monitor signs of incipient earthquakes (e.g., anomalous behavior in animals). This phenomenon is in Fan's view not a case of Chinese exceptionalism—once the default view—but the manifestation of one version of a concept of citizenship that in varied forms is an essential operating principle of all modern (and modernizing) nation states. The case of science in Mao's China thus leads us to rethink science-state relations in the West not as a universal model but as a manifestation of the West's singular conceptions and practices of citizenship and state building. In an Old World picture a New World meaning opens up, and with the category of citizenship at the center of both it is no great leap from one to the other.

Nyhart makes a similar move in reopening the history of the concepts of general and special in nineteenth-century German zoology. These uncontentious epistemic categories became more freighted and urgent in the age of university reform and discipline building, when the ideology of *Wissenschaft* valued general knowledge above all else, while the imperative of deep empirical research demanded and valorized specialization. Concepts of general and special thus guided the competition for top rungs of the academic pecking order. And changes in these categories between the 1870s and 1910s were foundational to the dramatic changes in the organization and practices of science that constitute modernity. Historiographic categories long accepted as defining that great transformation—laboratories, experimental method, reductive theory, and so on—become in Nyhart's view secondary aspects of the master principles of general and special. If she is right, then familiar histories of modern science stand to be revisited and revised.

Strasser revisits another long-accepted historiographic view: that the experimental sciences are a separate domain owing nothing to natural history. From his special case—of molecular biologists adopting natural-history practices of collecting and curating to create vast databases of DNA sequences and protein structures—he draws a general view: that the categorical distinction between lab science on the one hand and society and the field on the other is not a separating wall but regulates an active two-way traffic between different domains of practice. There is more of the field in lab science than has been thought, and standard histories of experimental life science, Strasser foresees, stand in need of substantial revision. What Nyhart does with biologists' conceptions of general and special, Strasser does with their practices of collecting and curating. The full consequences of these revisions remain to be worked out.

Kaiser's case study of manpower bubbles in Cold War physics arrives at a similar view by a third approach, via the social dynamics of demographic growth in science. How these demographic cycles work is no mystery: like all economic bubbles they are driven by exuberance (and panic) and media amplifiers, way beyond the realities of supply and demand. Less understood is how these cycles of recruitment affect the intellectual life of science. Bubbles foster hyperspecialization—that is clear enough. Kaiser suggests that they also powerfully influence individuals' choices of subjects and subfields. In times of expansive growth, subjects will be preferred that provide a ready supply of topics suitable for PhD dissertations and publications, Kaiser argues, citing the cases of particle and solid-state physics—as well as of historians' flight from intellectual to social history. The serial fashions of big-picture histories of science may be another such case, and it is not hard to think of many more. Midlevel social phenomena like bubbles, Kaiser concludes, are broad roads back to "mesoscopic" histories of science that depend on, yet transcend, intensive local study. In their varied ways, all the articles in this volume do that.

THE CHALLENGES OF HISTORY

Clio Meets Science joins a chorus of scholars addressing the challenges of history, especially the challenge of thinking more broadly *and* more historically. Lynn Nyhart's opening statement as President of the History of Science Society expressed the challenge in terms of expanding our tolerance of ways of representing the past: to adopt more accessible ways of writing, to prepare ourselves to address the significance of the history of science to larger and different audiences, and to accept the polycentric intellectual universe in which we live.[14] Across the pond, where there has been a special interest in the popular success of Dava Sobel's *Longitude,* Sophie Waring has warned of the dangers of criticizing Sobel for her shortcomings as a historian. Waring advocated an "if you can't beat 'em, join 'em" attitude, arguing that historians of science had a lot to gain by adopting the storytelling elements of a narrative, if held in check by scholarly historical standards.[15] Especially germane to the historiographical concerns of this volume are the broad-based and penetrating historiographical contributions to the 2011 meeting of the Gesellschaft für Wissenschaftsgeschichte, which are certain to cause ripples throughout the field. One in particular stands out. Katherina Kinzel cast a critical eye on historical epistemology as practiced largely, but not exclusively, at the Max Planck Institute for the History of Science in Berlin. She condemned its near total abandonment of historical context; its sharp separation of micro- and macrohistorical perspectives; and its neglect of political, social, ethical, and other historical dimensions of science. The result, she claimed, is a type of historical explanation that has "problematic consequences," chief among them the reproduction of an internal history of science and the production of a history that does not address causal factors.[16] Not everyone will agree with her analysis. But her stance just might be indicative of the transparency and gravity with which a younger

[14] Nyhart, "History of Science Unbound," *Newsletter of the History of Science Society* 41 (January 2012), 1–2, 4.

[15] Waring, "Who Should Tell the Story of Longitude?" *Viewpoint: Newsletter of the British Society for the History of Science,* no. 95 (June 2011), 6; Sobel, *Longitude: The True Story of a Lone Genius Who Solved the Greatest Scientific Problem of His Time* (New York, 1995); Gascoigne, "'Getting a Fix': The *Longitude* Phenomenon," (cit. n. 6), and more broadly on the historical sins of popular writers without training in the history of science, Charles Gillispie, "The Distorted Meridian," *Isis* 98 (2007): 788–95, on Denis Guedj's *The Measure of the World: A Novel,* trans. Arthur Goldhammer (Chicago, 2001).

[16] Kinzel, "Geschichte ohne Kausalität: Abgrenzungsstrategien gegen die Wissenschaftssoziologie in zeitgenössischen Ansätzen historischer Epistemologie," *Ber. Wissenschaftsgesch.* 35 (2012): 147–62. We thank Paul Forman for bringing this article to our attention.

generation will take up historiographical principles as a matter of course in the practice of a history of science that is more historically inclusive.[17]

Historians of science are not alone in sounding the clarion call for higher levels of synthesis. The growing popularity and relevance of environmental history have contributed most to making synthesis the sine qua non of historical scholarship, as demonstrated by the works of historians like Mark Fiege, John McNeill, and Joachim Radkau.[18] Historians of science who also work in the field of environmental history invoke global themes necessarily and naturally, and their practices are sure to fold back into the history of science.[19] Indeed environmental historians have been vociferous in their promotion of large-scale histories, yet not to the point of denying the importance of microstudies. As William Cronon, dean of American environmental history and president of the American Historical Association, sagaciously put it: "The very best histories take their flight on both wings equally."[20] And if the predilections of environmental historians do not instill in us the virtues of thinking more broadly, those of general historians will. Of late the latter have discovered and exploited the significance that science, scientists, and engineers can add to their own visions of history, as have the Germanist Martin H. Geyer in his study of universal time, the Middle Easternist Osama Abi-Mershed in his examination of colonial rule in Algeria, and the Russianist Catherine Evtuhov in her imaginative exploration of how the "idea of province" emerged in Russia.[21] Can we not meet them halfway, by embedding our own work more directly in more expansive historical contexts, before history departments draw the conclusion that it might not be necessary to hire a historian of science when so many members of the department incorporate science in their writings as a matter of course?

It is thus timely that this volume considers the fundamentals of historiography and examines some of the most important conceptual foundations and assumptions in the field. Scaling up from the microstudy presents historiographical challenges, most obviously the reorganization and reshuffling of historical context. Three issues in particular beg for deeper examination. How should categories of historical analysis be

[17] Other recent articles on the topic notwithstanding. Dear and Jasanoff, "Dismantling Boundaries in Science and Technology Studies" (cit. n. 1).

[18] Fiege, *Republic of Nature: An Environmental History of the United States* (Seattle, 2012); McNeill, *Mosquito Empires: Ecology and War in the Greater Caribbean, 1620–1914* (Cambridge, 2010); Radkau, *Nature and Power: A Global History of the Environment* (Cambridge, 2008). Indeed the leading international center for environmental history, the Rachel Carson Center in Munich, actively cultivates connections with historians of science. The exchange is likely to have consequences, all good, for both sides.

[19] James Rodger Fleming, Vladimir Jankovic, and Deborah R. Coen, eds., *Intimate Universality: Local and Global Themes in the History of Weather and Climate* (Sagamore Beach, Mass., 2006); Fleming and Jankovic, guest eds., *Klima*, vol. 26 of *Osiris* (2011); Naomi Oreskes and Erik M. Conway, *Merchants of Doubt: How a Handful of Scientists Obscured the Truth on Issues from Tobacco Smoke to Global Warming* (New York, 2010).

[20] Cronon, "Breaking Apart, Pulling Together," *Perspectives on History* 50, no. 5 (2012): 5–6, on 6.

[21] Geyer, "Prime Meridians, National Time, and the Symbolic Authority of Capitals in the Nineteenth Century," in *Berlin–Washington, 1800–2000: Capital Cities, Cultural Representation, and National Identities*, ed. Andreas W. Daum and Christof Mauch (Cambridge, 2001), 79–100; Abi-Mershed, *Apostles of Modernity: Saint-Simonians and the Civilizing Mission in Algeria* (Palo Alto, 2010); Evtuhov, *Portrait of a Russian Province: Economy, Society, and Civilization in Nineteenth-Century Nizhii Novgorod* (Pittsburgh, 2011). Our impression is, too, that historical journals like the *American Historical Review, Journal of Interdisciplinary History, Journal of Modern History,* and others have in recent years reviewed more books in the history of science than a decade or more ago. If correct, this trend is a further indication of the penetration of history of science into general history.

defined and constructed? What is the appropriate range for historical periodization? And finally: What kinds of historical subjects can be selected, and how should they be framed contextually? These three issues—categories of analysis, periodization, and subjects—are the métier of reflective historiography. Although general historians have examined them for generations, only recently have historians of science taken them up consistently.

Take categories of analysis, a tool frequently discussed by general historians.[22] A landmark example of this genre is Joan W. Scott's examination of the term "gender" when women's history was taking hold, but still challenged, in the historical profession.[23] As categories of historical analysis, gender and science are similar to one another. Both demand a rejection of essentialism that stems from a limited descriptive approach and that promotes a deterministic point of view. And both, at one time or another, were strongly shaped by two major ideologies, Marxist and Weberian, that predetermined causal elements (economics and rationality, respectively) as well as the contours of their historical narratives (class struggle and the march of progress, respectively).

The advantages for historical scholarship that Scott attributed to refining gender conceptually apply as well to the examination of science as a category of analysis. Such an analytical exercise turns historical scholarship on its head by forcing "a critical reexamination of premises and standards of existing scholarly work," changing the terms and bases for judging historical significance and even asking historical questions, and finally, offering a "synthesizing perspective that can explain continuities and discontinuities."[24] More bluntly: reexamining categories of analysis is essential for the continued health of historical scholarship because it compels us to rethink how we do things. Like women's history in the 1980s, the history of science still plays a marginal role in the historical field as a whole, as demonstrated by its continued poor integration into history textbooks and syllabi. To paraphrase Scott, it is simply not enough for historians of science to demonstrate either that science has a history or that science and scientists were important factors in the history of civilizations.[25] Nor is it entirely a matter of defining what "science" is in itself as a body of ideas and methods or as a framework for certain kinds of institutional practices. It is also a matter of asking what science is in relation to the other elements of human life, including power, social relations and organization, economic development, institutional frameworks, international affairs, cultural production, and—to hark back to environmental history—even humanity's ecological relationship to nature.

What, now, about the category "science"? We no longer apply the term "science" to all we study—there's alchemy, chymistry, natural philosophy, and pseudoscience,

[22] As can be seen by perusing recent decades of the *American Historical Review*. We certainly do not claim here that historians of science have not created or analyzed categories of historical analysis (for they have), only that the practice is more common in general history, where perforce the relationships between categories and historical contingencies are more explicitly developed. Historians of science at the Max Planck Institute for the History of Science in Berlin have contributed significantly to the articulation of certain categories, such as scientific persona, rationality, objectivity, and others, but connections to historical contingencies are less in evidence. For a lucid and accessible discussion of categories of analysis in the history of science, see the CBC radio series "How to Think about Science," http://www.cbc.ca/ideas/episodes/2009/01/02/how-to-think-about-science -part-1---24-listen/ (accessed 15 July 2012).

[23] Scott, "Gender: A Useful Category of Historical Analysis," *Amer. Hist. Rev.* 91 (1986): 1053–75.

[24] Ibid., 1054, 1055, 1066.

[25] Ibid., 1055.

to name just a few of the alternative choices—and yet it remains the core concept of the field. Over the twentieth century the field moved considerably far away from iconoclastic and rigid definitions of science. Yet what is "science" as a category of historical analysis?

In a fashion similar to Scott, Jan Golinski uses the history of the field to explicate the origins, trajectory, and eventual demise (but not disappearance) of the idea of a "singular science": the idea that there are universal principles, beliefs, values, and assumptions that are common to all branches of science. The long-term continuity in the idea of a singular science is a central thread in his strategy, a thread woven with the strands of positivism that, far from disappearing at the end of the nineteenth century, were resurrected and rehabilitated by founding fathers of the field, especially George Sarton, and by prominent commentators on science in the modern world, like C. P. Snow, who tied science's unity to the narrative of incessant progress. Hence tightly woven together were both the assumptions and definition of the central category of analysis, science, and the framework and message of the narrative that mapped out its historical path.

But at least from the first half of the twentieth century, Golinski argues, that thread was tangled. In the interwar period Gaston Bachelard and Ludwik Fleck, among others, wrote about the sciences in ways that exposed their cultural dependencies, and so their local natures. And although Marxist historians generally maintained a belief in the singularity of science, Joseph Needham ironically acted as an intellectual beacon, as he promoted the diversity and cultural nature of the sciences. His landmark studies of Chinese science challenged the idea of European cultural hegemony in the sciences, weakening the related assumption that science was an instrument of power in the European arsenal. Studies like Needham's thus opened up a path to a set of beliefs associated with multiculturalism at the end of the twentieth century when local studies—inspired by intellectual innovations like Thomas Kuhn's *Structure of Scientific Revolutions,* social studies of scientific knowledge, and social constructivism—shattered the notion of a singular science. In the end, however, we are left with a quandary: local studies are popular and they have shattered the idea of a singular science, yet the specter of a unified science lingers in the corners. Golinski claims we still have unfinished business: to explain the geographical range of the sciences, their institutional diversity, and the nature of their relation to technology.[26]

Peter Dear, too, admits that there is no singular science. But rather than using the history of the field as a reservoir for understanding what science is, Dear turns to the deeper past of the early modern period because it offers both an "anthropological strangeness" and a venue for appreciating and gauging historical contingency. The result, he claims, can be described in terms similar to Niels Bohr's notion of complementarity. Science in its modern form appears to be a duality: it is both natural philosophy (in the sense of a contemplative understanding of nature) and a form of instrumentality (with technical and applied capabilities). Dear is careful, however, not to succumb to either a nominalist or essentialist point of view. Just as the port-

---

[26] To this list might be added the need to take up more closely definitions of science in the eighteenth century, particularly on the continent. A foray into this issue that is complementary to Golinski's approach and argument is Paul Ziche, "Science and the History of Science: Conceptual Innovations through Historicizing Science in the Eighteenth Century," *Ber. Wissenschaftsgeschichte* 35 (2012): 99–112. Ziche argues that writing the history of science in the eighteenth century interacted dialectically with the evolution of the definition and constitution of science.

manteau concept "science" is dependent upon historical contingencies, so too are its components.

Yet in this essay Dear is less concerned with articulating those contingencies than with understanding their relationship over time. There are deep and complex historical problems associated with each. They are characterized by opposed purposes, values, and discourses that beg for elaboration. Their apparently peaceful coexistence has to be analyzed and explained historically (perhaps by what Dear calls a "situated functionalism," but certainly not by "technoscience," which he claims obscures the historical processes at work). They exist in a tense relationship with one another, and that tension is captured in the historical record and must be acknowledged and analyzed by historians. Most importantly, like the elusive electron with its wave and particle properties, they can only be explained together: we simply cannot escape one or the other. Whether this complementarity is culturally grounded in the West or characterizes all scientific cultures is an open question for Dear; but the act of isolating these components of science allows us to approach the historical embeddedness of science from a new angle. This complementarity provides an axis along which historical evidence can be identified, weighed, and analyzed; generates questions that structure historical inquiry; and supplies the driving elements of a narrative wedded neither to a notion of progress nor to present definitions of science. The relationship between the two is historically contingent, as is their composition, and both change over time. Thus, in Dear's view, the history of science is the history of an ideology, the ideology of modern science.[27]

Paul Forman's essay both runs against the grain of and sits comfortably within the message of this volume. The latest in a series of essays addressing modernity and postmodernity as historical epochs, his essay unpacks—sometimes in jarring ways—the beliefs, values, and assumptions that undergird these epochs, currently the last two in a chain that runs from one master trilogy—antiquity, the Middle Ages, and the Renaissance—to the next—early modern, modern, postmodern. In one sense his essay is about periodization—about how we divide time spans meaningfully by demonstrating the differences in what makes them up. On a small scale we already do this automatically. For instance, the dates 1933, 1945, 1956, 1968, 1989, and 9/11 conjure up a set of cultural associations, political forces, and overall, a sense of historical significances and a confidence in them that are indisputable. Some dates are more complex, however. What is the nineteenth century? What is the twentieth? Both centuries have both long (1789–1914, 1860s–1989) and short (1806–1890s, 1914–1989) designations. Which one is correct? Why choose one range of dates over another? What historical factors hold a time span together? Historians debate these boundaries and must justify their choices with evidence and convincing arguments. So for David Blackbourn, the long nineteenth century for Germany is 1780 to 1914 for political reasons; for Charles Maier, the long twentieth century runs from the

---

[27] A position similar to one Paul Forman has argued. Forman, "The Primacy of Science in Modernity, of Technology in Postmodernity, and of Ideology in the History of Technology," *Hist. & Tech.* 23 (2007): 1–152; Forman, "(Re)cognizing Postmodernity: Helps for Historians—of Science Especially," *Ber. Wissenschaftsgesch.* 33 (2010): 157–75; and Forman's essay in this volume. A growing challenge to the field, not addressed in this volume, is how the history of science will fare as the categories of a "knowledge economy" and a "knowledge society" take hold. For early discussions of both, see Joel Mokyr, *The Gifts of Athena: Historical Origins of the Knowledge Economy* (Princeton, N.J., 2004), and Jakob Vogel, "Von der Wissenschafts- zur Wissensgeschichte: Für eine Historisierung der 'Wissensgesellschaft,'" *Gesch. Gesell.* 30 (2004): 639–60.

1860s to 1989 for territorial reasons; and for Giovanni Arrighi, the long twentieth century runs over seven hundred years for economic reasons.[28] The dates all depend on what kinds of problems you want to solve.

Yet periodization—the naming of historical epochs—is less about numbers than what these periods constitute. They are constructed by historians. They are ideologically loaded, and more often than not, they include unacknowledged assumptions. Speak of the medieval, and the sacred and the feudal come to mind; for the modern, the association is with secularism and capitalism. Epochs are about the politics of time in the broad sense: the medieval/modern distinction, for instance, has been used for generations to justify imperial civilizing missions.[29] Forman addresses the temporal and cultural distinctiveness of modernity and postmodernity (no easy task) by discerning both a temporal divide (1960s–1970s) and a set of traits and values that distinguish them: from discipline (in the characterological sense), proceduralism (an emphasis on means), disinterestedness, autonomy, and solidarity in modernity to the rejection or inversion of all of these in postmodernity. His evidence is sweeping, and his argument encapsulates a good part of the intellectual history of the United States in the postwar period.

The shift from modernity to postmodernity, however, did not involve merely a transformation in personal values, although that transformation is in itself worth including in the historian's tool box for the analysis of intellectual actors in the twentieth century. The deeper levels of his argument concern historians of science and what they do—or do not do. For in the transition from modernity to postmodernity, Forman finds the reasons for why we practice history of science the way we have, from Sarton's sequestering of the field's boundaries for professional reasons to our acceptance in the early twenty-first century not only of what science popularizers do, but also of the way they write. In that transition he isolates a momentous intellectual shift: from the focus on disciplinarity in modernity to an antidisciplinarity in postmodernity. Yet his analysis runs deeper still and addresses directly why we use or do not use certain categories of historical analysis, thus providing the historical context and reasons for our own practices. He chides us for failing to take up the distinction between discipline and profession in any great detail (we who were so suited to do so) and for several other missed opportunities that he believes did a disservice to the histories we write. His bottom line is of relevance to this volume and to the general thrust of recent issues of *Osiris*. There is a conflict, he claims, between history (the umbrella profession of which we are a part) and science (the object of our analysis). History as a practice is broad ("a seamless web," he calls it), while science (as we see it and as it is most often practiced) has for the most part signified discipline, autonomy, and boundary. Cognitive dissonance and incompatibility were inevitable, Forman concludes. To which we might add: so was a low, or at least belated, integration of historical scholarship into history of science, and vice versa.

---

[28] Blackbourn, *The Long Nineteenth Century: A History of Germany, 1780–1918* (Oxford, 1998); Maier, "Consigning the Twentieth Century to History: Alternative Narratives for the Modern Era," *Amer. Hist. Rev.* 105 (2000): 807–31; Arrighi, *The Long Twentieth Century: Money, Power, and the Origins of Our Times* (New York, 1994).

[29] For a deeper, if not dense, examination of the medieval/modern distinction, see Kathleen Davis, *Periodization and Sovereignty: How Ideas of Feudalism and Secularization Govern the Politics of Time* (Philadelphia, 2008). On modernity, see Michael Saler, "Modernity and Enchantment: A Historiographic Review," *Amer. Hist. Rev.* 111 (2006): 692–717, and the "*AHR* Roundtable: Historians and the Question of 'Modernity,'" *Amer. Hist. Rev.* 116 (2011): 631–751.

As historical subjects of investigation have changed or been recast, so have strategies for contextualizing them and grounding them in historical contingencies. The task is often complex. Take the recent growth in interest in the history of the senses, a topic of relevance to the contemporary world where our senses interact much more often with technology than with nature.[30] Karl Marx's assertion that the senses were alienated under capitalism serves as one possible starting point for writing a history of them, but his historical framework, grounded in the economy, ignores the physiological and cultural contexts in which the operation of the senses is grounded. To define what the senses are as subjects of history is to take up the history of the body, its interaction with the world, and the meanings we give to sense impressions. We are taught how to interpret our senses, but evidence for that cultural transmission comes from linguistic analyses, everyday life, cultural studies (especially material culture), and even environmental history, rather than from texts. Even with an ample evidentiary foundation for the history of the senses, we still have to deal with the difficult tasks of periodization and narration, not to mention the determination of the level at which their story is to be told. Similar intellectual tasks lie before historians in the new histories of sleep, comfort, and pain.[31] Historians of science have dealt with similar historiographical problems when reframing the historical significance of Copernicus in an era that believed in prognostication; when recasting our understanding of book 3 of Newton's *Principia Mathematica* in terms of the global information order; when trying to understand the meaning of science and technology in the nineteenth century; in challenging teleological narratives of the Soviet space program; and in other areas.[32]

Two essays in this volume demonstrate the complexity of defining and contextualizing historical subjects. In many respects the essays by Harold J. Cook and Edward Grant are polar opposites in that Cook's essay on early modern Dutch culture during the Scientific Revolution embraces a deep engagement with and even debate over questions of historical contextualization and contingencies, while Grant's on medieval history of science is much more reluctant to reach beyond textual evidence. Historiographical traditions in part account for the differences between them, but so does the variance between the focal points of their historical lenses, with Cook focused on macrophenomena and Grant on detailed textual analyses. Grant offers us the long-term view of a senior scholar who has had a deep engagement with intellectual history and who, consequently, has remained steadfast in his conviction that doing the history of science requires some scientific expertise. In Grant's view the exact sciences were immune to external influences in the Middle Ages, and natural philosophy remained secular. Yet as his essay and several successive editions of his books demonstrate, Grant comes to accept, if only in a limited way, the importance of broader historical contexts for understanding medieval science: the church, the university, theology, religion, and the role of the Greco-Arabic tradition. And while

---

[30] "*AHR* Forum: The Senses in History," *Amer. Hist. Rev.* 116 (2011): 307–400.

[31] A. Roger Erich, *At Day's Close: Night in Times Past* (New York, 2005); John E. Crowley, "The Sensibility of Comfort," *Amer. Hist. Rev.* 104 (1999): 749–82; Esther Cohen, "The Animated Pain of the Body," *Amer. Hist. Rev.* 105 (2000): 36–68.

[32] Robert Westman, *The Copernican Question: Prognostication, Skepticism, and Celestial Order* (Berkeley, 2011); Simon Schaffer, "Newton on the Beach: The Information Order of the *Principia Mathematica*," *Hist. Sci.* 47 (2009): 243–76; Iwan Rhys Morus, "Essay Review: Working Out in the Nineteenth Century," *Stud. Hist. Phil. Sci.* 38 (2007): 605–9; Asif A. Siddiqi, *The Red Rockets' Glare: Spaceflight and the Soviet Imagination, 1857–1957* (Cambridge, 2010).

he claims that social influences have little to do with texts, he admits they have everything to do with the life of the author. The nature of his subject, and the close range within which he examines it, keeps Grant himself immune from considering larger historical contexts and the contingencies that accompany them. Both are none-theless present in his historical method, albeit in foreshortened ways. Historiographi-cal traditions are often difficult to overcome.

Cook, on the other hand, creates his subject—how the economy, material culture, and knowledge systems were coproduced—by triangulating between different his-toriographical traditions. While his primary analytical framework is derived from economic history, he modifies it by using science and development paradigms (from international relations and political theory), which enable him to address simulta-neously the uses of knowledge, growth in material wealth, and the transfer of political goods. Not only is he able then to criticize and revise the ways in which economic historians have treated scientific knowledge; he is also able to recast old problems in the history of science with his conclusion that early modern economic expansion and the Scientific Revolution were codependent phenomena.

Interesting from a historiographical perspective is how Cook's definition of "science" adjusts to the contours of his engagement not only with empirical historical evidence, but also (and perhaps more importantly) with economic history. The kinds of knowledge that contributed to economic growth, he argues, were prescriptive and informational, and they dealt with material processes. In contradistinction to scholars who have used local contexts to explain the configuration of scientific knowledge, Cook points out that because knowledge has to move easily across boundaries in order to account for economic growth, it has to be universal. Cook thus addresses one facet of Golinski's "unfinished business" by accounting for the geographical range of scientific knowledge. Through his artful combination of economic history, Dutch history, and history of science, Cook provides an alternative perspective on why uni-versal knowledge was so prized in early modern Europe. The Dutch urban elite quite simply valued matters of fact about the natural and material worlds because they fit so well into their commercial way of life. They created an "information economy" that resonated with the exchange of information in later centuries.[33]

Essays in this volume deepen our concern for the tools of the trade, and appropri-ately so. After all, we still call ourselves *historians,* and we continue to call the field the *history* of science even as the number and kind of scholars analyzing science have expanded considerably beyond the philosophers and sociologists who have always been at our side. It behooves us, then, to think historiographically. Join us in the dia-logue between Clio and science.

---

[33] In this connection, see Cook's larger study, *Matters of Exchange: Commerce, Medicine, and Science in the Dutch Golden Age* (New Haven, Conn., 2007), and Schaffer, Roberts, Raj, and Del-bourgo, eds., *The Brokered World* (cit. n. 9).

REFLECTIONS

Is It Time to Forget Science?

Reflections on Singular Science and Its History

*by Jan Golinski**

ABSTRACT

The name *history of science* reflects a set of assumptions about what science is. Among them is the claim that science is a singular thing, a potentially unified group of disciplines that share a common identity. Long promoted by scientists and philosophers on the basis of a supposedly universal scientific method, this claim now looks very embattled. I trace its development from the early nineteenth century and the growth of the positivist movement to its various manifestations in the twentieth century. Recently, some historians have called for the term *science* to be relinquished, and for adoption of a more relaxed pluralism. Yet the complex legacy of the notion of singular science cannot be so easily abandoned.

The history of science has a paradoxical relationship to the thing it studies. Although the name of the discipline embeds within it an assumption that a singular thing called science is the object of its attention, that object has become harder to pin down as historical and other studies have gone in search of it. Attempts to demarcate the boundaries of the activity we call science have been repeatedly frustrated; claims to have isolated its essence or to have traced its continuity through the centuries have been seriously challenged. The historicization of the category of science has ended up by fragmenting the entity in question. To speak of science as a single thing suggests a degree of unity, exclusivity, and long-term continuity that the historical record does not seem to manifest. Science as we know it today is harder to find the farther we recede from the present, as has often been observed by those who study the premodern or early modern periods. And when non-Western cultures are brought into focus, it is even more difficult to locate in its familiar form.

It seems, then, that the concept of science is ripe for critical examination. The term is central to the cultural capital of our field, in other words to the symbolic vocabulary by which it has secured its authority. It is part of a venerable tradition, and as such—as historians above all should understand—is not to be dispensed with by mere wishful thinking. On the other hand, there has been a growing realization that this word *science* is a peculiar and rather fraught inheritance. It embraces the specific ideas and practices of the scientific disciplines, along with a complex set of metascientific beliefs and values. What we take to be science includes, along with the

---

* Department of History, 20 Academic Way, University of New Hampshire, Durham, NH 03824; jan.golinski@unh.edu.
I would like to thank the volume editors for their advice and support.

actual contents of scientific knowledge, a whole range of peripheral assertions about it. These include the claims that science has a privileged grasp of the truth, that it constitutes the only path to sound knowledge, that it enjoys a special freedom from social and political influences, that it is objective in its findings, and that it is universal in its application. Many of these claims have been subjected to fairly stringent criticism in recent years. It seems that such ideals as truth, value-neutrality, and objectivity are neither eternally unchanging nor universally accepted. Rather, they are historical constructs, interpreted in a range of different ways, and coming into prominence at particular times for particular reasons. Several important works of historical scholarship have mapped this terrain, and I do not need to reiterate their conclusions here.[1]

Instead, I want to see what can be brought into focus by pursuing a different but related metascientific theme: the notion that science is a singular entity, that all scientific beliefs and practices are components of one thing. This notion of "singular science" (as I shall call it) may be broken down into several component parts. It may rely on the assertion that all scientific disciplines adhere to one kind of method, for example, or at least that they draw upon a common set of approaches. It may invoke a kind of hierarchy of the sciences, insofar as they are thought to have advanced to different degrees in pursuit of their common goals. Singular science may involve an aspiration to unify all of the scientific disciplines at some point in the future. And it may also involve an insistence that science—as practiced in European civilization in the modern era—is a uniquely effective means of understanding and controlling the natural world.

I do not intend to disentangle all of the ancillary assumptions connected with the idea of singular science, as one might for the purposes of philosophical analysis. My aim is rather to explore the history of their entanglement.[2] In doing this, I will show that the notion of singular science was advanced in response to the rise of a large number of new disciplines in the early nineteenth century.[3] The fragmentation of the field of scientific inquiry called forth a vision of its potential unification. This ideal was particularly characteristic of the positivist philosophy, with its model of a scale of perfection on which all the scientific disciplines could be arranged. The positivist assertion that the disciplines had a common goal, which they would all eventually attain, allowed science to establish its credentials vis-à-vis other cultural domains in the period, such as religion and technology. Positivism also projected its vision of singular science globally, insisting that it was the only knowledge of nature that offered a path to material and social progress. Singular science was asserted as a component of European cultural hegemony in the nineteenth and early twentieth centuries.

---

[1] On the historical formation of notions of truth, value-neutrality, and objectivity, see Steven Shapin, *A Social History of Truth: Civility and Science in Seventeenth-Century England* (Chicago, 1994); Robert N. Proctor, *Value-Free Science? Purity and Power in Modern Knowledge* (Cambridge, Mass., 1991); Lorraine Daston and Peter Galison, *Objectivity* (New York, 2007).

[2] This inquiry draws upon the work of several other historians, including Peter Galison and David J. Stump, eds., *The Disunity of Science: Boundaries, Contexts, and Power* (Stanford, Calif., 1996); Thomas F. Gieryn, *Cultural Boundaries of Science: Credibility on the Line* (Chicago, 1999); and Peter Dear, *The Intelligibility of Nature: How Science Makes Sense of the World* (Chicago, 2006).

[3] On this, see Richard G. Olson, *Science and Scientism in Nineteenth-Century Europe* (Urbana, Ill., 2008); Andrew Cunningham and Perry Williams, "De-centring the 'Big Picture': *The Origins of Modern Science* and the Modern Origins of Science," *Brit. J. Hist. Sci.* 26 (1993): 407–32.

The history of singular science I shall present will of course be a selective one, reflecting the limits of my own knowledge. But I hope it will illuminate some aspects of the changing concepts of science and its history in the last two centuries. Positivism was not the only source of the aspiration to unify the sciences during this period, but a focus on the positivist movement will allow me to connect that aspiration to a specific historical consciousness that flourished in the same era. The positivist outlook included a place for historical studies, albeit a limited one. History was assigned a role that subordinated it to the sciences themselves, a role dictated by the vision of their eventual unification. As the influence of positivism has diminished in recent decades, history has assumed a more critical function. Historical and other studies have emphasized the differences between the disciplines, calling into question the purported singularity of the scientific enterprise and its tendency toward unity. At the same time, the rise to prominence of the biological and information sciences has undermined positivist assumptions about the hierarchy of the disciplines. This has coincided with important social and technological changes that have significantly altered the character of the scientific community and the perceived relationship between knowledge and technical practices. And the same period has also witnessed the diminished influence of European powers in global affairs, with consequent criticism of the idea that science has been uniquely a feature of Western civilization. All of these factors have undermined the traditional concept of singular science.

For historians of science, this situation raises an important set of questions. Should we cut ourselves loose from a concept of singular science that now looks decidedly threadbare? Can history of science simply rebrand itself "history of *the sciences*," acknowledging the plurality and diversity of its subject matter, and leave it at that? Or are there things we need to hang on to in the notion of singular science? Do we still require such an idea to secure the credentials of the field and provide it with a common agenda? My hope is that, by reviewing how we found ourselves at this point, we can reach toward some answers to these questions.

NINETEENTH-CENTURY ORIGINS

To begin, we need to go back to the early nineteenth century, since it was then that the modern conception of science as a singular and potentially unified entity was forged. This was a time of spectacular new discoveries and the creation of new disciplines. What was later called "classical physics" emerged at this time, centered on the phenomena of heat, light, and electricity, and on the concept of energy that linked them all. Chemistry underwent its own revolutionary developments, with crucial new findings about chemical composition. Biology arose from an intensified focus on the fundamental processes of life. The era also witnessed the growth of new institutions, including those specializing in the newly recognized disciplines. In response to these changes, a good deal of attention was devoted to trying to define science, to demarcate its boundaries and secure its standing in society. In the English-speaking world, the crucial metascientific works included John Herschel's *Preliminary Discourse on the Study of Natural Philosophy* (1830), Mary Somerville's *On the Connexion of the Physical Sciences* (1834), William Whewell's *History of the Inductive Sciences* (1837) and *Philosophy of the Inductive Sciences* (1840), and John Stuart Mill's *System of Logic Ratiocinative and Inductive* (1843). Looming over them was the massive oeuvre of Auguste Comte, the founder of positivism, whose *Cours de philosophie positive* was published between 1830 and 1842.

What was new at this time was not the basic idea that all knowledge might be interconnected. That vision could be traced back to classical antiquity, and it had already inspired the encyclopedic projects of the eighteenth-century Enlightenment. What was new was the notion that science—meaning specifically the study of the natural world—could serve as the foundation for that unified knowledge. It was at this point that the English word *science* acquired a somewhat narrower meaning than it had had before, coming to designate the body of natural and physical sciences as opposed to knowledge more generally. The *Oxford English Dictionary* gives 1867 as the first date for this usage, but the passage adduced in illustration declares that the word already bears this meaning in common speech. Indeed, Whewell pointed toward this more restrictive meaning in his *History* of 1837, when he wrote that, by using the criterion of correct method, "some portions of *knowledge* may properly be selected from the general mass and termed SCIENCE."[4]

Although the meaning of the word *science* was apparently narrowed in this way, the influence of positivism and other European philosophical movements also led to a considerable expansion in its scope of application. Science came to designate a particular kind of knowledge, but it was also hailed as a model for all fields of inquiry and a focus for their potential unification. At the end of the nineteenth century, the historian J. T. Merz claimed that the vision of the unification of all human thought was the main achievement of the epoch, or "the higher work of our century."[5] As champions of this vision, he mentioned both philosophers, such as Comte and Herbert Spencer, and scientists, such as Hermann von Helmholtz and Emil Du Bois-Reymond, who shared the positivist notion of a universal scientific method. The positivists saw science as the summation of all human endeavor, the final project in the historical progress of humanity. Far from being a narrowing of the conception of science, this amounted to a very grandiose vision of its historical importance.[6]

In Britain, it was Whewell who assumed the role of creating a study of science that would be external to scientific investigation as such but would secure its philosophical and historical credentials.[7] Whewell not only coined the term *scientist* in 1833, but gave substantial theoretical support to the project of the British Association for the Advancement of Science (BAAS), to whom he suggested this neologism. As a result, he gained a lasting reputation—in certain quarters at least—as a founder of the history and philosophy of science. For Whewell, as for Comte and Mill, the crucial idea was that the different fields of scientific inquiry were potentially unified by a shared method, but that they had to be distinguished in terms of the progress

---

[4] William Whewell, *Selected Writings on the History of Science*, ed. Yehuda Elkana (Chicago, 1984), 5; emphasis in the original. I am aware, of course, that the same connotations do not surround the words normally translated as "science" in other European languages. The cognate terms generally cover a wider range of disciplines, and this suggests that the tendency to impute a singular identity to science may have been articulated more stridently in English-speaking countries than elsewhere.

[5] John Theodore Merz, *A History of European Scientific Thought in the Nineteenth Century*, 4 vols. (London, 1904–12), 1:28–45, on 33.

[6] On positivism and the singularity of science in nineteenth-century Germany, see Peter Galison, "Introduction: The Contexts of Disunity," and Ian Hacking, "The Disunities of the Sciences," in Galison and Stump, *Disunity of Science* (cit. n. 2), 1–33 and 37–74.

[7] Richard Yeo, *Defining Science: William Whewell, Natural Knowledge and Public Debate in Early Victorian Britain* (Cambridge, 1993), 32–8, 145–75.

they had made to date in implementing it. The notion of an ordering of the disciplines according to their different rates of progress was taken to be a distinct advance beyond earlier attempts—such as those of Herschel and Somerville—to relate the various sciences to one another. As G. H. Lewes later put it, Comte had shown that all sciences are "branches of one Science, to be investigated on one and the same Method."[8] As Jack Morrell and Arnold Thackray have noted, the idea of singular science did important ideological work for the BAAS, helping to strengthen its relations with engineers, manufacturers, and professionals, and sharpening its appeal to the government from which it sought patronage.[9]

The fortified image of science was grounded in a sense of history. Whewell had an eye on the role played by contemporary political historians as he claimed the prerogative of legislator of science. Bound by a common method, the sciences were expected to look to history to measure their position along the path of progress. Some were recognizably more developed, and this was assumed to be because they dealt with simpler kinds of phenomena, though Whewell categorically denied that they could ever free themselves from metaphysical assumptions, as Comte had hoped. Nonetheless, Comte's classification of six fundamental disciplines—mathematics, astronomy, physics, chemistry, physiology, and sociology—was mirrored fairly closely in the six sections of the BAAS after 1833. And the order of the disciplines, as Lewes later explained, had a "necessary conformity to the actual order of the development of natural philosophy." He went on: "This is verified by all we know of the history of the sciences, particularly during the last two centuries."[10] Debate continued as to how rapidly and how exactly the moral sciences would catch up with planetary astronomy, which offered the prime example of the reduction of natural phenomena to mathematical laws. Whewell was skeptical that they would ever make the grade, and the BAAS accordingly confined social and political inquiries to the domain of "Statistics" in Section VI (later known as Section F). Mill agreed that it remained uncertain whether studies of mind or society were "capable of becoming subjects of science in the strict sense of the term." But he drew encouragement from the case of meteorology, which everyone seemed to expect would yield mathematical laws eventually, though it had failed to do so yet. As Mill explained: "The science of human nature . . . falls far short of the standard of exactness now realised in Astronomy; but there is no reason that it should not be as much a science as Tidology is, or as Astronomy was when its calculations had only mastered the main phenomena, and not the perturbations."[11] Questions of scientific standing were thus brought before the tribunal of history, though the historical investigation that would resolve them was strictly constrained by a very specific theory of progressive development. In this manner, history became an auxiliary to the project of establishing singular science.

Conjoined in this way with a theory of scientific method, history was used to secure the authority of science in the nineteenth century. Positivism continued to be influential, and the later prominence of evolutionary thinking also strengthened the

---

[8] Lewes, *Comte's Philosophy of the Sciences: Being an Exposition of the Principles of the "Cours de philosophie positive" of Auguste Comte* (London, 1890), 10.

[9] Morrell and Thackray, *Gentlemen of Science: Early Years of the British Association for the Advancement of Science* (Oxford, 1981), 96, 224, 259–60.

[10] Lewes, *Comte's Philosophy of the Sciences* (cit. n. 8), 46.

[11] John Stuart Mill, *A System of Logic, Ratiocinative and Inductive*, people's ed. (London, 1896), 546, 553–4.

historical mode of legitimation. The great Victorian sages of science, including T. H. Huxley, John Tyndall, and Karl Pearson, repeatedly invoked the historical vision of scientific progress in their writings.[12] Though each of them recognized important differences among the individual disciplines, they deployed the notion of singular science strategically to uphold the credentials of science as a whole in relation to other cultural fields. In the later decades of the century, an evolutionary model of cultural progress strengthened the positivist expectation that science was destined to succeed theology and metaphysics. The theological outlook that had been so important earlier to the gentlemen of the BAAS was called into question. Instead of the framework of Christian belief being taken for granted, it became possible to conceive of religion as an observable feature of human cultures in general and a counterpart to science as a basic way of understanding the world. It has been noted that the term *religion*, applied in the eighteenth century to Judaism, Christianity, and Islam, began to be extended to cover the major Asiatic faiths at just the same time as *science* was acquiring its modern meaning. Thereafter, it became possible to think of religion and science as complementary and even opposed systems of thought.[13] The end of the nineteenth century saw the emergence of some celebrated works arguing that science was inherently in conflict with religion. In their writings, J. W. Draper and A. D. White adopted the positivist assumption that science and religion were necessarily rivals, with the former destined to supersede the latter in the course of historical development.

Not all writers on science and religion in the period endorsed the "conflict thesis" concerning their relationship. The point is that it only became possible to conceive of the relationship—in these or other terms—when science had been defined as a singular entity, with which religion (similarly defined) could be compared. The consolidation of the notion of singular science also enhanced its claims for intellectual authority over material practice. In comments at the same meeting of the BAAS at which he proposed the word *scientist*, in Cambridge in 1833, Whewell explained how he saw the difference between scientific knowledge and that of the practical arts. The artisan's knowledge was inarticulate and restricted in its circulation, he claimed, whereas that of the scientist was explicit and public. The practitioners of the arts should therefore be guided by those who had a grasp of scientific theory.[14] The differentiation of mental from manual labor, with its associated assumption that the mind should rule over the hand, was rooted in the thinking of the eighteenth-century Enlightenment, but it found a ready acceptance in the Victorian era. It was flattering both to those who saw themselves as theorists and to engineers and mechanics who aspired to the cultural validation that theoretical knowledge conferred. A few decades later, Tyndall defended a similar demarcation of science from mechanical skill in lectures at the Royal Institution in London. Somewhat more controversially, he also

---

[12] Paul White, *Thomas Huxley: Making the "Man of Science"* (Cambridge, 2003), 67–99; Theodore M. Porter, *Karl Pearson: The Scientific Life in a Statistical Age* (Princeton, N.J., 2006).

[13] Peter Harrison, "'Science' and 'Religion': Constructing the Boundaries," in *Science and Religion: New Historical Perspectives*, ed. Thomas Dixon, Geoffrey Cantor, and Stephen Pumfrey (Cambridge, 2010), 23–49.

[14] Morrell and Thackray, *Gentlemen of Science* (cit. n. 9), 259–60; Yeo, *Defining Science* (cit. n. 7), 224–30. On changing notions of the relationship between scientific knowledge and material practice, see Lissa Roberts, Simon Schaffer, and Peter Dear, eds., *The Mindful Hand: Inquiry and Invention from the Late Renaissance to Early Industrialization* (Amsterdam, 2007).

sought to redraw the lines dividing the domains of science and religion.[15] Both of these strategic moves relied on the identification of singular science introduced by positivism.

Positivism also had important implications for the history of science. It promoted historical investigation while simultaneously tying it to the philosophical vision of the unity of the sciences. Comte called for the establishment of a chair in "general history of the sciences" at the Sorbonne as early as 1832, perceiving its importance in advancing his metascientific mission. History was to be enlisted in defense of the authority of singular science, which was defined by adherence to a common method. The historical record was to be interpreted in accordance with the positivist theory of progress, thereby contributing to the unification of the sciences. Although historical study was valuable in the metascientific realm, it had no necessary educational role for scientific practitioners themselves; in the mature sciences it should give way to a systematic presentation of the established facts. It would be quite incorrect, in Comte's view, to teach a science such as astronomy historically.[16] In a sense then, the position of the history of science was both privileged and dependent. It was accorded primacy as the key to revealing the underlying method of the sciences and assigning each of them its correct place in the order of development. But history itself was not a science, and could not be one as long as it retained the narrative method and had not yet discovered its own laws. It was confined to a metascientific role, part of the positive philosophy that was itself to be transcended in the course of transition to a fully scientific era of human development.

TWENTIETH-CENTURY VARIATIONS

This paradoxical situation clearly irked George Sarton, the Belgian American who founded the journal *Isis* in 1913 and agitated consistently for the institutionalization of history of science in the first half of the twentieth century. Sarton's campaign was launched under the aegis of positivism, in which he had been interested from his early days as a student of philosophy. He professed admiration for the movement's founder, though he blithely acknowledged Comte's periodic bouts of mental instability. He even recorded having "communed" with the great man's spirit during a visit to his Paris home in the 1940s.[17] Comte was worthy of veneration, Sarton maintained, though his historical scholarship never went beyond the superficial level. Sarton insisted on more rigorous scholarly standards in the history of science, in order to buttress its claim to independent academic standing. His perspective could still be described as positivist, however, with its emphasis on collecting and organizing factual information, compiling bibliographies, and establishing the institutional basis for the discipline. As Peter Dear has noted, Sarton also insisted on the singularity of the science whose history he was studying. While noting Comte's call for institutionalization of the discipline, he changed its designation from "history of the sciences"

---

[15] Thomas F. Gieryn, "John Tyndall's Double Boundary-Work: Science, Religion, and Mechanics in Victorian England," in Gieryn, *Cultural Boundaries of Science* (cit. n. 2), 37–64.

[16] Gertrud Lenzer, ed., *Auguste Comte and Positivism: The Essential Writings* (New York, 1975), 91–3.

[17] George Sarton, "Auguste Comte, Historian of Science: With a Short Digression on Clotilde de Vaux and Harriet Taylor," *Osiris* 10 (1952): 328–57.

to "the history of science."[18] The reformulation was crucial. As far as Sarton was concerned, the unity of science would emerge from properly conducted historical scrutiny. To encourage this, it was necessary to step back from the small-scale details of scholarship and take the long view. The minutiae of developments in the different scientific disciplines were less relevant to Sarton than the consolidated knowledge formalized in textbooks. Stories of individual accomplishment should ideally be synthesized into general accounts of eras and cultures. Then, the overall pattern of progress would be discerned and the unity of scientific knowledge correctly grasped. Sarton's *Introduction to the History of Science* was projected as a massive and comprehensive work to convey this unified vision. It was never more than partially completed, two volumes being published in 1927 and one more in 1948.

Sarton's vision of the discipline combined his positivist outlook with a deeply held commitment to pacifism and international solidarity. He believed that the codification and unification of all human knowledge would form the basis of global harmony and point the path toward the ending of military conflict. Launching his journal *Isis* on the eve of World War I, Sarton insisted that the positivist ideal of the unity of science was a buttress against national and ethnic rivalries, a prop for international cooperation, and a spur toward human unity. The grim alternative was intellectual and social fragmentation. As he later put it: "The fact that the building up of science has been done in the past and is done today by men of various races and many nationalities, inspired by different faiths, speaking different languages, proves that these men have the same needs and aspirations, reason in the same way, and, as far as they collaborate in the essential task of mankind, are united."[19] In Sarton's vision, then, the history of science was the heritage of the human species as a whole, a testimony to its common investment in the essential task of mankind. Sarton thus gave singular science its most profound significance, hailing its development as the central theme of human history and the brightest hope for the future.

The vision of singular science as a collective enterprise of humanity as a whole was shared by other inheritors of the positivist outlook in the early twentieth century. As Lewis Pyenson and Christophe Verbruggen have shown, Sarton drew upon the work of the Belgian bibliographers Paul Otlet and Henri-Marie Lafontaine, whose attempts to categorize and index all knowledge reflected their internationalist vision.[20] Sarton's activities in turn inspired those of the Italian scholar Aldo Mieli, founder of the *Archivio di storia della scienza* in 1919 (renamed *Archeion* in 1927) and of the International Academy of the History of Science in 1929.[21] Similar political ideals informed the work of the Unity of Science movement, associated with the logical positivists of the Vienna Circle. Members of this group, including Otto Neurath, Philipp Frank, Charles Morris, and Rudolf Carnap, emphasized the methodological features common to all of the sciences. Their approach was philosophical rather than historical, but it issued in an encyclopedic project similar in scope to Sarton's and more

[18] Dear, "The History of Science and the History of the Sciences: George Sarton, *Isis*, and the Two Cultures," *Isis* 100 (2009): 89–93.

[19] George Sarton, "Four Guiding Ideas," in *Sarton on the History of Science*, ed. Dorothy Stimson (Cambridge, Mass., 1962), 15–22, on 15. This essay was originally published in 1947.

[20] Pyenson and Verbruggen, "Ego and the International: The Modernist Circle of George Sarton," *Isis* 100 (2009): 60–78.

[21] Robert Fox, "Fashioning the Discipline: History of Science in the European Intellectual Tradition," *Minerva* 44 (2006): 410–32.

extensively realized: the *International Encyclopedia of Unified Science*, of which twenty monographs were published by the University of Chicago Press between the mid-1930s and the early 1960s. For Neurath, the unification of scientific knowledge was a distant prospect rather than an immediate goal; the encyclopedia was intended to facilitate the communication among scientists that could gradually bring it about. His campaign for this kind of unification was linked to his belief in socialist economic planning, which he attempted to put into practice in the short-lived Bavarian revolution after World War I, in leftist administrations in Vienna, and in work with the Bauhaus movement in the Weimar Republic.[22] Frank, an Austrian physicist and philosopher of liberal outlook, and Morris, an American pragmatist philosopher, shared to some degree Neurath's social vision for the Unity of Science project.

The social dimension of the movement was largely lost sight of after its leading members relocated to the United States in the 1930s, especially in the face of the concerted anticommunist campaign in the years after World War II. During the Cold War, American suspicion of internationalist and socialist political programs led to the rise of a purely technical version of logical positivism, associated with Carnap, Hans Reichenbach, and Herbert Feigl. As George A. Reisch has recently documented, Neurath's death in 1945 left the movement without the most articulate spokesman for its political vision.[23] Morris faced pressure from anticommunist agitators in Chicago, while Frank and Carnap were subjected to FBI investigation as suspicious foreigners. At this juncture, logical positivism was refocused on technical issues of epistemology, abandoning the popular and collectivist social aims originally invested in the Unity of Science movement.

The idea of singular science nonetheless lived on, accorded a common respect—though understood differently—by each side in the fierce ideological conflicts of the middle decades of the twentieth century. Despite deep divisions between Marxists and anti-Marxists in this period, both sides sought to borrow the cultural capital of science. In some respects, twentieth-century Marxism nurtured a continuation of the positivist tradition, though positivism was also fiercely criticized in some communist countries for its opposition to materialist metaphysics. Many Marxists embraced the notion of singular science, often claiming that dialectical materialism itself had scientific status and had revealed the long-sought laws of historical development.[24] The internationalist outlook associated with this was upheld by such leftist scholars as Joseph Needham and Benjamin Farrington, both of whom were involved with the UNESCO (United Nations Educational, Scientific, and Cultural Organization) Committee for History and Social Relations of Science in the 1940s.

On the other hand, the anti-Marxists who played an important role in institutionalizing the history of science in the West conceded nothing in their admiration of science, though they often understood it in a pointedly antipositivist manner. Thus, the Russian émigré Alexandre Koyré set out his idealist model, in which scientific knowledge was said to rely on metaphysical foundations adopted prior to sensory

---

[22] Richard Creath, "The Unity of Science: Carnap, Neurath, and Beyond," and Jordi Cat, Nancy Cartwright, and Hasok Chang, "Otto Neurath: Politics and the Unity of Science," in Galison and Stump, *Disunity of Science* (cit. n. 2), 158–69, 347–69.

[23] Reisch, *How the Cold War Transformed Philosophy of Science: To the Icy Slopes of Logic* (Cambridge, 2005), esp. 1–26, 259–76.

[24] Terrell Carver, "Marx and Marxism," in *The Cambridge History of Science*, vol. 7: *The Modern Social Sciences*, ed. Theodore M. Porter and Dorothy Ross (Cambridge, 2003), 183–201.

experience. Koyré was strongly antipathetic to Marxism or anything tainted by association with it; his views had a deep influence on conceptions of the "Scientific Revolution" of the sixteenth and seventeenth centuries. For Koyré, the intellectual transformation of that era was identified with the rise of a new theory of motion and the "geometrization" of space, especially in the works of Galileo and Descartes.[25] Koyré's antipositivist and antiempiricist conception of science was influential with a whole generation of American and British historians, including Herbert Butterfield, A. Rupert Hall, and Richard S. Westfall.[26] On a more popular level, the books of the Hungarian writer Arthur Koestler also advocated an idealist view of science, in which individual geniuses such as Johannes Kepler assumed the role of intellectual revolutionaries. A recent biography of Koestler has shown how he was drawn to metaphysical and psychological studies of science after gaining fame as an anticommunist novelist and polemicist in the postwar period.[27]

Thus, singular science enjoyed veneration on both sides of the ideological gulf of these years. Marxist historians, from the Soviet physicist Boris Hessen in the 1930s to the Irish crystallographer J. D. Bernal in the 1950s, tended to accept a rather similar characterization of the intellectual contents of science as their anti-Marxist opponents. They set out to uncover the material and economic needs that had given rise to the intellectual structure in question.[28] A recent commentator has written of Bernal, "He was quite as 'idealist' a historian of scientific ideas as anyone can imagine in the 'bourgeois camp.'"[29] Marxists shared with their ideological enemies a tendency to view science as a single thing, primarily identified with a set of metaphysical assumptions that could change dramatically in quite a short period. Hence the common interest in the Scientific Revolution of the early modern era. The gaps appeared in other areas of history, such as the Enlightenment, which was not properly brought into focus by either camp. Idealists tended to see the eighteenth-century movement as a mere aftershock of the intellectual revolution of the previous century, while Marxists viewed it as a bourgeois diversion prior to the more important Industrial Revolution.[30]

Nobody better demonstrates the ambiguities of Marxist historiography in the twen-

---

[25] Roy Porter, "The Scientific Revolution: A Spoke in the Wheel," in *Revolution in History*, ed. Porter and Mikuláš Teich (Cambridge, 1986), 290–316; David C. Lindberg, "Conceptions of the Scientific Revolution from Bacon to Butterfield: A Preliminary Sketch," in *Reappraisals of the Scientific Revolution*, ed. Lindberg and Robert S. Westman (Cambridge, 1990), 1–26.

[26] David A. Hollinger, "Science as a Weapon in *Kulturkämpfe* in the United States during and after World War II," *Isis* 86 (1995): 440–54; Anna K. Mayer, "Setting Up a Discipline: Conflicting Agendas of the Cambridge History of Science Committee, 1936–1950," *Stud. Hist. Phil. Sci.* 31 (2000): 665–89.

[27] Michael Scammell, *Koestler: The Literary and Political Odyssey of a Twentieth-Century Skeptic* (New York, 2009).

[28] On Hessen, see Loren R. Graham, "The Socio-political Roots of Boris Hessen: Soviet Marxism and the History of Science," *Soc. Stud. Sci.* 15 (1985): 705–22; Gary Werskey, *The Visible College: A Collective Biography of British Scientists and Socialists of the 1930s*, 2nd ed. (London, 1988), 138–49; Anna K. Mayer, "Setting Up a Discipline II: British History of Science and the 'End of Ideology,' 1931–1948," *Stud. Hist. Phil. Sci.* 35 (2004): 41–72. On Bernal, see Andrew Brown, *J. D. Bernal: The Sage of Science* (Oxford, 2005).

[29] H. Floris Cohen, *The Scientific Revolution: A Historiographical Inquiry* (Chicago, 1994), 220.

[30] For more on the neglect of Enlightenment science, see William Clark, Jan Golinski, and Simon Schaffer, "Introduction," in *The Sciences in Enlightened Europe*, ed. Clark, Golinski, and Schaffer (Chicago, 1999), 3–31.

tieth century than the English Sinologist Joseph Needham. Needham's monumental studies of the history of Chinese science opened up the issue of the comparative study of non-European cultures, with important implications for the notion of singular science. Needham had a famously heterodox intellectual formation, in which Marxism, Christianity, and Taoism jostled in sometimes uneasy proximity, and his massive scholarly labors gave him an unrivaled knowledge of Asian and European cultures. With his training in biochemistry and developmental biology, he was uncomfortable with the positivist hierarchy of the sciences. He disputed the right of physics to be considered the model for all science, though he sometimes also paid homage to the idea that science was a singular entity. He said the initial spur to his research on China was the question of why modern science had arisen only in Europe—a formulation that took for granted that science was defined by the European experience. But, as Needham's research accumulated, it succeeded in showing the magnitude and range of the Chinese accomplishment and implicitly made the case for the diversity of sciences in different cultures.[31] So, although the famous "Needham question" was premised on an assumption of the singularity of science, in the end his work undermined the whole idea that science was a uniquely European enterprise. As H. Floris Cohen has shown, Needham gave no definitive or unambiguous answer to his own question. While he cited Koyré's description of the Scientific Revolution as a transformation in metaphysics, he denied its completeness because it ignored the biological sciences.[32] Needham also included technology and medicine within his capacious vision, unlike Koyré, who firmly segregated such applied fields from science's intellectual core.

In these ways, Needham pointed forward to the new world in the last few decades of the twentieth century, in which notions of singular science had to be reconsidered. Assumptions of the uniqueness of European modes of understanding were increasingly called into question in an era of increased global communications. Progressive politics in the West, which used to be identified with the unification of the sciences and the singularity of the European achievement, now came to embrace multiculturalism, which implied that the sciences might take forms other than just the canonical one. Needham was a critical figure in this transformation. Sometimes he emphasized the singularity of science as the common enterprise of humankind as a whole, in the internationalist manner of Sarton, Neurath, and Bernal. At other times, he acknowledged its diversity in relation to cultural differences and historical change. In this way, he foreshadowed the trouble that lay ahead for the idea of singular science.

Needham's farsightedness in this connection is illuminated by a comparison with his Cambridge contemporary C. P. Snow. Snow's much-discussed lecture of 1959, "The Two Cultures and the Scientific Revolution," confidently reiterated the notion

[31] Gregory Blue, "Joseph Needham," in *Cambridge Scientific Minds*, ed. Peter Harman and Simon Mitton (Cambridge, 2002), 299–312; Simon Winchester, *The Man Who Loved China: The Fantastic Story of the Eccentric Scientist Who Unlocked the Mysteries of the Middle Kingdom* (New York, 2008); Needham, *The Grand Titration: Science and Society in East and West* (London, 1969); Needham, "The Making of an Honorary Taoist," in *A Selection from the Writings of Joseph Needham*, ed. Mansel Davies (Jefferson, N.C., 1990), 29–54; Needham, *Science and Civilization in China*, ed. Christopher Cullen, 24 vols. (Cambridge, 1954–).

[32] Cohen, *Scientific Revolution* (cit. n. 29), 418–82, esp. 445.

of the singularity of science.[33] By "the scientific revolution," Snow meant not the early modern shift in metaphysical outlook described by Koyré, but the ongoing technological transformation of modern society since industrialization. The key to this process, in Snow's view, was the knowledge base of Western science. He proclaimed that European scientific education was the sine qua non of material progress. Countries elsewhere in the world had simply to educate scientists in the approved way and industrial development would follow. Snow's outlook was thoroughly imbued with the assumption of the uniqueness of European science and the particular version of internationalism that derived from it. And, of course, he also trumpeted the singularity of scientific culture when contrasting it with that of literary intellectuals. The scientific and the literary spheres constituted the "two cultures" for which his lecture became famous. Comparing science with a degenerate literary modernism, he declared that the former was superior both as an intellectual outlook and as a repository of moral values. He subsequently resisted proposals that the two cultures should be increased to three or more to recognize the social sciences or other academic fields. Singular science remained crucial to Snow's whole perspective. And because his lecture was so influential—not least in inspiring the establishment of programs in history of science and allied subjects to "bridge" the two cultures—he breathed new life into the notion.

Just three years later, however, Thomas S. Kuhn's *Structure of Scientific Revolutions* provided the occasion for more intensive scrutiny of singular science.[34] This is not to say that Kuhn was setting out to completely subvert established ideas. In fact, it has been plausibly argued that his aims were not really radical at all, and his influence was out of all proportion to his intentions. In many respects, he stuck closely to the classical notion of singular science. His idea that certain "paradigms" guide the development of the "mature" scientific disciplines suggested a uniform standard of method. And he closely followed the positivist hierarchy of the disciplines in marking when different fields had achieved paradigmatic form: the mathematical sciences in antiquity, experimental physics in the seventeenth and eighteenth centuries, chemistry somewhat later, and the biological sciences only in the nineteenth century and since. When Kuhn declared that the social sciences had yet to create their paradigms, he was echoing Comte's sentiments of more than a century before. Furthermore, mature, paradigmatic, or "normal" science was characterized by Kuhn as a self-contained world, largely segregated from those large-scale social forces he classified as "external."[35] As such commentators as Steve Fuller have pointed out, this clearly reflected the assumptions of the Cold War scientific intelligentsia from which Kuhn emerged.[36] It was taken for granted in those circles that scientific progress would take the form of increasing specialization and professionalization of the various fields of research. Scientists were expected to focus on ever-narrower topics of study as they isolated themselves institutionally, organizing their own social arrangements

---

[33] Snow, *The Two Cultures* (Cambridge, 1993).

[34] Kuhn, *The Structure of Scientific Revolutions*, 2nd ed. (Chicago, 1970).

[35] Ibid., 69. See also Steven Shapin, "Discipline and Bounding: The History and Sociology of Science as Seen through the Externalism-Internalism Debate," *Hist. Sci.* 30 (1992): 333–69.

[36] Fuller, *Thomas Kuhn: A Philosophical History for Our Times* (Chicago, 2000), 1–37. See also Reisch, *How the Cold War* (cit. n. 23), 229–33; Philip Mirowski, "What's Kuhn Got to Do with It?" in *The Effortless Economy of Science?* (Durham, N.C., 2004), 85–96.

independently of forces in the wider society. Again reflecting the assumptions of his milieu, Kuhn remarked that this form of social organization had originated in early modern Europe and was to be found exclusively in the cultures that descended directly from it.[37]

On the other hand, Kuhn's book was received in an intellectual climate that was growing increasingly skeptical about the idea of singular science.[38] Many contemporaries read his work as opening the door to a more discriminating and rigorously historical account. Kuhn was more a symbol of these changes than a leader of them; he subsequently disavowed what he took to be the wildest misinterpretations of his work. But that does not mean that those who perceived radical implications in his writings were altogether mistaken.[39] In certain respects, Kuhn's model did undermine the program of singular science, despite the fact that the book appeared as a volume in the *International Encyclopedia of Unified Science*, founded by Neurath and his colleagues from the Vienna Circle. Kuhn noted that scientific methods and practices were not universal, but localized within quite tightly bounded communities of practitioners. He acknowledged that phenomena could not be observed raw, but were always interpreted through a framework of preconceptions and according to assumptions bound up with the use of certain instruments. And he recognized that, while paradigms guided scientific research, they did not determine what sense could be made of new experiences. Concepts did not come prepackaged with all of their possible applications; they were fitted to the findings of inquiry by a process of interpretive fudging and creative extrapolation. If one took seriously these aspects of Kuhn's perspective, then science appeared as a methodologically looser and much more fragmented activity than the positivist tradition had allowed.

One group of his interpreters who developed Kuhn's insights in this direction were the members of the Edinburgh school, who advanced the so-called Strong Programme in science studies in the 1970s and early 1980s. The philosopher David Bloor and the sociologist Barry Barnes seized particularly on Kuhn's debts to the philosophy of Ludwig Wittgenstein.[40] Kuhn's vision of normal science as the creative and undetermined extension of existing ideas echoed Wittgenstein's insistence that rational thought was not governed by logical rules and that language was always changing in response to practical needs. This suggested that the meanings of scientific ideas were related to the local circumstances and immediate needs of investigators; they were not determined by logical deduction from some more general theory. To the Edinburgh school, the fundamental implication of this was clear, though Kuhn himself had not perceived it. Science was a social enterprise at its very core, since the concepts and methods that constituted paradigms were the attributes of small

[37] Kuhn, *Structure of Scientific Revolutions* (cit. n. 34), 168.

[38] For background, see Michael Aaron Dennis, "Historiography of Science: An American Perspective," in *Companion to Science in the Twentieth Century*, ed. John Krige and Dominique Pestre (London, 2003), 1–26; Hollinger, "Science as a Weapon" (cit. n. 26).

[39] For a more developed argument on this point, see Jan Golinski, "Thomas Kuhn and Interdisciplinary Conversation: Why Historians and Philosophers of Science Stopped Talking to One Another," in *Integrating History and Philosophy of Science: Problems and Prospects*, ed. Seymour Mauskopf and Tad M. Schmaltz (Dordrecht, 2012). A contrasting account of recent trends in science studies, which roots them in a series of "misunderstandings" of the works of postpositivist philosophers, is given by John H. Zammito, *A Nice Derangement of Epistemes: Post-positivism in the Study of Science from Quine to Latour* (Chicago, 2004).

[40] Barnes, *T. S. Kuhn and Social Science* (London, 1982); Bloor, *Wittgenstein: A Social Theory of Knowledge* (London, 1983); Bloor, *Knowledge and Social Imagery*, 2nd ed. (Chicago, 1991).

groups united by their focus on a common task. Social relations within such groups had an immediate bearing on how science was done; they were not at all external to its practice.

The most prominent application of this claim was in the studies of scientific controversies that flourished among historians and sociologists of science in the 1980s. The sociologists Harry Collins and Trevor Pinch led the way, documenting the disputes over physicists' attempts to detect gravity waves or solar neutrinos.[41] Historical studies followed in their footsteps, by Steven Shapin, Simon Schaffer, Martin Rudwick, James Secord, and others. They scrutinized the controversies surrounding Robert Boyle's experiments with the air pump, Isaac Newton's work on light and colors, and Victorian geologists' mapping of the Earth's strata.[42] When such debates were studied with due attention to both sides, they seemed to be not just about facts, but about methods, instruments, competences, and indeed the whole direction science should take. What Kuhn had called "incommensurable paradigms" seemed to be squaring off against each other. In Kuhn's own phrase (echoing Wittgenstein), "incompatible modes of community life" were at stake.[43] The methods and practices of the sciences were found to differ, not just in different civilizations and nations, but even at the level of individual researchers. With the exposure of such apparently unbridgeable chasms between practitioners in the same field, a sharp blow was dealt to the notion of singular science.

CONTEMPORARY PROSPECTS

The exposure of deeply rooted controversies at the leading edges of scientific research is but one of the ways in which the idea of singular science has been eclipsed in recent years. The themes of diversity, disunity, and localism have come to overshadow those of consistency, unification, and universalism in analyses of scientific practice. There are many factors that have shaped the current situation, and it is clear that only a small part of the responsibility can be laid at the door of Thomas Kuhn. In the last few decades, the field of science studies has come to embrace political, economic, and literary disciplines, as well as the traditional triad of history, philosophy, and sociology. These academic interests have intersected in complex ways with broader cultural and intellectual changes, often summarized under the label *postmodernism*, and with remarkable developments in the sciences themselves. As individuals of all nations have become scientists, questions of cultural diversity have been raised much more directly than C. P. Snow could ever have imagined. The entry of more women into professional positions has both inspired and been encouraged by

[41] Collins, *Changing Order: Replication and Induction in Scientific Practice* (Beverly Hills, Calif., 1985); Pinch, *Confronting Nature: The Sociology of Solar Neutrino Detection* (Dordrecht, 1986).

[42] Shapin and Schaffer, *Leviathan and the Air-Pump: Hobbes, Boyle, and the Experimental Life* (Princeton, N.J., 1985); Schaffer, "Glass Works: Newton's Prisms and the Uses of Experiment," in *The Uses of Experiment: Studies in the Natural Sciences*, ed. David Gooding, Pinch, and Schaffer (Cambridge, 1989), 67–104; Rudwick, *The Great Devonian Controversy: The Shaping of Scientific Knowledge among Gentlemanly Specialists* (Chicago, 1985); Secord, *Controversy in Victorian Geology: The Cambrian-Silurian Dispute* (Princeton, N.J., 1986). Also important in directing historians' attention to the value of the Edinburgh perspective was Shapin, "History of Science and Its Sociological Reconstructions," *Hist. Sci.* 20 (1982): 157–211.

[43] Kuhn, *Structure of Scientific Revolutions* (cit. n. 34), 94.

increased academic attention to the role of gender in science. These changes have enhanced the appeal of a model of science that emphasizes pluralism rather than unification, and that recognizes how knowledge remains tied to the attributes of local cultures rather than entirely escaping from them. In addition, scientific knowledge has been implicated as both cause and effect in the staggering technological advances of recent decades. The ubiquity of information technology in all kinds of scientific research has made it impossible to maintain the fiction that technological change is always dependent on prior advances in "pure" science. The rise of the biological sciences has been especially notable in this period, undermining the positivist claim that physics is the preeminent discipline. Additional sciences have emerged with bewildering rapidity, and interdisciplinary fields have proliferated as the foci of research and pedagogy. All of this has tended to confirm the impression that the traditional notion of a singular and prospectively unified science is obsolete—that the diverse fields of contemporary inquiry and innovation cannot all be subordinated to a single method or ideal.

In this situation, historical studies have turned for inspiration to philosophical traditions other than positivism, in which the singularity of science has not been so readily accepted. In a recent account of these traditions, Hans-Jörg Rheinberger has argued that Kuhn did no more than synthesize the conclusions of "a protracted effort that took various forms over a good half-century, despite the contemporary philosophical dominance of logical positivism."[44] Rheinberger locates the roots of scientific pluralism in the turn toward history by certain philosophers of science in the early twentieth century. He points in particular to the importance of the 1920s, when ideas about the plurality of scientific cultures were provoked partly by the revolutionary developments in physics and partly by reflections on the biological sciences. The former were of interest to the French philosopher Gaston Bachelard, who drew the conclusion that scientific ideas were tied to very localized arrangements of experimental apparatus and the phenomena they exhibited. Working in the biological sciences at the same time, the Polish immunologist Ludwik Fleck drew similar conclusions about the local specificity of scientific cultures. Fleck wrote of the styles of thinking characteristic of particular laboratories and of the material practices that lay behind them. The writings of these two authors, along with the more general philosophical perspectives of Edmund Husserl, Martin Heidegger, and Ernst Cassirer, are seen by Rheinberger as having laid the foundations for the more widespread emergence of pluralistic notions of science in the years after World War II.

Rheinberger's narrative can be read as a counterpart to my own. While I have traced the long legacy of nineteenth-century positivism, he has shown that it was nurturing the roots of postpositivism even during its heyday. At the least, this serves as a salutary corrective to the temptation to see pluralistic ideas about science as entirely due to very recent cultural trends. Taking that line, some commentators have sought to pin the blame on the liberalism of the 1960s, on the feminism and postcolonialism that emerged from that decade, or on the postmodernism that came to the fore in the 1980s. These were the targets in the so-called science wars of the 1990s, when what was thought to be an attack on science drew forth a vituperative response from

---

[44] Rheinberger, *On Historicizing Epistemology: An Essay* (Stanford, Calif., 2010), 79–80.

its self-appointed defenders.[45] The clamor of that battle should not obscure the fact that criticisms of the positivist vision of science had been brewing for several decades already. And a central feature of these criticisms was the theme that science was not a singular thing. To those who recognized this, the science wars seemed like a noisy distraction, a hubbub of mutual incomprehension. If science is not a single entity, then it is not really possible either to attack or to defend it as such. Critics of the notion of the singularity of science were not, of course, attacking the sciences, or even any part of them. Nor were they likely to be persuaded by reiterations of the traditional positivist line by those who claimed to be defending science itself.

If the polemics of the 1990s confirmed that the idea of singular science had taken a beating by the end of the twentieth century, they also showed that it was not entirely extinct. Indeed, some of those who sprang to the defense of science enthusiastically reasserted the possibility of the unification of scientific knowledge. A few physicists prophesied the imminent reduction of natural phenomena to a single all-encompassing law, or the unification of physical forces through new discoveries of subatomic particles. Some biologists foresaw the expansion of Darwinian natural selection beyond plants and animals to cover the evolution of the inorganic world and human culture as well. Visions of the singularity of science, it seemed, could themselves take multiple forms. On the other hand, even while the onslaughts of the critics were being vehemently repulsed, the journalist John Horgan found several prominent scientists who acknowledged that singular science as it had traditionally been understood was coming to an end.[46] In the course of a series of interviews with leading practitioners, he uncovered what he called an "ironic" attitude of contemporary scientists to their past. Many of them were highly conscious of the complexities and discontinuities in the historical record, hesitant to predict the future, reluctant to prophesy that the truth would finally be found, and aware that—in the long term—current ideas would pass away. As Horgan pointed out, this kind of ironic sophistication about the history of science is also a widespread characteristic of our age.

This complex situation frames current debates about the central issue I have been discussing: the relationship between the belief that science is a singular entity and the consciousness of its history. In the positivist tradition, the history of science was read as a narrative of progress, of the accumulation of knowledge and the gradual reduction of phenomena to regular laws. This account provided a guarantee that scientific knowledge would in time be unified, as the different disciplines in turn attained the form of a true science. The scientists Horgan interviewed seem to have realized that that guarantee has now been withdrawn. Singular science, whose unity was forecast on the basis of a specific interpretation of the past, has lost the warrant of history. In this sense, if in no other, the end of science is upon us.

Professional historians, at least, seem unworried by this. Indeed, they have experienced the release from positivist expectations as something of a liberation. Few historical narratives now trumpet unidirectional progress and the march toward the unification of knowledge. On the contrary, most of them stress fragmentation and

[45] For illuminating commentary on the science wars, see Gieryn, *Cultural Boundaries of Science* (cit. n. 2), 336–62; Arkady Plotnitsky, *The Knowable and the Unknowable: Modern Science, Nonclassical Thought, and the "Two Cultures"* (Ann Arbor, Mich., 2002), 157–99; John Guillory, "The Sokal Affair and the History of Criticism," *Crit. Inq.* 28 (2002): 470–508.

[46] Horgan, *The End of Science: Facing the Limits of Knowledge in the Twilight of the Scientific Age* (New York, 1996).

localization. In this manner, research on the history of the sciences has largely freed itself from the notion of singular science, which was its primary source of legitimacy when positivism held sway. In addition, recognizing the plurality and disunity of the sciences makes it easier to claim a place for history among them. No longer striving to model itself after the physical sciences, history can escape from the subordinate position to which positivism confined it. One can even argue, as Roger Smith has recently, that all humanistic study is historical to one degree or another, and hence that history has a fundamental role in all of the "human sciences."[47]

On the other hand, the retreat of the idea of singular science leaves unfinished business in its wake. The traditional notion presented answers to a series of questions—answers that we no longer find adequate but have yet to replace fully. Science was supposed to present a single path to the truth, a truth that would necessarily prevail everywhere and would yield intellectual command over material practice. Science was supposed to be inherently egalitarian and public, yet unencumbered by economic interests or political forces. If we no longer find this vision convincing, we have to derive new ways to respond to the questions it was originally intended to answer. We need new explanations for the extraordinary geographical range over which scientific knowledge has proven itself, its intimate relationship with technology, and its institutionalization in diverse cultures. These questions of the character and context of scientific knowledge remain fundamental, even if we jettison the truisms by which they were traditionally answered. They constitute key issues surrounding the historical role of the sciences in the modern era. And it is the heritage of engagement with these issues that distinguishes what has been called the history of science from other fields of historical study.

A lot has been done along these lines, though much still remains to be done. As singular science was deconstructed into a patchwork of diverse and localized forms of knowledge, the pressing issue arose of explaining how that knowledge escaped the local circumstances of its origin and was given more general acceptance. Answers have been sought in the various social mechanisms through which artifacts, people, and texts are mobilized. Practices of translation, replication, and metrology have taken the place of the universality that used to be assumed as an attribute of singular science. When science was thought of as the same everywhere, its uniformity mapped directly onto that of nature itself. In the new view, science achieves its singularity—to the extent it does—as the result of practices that are socially and historically located. These practices have been brought under scrutiny by historians, sociologists, philosophers, and others, whose task has been to show how scientific knowledge is made to seem universal and consistent—in other words, to fill the gap left by the disappearance of the idea of singular science. Through their efforts, a fair amount has been revealed about the mechanisms by which scientific knowledge has circulated through modern society and across geographical space.[48] A picture is emerging that stresses the role of networks and circulation in allowing knowledge to

---

[47] Smith, *Being Human: Historical Knowledge and the Creation of Human Nature* (Manchester, 2007).

[48] Surveys of this work include Steven Shapin, "Here and Everywhere: Sociology of Scientific Knowledge," *Annu. Rev. Sociol.* 21 (1995): 289–321; Jan Golinski, *Making Natural Knowledge: Constructivism and the History of Science* (Chicago, 2005); David N. Livingstone, *Putting Science in Its Place: Geographies of Scientific Knowledge* (Chicago, 2003); James A. Secord, "Halifax Keynote Address: Knowledge in Transit," *Isis* 95 (2004): 654–72.

transcend the cultures, languages, and communities within which it originates. We are beginning, in other words, to formulate an alternative model. But, as long as this remains a collective work in progress, we can still expect to have it measured against the all-too-easy answers provided by the notion of singular science. The legacy of that notion still bears heavily upon us, and is not easily to be shrugged off. We have not yet arrived at the point when we can forget science.

Science Is Dead; Long Live Science

by Peter Dear*

ABSTRACT

Scholars nowadays generally accept that there is no single enterprise of "science"—
that there is a multiplicity of special studies that are grouped together under that
label. This article discusses a possible route by which to trace certain common cultural characteristics of modern science, rooted in a particular (and complex) relationship between discourses of philosophical understanding and utilitarian practice.
The issue is examined in relation to recent discussions of the scope of the history
of science, metaphors of "networks" and "circulation" of knowledge, present-
centeredness, and the temporal continuity of historical accounts. The article illustrates how the two entwined themes of natural philosophy and instrumentality may
be traced in forms characteristic of an ideology of modern science.

SCIENCE OR "SCIENCE"?

Is the history of science a history of science, or a history of "science"? If the former,
it may comprise any sort of knowledge or human activity to do with the world that we
regard as serious, formally organized, and respectable. It could range from gnomons
to genomics; from satellites to stalactites; from ancient Kenyan iron-ore smelting to
Polynesian navigation. Very little would be off-limits, and a broad vision to encom-
pass it would have little real coherence. If the latter, however, familiar tools of social
and intellectual history could serve to delineate the history of the idea, or ideology, of
"science" that has so possessed modern culture.

There is nothing novel in proclaiming the diversity of those activities and bodies of
knowledge that we call *science*; the theme of the "disunity of science" is nowadays a
familiar one.[1] In 1962, in the optimistically titled *Foundations of the Unity of Science*,
volume 2, number 2, Thomas Kuhn wrote of science as a "rather ramshackle structure with little coherence among its various parts."[2] And yet little has been done about
it. In the early 1990s the watchword *practice* promised to usher in a new approach
to understanding science that shook free of previous, and philosophically motivated,
idealist conceptions of science[3]—those conceptions having been shared in important ways by the approaches of the 1970s and '80s associated with the Sociology of

---

* Department of History, Cornell University, 435 McGraw Hall, Ithaca, NY 14853; prd3@cornell
.edu. Thanks to Carin Berkowitz for valuable suggestions regarding argumentative structure.

[1] E.g., Peter Galison and David J. Stump, eds., *The Disunity of Science: Boundaries, Contexts, and
Power* (Stanford, Calif., 1996).

[2] Kuhn, *The Structure of Scientific Revolutions*, 2nd ed. (Chicago, 1970), 49. This quotation is un-
changed from the first edition of 1962 (51).

[3] Andrew Pickering, ed., *Science as Practice and Culture* (Chicago, 1992); Jan Golinski, "The
Theory of Practice and the Practice of Theory: Sociological Approaches in the History of Science,"
Isis 81 (1990): 492–505.

Scientific Knowledge (SSK). In the event, however, this failed to address the problem of disunity: instead, the trick was simply to study practices in any area of activity deemed part of science, not to investigate why that area was so deemed.

Historians of science, then, have not worried overmuch about the possibility that those multifarious activities and bodies of knowledge that they, or others, have designated by the name *science* might in fact be an assemblage created by historical contingencies, rather than being bound together by necessary relations, whether cognitive or methodological. Why do we not have thriving specialties in the history of mortuary science in our university departments of the history of science?[4] No one seems likely to be able to give a principled answer as to why some activities that use the name *science* should be admitted, but not others. In order to interrogate this problem more closely, I propose to consider what happens when the historian attempts to treat the category *science* with a degree of anthropological strangeness, accounting for what is and is not seen as falling within its scope in the terms of historical contingency, with no presuppositions, implicit or explicit, about what ought really to count.

Such a vision of the history of science might, of course, reduce to a nominalism that would drain that historical enterprise of all meaning—pure historical granularity with nothing to link the grains except the vagaries of linguistic practice. If there is to be a broad vision in the history of science at all, the first question to ask, as Georges Canguilhem put it, "is what it is that the history of science claims to be the history of."[5] Canguilhem failed to shrug off certain lingering remnants of Comtean positivism in addressing the point: he characterized science as a "certain cultural form," the complexity of which necessitated the historical tracing of various quasi-disciplinary "ideologies" that in each case preceded "the institution of science."[6] But his question invites a consideration of precisely the distinction with which we have opened: are we concerned with the history of science or with the history of "science"?

Given the problem of disunity, it may be that the writing of the first history is impossible, or else possible only as a thin and miscellaneous account with little intellectual value. But the second sort of history is much more promising. A history of an ideal of inquiry, of the vicissitudes of its applications and associations, need not rely on an acceptance of its privileged epistemological character, its literal efficacy, or the propriety of its uses. It would be a history of "science" as a sort of ideology, but not in the way that Canguilhem suggested, where each scientific field is originally birthed as what Kuhn, in 1962, called a preparadigmatic stage. Instead, the idea of science itself is the ideology, striving to account for the apparent successes of many diverse knowledge enterprises.

One of the central virtues of the history of science is its ability to shed light on this most characteristic aspect of the modern world (modern in the sense of both modernist and, where appropriate, postmodernist). The reality of what science truly is resides less in essential similarities between its assorted branches than it does in the

---

[4] There are numerous programs and departments of mortuary science at US institutions of higher education, mostly state colleges and community colleges.

[5] Canguilhem, "What Is a Scientific Ideology?" (1970), in *Ideology and Rationality in the History of the Life Sciences*, trans. Arthur Goldhammer (Cambridge, Mass., 1988), 27–40, on 27; see also C. Hakfoort, "The Missing Syntheses in the Historiography of Science," *Hist. Sci.* 29 (1991): 207–16. Cf. Peter Dear, "What Is the History of Science the History *Of*? Early Modern Roots of the Ideology of Modern Science," *Isis* 96 (2005): 390–406, written in ignorance of Canguilhem's article.

[6] Canguilhem, "What Is a Scientific Ideology?" (cit. n. 5), 27, 38.

ideology that binds them. This ideology presents the appearance of unity where none truly exists.[7] A broader vision of the development of this ideology, integrated with some of the more signal examples of the knowledge activities and practices that have come to constitute it, may help to exorcise the totalizing ghost of "science" while still telling us what it is.

NATURAL PHILOSOPHY AND INSTRUMENTALITY

The dominant, but usually covert, ideology of science amounts to a self-denying perception according to which "science" is fundamentally about natural philosophy—contemplative understanding of a world independent of the knower—while at the same time, and paradoxically, it is also fundamentally about instrumental capabilities. The latter include the instrumental tools of science and the employment of such tools for desired purposes. The ideology of modern science, according to this argument, is one in which natural philosophy and instrumentality (the manipulation of nature) play complementary but generally mutually invisible roles in supporting the cultural dominance of science. Each perspective relegates the other to a subsidiary role, and itself to a position of primacy. When talk of one is paramount, the other tends to be hidden, and which is at the forefront in any particular instance depends on local contextual convenience.[8] As for the question of the supposed relation between science and technology, it can often be expedient to skirt the term *technology* entirely, given the immensely broad senses in which it is used by all sorts of scholars, including most historians of technology.[9] The now-common term *technoscience* and the grouping of science with technology and medicine, as if the three were naturally the same endeavor, help to perpetuate, while also obscuring, the ideological framework from which science derives its cultural power.

As in Bohr's principle of complementarity, according to which whether you speak in terms of particles or waves depends on the experimental situation, so whether you speak in terms of science as natural philosophy or as instrumentality depends on what best serves the particular purpose. But in the latter case, the distinction between natural philosophy and instrumentality is never actually addressed; doing so would undermine the discursive efficacy of each. When, from another ideological standpoint, Jürgen Habermas has attempted to represent scientific work as instrumental rationality, he too has unwittingly reinforced the usual view insofar as, at the same time, he has neither argued in detail for the reduction of all scientific theory to its instrumental efficacy nor rejected outright the existence of any natural-philosophical dimension to science.[10]

Drawing an analytical distinction between natural philosophy and instrumentality does not, then, amount to offering a choice between two alternatives. It is not a

---

[7] Dear, "What Is the History of Science?" (cit. n. 5).

[8] See the fuller account ibid. and in Peter Dear, *The Intelligibility of Nature: How Science Makes Sense of the World* (Chicago, 2006), intro.; cf. Paul Forman, "The Primacy of Science in Modernity, of Technology in Postmodernity, and of Ideology in the History of Technology," *Hist. & Tech.* 23 (2007): 1–152, sec. 3.

[9] An important exception is Eric Schatzberg, "*Technik* Comes to America," *Tech. & Cult.* 47 (2006): 486–512.

[10] Habermas, *Technik und Wissenschaft als "Ideologie"* (Frankfurt, 1968); partially translated in Habermas, *Toward a Rational Society: Student Protest, Science, and Politics*, trans. Jeremy Shapiro (Boston, 1970).

choice between (say) realism and instrumentalism. It is instead a way of seeing the underlying ideological structure of a hybrid enterprise, by directing attention to the difficulty of representing science as either one of those components or the other in isolation.

The most challenging feature of this picture from an empirical perspective is perhaps the adequate framing of the ideology of modern science itself, without which a corresponding historical account of its emergence would be otiose. The central idea of two discrete ideals of inquiry, frequently with opposed purposes and moral values, operating as mutually supportive alternative discourses requires a working ideological construct that can systematically obscure the contradictions. That this is how the practices of modern science are understood, by the practitioners themselves as well as by other cultural institutions, is not itself a demonstrable proposition; like any other broad generalization in the qualitative social sciences or the empirically based humanities (admitting history among one or the other of those groupings), the proposition is a conjectural hypothesis. But, like a temporal model that purports to show how that state of affairs came about historically, this characterization of modern science, as ideology and as concomitant social institution, must be made plausible as a correct description of how modern science really is, and must also provide indications of why it should be so, and how that structure is maintained.

The last point is the most difficult. I take it as a generally acceptable statement that specialties in the contemporary sciences frequently use self-representations that stress the practical benefits potentially or actually accruing to their research.[11] In the decades following the Second World War, the rhetorical stress on "basic research" was itself justified by the metaclaim that practical benefits emerge most effectively from fundamental, disinterested research rather than from explicitly targeted "mission-oriented" science; claims of instrumental effectiveness applicable to socially desirable purposes were even then rhetorically crucial.[12] At the same time, it would seem to be almost impossible to find a self-identifying science that did not also purport to discover truths about an independently existing world, regardless of their usefulness. Hence the two categories of instrumentality and natural philosophy. How the two are interwoven then becomes the crucial issue, one whose nature remains, to me, rather obscure. A tempting version would connect them piecemeal, through a kind of situated functionalism: whenever it suits the purposes of particular scientists or groups of scientists to emphasize the instrumental virtues of their work, they will do so, and whenever it benefits them to emphasize disinterested pure science, they will do so. Consequently, the same piece of work or research program could be presented and discussed in either of those two principal modes. Furthermore, the two alternatives may even formally contradict one another.

DUELING DISCOURSES

A sterling example appears in Adelheid Voskuhl's account of her work as a participant-observer on the development of speech-recognition (i.e., transcription)

---

[11] Indeed, such representations are usually explicitly required by governmental granting agencies such as the National Science Foundation, and not only in the United States.

[12] Michael Aaron Dennis, "Reconstructing Sociotechnical Order: Vannevar Bush and US Science Policy," in *States of Knowledge: The Co-production of Science and Social Order*, ed. Sheila Jasanoff (London, 2004), 225–53.

software.[13] The researchers Voskuhl studied were members of a medical physics research unit, concerned with the processes of human auditory signal processing; that is, with understanding human hearing. At the same time, the group gauged its success by the performance of its speech-recognition machine in an annual international competition, in which the victor was determined solely by the level of accuracy of its transcriptions. Voskuhl discusses how the researchers talked about and negotiated the boundary that they themselves recognized between "engineering" and "medical" traditions in their work. The engineering aspect was solely concerned with creating a machine that would perform well at automatic speech recognition, regardless of how implausible the system's features might be as representative of human speech recognition. The medical aspect, to which the group was organizationally committed, concerned precisely this issue—how well its machine modeled the way that human speech recognition operates. Rather than perceiving a sharp demarcation between these two endeavors, however, the researchers in effect fudged the differences. One of them reflected on the matter in this way:

> An ideal for me would be some kind of a circular process. So—I look, "hey, what's going on physiologically in human hearing and, on higher levels, in the auditory cortex? What's actually happening there?" Then I try to model it, replicate it, and to solve certain tasks with [that model]. And my ideal would be that I see "oops, that functions really well," so that I have on one hand a nice application and on the other hand I may have progressed a *tiny* little bit *possibly* in understanding what really is happening when humans recognize speech. But that's an ideal. There's so much—yeah—so many technical peculiarities and restriction [*sic*] in there, that the inference back to humans probably doesn't work. At any rate, one has to be *very*, *very* cautious with that.[14]

The faltering attempt here to generate an Aristotelian nonvicious circle indicates the embarrassment caused by having to respond on the issue in a self-consciously methodological register.[15] Practice (and practical reasoning) usually took care of things quite adequately for these researchers, as long as an ethnographer refrained from asking awkward questions.

A case from the history of chemistry shows the empirical value of this perspective for the historian of science. The acceptance of Daltonian chemical atomism in the first half of the nineteenth century, as we know especially from the work of Alan Rocke, was only partial and qualified.[16] In fact, its discussion displayed exactly the wavering situational focus between instrumentality and natural philosophy that one might by now expect, in the very period from which "modern science" is most often dated. In his 1825 textbook *An Attempt to Establish the First Principles of Chemistry by Experiment*, Thomas Thomson, Regius Professor of Chemistry at Glasgow, first addresses his book to medical students, an important market for him at Glasgow. So he explains to them "the great advantage which medical practitioners will derive from a knowledge of the atomic weights of bodies, and of the weights of the integrant particles of the salts, &c., which they have occasion to employ in their

---

[13] Voskuhl, "Humans, Machines, and Conversations: An Ethnographic Study of the Making of Automatic Speech Recognition Technologies," *Soc. Stud. Sci.* 34 (2004): 393–421.

[14] Ibid., 410, as there translated from the German.

[15] See Aristotle, *Posterior Analytics*, bk. 1, chap. 13, on planets and their failure to twinkle.

[16] Rocke, *Chemical Atomism in the Nineteenth Century: From Dalton to Cannizzaro* (Columbus, Ohio, 1984).

prescriptions."[17] He also talks about the potential value of chemistry for physiology. In the preface itself, he draws on another aspect of the instrumentality of his book's atomic first principles:

> Several chemical manufacturers in this country have already availed themselves of the atomic theory to rectify their processes. . . . One of the great objects of the present work is to reduce the whole doctrine of atoms to the utmost degree of simplicity and accuracy; so as to put it in the power of chemical manufacturers in general, to avail themselves of it in order to bring their processes to the requisite degree of precision.[18]

And similarly for "apothecaries and druggists."[19]

The first chapter provides a historical account of the work of chemists starting with Wilhelm Homberg in the late seventeenth century, focusing on the chemistry of salts and the issue of combining weights. It concludes with discussion of William Wollaston and Jacob Berzelius, and finally of William Prout's conjectured modification of Dalton's atomic theory—including, of course, Thomson's own work to determine atomic weights, which he says supports Prout's hypothesis. The chapter's discussion focuses on experimental work and results, devoting particular attention to the pragmatic advantages of using the atomic weight of oxygen as the system's unity rather than Dalton's preferred hydrogen.[20]

Thomson's "Historical Introduction" is followed by a chapter on the atomic theory, which one might have thought had just been covered. Thomson's themes in this chapter comprise the divisibility or indivisibility of matter, concentrating on G. W. Leibniz and Roger Boscovich, and on Thomson's own version of Daltonian atomic theory, being principally concerned with proper terminology and with establishing the composition and relative atomic weights of various acids, bases, and salts, as well as of the "simple bodies" that compose them. Thomson knits these themes together through a conventional expression of nescience: after identifying the "maze of metaphysical subtilities" concerning the ultimate divisibility of matter, he brushes the question off, while still taking advantage of atomic language.

> But such discussions being quite unsuitable to a system of chemistry in its present state, it will be better to avoid them altogether. By *atom*, then, in the following pages, I would be understood to mean, the ultimate particles of which any body is composed, without considering whether the farther division of these particles be possible or not. . . .
> The weights attached to the atoms of bodies, and deduced from the combinations into which they enter, merely express the ratios of the atomic weights, and have no reference whatever to the size or specific gravity of the atoms to which they are attached.[21]

Thomson even allows himself to speak on occasion of "half-atoms" as being involved in combinations, on the grounds that they can always be eliminated by doubling all the numbers involved: "But I would not be understood to support the ab-

---

[17] Thomson, *An Attempt to Establish the First Principles of Chemistry by Experiment*, 2 vols. (London, 1825), 1:viii.

[18] Ibid., xiii–xiv.

[19] Ibid., iv.

[20] On oxygen as unity: ibid., 14–7. On Thomson and Prout, see W. H. Brock, *From Protyle to Proton: William Prout and the Nature of Matter, 1785–1985* (Bristol, 1985), esp. chaps. 5, 7.

[21] Thomson, *Attempt to Establish* (cit. n. 17), 1:31–2.

stract idea of the possibility of the existence of half-atoms; but to have been actuated by the arithmetical facilities derived from the method which I have adopted."[22] When other chemists avoid the term *atom*, they are being, he says, "squeamish," because the term is "more convenient, shorter, and more distinct."[23]

Most of Thomson's book consists of experimental procedures, reports, and results concerning the measurement of specific gravities and atomic weights.[24] The characteristic feature of the work is, of course, the intermingling of natural-philosophical concerns—the accessibility or otherwise of information about atoms—with up-to-date stoichiometric procedures of a self-evidently operational or instrumental kind. Thomson's insistence on the use of the term *atom*, while certainly not typical of chemical discourse in this period, reflects a concern that his accounts of laboratory manipulations be seen as philosophically respectable, as bearing on issues of matter theory even if only at several conditional removes. Hence his plain acceptance that other terms might have been used, such as Wollaston's *equivalent*, or, in the case of some of Thomson's own usages, *integrant particle*, and his denial that chemical atoms are necessarily literally indivisible.[25] Mi Gyung Kim has dubbed the subject of such useful evasions the "stoichiometric atom."[26]

In fact, the virtual hybridity of Thomson's endeavor is already indicated in his book's title, which refers to both "first principles" and "experiment." He wants to have it both ways, which is the feature to look out for in tracking the slippery ideology characteristic of modern science.

SCIENCE AND TECHNOLOGY AND MEDICINE

A fairly recent serious attempt at addressing the wide scope of science as a product of historical contingencies is John Pickstone's *Ways of Knowing: A New History of Science, Technology and Medicine*. One of the book's striking features is that it speaks of "science, technology, and medicine" (STM) as a proper grouping for study without providing any analytical justification for treating the subjects as intimately related, or else for treating them as fundamentally distinct from one another. Pickstone often uses *technoscience* to gesture at aspects of some of the relations involved, but for the most part he traces particular "ways of knowing" across the putative boundaries comprised by STM without asking what the apparent boundaries themselves mean, or why they appear to exist.[27] His overall approach is therefore

[22] Ibid., 32–4, quotation on 34.

[23] Ibid., 36.

[24] A classic discussion of Thomson's laboratory procedures is Jack Morrell, "The Chemist Breeders: The Research Schools of Liebig and Thomas Thomson," *Ambix* 19 (1972): 1–46; see also Catherine M. Jackson, "Visible Work: The Role of Students in the Creation of Liebig's Giessen Research School," *Notes Rec. Roy. Soc. Lond.* 62 (2008): 31–49.

[25] Thomson, *Attempt to Establish* (cit. n. 17), 1:34, 36.

[26] Kim, "The Layers of Chemical Language, I: Constitution of Bodies v. Structure of Matter; II: Stabilizing Atoms and Molecules in the Practice of Organic Chemistry," *Hist. Sci.* 30 (1992): 69–96, 397–437, esp. pt. 2, e.g., 398; see also Kim, *Affinity, That Elusive Dream: A Genealogy of the Chemical Revolution* (Cambridge, Mass., 2003), 434, attributing such stoichiometric atomism to Thomson and Berzelius.

[27] Pickstone, *Ways of Knowing: A New History of Science, Technology and Medicine* (Chicago, 2001); see also Pickstone, "Working Knowledges before and after circa 1800: Practices and Disciplines in the History of Science, Technology, and Medicine," *Isis* 98 (2007): 489–516.

analogous to Bruno Latour's methodological trope of "following scientists and engineers through society,"[28] where an implied no-nonsense empiricism overrides picky categorical distinctions between different kinds of activity. Pickstone's categories are developed in a sort of orthogonal relationship to those that make up STM: they instead comprise various "ways of knowing," such as natural history, analysis, and experimentalism, which themselves admit of quite specific characterizations in Pickstone's account, and which trace each its own historical path.[29] For some purposes, and in order to answer certain kinds of questions, Pickstone's way of tackling the big picture of the history of science (and technology, and medicine) has many virtues; what it cannot do, however, is to justify or explain the juxtaposition of S and T and M, or the circumstances in which such a juxtaposition has historically been avoided.

The issue is familiar in the field of science and technology studies, where it has long been investigated in cases of historical controversy over whether particular areas of study or pieces of research should be admitted as science or else denied that status.[30] One of the central themes in such work is that of cultural status: most cases involve a determination of whether some kind of intellectual work should be valued highly or not, and the label *science* successfully attached to the work usually signals its high value, with attendant social prestige and related characteristics deemed desirable by sociologists. All this seems fairly uncontroversial. But an analogous set of themes should appear in the drawing of any kind of boundaries, such as demarcations between science, technology, and medicine. What sets these alternative boundary negotiations apart from that between science and nonscience is that the relative value of the labels is not always so obvious. This lends an analytical richness to the study of any specific instance, since nuance and contrasting evaluations of the labels themselves add extra dimensions to the contested boundary lines.

Yet there are still a few generalizations to be made. In European traditions, the modern demarcation between science and technology has often carried with it a head/hand distinction of social status: better to be an intellectual scientist than a manual artisan.[31] A general scholarly recognition that there is no in-principle line to be drawn that will cleanly separate science from technology (or pure from applied science) has resulted, however, in a makeshift solution that evades the crucial historical questions entirely. The term *technoscience* is now widely used to deny the legitimacy of a distinction that continues to be made everywhere but in specialized scholarly science-studies communities. But it is not in itself a particularly insightful or useful denial; human endeavors generally tend to merge into each other at critical points, and the historian (or contemporary social analyst) might like to remain alert

[28] Latour, *Science in Action: How to Follow Scientists and Engineers through Society* (Cambridge, Mass., 1987).

[29] Cf. A. C. Crombie, *Styles of Scientific Thinking in the European Tradition: The History of Argument and Explanation Especially in the Mathematical and Biomedical Sciences and Arts*, 3 vols. (London, 1994).

[30] Examples are examined in Thomas F. Gieryn, *Cultural Boundaries of Science: Credibility on the Line* (Chicago, 1999); Gieryn, "Boundary-Work and the Demarcation of Science from Non-science: Strains and Interests in Professional Ideologies of Scientists," *Amer. Sociol. Rev.* 48 (1983): 781–95; Gieryn, "Boundaries of Science," in *Handbook of Science and Technology Studies*, ed. Sheila Jasanoff et al. (Thousand Oaks, Calif., 1995), 393–443; see also Steven Shapin, "Discipline and Bounding: The History and Sociology of Science as Seen through the Externalism-Internalism Debate," *Hist. Sci.* 30 (1992): 333–69.

[31] Lissa Roberts, Simon Schaffer, and Peter Dear, eds., *The Mindful Hand: Inquiry and Invention from the Late Renaissance to Early Industrialization* (Amsterdam, 2007).

to questions regarding when and why people choose to distinguish between them and to taxonomize them.[32] Casually speaking of "technoscience" ignores those realities.

In the case of medicine, similar categorical relations are slightly more complex. As an intellectual, university-taught specialty, medicine properly so called was once of higher status than the butchery of the surgeon or the recipes of the apothecary; a closer formal unification of the three specialties had to wait until the early nineteenth century, well after the antiestablishment Paracelsian medicine of the sixteenth and seventeenth centuries had raised the social stakes of different approaches to medical treatment. The nineteenth century's concern with developing so-called scientific medicine indicates the broad direction of medical reform: just as medieval medicine had been a learned profession, while surgery and the concoction of drugs emphatically had not, so the reformed medicine of the nineteenth century strove for the validating label *scientific*; medicine that was not scientific was mere quackery.[33] In this way, the demarcating boundaries between science and medicine may have possessed more significance for the new physician than for the nonmedical scientist. At the present day, the term *biomedicine*, relating medicine to the scientific credentials of biology, seems to enshrine a desired relationship between science and medicine. In practice, however, the word is used not simply to designate medical research of a physiological and biochemical kind, but also more generally to represent both the medical knowledge and the medical practice that used to be called, just as imprecisely, "Western" medicine. Evidently, a vaguely "scientific" terminology can serve to globalize a specific medical culture.

An adequate, integrated history of science needs to confront many preconceptions about what science is and is not: whether it is one or many (methodologically or otherwise), or, indeed, whether it is nothing more than a culturally prestigious label to be attached to absolutely anything whenever one can get away with it. Everyone has a favored belief on what science really is, perhaps, but to tell a convincing story about the history of science, one that will be useful and convincing to many other people, seems to require a common ground. A strong and fundamental link is to be found in the various ways by which people have represented science, both to themselves and to others, and in the institutional establishment of those representations. The twin discourses of natural philosophy and instrumentality are good ones to track in their interwovenness, and the codependence that emerges between the two is the hallmark of the ideology of modern science—an ideology that shapes and is formed by the practices and institutions that thereby count as scientific.

If a narrative can be constructed along those lines, several implications become apparent. One of the most salient concerns the issue of "Western" versus "non-Western" science. If modern science is the legacy of historically specific developments whereby natural philosophy became entangled with instrumentality to produce a characteristic hybrid, then science in that sense is a particular ideologically understood set of practices emergent from a particular cultural matrix. In brief, it is an endeavor created in a particular (broad) cultural context in a particular (broad) period, and to the extent

---

[32] Cf. esp. Barry Barnes, "Elusive Memories of Technoscience," *Perspect. Sci.* 13 (2005): 142–65; Paul Forman, "(Re)cognizing Postmodernity: Helps for Historians—of Science Especially," *Ber. Wissenschaftsgesch.* 33 (2010): 157–75. On related issues of classification, see Geoffrey C. Bowker and Susan Leigh Star, *Sorting Things Out: Classification and Its Consequences* (Cambridge, Mass., 1999).

[33] Roy Porter, *Health for Sale: Quackery in England, 1660–1850* (Manchester, 1989).

that it has become a worldwide enterprise, it has done so through export and adoption in other cultural contexts. The idea of inherently non-Western science becomes as strange a notion as non-Western contract bridge; the specificity is part of its cultural meaning. Only the rather nineteenth-century presumption that science, both as knowledge and as activity, is a mark of progressive cultural achievement impedes this perception of the mundane and contingent contents of the ideological box marked *science*. Everyone wants credit for making scientific discoveries; everyone seems to have adopted them as signs of intellectual attainment.[34] But discoveries (innovations of some kind) can only be called scientific if they are seen as exemplifying the ideology of science. This is a clear form of cultural hegemony.

METAPHYSICAL REDUCTION: NETWORKS AND CIRCULATION

Among the most popular metaphors, or full-fledged theories, used to understand science and its history have been those postulating varieties of "networks." The Cambridge philosopher of science Mary Hesse garnered attention in British SSK circles in the 1970s with her proposal that the "entrenchment" of concepts in a scientific theory was achieved by their mutual interrelations in a network.[35] This was an intellectual model of scientific ideas, in which the slightest alteration of any single node inevitably affected all the others by a sort of mutual adjustment of the connections throughout the web. The nodes of this network were concepts, while the links between them Hesse represented as relations of mutually implicated probabilities, particularly probabilities relating to empirical substantiation.

Hesse's model was adopted in a significantly modified form by the sociologist of science Barry Barnes.[36] Barnes took the network metaphor (dubbing it the "Hesse net") and reconceived the links between the nodes as social conventions rather than probabilities. This enabled him, and others in SSK, to attempt a sociological understanding of scientific theories, whereby the interrelated ideas of a theory would owe their relative entrenchment in the network of the theory to the social conventions that established their mutual implications.

That particular web or network had both advantages and drawbacks. Its drawbacks lay in the limited explanatory palette that it offered to the science-studies scholar. Because the nodes were taken to be concepts, the network functioned as a means of accounting for science as a knowledge system consisting of statements about the world that were held to be true or false (or relatively likely). It was therefore unsuited to—because not directed toward—understanding science as, for example, an organizational system or a set of interrelated practices. Its chief advantage, however, like that of Hesse's original version, lay in its degree of ontological specificity: instead of being a vague metaphor of interconnectedness, Barnes's model indicated what kinds of things constituted the nodes (concepts) and how the links between them were to be understood (as social conventions). That advantage set it apart from the most widely

---

[34] See, for one among many popular examples, the recent exhibition "1001 Inventions," originally presented in London, in which much serious "golden age" Islamic and Arabic work is reduced to anticipations of later European science.

[35] Hesse, *The Structure of Scientific Inference* (London, 1974), chap. 1, "Theory and Observation," 9–44; see esp. Barry Barnes, "On the Conventional Character of Knowledge and Cognition," *Phil. Soc. Sci.* 11 (1981): 303–33.

[36] Barnes, "On the Conventional Character" (cit. n. 35); see also H. M. Collins, *Changing Order: Replication and Induction in Scientific Practice* (London, 1985), 17.

adopted of network models in the study of science, the actor-network theory (ANT) propounded by Bruno Latour and Michel Callon.[37] The network imagery has owed much of its popularity, no doubt, to its adaptability to many different projects; but that adaptability comes at a price. ANT uses the idea of a so-called heterogeneous network, which in practice seems to forbid almost nothing. Its nodes can be human agents, or anything else whatever, from elephants to electrons; *things*, one might say, but so generously construed, in accordance with a semiotic theoretical justification,[38] that apparently abstract concepts themselves could also count as legitimate nodal points in the network. Because of this ontological promiscuity, ANT provides very few, if any, theoretical constraints on what the scholar might do with it. The absence of constraints applies not just to the nature of the network's nodes, but also to the character of the links between them.[39]

Work in the history of science has from time to time made use of network conceptualizations that take their theoretical lead from Latour.[40] However, useful as these attempts have sometimes been, their value has resided more in a generic model of a network of human or social actors, as in cases of correspondence networks or centrally coordinated Jesuit institutional and missionary networks, than in their exploitation of the unique features of ANT. Some recent historiographical endeavors to write the biographies of scientific objects (such as the electron), rather than of human actors, have attempted to follow, in revised forms, the ANT notion of nonhuman "actants" as playing quasi-causal roles in passages of scientific activity.[41] The principal risk here has long been seen as that of renovating a naive form of scientific realism, in concert with teleological accounts of the development of currently accepted scientific ideas, including current beliefs about the reality and properties of the nonhuman actants themselves.[42]

Latour's related model of "centers of calculation," with its focus on accumulation at a rapacious metropolitan center, has also attracted notice.[43] Although Latour's explicit intent was to dissolve the "Great Divide" between the West and the rest of the world,[44] the net effect of his work has been to reinforce the view that science and its associated physical power was exported wholesale as a fundamentally European commodity. It is that view of science and colonialism that recent scholarship in

[37] Latour, *Science in Action* (cit. n. 28); Latour, *Reassembling the Social: An Introduction to Actor-Network-Theory* (New York, 2007); Callon, "Some Elements of a Sociology of Translation: Domestication of the Scallops and the Fishermen of St. Brieuc Bay," in *Power, Action and Belief: A New Sociology of Knowledge?* ed. John Law (London, 1986), 196–229.

[38] This semiotic approach is most clearly expressed in Bruno Latour and Françoise Bastide, "Writing Science—Fact and Fiction," in *Mapping the Dynamics of Science and Technology*, ed. Michel Callon, John Law, and Arie Rip (London, 1986), 51–66.

[39] See Steven Shapin's review of *Science in Action*, "Following Scientists Around," *Soc. Stud. Sci.* 18 (1988): 533–50, which argues that Latour's links are usually tantamount to the sociologist's "social interests."

[40] E.g., Steven J. Harris, "Confession-Building, Long-Distance Networks, and the Organization of Jesuit Science," *Early Sci. & Med.* 1 (1996): 287–318; David S. Lux and Harold J. Cook, "Closed Circles or Open Networks? Communicating at a Distance during the Scientific Revolution," *Hist. Sci.* 36 (1998): 179–211.

[41] See, for examples, contributions to Lorraine Daston, ed., *Biographies of Scientific Objects* (Chicago, 2000).

[42] Latour's *Pasteurization of France*, trans. Alan Sheridan and John Law (Cambridge, Mass., 1988), was effectively critiqued along these lines in Simon Schaffer's review, "The Eighteenth Brumaire of Bruno Latour," *Stud. Hist. Phil. Sci.* 22 (1991): 174–92.

[43] Latour, *Science in Action* (cit. n. 28), chap. 6.

[44] See also Jack Goody, *The Domestication of the Savage Mind* (Cambridge, 1977).

the history of science challenges, shifting "Western" into "modern" science instead. Thus Kapil Raj stresses the theme that modernity was not an exported European invention, but something spawned in the interactions between Europeans and peoples from elsewhere (such as the multifarious populations of early modern Calcutta) in what he calls "contact zones."[45] While acknowledging the considerable imbalances of power between European rulers and those whom they ruled, Raj details ways in which theoretical knowledge and practical techniques were generated in such contact zones to create cultural practices formative of modern science. European states, in other words, were never able to be cut off from the rest of the world when they made the "rational" knowledge that they increasingly represented as their own characteristic achievement.

The use of networks and web metaphors in the history of science, Latourian or otherwise, has tended to be imprecise and allusive rather than theoretically specific. In the history of early modern science, however, the allusions have also invoked notions familiar from economic historians of the period, like John Brewer. For some time economic and other historians have been concerned with global networks, impelled most notably by the work of Immanuel Wallerstein.[46] In turn, such interests have begun to reinvigorate particular areas of the history of science concerned with European colonialism and the global growth of scientific knowledge systems since (at least) the eighteenth century. Such work has augmented the network view with the somewhat amorphous concept of "circulation."

This figure emphasizes the movement of material objects as well as of instrumental practices (including such items as plants, instruments, books, astronomical data, and ethnographic reports) along the trade routes of the early modern world, especially those of the Atlantic and Indian Oceans.[47] Other scholars, most notably Harold J. Cook, have interrogated issues of global commerce without employing the specific notion of circulation;[48] the significance of those studies that do so, however, lies in their interpretive assumptions about the "matter" of science. The metaphor is partly drawn from the notion in economics of the circulation of money and goods. Its adoption by historians of science in the form of the "circulation of knowledge" and the "circulation" of scientific (natural-philosophical and mathematical) practices seems to depend on the implied technical content of the economic models themselves, where commercial networks instantiate economic theories of wealth as generated by

---

[45] He uses this term in a somewhat more concrete sense than Peter Galison's well-known *trading zone*. See Raj, *Relocating Modern Science: Circulation and the Construction of Scientific Knowledge in South Asia and Europe* (Basingstoke, 2007).

[46] Wallerstein, *The Modern World-System*, 3 vols. (New York, 1974–88).

[47] For a very few examples, see James Delbourgo and Nicholas Dew, eds., *Science and Empire in the Atlantic World* (New York, 2008); Raj, *Relocating Modern Science* (cit. n. 45); see also the discussion of circulation in Pamela H. Smith, "Science on the Move: Recent Trends in the History of Early Modern Science," *Renaiss. Quart.* 62 (2009): 345–75; Dew, "Scientific Travel in the Atlantic World: The French Expedition to Gorée and the Antilles, 1681–1683," *Brit. J. Hist. Sci.* 43 (2010): 1–17. Ongoing work by Mary Terrall on Réaumur also uses the theme of circulation. The term is of long standing among anthropologists; it seems to have little clear definition outside economics. Newspapers have long been said to circulate rather than being distributed, but the origin of this usage is obscure: the *Oxford English Dictionary*'s first record of this sense (s.v. "circulation") is from 1847; many earlier cases may be found through Google Books.

[48] Cook, *Matters of Exchange: Commerce, Medicine, and Science in the Dutch Golden Age* (New Haven, Conn., 2007).

trade—wealth created, therefore, through circulation, especially the circulation of money: in the ideal (classical) economic case, of course, money itself really does circulate rather than just travel. But the application is less clear for the history of science.

The phrase *circulation of knowledge* no doubt resonates with Joel Mokyr's arguments on the rise of the "knowledge economy."[49] But for early modern science, the existence of circulation may be only loosely figurative: there was certainly long-distance travel, and the dispersal of ideas, but circulation analogous to that of money seems to have been absent. Artifacts and instruments moved across the North Atlantic, for example (one thinks of Benjamin Franklin receiving his glass electrical tubes from Peter Collinson in London),[50] with experimental reports involving their use coming back again; astronomical, meteorological, and other instruments, together with their users, certainly traveled on expeditions sent out from Europe, and often returned;[51] but ideas themselves did not circulate and proliferate as a result, because ideas are not like money—not even like paper money.

Techniques, practices, and skills also travel, of course, generally embodied in the people who command them. But here again, *circulation* fails to capture the process. As such things travel, to be locally communicated to people in new locales, they seem to elaborate networks, rather than circulating around in feedback loops. At the same time, as studies of standardization in the sciences during the nineteenth century have made clear, the establishment of an authoritative center or privileged node in the network is of essential importance.[52] The whole point in such cases is not to be dynamic, but to have authoritatively enforced standards, policed in various ways, that try to prevent slippage within the network; controversy should not be allowed to break out.

In that sense, the picture of networks is opposed to the economic model of circulation. In the latter, it is important that not everywhere be the same as everywhere else (e.g., that there be different demand for particular commodities in different places). In the former, however, it is important that standards be uniform, on a level, universal. And in that light, there is no point to circulation, only to distribution and communication.

If the economic model of circulation is to be applicable to science, especially to premodern science, it will require both the equivalent of money and, especially, the equivalent of wealth. The economic model was designed so that the concept of circulation could perform explanatory work—namely, to show how trade creates wealth. There seems to be no functional parallel for circulation in science, which makes it unsurprising that such circulation (as opposed to travel, or communication) is hard

[49] Mokyr, *The Gifts of Athena: Historical Origins of the Knowledge Economy* (Princeton, N.J., 2004); see, e.g., Dew, "Scientific Travel" (cit. n. 47).

[50] See James Delbourgo, *A Most Amazing Scene of Wonders: Electricity and Enlightenment in Early America* (Cambridge, Mass., 2006), which also speaks of circulation.

[51] Dew, "Scientific Travel" (cit. n. 47); see also Florence C. Hsia, *Sojourners in a Strange Land: Jesuits and Their Scientific Missions in Late Imperial China* (Chicago, 2009), which does not speak of circulation.

[52] See Joseph O'Connell, "Metrology: The Creation of Universality by the Circulation of Particulars," *Soc. Stud. Sci.* 23 (1993): 129–73 (naturally, I regret O'Connell's choice of the word *circulation*, although his use of it is innocuous); Simon Schaffer, "Late Victorian Metrology and Its Instrumentation: A Manufactory of Ohms," in *Invisible Connections: Instruments, Institutions and Science*, ed. Robert Bud and Susan E. Cozzens (Bellingham, Wash., 1992), 23–56; Theodore M. Porter, *Trust in Numbers: The Pursuit of Objectivity in Science and Its Public Life* (Princeton, N.J., 1995); and much else.

to find. The prize in science is, rather, the establishment of uncontested normativity (i.e., standardization), or what is usually called reason.

The very plausibility of metaphors of circulation in the historiography of science reflects once again the confusion between natural philosophy and instrumentality that characterizes modern science itself. Trade routes as conduits for exchange, mediated by money (whether specie or credit), surely moved objects and people around.[53] The instrumental role of ships and (more abstractly) trading companies in effecting that movement usefully reifies and materializes this aspect of the development of global science. Ships also conveyed letters and papers—intelligence—from place to place;[54] hence the association of trade routes with distributed knowledge and the instrumentality of technical capabilities. The universality of Linnaean taxonomy in the eighteenth century and after was presupposed as a dimension of the worldwide journeys by naturalists to collect and describe the specimens that were fitted into that taxonomy.[55] Such universality also encouraged and underwrote attempts at the naturalization of useful species in alien locales. But universal taxonomies, as well as having instrumental aspects, always embodied the philosophical ambition to be *natural* classifications, whether based on a divine plan, on functional parallels, or on descent with modification. Ships and their naturalists permitted the nineteenth-century work on geographical distribution that proved so crucial to establishing that third natural-philosophical alternative.[56]

The ships and the naturalists (usually) came back; they literally circulated. The knowledge, to the extent that it was a commodity that could be recorded in accounts, accumulated. The natural philosophy, however, was never a *thing* to be collected, even though it was crucial.

But how do we know what was taken to be crucial at any given period, and how do we know what the enterprises and categories for which this cruciality obtained were taken to be? Whenever we consider these sorts of models for the operation of modern science, we must not forget that history is also about change, and that change creates difference—the source of anachronism.

HISTORICAL RELATIVISM AND HISTORICAL CONTINUITY

There is no way to avoid the cardinal sin of anachronism in the history of science; anachronism may well be the original sin of historical scholarship in general. It would be easier if the sin had only concerned knowledge of good and evil, but our sin is a determination to know the past as we claim to know the present. Historians

---

[53] Carl Wennerlind, "Credit-Money as the Philosopher's Stone: Alchemy and the Coinage Problem in Seventeenth-Century England," in "Oeconomies in the Age of Newton," supplement, *History of Political Economy* 35, no. S1 (2003): 234–61.

[54] Simon Schaffer, "Golden Means: Assay Instruments and the Geography of Precision in the Guinea Trade," in *Instruments, Travel and Science: Itineraries of Precision from the Seventeenth to the Twentieth Century*, ed. Marie-Noëlle Bourguet, Christian Licoppe, and H. Otto Sibum (London, 2002), 20–50; Schaffer, *The Information Order of Isaac Newton's "Principia Mathematica,"* Hans Rausing Lecture (Uppsala, 2008); Lissa L. Roberts, *Technology out of Context,* Inaugural Professorial Lecture, University of Twente (Enschede, 2010).

[55] Lisbet Koerner, *Linnaeus: Nature and Nation* (Cambridge, Mass., 1999).

[56] Among such naturalists were, of course, Charles Darwin, A. R. Wallace, Henry Bates, and Joseph Hooker; on Hooker, see the recent study by Jim Endersby, *Imperial Nature: Joseph Hooker and the Practices of Victorian Science* (Chicago, 2008).

of science in particular, for reasons that are slightly obscure, eagerly adopted Herbert Butterfield's condemnation of "whig history," or whiggishness, in our own field, believing that science is especially liable to it. The positive remedy has been less clear, however: broadly speaking, there is general agreement that one should treat the past "in its own terms," but there remains much anxiety about admitting that this solution might also be called relativism.

In areas of cultural history, especially concerning early modern Europe, the anthropological turn that became so popular in anglophone scholarship in the 1980s left a strong mark on the history of science. The discipline, eager to distance itself from whiggishness, could adopt cultural relativism through anthropology at one remove, as it were. The results have tended to take two principal forms. One is the history of early modern science without the science: that is, historically appropriate contemporary terms, such as *natural philosophy*, are used instead of the anachronistic *science*. The other form is a carefree, loose use of the term *science*, implicitly justified by the very knowledge that no analytically precise sense is intended or attainable.[57] Both approaches effectively deny the coherence of the history of science as a scholarly specialty. The latter merely does so by omission, and is therefore either more innocent or more culpable, depending on how important one considers the history of science as a topic. The former, because its denial is more principled, gives rise to a number of problems that are not restricted to the history of science alone and that invite us to consider the nature of our work as historians more generally.

Thoroughgoing historicism (in its usual contemporary sense)[58] suggests that differences in basic categories of understanding and action render people living in past worlds, such as that of last week, wholly other than ourselves and not to be explicated in our necessarily anachronistic terms. They must be understood, we say, in their own terms, as early modern courtiers or natural philosophers, or Victorian "men of science," rather than as modern scientists. All this is well and good, and a standard presupposition in the history of science. But fears of anachronism, or of loosely defined whiggishness, while they have been crucial to creating sensitive and insightful historical studies, cannot adequately define what historians of science do and, in particular, what they are good for in the enterprise of science studies. There remains the issue of what kinds of questions, originating from what foundations, and subject to what social or material constraints, drive historical inquiries.

A notable discussion of this issue, the relationship between the historian's present and history's past, remains the study by Adrian Wilson and Trevor Ashplant, who identified the cardinal sin of "present-centeredness." They meant by this term, quite specifically, the use of present-day conceptual categories in making sense of the past.[59] Wilson and Ashplant provide some illustrations of the fallacies that can result from such hermeneutic circularity; one example concerns Keith Thomas's *Religion and*

---

[57] See, e.g., Paula Findlen, *Possessing Nature: Museums, Collecting, and Scientific Culture in Early Modern Italy* (Berkeley and Los Angeles, 1994); Deborah E. Harkness, *The Jewel House: Elizabethan London and the Scientific Revolution* (New Haven, Conn., 2007).

[58] In other words, not Hegelian historicism of the sort against which Karl Popper inveighed in his *Poverty of Historicism* (London, 1957).

[59] Wilson and Ashplant, "Whig History and Present-Centred History," *Hist. J.* 31 (1988): 1–16; Ashplant and Wilson, "Present-Centred History and the Problem of Historical Knowledge," *Hist. J.* 31 (1988): 253–74.

the Decline of Magic (1971) and targets Thomas's scissors-and-envelopes approach to categorizing exemplary quotations.[60] But besides drawing attention to an idol of the understanding, they also acknowledge the need for practical solutions to the problem of being trapped by one's anachronistic envelopes. The unavoidability of the hermeneutic circle does not, in their view, excuse the historian from trying to escape it. Their solution, pragmatic rather than epistemological, is to address one's source materials from the initial perspective of understanding the source-generating process itself—that is, by asking why such materials (documents of various sorts, usually) were created or exist at all.[61] Generally speaking, they were not made to assist historians, although their archival preservation may have occurred because of certain assumptions at the time about what future historians might want.

The virtue of Wilson and Ashplant's idea lies in its pragmatism. They are quite aware that they have not in principle solved the problem of present-centeredness, but have merely shifted it up a level, where it will continue to threaten understandings of the source-generating process itself. But they attempt to lend a practical hand to the historian by suggesting a concrete question that can redirect attention away from, or around, some of the more insidious pitfalls. Not only is it strictly impossible to avoid using our own categories in understanding the past; it is often undesirable. Anachronism is a form of advocacy, and usually a suspect form, but advocacy is an integral part of what all historians do, whether deliberately or not. In representing the past, whether narratively or analytically, we must always translate, because we have our own things to say about it and must use our own categories of understanding in saying them. The anthropological strangeness promoted by the practices of early modern cultural history is immensely valuable intellectual therapy, especially in studying the pasts of modern science. But history is also about understanding continuities, and processes of change, that serve to connect the past to the present;[62] it is more than temporal anthropology.

A makeshift way around this challenge has been the norm at Cambridge for a decade or two by now: the use of the term *history of the sciences* instead of *history of science* for our field. The term has the virtues of rendering explicit the notion of the disunity of science and acknowledging the wealth of potential anachronisms that attend too easy a use of the monolithic label *science*, but it achieves them while neglecting the significance of calling each and every constitutive enterprise a science (this can still be attended to piecemeal, of course). The problems of present-centeredness that plague the history of science are not thereby actually solved, merely obscured; they are multiplied and distributed among the various putative "sciences" so considered.

Quentin Skinner, in influential work on the conduct of intellectual history (more specifically, the history of political thought), addressed precisely this problem in a famous article of 1969.[63] Skinner's advocacy of a "speech-act" approach to the history of ideas was, of course, widely noticed and helped to establish practices of local

---

[60] Wilson and Ashplant, "Present-Centred History" (cit. n. 59), 257–61. For a recent description of his own working methods, see Keith Thomas, "Diary," *London Review of Books* 32 (2010): 36–7.

[61] Wilson and Ashplant, "Present-Centred History" (cit. n. 59), 268–73.

[62] This last point is emphasized in A. Rupert Hall, "On Whiggism," *Hist. Sci.* 21 (1983): 45–59.

[63] Skinner, "Meaning and Understanding in the History of Ideas," *Hist. & Theory* 8 (1969): 3–53; J. L. Austin, *How to Do Things with Words*, 2nd ed. (Cambridge, Mass., 1975); also discussed in Wilson and Ashplant, "Present-Centred History" (cit. n. 59).

contextualism in the history of science itself.[64] Another aspect of Skinner's discussion was less remarked, however, at least among historians of science: he noted that intellectual historians are, of necessity,

> committed to accepting *some* criteria and rules of usage such that certain performances can be correctly instanced, and others excluded, as examples of a given activity. Otherwise we should eventually have no means—let alone justification—for delineating and speaking, say, of the histories of ethical or political thinking as being histories of recognizable activities at all.[65]

For the student of premodern science in particular, Skinner's observation nicely captures the instability that one always tends to feel in asking whether the endeavors that make up one's subject matter really are science, or should be understood as parts of entirely distinct enterprises. Skinner recognizes the epistemological abyss that confronts the historian whenever basic questions of relevance and continuity are raised and counsels a knowing and pragmatic turning away from it.

There is, nonetheless, an apparently easy way around Skinner's problem. One might claim that one's premodern historical materials, while not part of modern science itself, are part of a tradition that "fed into," and thus helped to constitute, modern science (whenever that might be taken to have begun).[66] Steven Shapin and Simon Schaffer, in one of the canonical texts of science studies, *Leviathan and the Air-Pump*, in effect make that very move, managing very slickly to insinuate into their account at the outset the proposition that everyone has always agreed that Robert Boyle's air-pump work was foundational for modern experimental science.[67] Since no potential critic of Shapin and Schaffer's approach was likely at that stage of the argument to raise an objection, they could take the point as if it were established and go on to show things about Boyle's work that, by implication, already applied to "modern experimental science" without their having to argue for the link of continuity themselves. This was clever, and, by and large, it worked.

But there is a limit to how often, and how effectively, that trick can be used. Eventually, as Skinner anticipated, all that can be done is to assert historical continuities and hope that others will be convinced. Skinner had abandoned (or rebelled against) a dominant approach in his field that took certain concepts or problems as effectively sempiternal, and therefore in play in all times and places, but he simultaneously felt the need to avoid a consequent historical relativism that would exclude him from participating in his own well-established scholarly specialty; hence his blunt assertion that there were in fact marks of a specific intellectual activity recognizable to us as

[64] See Steven Shapin, "Social Uses of Science," in *The Ferment of Knowledge: Studies in the Historiography of Eighteenth-Century Science*, ed. George S. Rousseau and Roy Porter (Cambridge, 1980), 93–139, on contextualism and his concomitant opposition to historical explanation in terms of influence (on which see also Skinner, "The Limits of Historical Explanations," *Philosophy* 41 [1966]: 199–215; cf. Forman, "(Re)cognizing Postmodernity" [cit. n. 32], attempting to rehabilitate talk of "influence" in the history of science).

[65] Skinner, "Meaning and Understanding" (cit. n. 63), 6.

[66] A particularly forthright expression of a related perspective is found in Andrew Cunningham and Perry Williams, "De-centring the 'Big Picture': *The Origins of Modern Science* and the Modern Origins of Science," *Brit. J. Hist. Sci.* 26 (1993): 407–32.

[67] Shapin and Schaffer, *Leviathan and the Air-Pump: Hobbes, Boyle, and the Experimental Life* (Princeton, N.J., 1985), 3–5. See now the authors' new introduction to the 2011 edition, "Up for Air: *Leviathan and the Air-Pump* a Generation On," xi–xl.

theoretical political discourse in the distant setting of early modern Europe. By call-ing these marks "criteria and rules of usage," but without enumerating them in any principled way, he in effect claimed that you recognize them when you see them—a very conservative stance for a then-young Turk to adopt.

A very similar formalization of this issue, independent of these specifically his-torical considerations, was proposed in 1993 for the case of science studies by my colleague Michael Lynch.[68] When one looks around at the enormously diverse set-tings that constitute the practices of science, there appear many recurring themes, spoken of by the scientists themselves, that are related to the making of scientific knowledge: theory, deduction, experiment, hypothesis, observation, and so on. These epistemic themes can be understood as having a historically transitive, as well as a general situationally transitive, application. Just as Skinner wanted to save the "his-tory of political thinking" as the history of a still "recognizable activity," so Lynch's epistemic themes serve to save science studies as the investigation of a recognizable activity in a wide range of different times and places. And that would seem to re-inforce the argument for doing the same in the history of science itself.

The problem remains, in practice, of how the historian is to identify and justify the temporally transitive categories (epistemic themes) that will serve to make some past scene a legitimate part of the history of science. Lynch's proposal relies on track-ing the use of characteristic terms and idioms among (typically, English-speaking) groups of scientists. Skinner had to allow a broader linguistic scope to his early modern political recognizable activities, but he could generally manage in practice, thanks to cognates as well as the widespread use of Latin, which allowed for stan-dardized contemporary translations of terms from European vernaculars. But what the parallels between Lynch and Skinner succeed in doing is underlining the point that the history of science is never the history of an ahistorical human activity, just as political history properly speaking is not the history of a universal human practice, or (to take other parallels) just as the history of art is subject to disputes over the nature of its subject matter, or religious studies to disagreements over what religion is. The European models of the nineteenth century can no longer be assumed as universal.

NATURAL PHILOSOPHY AND INSTRUMENTALITY
AS THE THEMES OF MODERN SCIENCE

In effect, therefore, those nineteenth-century European models have a consequential history, and science is arguably the most consequential of them all. Modern science, understood as an ideology, may be breaking apart, rather than shifting from an enter-prise seen by scholars as basic science to one seen as fundamentally "technology," as Paul Forman has it. The original composition of this ideology *was* the formation of modern science. The dissolution of the ideology will perhaps involve a reformulation of what science itself is taken to be, but the ideology should not be counted out yet.

Adopting Skinner's broad historicist sensibilities while at the same time attempt-ing to retain his sense of historical continuity, of links to the present, depends in this case upon justifying the categories of natural philosophy and instrumentality, discussed above, as applicable both to contemporary science and to earlier periods

---

[68] Lynch, *Scientific Practice and Ordinary Action: Ethnomethodology and Social Studies of Science* (Cambridge, 1993), esp. 280, 300.

in (primarily European) history. There remains, however, something of a difference from Skinner's specific problem. Having determined the characteristic applicability of the two categories to modern science, and located the time and sociocultural region in which modern science in that sense first became established, it still remains necessary to examine, understand, and explain the earlier careers of those categories and how the two became, as it were, symbiotic so as actually to create it.

That tripartite historical narrative would, of course, be necessarily present centered, insofar as it would be premised on the eventual emergence of a familiar modern science with a particular ideological form. There is nothing epistemologically suspect in this kind of present-centeredness, however. It is not teleological, since it does not use the (historically contingent) eventual appearance of modern science as a causal agent to account for earlier events; it is not, by the same token, "historicist" in the older sense. Furthermore, the categories that the narrative uses, natural philosophy and instrumentality, would have to be identifiable with equivalent categories in the cultural settings constituting the prehistory of modern science; if they were not, they could have no explanatory efficacy in this particular story. The entire model is theoretical, in that it is conjectural and to be judged by the work it does in organizing and making sense of historical accounts, much like many explanatory models in the sciences themselves.

Only the presence of this ideology can identify "science" in its modern sense: not particular practices, or specific ideas, but a self-effacing ideological construct that makes claims going beyond what it can fully deliver. The history of this ideology, made up of cultural, intellectual, and social reifications, is a coherent project for writing the history of modern science. The only alternative seems to be a history of "the technical," of STM, a grouping so amorphous, once released from quasi-positivist assumptions about the nature of science that few in our field would now willingly admit, that no intellectually respectable discipline could be formed around it.

On the Historical Forms of Knowledge Production and Curation:

Modernity Entailed Disciplinarity, Postmodernity Entails Antidisciplinarity

by Paul Forman[*]

ABSTRACT

This article continues and extends my previous efforts to characterize modernity and to map the change in state of mind constituting the transition to postmodernity. Here I make a case for marking that transition by the dramatic fall, between the early 1960s and the early 1970s, in the cultural valuation of professions and of disciplines. I show that four culturally presupposed values—proceduralism, disinterestedness, autonomy, and solidarity—along with discipline, in a characterological sense, were characteristic for modernity and indispensable to the conception and sustentation of disciplinarity. Then I evidence the inversion in contemporary culture of each of those five value presuppositions. In this way I show postmodernity to be antithetical to disciplinarity.

> Students used to come to me saying things like, "thank you for telling us about paradigms—now that we know what they are we can get along without them." All seen as examples of oppression. That wasn't my point at all! —Thomas S. Kuhn, 1995[1]

INTRODUCTION

"The fate of the disciplines" has rightly been a matter of urgent concern in academia for more than a generation. Indeed, it is a matter of far more than academic import: as Perry Anderson observed in *The Origins of Postmodernity*, "there could be no more ominous symptom of some cracking in the modern" than the rise of discipline-

[*] MRC-631, Smithsonian Institution, Washington, DC 20013–7012; formanp@gmail.com.
This paper has been improved though the criticism of many colleagues: Robert Adcock, Aitor Anduaga, Joan Bromberg, John Burnham, Martin Collins, Charles Gillispie, Tal Golan, John Heilbron, Karl Hufbauer, Anne Marcovich, Allan Needell, Lewis Pyenson, Sam Schweber, Charles Thorpe, Zuoyue Wang, and Peter Westwick. Their interest and their time, generously given, are evidence against my thesis.
[1] Aristides Baltas, Kostas Gavroglu, and Vassiliki Kindi, "A Discussion with Thomas S. Kuhn," i.e., the "autobiographical interview" referred to in the title of Kuhn, *The Road since "Structure": Philosophical Essays, 1970–1993, with an Autobiographical Interview*, ed. James Conant and John Haugeland (Chicago, 2000), 253–323, on 308. The exclamation point is present only in the original publication: *Neusis* 6 (1997): 145–200, on 187–8.

disregarding discourse. It is thus passing strange that, while the commentaries on this circumstance reach into the thousands, there does not exist any extended historical discussion treating that dedifferentiation of scholarly discourse, and the breakdown of disciplinarity, as a manifestation of the broad cultural transformation to which Anderson points; namely, the cracking-up of modernity and the onset of an essentially different cultural-historical epoch: postmodernity.[2]

Although it is widely understood that, as David Hollinger put it ten years ago, "universities with the structure and functions we take for granted are the products of a particular historical moment long since gone," those within our institutions of higher learning who wish them restructured and repurposed to comport with the present historical moment are generally to be found only in its topmost administrative ranks and bottommost academic ranks.[3] Professors understand that such restructuring and repurposing would bring complete intellectual chaos—not to mention the elimination of almost every gratifying feature of the professorial profession. The substance of the matter is thus never treated more lengthily than with a few paragraphs, and the demise of disciplinarity is almost always approached from the perspective of postmodern pragmatism; namely, how do we academics manage to keep our mills turning, ensure careers for our graduate students, and otherwise just get along, while studiously ignoring the collapsing cultural foundations for disciplinary production and curation of knowledge?[4]

[2] Anderson, *The Origins of Postmodernity* (New York, 1998), 61; "The Fate of the Disciplines," ed. James Chandler and Arnold I. Davidson, special issue, *Crit. Inq.* 35, no. 4 (2009). The lead article provides an introduction to the literature: Robert C. Post, "Debating Disciplinarity," 749–70. Measured by the amount of literature cited, Post's is, to my knowledge, the closest approach to a scholarly discussion of the fate of the disciplines. However, law professor Post, innocent of historical sensibility, considers that literature without regard to its date of publication. Moreover, like most writers on disciplines and disciplinarity in the past thirty to forty years, Post combines hostility to disciplinarity with a presumption that disciplines are here to stay. Similar to Post is Louis Menand, who for years has enjoyed heavy funding to do better, but has produced little; *The Marketplace of Ideas* (New York, 2010). Peter Novick's *That Noble Dream: The "Objectivity Question" and the American Historical Profession* (New York, 1988), an impressive accomplishment, comes closest to being such an account of the breakdown of disciplinarity for one discipline, but suffers from Novick's refusal explicitly to treat historians as moving with their wider cultural milieu, albeit he provides much evidence of this. John Ziman's *Real Science: What It Is, and What It Means* (New York, 2000) is an informed, insightful account of the transformation of scientific life that postmodernity has brought about, and specifically in its discipline-disregarding aspects. Peter Weingart's *Die Stunde der Wahrheit? Zum Verhältnis der Wissenschaft zu Politik, Wirtschaft und Medien in der Wissensgesellschaft* (Weilerswist, 2001) also provides an overview of the manifold antidisciplinary forces at work on and in the sciences at the end of the twentieth century. The ever more widespread and implicitly antidisciplinary rhetoric of interdisciplinarity, and its vast literature, I leave entirely aside.

[3] Hollinger, "Faculty Governance, the University of California, and the Future of Academe," *Academe* 87, no. 3 (2001): 30–3, on 33, available at http://www.aaup.org/AAUP/pubsres/academe/2001/MJ/ (accessed May 2008). Outside the professorial ranks advocates are plentiful—above them in the ranks of university administrators (most prominently, Michael Crowe), and below them among the providers of remedial instruction; e.g., Randi Gray Kristensen and Ryan M. Claycomb, eds., *Writing against the Curriculum: Anti-disciplinarity in the Writing and Cultural Studies Classroom* (Lanham, Md., 2010). Within the professorial ranks, most prominent is Mark C. Taylor, "End the University as We Know It," op-ed, *New York Times*, April 27, 2009, http://www.nytimes.com/2009/04/27/opinion/27taylor.html; Taylor, *Crisis on Campus: A Bold Plan for Reforming Our Colleges and Universities* (New York, 2010).

[4] Exemplary are Allan Kulikoff, "A Modest Proposal to Resolve the Crisis in History," *Journal of the Historical Society* 11 (2011): 239–63, and, again, Menand, *Marketplace of Ideas* (cit. n. 2), 55, 80, 102–5. As the natural scientists require outside support, they have no choice but to go with the flow, and thus have all the less reason to overcome their disinclination to such discussion.

The widely shared illusion that such a head-in-the-sand coping strategy is legitimate and effective arises, in part, from a near-universal confusion of postmodernism and postmodernity—that is, of the self-consciously held ideology and associated literary practices that constitute postmodern*ism*, and postmodern*ity*, a historical epoch characterized by a set of unreflectively held cultural presuppositions.[5] When postmodernism and postmodernity are confused, the indications of a general recoiling from postmodern*ism* in the academy since the 1990s can be, and commonly are, taken as evidence that postmodern*ity* is going away and therefore need not be confronted in any but a pragmatic, makeshift way.[6] Thus we continue to speak blithely of this and that discipline, as though there still exist—as though there *could* still exist—meaningful instantiations of the ideals and institutions signified by disciplinarity. But, I here contend, the ideals and institutions so signified were expressive of, and consequently dependent upon, key cultural presuppositions of modernity, presuppositions that today, in postmodernity, as postmodernity, no longer hold sway. Indeed, in important respects the presuppositions of our (now increasingly global) common culture are inversions of those distinctive of modernity. Consequently, disciplinarity is now an impossibility, both ideologically and practically.

In a previous essay at locating and characterizing the modern-to-postmodern transition I took the inversion of the cultural valuation of technology relative to science as a marker of that transformation. I contended that circa 1980 technology and science exchanged the rank and role that had been ascribed to them, respectively, in modernity, by modernity; that the primacy of science to and for technology, everywhere presupposed in modernity, has, since circa 1980, been superseded by a presupposition of the primacy of technology.[7] However, a fact on the ground of culture so

---

[5] Clearing up that confusion of postmodernity with postmodernism would be a hard job. Even Anderson (*Origins of Postmodernity*, cit. n. 2), whose knowledge is broad enough to have allowed him to avoid that confusion, fell into it, placing *postmodernity* in his title but devoting his text to postmodernism. Early on, Zygmunt Bauman sought to maintain a distinction between the ideological *-ism* and the historical *-ity*. But facing the difficulty of bringing this distinction home to his ever-wider, historically uninterested, audience, and fearing that any use of any form of the word would inevitably tar him as a postmodernist, Bauman abandoned *postmodernity*; Michael Hviid Jacobsen, Sophia Marshman, and Keith Tester, *Bauman beyond Postmodernity: Critical Appraisals, Conversations and Annotated Bibliography, 1989–2005* (Aalborg, 2007), 12–3. I, having much less to lose, and having a stronger commitment to clarity and consistency, have stuck to that distinction; Paul Forman, "Recent Science: Late-Modern and Post-modern," in *Science Bought and Sold: Rethinking the Economics of Science*, ed. Philip Mirowski and E.-M. Sent (Chicago, 2002), 109–48; Forman, "(Re)cognizing Postmodernity: Helps for Historians—of Science Especially," *Ber. Wissenschaftsgesch.* 33 (2010): 157–75.

[6] In "What If They Gave a Science War and Only One Side Came? Ask the American Anthropological Association," *Chronicle of Higher Education*, January 14, 2011, sec. B: *Chronicle Review*, B10–B11, available at http://chronicle.com/article/What-if-They-Had-a-Science-War/125828/ (accessed January 2011), Hugh Gusterson evinces such a confusion of postmodernity with postmodernism, combined with a dismissal of postmodernism as passé, all in the service of disciplinary damage control. It is also an example of the phobia toward all forms of the word *postmodern* that is common to scholars in science and technology studies (STS), stemming in large part from the science wars of the late 1990s; Forman, "(Re)cognizing Postmodernity" (cit. n. 5), 159, 168 nn. 8–9, 175 n. 52.

[7] Paul Forman, "The Primacy of Science in Modernity, of Technology in Postmodernity, and of Ideology in the History of Technology," *Hist. & Tech.* 23 (2007): 1–152. As the method and claims of that article have not always been understood, I emphasize here that it does not consider the actual, factual relations between science and technology, but only the putative, culturally presumed relations. If this rigorous restriction to cultural description is understood, then it is less likely that I be misunderstood as sharing the wide scholarly prejudice in favor of economic explanations of such cultural transformations as are described in that article and this. Thus Robert Kohler ("Lab History: Reflections," *Isis* 99 [2008]: 761–8, on 765), in placing his considerable authority behind the view that modern, disciplinary science was "an interlude" in world history, intimates that I am with him in thinking that the proper

specific as the valuation of technology relative to science cannot possibly be an isolated fact; it must necessarily be a particular manifestation of a more general, more comprehensive system of cultural values having wider implications for the valuation of science—and not merely science. I proposed in that previous article that the more comprehensive change in cultural presuppositions underlying the upward valuation of technology and downward valuation of science was the inversion of primacy between means and ends. This inversion was a world-historical reversion—a reversion from that primacy of means, that proceduralism and the esteem of disinterestedness entailed by proceduralism, that distinguished modernity among all geotemporal cultural domains—a reversion to the primacy of ends, the presumed principle from which action proceeded in all other times and all other climes.[8]

The present article continues and extends that endeavor to locate temporally, and discriminate culturally, the modern and postmodern eras. Here, in lieu of the valuation of technology relative to science, the discriminating characteristic is the valuation of disciplinarity relative to any and all other ways of generating and instituting knowledge. *Disciplinarity* is an abstract noun referring to a cultural ideal, to a set of presuppositions about where the value of knowledge lies and what sorts of knowledge possess highest value, about the morally charged behavioral norms that producers and curators of knowledge must satisfy, and about the proper embodiments of knowledge in formal institutions. As a distinct cultural constellation disciplinarity began to take shape only toward the end of the eighteenth century. It attained clear articulation and concerted implementation only in the nineteenth century, and even then was realized only slowly and imperfectly. The triumph of disciplinarity as a hegemonic cultural ideal came about during the fifty years following the First World War. Toward the end of that half century, in the two decades following the Second World War, disciplinarity was almost universally regarded as the inevitable, as well as the most estimable, mode of knowledge production. Once attained, it was supposed necessarily to remain—in perpetuity, the end of history.

Yet, in a volte-face then inconceivable, the next decade, from the midsixties through the early seventies, saw the wide and rapid rise of a conviction that the institutions

---

explanation of the inception of that interlude, the content of that interlude, and the ending of that interlude "is more economic than cultural." I, however, do not think that the rise of disciplinary science can be explained economically. I am quite certain that the world-historical transformation from modernity to postmodernity that caused the Icarian fall of disciplinarity was too broad, too fast, and involved too radical a reversal of modern perspectives to be explained by economic circumstances. I welcome the recent work of Philip Mirowski, *Science-Mart: Privatizing American Science* (Cambridge, Mass., 2011), in which this historian of economic thought assigns the determinative role not to economic realities but to an economistic ideology.

[8] Transcendence is the most important aspiration or illusion of modernity neglected in this exposition of those determinative of disciplinarity. My warrant for here neglecting the aspiration to transcendence is that although characteristic of modernity, it is not distinctive of modernity. That aspiration is also characteristic of significant sectors of knowledge production and curation throughout the length and breadth of premodernity. The preposterousness yet strange durability of this aspiration or illusion of transcendence was the theme of all of David F. Noble's scholarly books from the mid-1980s to the untimely end of his life: *A World without Women: The Christian Clerical Culture of Western Science* (New York, 1992); *The Religion of Technology: The Divinity of Man and the Spirit of Invention* (New York, 1997); *Beyond the Promised Land: The Movement and the Myth* (Toronto, 2005). Writing this last, it seemed to Noble that contemporary global environmental justice activists, but so far they only, had "come to the understanding that beyond the promised land is the here and now" (1). This understanding might, however, be seen as presumed by postdisciplinary knowledge production—which is to say most postmodern knowledge production. Meanwhile, elsewhere in postmodern life and thought we cling to transcendence in ever more preposterous forms.

and presuppositions of disciplinarity are nefarious impositions whose structures and strictures must needs be transgressed. Like the sudden reversal circa 1980 in the valuation of technology relative to science, the sudden reversal in the cultural valuation of disciplinarity provides an index of the modern-to-postmodern transition—an index both more sensitive and more comprehensive than the valuation of technology relative to science; an index that points, as we should rather expect, to a slightly earlier date for the onset of that epochal cultural transformation. Thus, even more than the devaluation of science relative to technology, the abrupt devaluation of disciplinarity, modernity's final and distinctive mode of knowledge production and curation, underscores both the uniqueness and the fragility of modernity and compels us to regard modernity not as the end but as the anomaly in human history.

The sensitivity and comprehensiveness of the valuation of disciplinarity as an index of the transition from modernity to postmodernity is necessarily a function of the centrality and fundamentality of the cultural presuppositions implicating and implicated by disciplinarity. I here identify four cultural values central and fundamental to modernity that seem self-evidently necessary for the conception and sustentation of disciplinarity as the ideal form of knowledge production and curation: the two already emphasized as underlying the valuation of technology relative to science—namely, the primacy of means and the esteem of disinterestedness—and two further principal presuppositions of modernity; namely, the high value of autonomy and of solidarity. These four values, along with the high value placed on discipline in a generalized characterological sense, are not only essential to disciplinarity, but spanned the cultural space of modernity, and their negations very nearly span the cultural space of postmodernity.[9] Consequently, the metamorphosis of scientific and scholarly disciplines from avatars of cultural ideals into cultural bêtes noires constitutes a very good indicator—perhaps the best indicator—of the great break in cultural history that is the modernity-to-postmodernity transition.

Disciplines versus Professions

Disciplines are institutions for production, validation, and transmission of knowledge. They were peculiar to modernity; indeed, to the short modernity dating from the second third of the nineteenth century. They are, however, commonly conflated with a different type of institution that long predated modernity: the profession. Professions are institutions oriented primarily to the provision of a practical service on the basis of possession—real, presumed, or pretended—of a distinctive body of knowledge. Disciplines are institutions oriented not to the provision of a practical service but to the production and curation of a distinctive body of knowledge. Although in relation to its distinctive epistemic basis any given profession abuts, overlaps, and draws upon one or more disciplines, professions and disciplines have different, sometimes antithetical, purposes and practices.[10] Although there is some similarity in

---

[9] None of these terms (method, disinterestedness, autonomy, solidarity, or discipline) appears among Raymond Williams's approximately 150 keywords of modern culture—neither in Williams's 1973, 1983, and 1985 editions, nor in the 2005 edition; Tony Bennett, Lawrence Grossberg, and Meaghan Morris, eds., *New Keywords: A Revised Vocabulary of Culture and Society* (Malden, Mass., 2005).

[10] Long ago, Dorinda Outram ("Politics and Vocation: French Science, 1793–1830," *Brit. J. Hist. Sci.* 13 [1980]: 27–43, on 28), introducing her critique of the imposition of the concept of professionalization upon the scientific "vocation," observed that professions "assume an applied component,

the formal institutions through which the differing purposes of professions and disciplines are pursued, it is not those formal similarities that are of prime importance for the history of science but the differences in purpose and practice. It is thus greatly to the discredit of the discipline of the history of science that it never elaborated this distinction between professions and disciplines, nor attended carefully to it.[11] Rather, an indiscriminate alternation between *discipline* and *profession* has long been characteristic of the writings of most historians and sociologists of science and now mars nearly all writings on the crisis of disciplinarity by scholars of every stripe.[12]

But if this conflation of disciplines with professions is deeply seated in the minds of those whose interests should lie in carefully discriminating these two sorts of institution, so different in purpose and practice, how could it be otherwise in the public mind? And if both clerisy and laity so conflate disciplines and professions, a

---

an ideal of service. In the natural sciences, however, it is difficult to identify the client served or the service provided." This pertinent point has been rather consistently ignored in the past thirty years. Of particular value as a review of the history of the preoccupation with professionalization in a closely related historical discipline is John C. Burnham, *How the Idea of Profession Changed the Writing of Medical History* (London, 1998).

[11] Circa 1980 the need for clarification of the difference between professionalization of science and discipline formation in science was widely felt; *Dictionary of the History of Science*, ed. W. F. Bynum et al. (Princeton, N.J., 1981), s.v. "discipline history," by Robert E. Kohler; Karl Hufbauer, *The Formation of the German Chemical Community, 1720–1795* (Berkeley and Los Angeles, 1982), 2–5, 150; Outram, "Politics and Vocation" (cit. n. 10). Recently, G. Matthew Adkins, in "The Renaissance of Peiresc: Aubin-Louis Millin and the Postrevolutionary Republic of Letters," *Isis* 99 (2008): 675–700, has noted that the large body of scholarship on that classic phase of French science suggests not a conjunction but an opposition between those two processes. The antithesis between profession and discipline, as well as the confused and inconsistent usage in the history of science literature, was stressed by Yves Gingras, *Physics and the Rise of Scientific Research in Canada*, trans. Peter Keating (Buffalo, N.Y., 1991), 4–8, 117–9. Alone, to my knowledge, Gingras's "L'institutionnalisation de la recherche en milieu universitaire et ses effets," *Sociologie et sociétés* 23 (1991): 41–54, on 41–3, made a suggestion for a consistent distinction.
The antithesis between a profession and a discipline seemed clear among the late nineteenth-century American cultural elite, as has recently been emphasized by Paul Lucier, "The Professional and the Scientist in Nineteenth-Century America," *Isis* 100 (2009): 699–732. It can be seen in the highly favorable connotations of *discipline* and the unfavorable connotations of *profession* in Herbert Croly, *The Promise of American Life* (New York, 1909; reprinted four times in the following five years and now available at http://en.wikisource.org/wiki/The_Promise_of_American_Life [accessed May 2010]), where one finds "professional millionaires," "professional bankrupt," "professional revolutionists and reformers," and "professional socialists." This, presumably, is what Charles Rosenberg had in mind when insisting on (but not stating or elaborating) that difference; "Toward an Ecology of Knowledge: On Discipline, Context, and History," in *The Organization of Knowledge in Modern America, 1860–1920*, ed. Alexandra Oleson and John Voss (Baltimore, 1979), 440–55. But by then insistence on that difference had been rendered passé by the onset of postmodernity, as reflected in Thomas L. Haskell's "Are Professors Professional?" *J. Soc. Hist.* 14 (1981): 485–93, an unsympathetic review of that volume largely devoted to Rosenberg's contribution. Although Rosenberg's commentary remains one of the most frequently cited articles in the history of science literature, postmodernity has ensured that we have heard almost nothing more of the discipline-profession distinction in the past two decades.

[12] So, e.g., the otherwise intellectually rigorous M. H. Abrams, in "The Transformation of English Studies: 1930–1995," *Daedalus* 126, no. 1, "American Academic Culture in Transformation: Fifty Years, Four Disciplines" (1997): 105–31, moves back and forth, indiscriminately, between "the profession of literature," "the literary discipline," "English Studies," and "English . . . as an established academic discipline." Jerry Z. Muller's "Discontent in the Historical Profession," *Society* 36, no. 2 (1999): 12–8, an account of the formation of the Historical Society in consequence of discontents in which, implicitly, the distinction between discipline and profession is central, has eleven appearances of a *d* word and eighteen of a *p* word, with not a word on the difference. Similarly, Steven Shapin has packed into the first two pages of "Hyperprofessionalism and the Crisis of Readership in the History of Science," *Isis* 96 (2005): 238–43, some twenty-three appearances of a *d* word, and eighteen of a *p* word, with, again, not a word on the difference.

historical examination of the cultural regard of disciplinarity must necessarily attend also to the cultural regard of professionality. Indeed, one by-product of the present inquiry is a better understanding of the grounds of this conflation; namely, that however fundamental from the perspective of the history of science the difference between disciplines and professions is, or ought to be, that difference counts little against the broad overlap in cultural presuppositions determining the cultural regard for professions and disciplines alike—to their common weal in modernity and their common woe in postmodernity.

THE INEVITABLE'S BELATED ARRIVAL

The history of science has a long history. From the late eighteenth through the late nineteenth century, histories unsurpassed in their amassment of material were compiled for various fields of inquiry, motivated in part by the aim of creating clear and distinct disciplinary identities. Yet by the early twentieth century a conception of what the history of science should be—a conception that judged all these prior historical accounts to be fundamentally flawed and requiring reconstitution—seemed uncontestable. George Sarton made that new instauration his mission. "The fundamental purpose" of his tireless efforts was, he said, "to establish the history of science as an independent and organized discipline." But how? Above all, a discipline demanded discipline; for Sarton and for his generation the duality of meaning was implicit and essential. "I believe," said Sarton, "that no greater service can be rendered to the history of science, at this juncture, than by relentlessly insisting upon the necessity of raising the standard of scholarship as high as possible." This, which Sarton declared publicly in 1918 in the pages of *Science*, he had earlier pled privately to the Carnegie Institution of Washington, where that perspective had already been firmly established with the appointment in 1905 of J. Franklin Jameson as director of the Department of Historical Research. "To set a standard of workmanship and compel men to conform to it" was the way Jameson had put it.[13]

Whence came Jameson's and Sarton's confidence that their disciplinary demands—so astonishing to us in their baldness—would be, must be, acceded to? At bottom, their confidence in the necessity and the inevitability of disciplinarity (which they shared with all their discipline-building contemporaries) rested on their presumption of a modern disciplinary personality. This was the personality that Weber identified as underlying the commercial success and capitalistic accumulation of Protestant, and more especially Calvinist, Europe in the seventeenth and eighteenth centuries. This

---

[13] Sarton, "Introduction to the History and Philosophy of Science (Preliminary Note)," *Isis* 4 (1921): 23–31, on 23, quoted by Arnold Thackray and Robert K. Merton, "On Discipline Building: The Paradoxes of George Sarton," *Isis* 63 (1972): 472–95, on 473; Sarton, "Review of Charles Singer, *Studies in the History and Method of Science*," *Science* 47 (1918): 316–9, on 318, quoted by John L. Heilbron in his lecture at the opening ceremony of the Max Planck Institut für Wissenschaftsgeschichte (MPIWG), March 31, 1995, published in MPIWG's *Annual Report* (1995), 183–201, on 188. Morey D. Rothberg, "'To Set a Standard of Workmanship and Compel Men to Conform to It': John Franklin Jameson as Editor of the *American Historical Review*," *Amer. Hist. Rev.* 89 (1984): 957–75, quoting Jameson, "The Influence of Universities upon Historical Writing" (1902). Jameson was then writing as head of the history department at the University of Chicago, from where he went, shortly after, to the Carnegie Institution of Washington. Likewise, Novick, *That Noble Dream* (cit. n. 2), 52; Lewis Pyenson, *The Passion of George Sarton: A Modern Marriage and Its Discipline*, Memoirs of the American Philosophical Society 260 (Philadelphia, 2007), 317–8.

was the personality indispensable to the success of self-governing nation-states, and in the nineteenth century all polities endeavoring to be such worked hard to inculcate this personality. This was the personality whose structure Freud mapped. Formed by internalization of an array of imposed disciplines, this personality was organized by and around self-discipline. This was thus a personality disposed to regard respectfully, and to derive self-respect from participation in, the endeavor to create knowledge production, knowledge curation, and knowledge application enterprises under the flag of discipline and on the basis of an individual and collective discipline.[14]

"Disciplines" versus "Disciplined"

By the middle of the twentieth century disciplinary production and curation of knowledge would become the most honored and emulated manifestation of modernity's high valuation of discipline—a valuation that had been rising and spreading through Western societies for a full four centuries. But disciplinarity, as such, had a much shorter history, being the last institutionalization of discipline to appear in modernity, and the briefest to reign. One can see how largely the modern conception of scientific and scholarly disciplines was built upon a cultural foundation of respect for discipline, especially self-discipline, and also how lately disciplinarity gained its place as modern cultural ideal, by tracing the predominant meaning of the word *discipline*, noting in particular both the tardiness and the suddenness of the word to come to prominence in the sense that we scholars now take so completely for granted and use so exclusively (except when speaking about other people's children).

Although the word *discipline* in the sense of a learned discipline goes back to classical antiquity and has a continuous history in Latin through the Western Middle Ages, and in all the European vernaculars from the early modern period, that sense did not become the predominant sense in which the word was used until the middle decades of the twentieth century—yes, the *twentieth* century. To be sure, Robert Boyle used the word *disciplines* in this sense in the title of an essay published in 1671. But there is not another such usage of this word in its plural form to be found among the titles in the catalogs of Harvard University libraries and the Library of Congress until 1932, when it appears in the title of Clifford Dobell's biography of Antoni van Leeuwenhoek.[15] This near absence of *discipline* in the sense of a scholarly

---

[14] Norbert Elias, *The Civilizing Process: Sociogenetic and Psychogenetic Investigations*, 2nd ed., rev., 2 vols. in one (Malden, Mass., 2000), 182, 190, 387; Philip S. Gorski, *The Disciplinary Revolution: Calvinism and the Rise of the State in Early Modern Europe* (Chicago, 2003). P. S. Atiyah, *The Rise and Fall of Freedom of Contract* (New York, 1979), on 273: "Englishmen were taught self-discipline by every possible means." (Thomas L. Haskell's writings drew my attention to Atiyah's important book.) William James, *The Moral Equivalent of War* [1906], *and Other Essays* (New York, 1971), 14. (By "moral equivalent" James meant "an equivalent discipline.") "Socialist discipline" was central to socialist thinking and practice from Comte forward; see n. 77 below. Freud's translators' liking for "self-discipline" is attested by *The Concordance to the Standard Edition of the Complete Psychological Works of Sigmund Freud*, ed. Samuel A. Guttman, Randall L. Jones, and Stephen M. Parrish (Boston, 1980–), 2:170. Novick (*That Noble Dream*, cit. n. 2, on 270) equates objectivity to work that is "inhibited by an objectivist superego."

[15] On the evidence of those catalogs, throughout the four centuries prior to the 1930s *disciplines* was used almost exclusively in the sense of religious disciplines. In its singular form, the noun *discipline* appears in many more titles, but the relative probability of finding it used in the sense of a scientific or scholarly discipline is even smaller. (The number of such titles is so great that I have not undertaken to examine each.) The shift in the predominant meaning of *disciplines* in the titles of works in the

or scientific discipline in the titles of books published prior to the twentieth century is also found, Aitor Anduaga kindly points out, in the catalogs of the national libraries of France, Spain, Portugal, and Italy.[16]

Even in the last decades of the nineteenth century and the first decades of the twentieth, as scientists and scholars were self-consciously engaged in construction of learned disciplines, that sense of the word seems to have remained rare in general cultural discourse, perhaps precisely because in those decades *discipline* in the sense that we now apply only to children was gaining strength as a cultural keyword. An indication of this centrality and high positive valence of *discipline* in a characterological sense is the number of appearances of the word in Herbert Croly's widely read and highly acclaimed *The Promise of American Life*, and more especially in its final, culminative, chapter, "The Individual and the National Purpose." There *discipline* or *disciplined* appears twenty-seven times—on every other page, on average—always in the sense of a beneficent effect of character or circumstance. Croly's usages approach the notion of learned discipline no closer than references to "educational discipline"; that is, they are still a long way off.[17]

Mugwump Croly, representing the cultivated classes, presented their conception of the basis for a national sociopolitical consensus. Yet as regards the centrality and valuation of *discipline*, even so self-consciously differently thinking and behaviorally nonconforming a scholar-intellectual as Thorstein Veblen shared Croly's view. Writing *The Higher Learning in America* in the second decade of the century, Veblen used the word *disciplines* only in conjunction with *practical* and *utilitarian*, a pejorative association in Veblen's mind. Nowhere in this book about the policies and practices of American universities did Veblen use that word to refer to learned disciplines. Five times more frequent than *disciplines* in that text are other forms of the word *discipline*—chiefly, *disciplined*—all of which refer, and refer favorably, to the character of some action, whether exerted or suffered, and not to fields of investigation or instruction, let alone a collective of learned persons devoted thereto.[18]

Harvard and Library of Congress catalogs began only around 1940. Over the following decades, those works treating of scientific and scholarly disciplines outnumbered those using that word in some other (usually religious) sense by ever-increasing factors. Doubtless a positive feedback effect contributed to this shift from the religious to the secular meaning.

[16] Of the 1,284 works in the catalog of the Bibliothèque Nationale de France published between 1500 and 1900 having *discipline* in their title, Anduaga found that not one uses the word in the sense of a learned discipline. Titles containing the plural form, *disciplines*, number only ten, and, again, not one uses the word in the sense of a learned discipline. Searching the catalogs of the Biblioteca Nacional de Portugal, the Biblioteca Nacional de España, and the Biblioteca Nazionale Centrale di Roma for works published between 1500 and 1900 having *disciplina* (the singular form in Latin, Portuguese, Spanish, and Italian) in their titles, he found that, in the first, among 634 works, only three use the word in the sense of a learned discipline; in the second, among 320 works, only four; in the third, among 262 works in Latin, only ten, and among 71 works in Italian, none. Again in every case, as in English, the plural form appears in titles far less frequently, but nonetheless there are within those much smaller numbers more pertinent uses of the word. Personal communication from Anduaga, August 15, 2011.

[17] Croly, *Promise of American Life* (cit. n. 11). This book "immediately established Croly as the most important theorist of American progressivism": James T. Kloppenberg, *Uncertain Victory: Social Democracy and Progressivism in European and American Thought, 1870–1920* (New York, 1986), 313–4. For the ambivalence that romanticism introduced into the valuation of discipline in these decades, see my examination of Lewis Mumford's attitudes, "Appendix: Mumford's 'Discipline,'" in Paul Forman, "How Lewis Mumford Saw Science, and Art, and Himself," *Historical Studies in the Physical and Biological Sciences* 38 (2007): 271–336, esp. 323–6.

[18] Veblen, *The Higher Learning in America: A Memorandum on the Conduct of Universities by Business Men* (New York, 1918). The closest Veblen comes to our use of *disciplines* is the following: "The university of medieval and early modern times, that is to say the barbarian university, was

Croly and Veblen were in this regard typical writers of books in English before 1920. Reference to the database created by Google as a by-product of its book-scanning operation shows that *disciplines* and *disciplined*, those two orthographically minimally differing grammatical forms of the noun-and-verb *discipline*, have had drastically different histories of currency in English. While *disciplined* has remained of roughly constant frequency of occurrence in books in English from 1880 to 2005, *disciplines* remained until 1920 at only about 15 percent of the frequency of *disciplined*, with most of its occurrences being, presumably, references to religious disciplines. Then, about 1920, *disciplines* began to rise—to rise exponentially, quadrupling in frequency by 1945, and quadrupling again by 1970.[19]

More surprising, perhaps, this general pattern holds also for writing by and for scientists. As figure 1 shows, although in the pages of *Science* the frequency of occurrence of *disciplines* relative to *disciplined* in the four decades 1880–1920 was five to ten times higher than in English books of every sort—that is, in *Science* the two word forms occurred roughly equally frequently[20]—the first faint signs of an increase of frequency of *discipline* in the sense of scientific discipline in *Science* discourse appear only a decade ahead of its takeoff in general discourse. In *Science* discourse the rise was even more rapid, with the frequency increasing by a factor of five by the 1940s and by a factor of almost five again in the next decade. Both the lateness and the steepness of this rise in the frequency with which *disciplines* appeared in writings in *Science* are astonishing. More important, however, for the principal thesis of this article is the shortness of its heyday. The pejoration of *discipline* taking hold at the peak of the sixties' cultural rebellion found synchronous, albeit tacit, expression as a marked drop in the frequency with which writers in *Science* chose the word—a disfavoring that increased in each succeeding decade.

The predominance of general cultural and characterological meanings of the word *discipline* among learned writers well into the twentieth century, and the absence of a reversal of such usage, even in writing by and for scientists, until the interwar period, is perhaps connected with the fact that however clearly the disciplinary ideal had taken shape in the minds of its academic proponents in the last decades of the nineteenth century, the realization of that ideal in academic and scientific practice remained far from that degree of completeness that came to be widely taken for granted

---

necessarily given over to the pragmatic, utilitarian disciplines, since that is the nature of barbarism" (34). Beyond this comic example there is only this: "The attention given to scholarship and the non-utilitarian sciences in these establishments has come far to exceed that given to the practical disciplines for which the several faculties were originally installed" (37) and several more references to "these utilitarian disciplines" in the following few pages. For *utilitarian* and *practical* as depreciative in Veblen's mind—and for the common misrepresentation of Veblen in this regard—see my discussion in "Primacy of Science in Modernity" (cit. n. 7), 86 n. 99.

[19] Jean-Baptiste Michel et al., "Quantitative Analysis of Culture Using Millions of Digitized Books," *Science* 331 (2011): 176–82. (Readers may construct their own graphical displays of these data at http://books.google.com/ngrams [accessed September 2011].) The graphs arising from the same searches over Google's American English database are essentially identical, both qualitatively and quantitatively. The graphs arising from the same searches over Google's English Fiction database are qualitatively very similar, except that the frequency of appearance of both terms is about one-third lower.

[20] This near equality of frequency of appearance of *disciplined* and *disciplines* should not be construed to imply equality of frequency of characterological usages of *discipline* and references to fields of inquiry. On the contrary, the appearances of *discipline* are far more frequent, and they are overwhelmingly characterological.

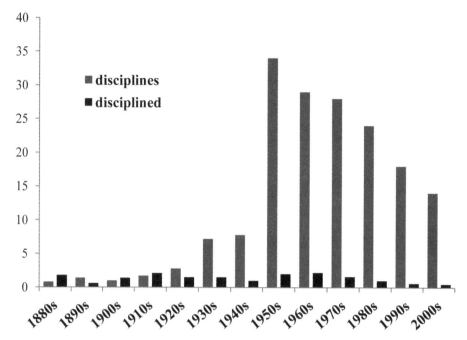

Figure 1. *Publications in Science, 1881–2010, containing the word* disciplines *or* disciplined: *number of such items per 1,000 pages published in each decade. Data obtained from "all item types" searches at JSTOR. Decades run from 1 to 0, not 0 to 9. Figure by author.*

in the middle decades of the twentieth century.[21] So, for example, until the 1930s the faculties of all the elite American universities were highly inbred. In Cornell's Physics Department, one of the best in the United States, thirteen of the fourteen men appointed to faculty posts between 1892 and 1927 had their doctorates from Cornell: good though it was, until the 1930s Cornell's Physics Department chose to be significantly less good, by purely disciplinary criteria, than it could have been.[22]

[21] One need only recall the pervasive departures from the demands of disciplinarity underscored by Weber in "Wissenschaft als Beruf" (1919). The surprisingly slow and incomplete development of the physics discipline in Germany and the sociology discipline in the United States are main themes of Christa Jungnickel and Russell McCormmach, *Intellectual Mastery of Nature*, 2 vols. (Chicago, 1986), and Andrew Abbott, *Department and Discipline: Chicago Sociology at One Hundred* (Chicago, 1999), 84–7. Novick (*That Noble Dream*, cit. n. 2, on 47–9) noted, and Robert Townsend has documented the fact, that until after World War II the American Historical Association (AHA) remained far from realizing the ideal characteristics of a disciplinary association; Townsend, "The Social Shape of the AHA, 1884–1945," *Perspectives on History* (2009): 36–40, available on the AHA website at http://www.historians.org/perspectives/issues/2009/0912/0912tim1.cfm (accessed 24 October 2011); Townsend, "Making History: Scholarship and Professionalization in the Discipline, 1880–1940" (PhD diss., George Mason Univ., 2009), esp. chap. 2, "Rethinking Assumptions: The Academic as Professional." Where I see a culturally driven, hence more or less inevitable, disciplinarization of knowledge production and curation, Townsend sees an arbitrary narrowing and preemption of the doing of history by one sector of a multifaceted "historical enterprise."

[22] Paul Hartman, *The Cornell Physics Department: Recollections and a History of Sorts*, rev. ed. (Ithaca, N.Y., 1993), 332–4. Roger L. Geiger, in *To Advance Knowledge: The Growth of American Research Universities, 1900–1940* (1986; repr., New Brunswick, N.J., 2004), 223–7, describes the inbreeding of US university faculties in this period—50 percent or more of senior faculty being typical of the best universities, reaching 70 percent in some cases. In some fields and institutions (e.g., the Harvard Business School) such inbreeding was maintained well into the post–World War II de-

Reflection on the historiography of science and scholarship in the first half of the twentieth century readily produces analogous examples of parochial solidarities trumping the demands of disciplinarity. The most notorious example is the repudiation by scientists and scholars of their transnational disciplinary solidarities following their country's entry into World War I, and, in varying degrees, their persistence through the decade following that war's end in their elevation of their nation's interests, as they saw them, above their disciplinary interests as scientists and scholars. This circumstance, now well known, still awaits adequate explanation. No less needful of explanation is the repression, in the middle decades of the twentieth century, of memory of this repudiation by those selfsame scientists and scholars, both individually and collectively. However one theorizes that abjuration of scientific internationalism during and after the First World War, the fact of that abjuration and the subsequent repression of memory of that fact are both evidence that over the course of the first half of the twentieth century disciplinarity's hold on the minds of scientists and scholars had become substantially firmer and more exclusive.[23]

Most Preposterous Presupposition

If in the decades following the Second World War the mandate of any scholarly discipline obliged it, qua discipline, to hold in mind how recently the demands of disciplinarity had attained cultural priority and academic institutionalization—and to recognize the recency of the repurposing of the word *discipline* in both common and scholarly parlance—that discipline was our discipline, the history of science. Yet if there is one discipline that contributed more than any other to obscuring and belying the recency of the institutionalization of the demands of disciplinarity, one discipline that did more to create and sustain the illusion that disciplines and disciplinarity were long since fully realized in all the well-established fields of knowledge and that disciplinary organization and operation would in the fullness of time be fully realized in each and every field of knowledge, that discipline so disserving of historical understanding was our discipline, the history of science. This signal disservice we accomplished chiefly, indeed overwhelmingly, through the unparalleled uptake across the spectrum of scholarly disciplines of Thomas Kuhn's *Structure of Scientific Revolutions* (1962).[24]

---

cades; Rakesh Khurana, *From Higher Aims to Hired Hands: The Social Transformation of American Business Schools and the Unfulfilled Promise of Management as a Profession* (Princeton, N.J., 2007), 306–8.

[23] The discoverer of both these circumstances was Brigitte Schroeder-Gudehus, *Deutsche Wissenschaft und internationale Zusammenarbeit, 1914–1928* (Geneva, 1966). Her most recent review of the literature is "Probing the Master Narrative of Scientific Internationalism: Nationals and Neutrals in the 1920s," in *Neutrality in Twentieth-Century Europe: Intersections of Science, Culture and Politics after the First World War*, ed. Rebecka Lettevall, Geert Somsen, and Sven Widmalm (New York, forthcoming), where she also draws attention to Sarton's contribution to repressing memory of that fracture of scientific solidarity. One would think that in the face of the large literature of the past four decades detailing and analyzing this failure of scientific internationalism, retrogression in historical interpretation to the amnesiac state that preceded Schroeder-Gudehus's 1966 publication would be impossible—and certainly impossible for a sometime president of the AHA, such as Akira Iriye, when writing a book titled *Cultural Internationalism and World Order* (Baltimore, 1997). But as the demise of disciplinarity in postmodernity has freed the historian to ignore the literature and indulge in wishful thinking, such retrogression is now an option.

[24] It is important, especially here, to emphasize that though the wider impact was almost entirely Kuhn's, the disciplinary perspective on modern science was common to all forward-looking historians

In that brief book, misrepresented to the scholarly world as the epistemic distillate of a large body of investigation by our still-fledgling discipline, Kuhn conjoined, on the one hand, a radical historicizing of scientific knowledge and, on the other hand, a radical dehistoricizing of the social-motivational structures through which scientific knowledge is produced—postulating disciplines and disciplinarity as the historically invariant institutional forms through which "mature" sciences create their histori-cally variant, noncumulative, incommensurable knowledges.[25] It never occurred to Kuhn that the abhorrence of whiggery that dominated his own practice and teach-ing of the history of scientific ideas should extend below ideas to the history of the institutional forms and motives for knowledge creation and propagation. But then, it never occurred to Kuhn's contemporaries either: whatever our reservations or objec-tions as historians to Kuhn's structuring of the succession of scientific ideas, as mod-erns we went docilely along with Kuhn's most preposterous presupposition, that dis-ciplinarity was meant to be, and would forever be.[26] That this historically so recent, and still so imperfectly realized, form of knowledge production and curation was

---

in the sixties and early seventies. An especially striking example of this programmatic conviction is in the editor's foreword to volumes 4 and 5 of *Historical Studies in the Physical Sciences*, prepared by Russell McCormmach in 1974. In his long list of questions needing investigation (quoted at length by Lewis Pyenson in "Editor's Foreword," *Historical Studies in the Physical Sciences* 37 [2007]: 189–204, on 198–9) every one of them sprang from, and fed into, the disciplinary conception. McCorm-mach was himself then close to completing the manuscript of what remains still today the longest history of a discipline (*Intellectual Mastery of Nature*, cit. n. 21), and may thus be excused from failing to see the handwriting that, by the time of his writing, had been on the wall for several years. Several years later still, Robert Kohler, opening *From Medical Chemistry to Biochemistry: The Making of a Biomedical Discipline* (New York, 1982), observed that "it is surprising, in view of its promise, how little discipline history has been done in the past 15 years" (2–3). He too, in view of his own invest-ment, may be excused for believing that he could see positive "signs of change" in this regard. Kathryn M. Olesko, in *Physics as a Calling: Discipline and Practice in the Königsberg Seminar for Physics* (Ithaca, N.Y., 1991), is not with that program. On the decline of interest in the sociology of professions during the 1970s, see Burnham, *How the Idea of Profession* (cit. n. 10), 119.

[25] Kuhn's conception that disciplinarity equals professionality equals the end of the prehistory and the beginning of the true history of a science—never clearly articulated in *The Structure of Scientific Revolutions* (Chicago, 1962; hereafter, *SSR*), but always understood by its readers—is explicit in the opening sentence of Kuhn, "The History of Science," in *International Encyclopedia of the Social Sciences* (New York, 1968), 14:74–83, as reprinted in Kuhn, *The Essential Tension* (Chicago, 1977), 103–26, on 103: "As an independent professional discipline, the history of science is a new field still emerging from a long and varied prehistory." In his "Postscript—1969" to the second edition, enlarged, of *SSR* (Chicago, 1970; 174–210, on 176), Kuhn said, "If this book were being rewritten, it would . . . open with a discussion of the community structure of science"—i.e., what everyone read into it. By 1974, in "Second Thoughts on Paradigms," reprinted in *Essential Tension* (293–319, on 297), Kuhn was ready to say that "less confusion will result if I instead replace it [*paradigm*] with the phrase 'disciplinary matrix'—'disciplinary' because it is the common possession of the practi-tioners of a professional discipline." Gad Freudenthal ("General Introduction," in Joseph Ben-David, *Scientific Growth: Essays on the Social Organization and Ethos of Science*, ed. Freudenthal [Berke-ley and Los Angeles, 1991], 1–25, on 17–8) alleged that Kuhn, taking the paradigm as determining *everything*, made the social dimensions of a mature science likewise a function of its ever-changing paradigm. This would indeed have been the logical position for Kuhn to take, but, for all that Kuhn's scheme was mainly an exercise in logic, he never took it. It is absent exactly where one would expect to find it, namely, in Kuhn's essay review "Scientific Growth: Reflections on Ben-David's 'Scientific Role,'" *Minerva* 10 (1972): 166–78. As Kuhn himself always insisted, a serious thinker's failure to follow the logic of his scheme is an indication of a blinding cultural commitment.

[26] The intensity of Kuhn's concern as a teacher with the crime of whiggery in respect of scientific ideas (but, I emphasize, ideas only) is dramatically conveyed in Errol Morris's reconstruction of his experience during his year as a graduate student with Kuhn at Princeton in the early 1970s, which he supports with the recollections of Norton Wise, included with his own; "The Ashtray: This Contest of Interpretation, part 5," Opinionator blog, *New York Times*, March 10, 2011, http://opinionator.blogs .nytimes.com/tag/incommensurability/ (accessed March 2011).

taken by Kuhn, and accepted virtually without question by his colleagues, as the end of history in science, is striking evidence that in those last few decades of modernity we were entirely in the thrall of the myth of disciplinarity, wholly certain of the naturalness, inevitability, and perpetuity of the disciplinary organization, production, and control of knowledge.[27]

This presupposition of our sometime discipline, the history of science, is less surprising (though no less embarrassing) if recognized as a particularization of the master concept of the discipline of sociology as it took form in the last decades of the nineteenth century and the first decades of the twentieth: the postulate of social, or institutional, differentiation. Sociology's assumption of a natural and inevitable process of increasing functional differentiation of social, cultural, and occupational roles, resulting in increasingly autonomous institutions operating to ever-higher functional standards, presumed, even as it explained, disciplinarity and professionality as the telos of social evolution. This postulate of social-institutional differentiation, articulated by Émile Durkheim (but implicitly assumed by most of Durkheim's sociological contemporaries), was elaborated in the 1930s by Talcott Parsons, the most influential sociological theorist of the middle decades of the twentieth century.[28]

It is, however, true that Robert Merton, although well familiar with Durkheim's work and in close contact with Parsons while preparing his Harvard dissertation, "Science, Technology, and Society in Seventeenth Century England" (1938), made in it only one, passing, footnote mention of Durkheim, and none of institutional differentiation.[29] As in all such cases, we may consider either that the omission implies the absence of any such perspective or, on the contrary, that the omission implies the taken-for-grantedness of that perspective in sociology in those decades. Taken-for-grantedness is proven by Merton's acknowledgment, when introducing the 1970 reissue of his dissertation, that the postulate of institutional differentiation—to be sure, still not so called, and still not crediting Durkheim—was "a principal assumption underlying the entire book."[30] Indeed, as Thomas Gieryn stressed in noting Merton's

---

[27] From this perspective it is easy to understand why the internal-external distinction—deplored by Steven Shapin ("Discipline and Bounding: History and Sociology of Science as Seen through the Externalism-Internalism Debate," *Hist. Sci.* 30 [1992]: 333–69, on 357) as the cardinal sin of "the founding fathers"—dominated methodological reflection in the history of science in those late modern decades. The history of science was faced more squarely than was any other historical discipline by the incompatibility between history's presupposition of a more or less seamless web and disciplinarity's presupposition of disciplinary autonomy, hence a boundary, a "skin" enclosing the "innards" of a discipline, in the vivid image of Thomas F. Gieryn, *Cultural Boundaries of Science: Credibility on the Line* (Chicago, 1999), 21. In *Professionalism: The Third Logic* (Chicago, 2001), Eliot Freidson, who turned advocate in his old age, put it more drastically: "*Without closure there can be no disciplines*" (199; emphasis in the original). Considering the irresolvable cognitive dissonance created by this incompatibility between the demands of history and those of the disciplinary conception of science, it is not surprising that, as Shapin observed, "much of what passed as debate was both diffuse and incoherent" (345).

[28] Thomas F. Gieryn, "Distancing Science from Religion in Seventeenth-Century England," *Isis* 79 (1988): 582–93, on 583; Gianfranco Poggi, *The Development of the Modern State: A Sociological Introduction* (Stanford, Calif., 1978), 13–5; Guenther Roth, reviewing Poggi, in *Contemporary Sociology* 8 (1979): 362–8.

[29] Merton, *Science, Technology and Society in Seventeenth Century England* (New York, 1970), 60 n. 10. Gieryn ("Distancing Science from Religion," cit. n. 28) says that Merton did not mention Durkheim, but he did not then have Google Books at his disposal to ferret out Merton's well-buried reference.

[30] Merton, "Preface 1970," in *Science, Technology and Society* (cit. n. 29), vii–xxix, on xix. There is of course a third alternative; namely, that the writer is simply avoiding acknowledging the sources of

retrospective admission, the postulate of institutional differentiation was the assumption underlying not only Merton's first work but of all Mertonian sociology of science.

Sarton's desideratum that those writing histories of science be held to ever-higher standards and Merton's Durkheimian assumption that the imposition and acceptance of such discipline was the result of the natural and inevitable increase in institutional differentiation were both foundational for the history of science when it did achieve disciplinary form in the decades after the Second World War. Sarton's desideratum must, from any disciplinarity-affirming perspective, be judged all to the good. Merton's—truly, modernity's—assumption, unquestionably functional for the creation and maintenance of the myth of disciplinarity, must be judged nefarious for the pursuit, practice, and understanding of history.[31] Kuhn, implicitly taking this Durkheim-Merton postulate of institutional differentiation as foundational for his fundamentally ahistorical account of how scientific conceptions change, found a resonance in the late modern intellectual world that is intelligible only by recognizing the breadth of implicit assent then given to that postulate.[32]

FOUR FURTHER PRESUPPOSITIONS OF MODERNITY— AND THEIR NEGATION IN POSTMODERNITY

Although I have provided some evidence of the high cultural rank and high degree of exigency that modernity attributed to discipline, disciplines, and disciplinarity, I have not offered a definition of disciplinarity, have not identified the behavioral norms or the institutional forms that constitute disciplinary knowledge production and curation. Implicitly, I have relied upon the reader carrying over into the reading of my text that same indulgence extended to authors contributing to the huge literature of recent decades that is uniformly critical and often condemning of disciplines and disciplinarity, a literature in which definitions are rare and rarely stand up to scrutiny. Careful definition has seemed supererogatory, precisely because we all sense that disciplinar-

---

his ideas. This was to no small degree the case with Merton. Thus his far more famous 1938 publication on anomie is a far more egregious failure properly to credit Durkheim.

[31] I have previously emphasized ("Independence, Not Transcendence, for the Historian of Science," *Isis* 82 [1991]: 71–86) that it is a mistake to suppose that what is good for disciplinary science is good for the discipline of the history of science. Lorraine Daston ("Science Studies and the History of Science," *Crit. Inq.* 35 [2009]: 798–813), in her contribution to the special issue "The Fate of the Disciplines" (cit. n. 2), makes that mistake (among several others): "Inexorably, immersion in the scientific practices that eventually created scientific disciplines led—by a kind of mimesis—to historical practices that turned the history of science into a discipline" (809). Inconsistently, but again mistakenly, Daston there also has this as the result of historians of science submitting themselves to the discipline of history, and that only recently. On this latter mistake, see n. 106 below and the concluding section generally.

[32] There is some irony in the fact that it seemed to sociologist Barry Barnes (*Scientific Knowledge and Sociological Theory* [London, 1974]) that the conception of science as a closed and bounded social-cultural system "is near enough taken for granted among sociologists . . . mainly due to the work of the historian of science T.S. Kuhn" (48). But so it was, and so it has remained: *SSR*, which simply reflected back to the sociologists their own presuppositions, is widely mistaken for an innovative work of sociological theory and discovery—so much so that Andrew Abbott, an undergraduate at Harvard in the later 1960s and entering graduate school at the University of Chicago at the end of that decade, could recall: "I grew up on the Cassirer-Langer-Mead philosophy of knowledge, the Kuhnian sociology of science . . ."; Abbott, *The Chaos of Disciplines* (Chicago, 2001), 4 n. 2. That what *SSR* offered was not sociology at all, but epistemology pretending to be sociology, ought to have been obvious, at least to sociologists, but was not, and still is not.

ity had an intimate affinity with modernity. Here, again, we are indebted to David Hollinger for drawing attention to that affinity evident in the set of norms for disciplinarity that Robert Merton articulated in the midst of the Second World War: communism, universalism, disinterestedness, and organized skepticism (since the late 1970s commonly referred to by the acronym CUDOS), a set of norms that mirrored the left-leaning, liberal-democratic, social-intellectual values that the Allies opposed to the Axis. More indicative, however, of the intimacy of that affinity between modernity and disciplinarity is the fact that Merton's norms, advanced in but a half-dozen pages without significant scholarly basis, and in an explicitly ideological context, were widely accepted in the postwar decades as a solid foundation for the field of sociology of science. So long as modernity continued—but only so long—CUDOS retained, on that slight but consensual foundation, wide acceptance not merely in the sociology of science but in the scholarly world generally.[33]

This absence of elaboration and of contestation does not argue for the accuracy of Merton's definition of disciplinarity, but, on the contrary, exemplifies the unimportance of accuracy in the definition of a cultural reality that is universally accepted as necessary and inevitable. How, then, to proceed in establishing that intimacy of connection between disciplinarity and modernity if we may not premise Merton's, or some other's, definition of disciplinarity? We must, I think, rely upon our general sense of what the functioning of scientific and scholarly disciplines requires—just as Merton relied on his, but with greater historical and personal distance than Merton had or could have had. More particularly, because our concern is with the intimacy of the connection between disciplinarity and modernity, we must seek the essence of disciplinarity not by socializing an epistemic position—as did Merton, and as would also the constructivists intent on displacing him—but rather by seeking the broadest and most fundamental value presuppositions of modernity that disciplinarity not only shared but required.[34]

---

[33] David A. Hollinger, "The Defense of Democracy and Robert K. Merton's Formulation of the Scientific Ethos," *Knowl. & Soc.* 4 (1983): 1–15; Hollinger, "Science as a Weapon in *Kulturkämpfe* in the United States during and after World War II" (1995), repr., with introductory notes, as chaps. 5 and 8 of Hollinger, *Science, Jews, and Secular Culture* (Princeton, N.J., 1998). The general thrust of the latter paper, first presented as a lecture to the History of Science Society, is entirely consistent with that of this article: that historians of science have been historically very shortsighted, projecting back into the early twentieth century cultural valuations of science that had become established in the United States only in the decade following World War II. In the former paper Hollinger (on 82) reviews the diverse titles under which Merton's original publication, "A Note on Science and Democracy" (1942), was reprinted, as indicative of the process whereby this short, unfocused, occasional essay was transmogrified into "theoretical foundations for a sociology of science." (I quote Nico Stehr, "The Ethos of Science Revisited: Social and Cognitive Norms," *Sociological Inquiry* 48 [1978]:172–96, on 173.) Stephen Cole, in "Merton's Contribution to the Sociology of Science," *Soc. Stud. Sci.* 34 (2004): 829–44, makes it clear that Merton's own theorization of CUDOS amounts to nothing more than that original essay.

[34] It is not out of place here to take issue with the "maxim of method" that Steven Shapin ("Hyperprofessionalism," cit. n. 12) avows as being that of his oeuvre, the more so as, on the one hand, it is the antithesis of the method pursued in this article and in my own work generally and, on the other hand, it has implicitly been the predominant perspective among historians of science in the past generation, to no small extent because of the influence of Shapin's works. "In my own line of work," Shapin explains, "a maxim of method—not an epistemological evaluation—has been to treat science as a typical form of culture. Whatever can be learned from the detailed, naturalistic study of a particular scientific practice may be applied to our overall understandings of knowledge and the conditions of its making" (242). It should be obvious, however, that Shapin's approach (supposing that it were indeed ever actually followed) is almost certain to lead the historian astray, for it provides no guidance in discriminating what is peculiar to "a particular scientific practice" and what is general in the culture

My aim, therefore, is to show that the taken-for-granted features of the common, consensual conception of disciplinarity are entailed by a set of cultural values enjoying general assent in modernity: proceduralism, disinterestedness, autonomy, and solidarity. Those four values, together with the high value placed on discipline itself, created disciplinarity as the ideal form of knowledge production and curation in modernity. Conversely, the negation of these five values in postmodernity, as postmodernity, has rendered disciplinarity odious, making disciplines, as Kuhn said with dismay, "examples of oppression," deplorable both for the character of the knowledge they produce and for their way of producing and propagating it.

Proceduralism

As Croly exemplifies the centrality of *discipline* in the U.S. intellectual's view of man and in society at the opening of the twentieth century, we should expect to find in *The Promise of American Life* also such other keywords of modernity as are connected with the notion of discipline. Among the keywords linked to discipline, that most distinctive of the long modernity extending from the mid-sixteenth century to the late twentieth century is *method*. Indeed, *method* appears twenty-five times in the concluding chapter of Croly's book, very nearly as often as *discipline*.

Historians of science, and historians of the philosophy of science, have never failed to stress the centrality of method for Bacon and Descartes and for the scientific movement going forward from their time. We have, however, never taken due account of the "frantic interest in the whole question of method which marks the two generations and more preceding Descartes"—in which period, Walter Ong pointed out, there emerged the modern concept of method; namely, "orderly procedure . . . a routine of efficiency" in practice, and not merely in discourse.[35] That every problem, every task, in realizing human purposes has an optimal method for its solution is a distinctively modern presumption. Only so can we understand that the evident successes of natural science were universally taken as demonstrations of the reality and effectiveness of a "scientific method." Indeed, the growing faith in scientific method through the eighteenth and nineteenth centuries, and the continuance of that faith into

---

of the period. That this is not perceived to be a problem is due to the presupposition—half denied in Shapin's interjection, "not an epistemological evaluation"—that science is in no way different. (To adapt his famous opener, "There's no such thing as science, and this is a history of it.") In articulating this anti-differentiationist maxim of method, Shapin identifies Barry Barnes as being the originator of it, citing Barnes's books in the 1970s. I think that Shapin misrepresents what was then, or is now, Barnes's point of view.

[35] Ong, *Ramus, Method, and the Decay of Dialogue: From the Art of Discourse to the Art of Reason* (1958; repr., with a foreword by Adrian Johns, Chicago, 2004), 225, 229. Likewise noting the absence in the medieval literature of *method* in the sense of *wissenschaftliche Vorgehen* or *Verfahren*: *Historisches Wörterbuch der Philosophie*, ed. Joachim Ritter (Basel, 1971–; henceforth, *HWP*), s.v. "Methode III: Mittelalter—2," by Ludger Oening-Hanhoff. On Ramus, method, and Calvinistic Protestantism, see Perry Miller, *The New England Mind in the Seventeenth Century* (1939; repr., Cambridge, Mass., 1983). David Simpson, in *Romanticism, Nationalism, and the Revolt against Theory* (Chicago, 1993), almost grasped that the regard of method is what distinguishes modernity. In the early modern period, the matter of method demanded attention not simply as concomitant to the general problem of replacing one scientific system by another, but equally because of the specific character of Aristotelian philosophizing, which gave primacy and centrality to ends. Thus the revolt against Aristotelianism almost inevitably, however gradually, carried with it the repudiation of the primacy of ends and the elevation of means, of method.

the middle of the twentieth century, had more affinity with religious faith than with the empirical sciences that it allegedly governed.[36]

Like disciplinarity, the scientific method is a good example of that unimportance of accuracy or acuity of definition where what is to be defined is a consensual cultural reality. The certainty felt by John Dewey, and those millions for whom and to whom he spoke, not merely of the existence of a scientific method, but, still more, that something was actually apprehended by his vague verbal gestures invoking said method, seems almost delusional from our early twenty-first-century perspective.[37] Wherefore Dewey's presupposition of "the supremacy of method" could by no means have been limited to those sharing Dewey's antimetaphysical metaphysics or Dewey's optimistic view of the prospects for human progress. On the contrary, the most often quoted articulation of a confident belief in the all-importance of method—and specifically, a method efficacious exactly where today we would most decidedly deny the possibility of such—came from a philosopher standing socially, affectively, and metaphysically opposite to Dewey. I refer, of course, to A. N. Whitehead and his claim that "the greatest invention of the nineteenth century was the invention of the method of invention. A new method entered into life."[38]

This faith in the existence of a uniquely efficacious method for realizing any goal of human action demands a name. *Methodism* suggests itself immediately. But as Microsoft insists on capitalizing this word wherever it appears, and as human readers too may confuse methodism with the beliefs of a mere sect in modernity, another name is needed.[39] I adopt *proceduralism* to denominate this nearly universal, quasireligious belief found only in modernity. I do so with trepidation because that word too is already in use, since the 1960s, in the field of political theory. There proceduralism, associated especially with the liberal tradition from Locke through Kant and Mill, does not carry an implication of practical efficacy so much as moral legitimacy. There proceduralism refers to one or more aspects of due process in the legal and

[36] Among the many voids in the literature of the cultural history of science, the lack of works on this subject is one of the largest. I know only Richard Yeo, *Defining Science: William Whewell, Natural Knowledge, and Public Debate in Early Victorian Britain* (1993; repr., New York, 2003), followed by Laura J. Snyder, *Reforming Philosophy: A Victorian Debate on Science and Society* (Chicago, 2006), and *The Philosophical Breakfast Club: Four Remarkable Friends Who Transformed Science and Changed the World* (New York, 2011)—this latter a cross between Menand and Dava Sobel achieving just that sort of success of which historians of science are now most desirous. Yeo recognized that those engaged in discussions of scientific method in Victorian Britain understood them as having implications wider than the production of reliable knowledge of nature, but he did not explore those implications. Still less did Snyder—less and less, from her scholarly work to her less scholarly work.

[37] Dewey's *The Quest for Certainty* (New York, 1929) contains his Gifford Lectures and is his most important epistemological statement to that date, with "The Supremacy of Method" its keystone chapter. The book contains no clear articulation, let alone examination, but only repeated postulations, of "the method" of the physical sciences. Dewey's *Logic, the Theory of Inquiry* (New York, 1938) is more technical—and more tautologously proceduralist. Opening his chapter there titled "Scientific Method and Scientific Subject-Matter," Dewey says that "since this body of subject-matter attains scientific standing only because of the methods that are used in arriving at them [*sic*], the systems of facts and principles of which science materially consists should disclose properties that conform to conditions imposed by the methods" (458). But what exactly the methods are, Dewey does not say.

[38] Whitehead, *Science and the Modern World: Lowell Lectures, 1925* (New York, 1925), 141.

[39] In "Primacy of Science in Modernity" (cit. n. 7), and again in "(Re)cognizing Postmodernity" (cit. n. 5), I used *methodism* and, occasionally, *procedurism* as monikers for this mind-set. Walter J. Ong, in "Peter Ramus and the Naming of Methodism," *J. Hist. Ideas* 14 (1953): 235–48, shows the close association of method with discipline in the early eighteenth century.

political spheres. Missing there, so far as I have seen, is an awareness that politico-legal proceduralism is but one manifestation of an inclusive perspective on social and cultural practice, and that our valuation of any particular manifestation depends on whether we accept or reject that inclusive perspective; that is, it depends on the historical epoch in which we stand and from which we inevitably draw our political presumptions.[40]

In that literature of liberal political theory, John Rawls's concept of "pure procedural justice," developed "as a basis of theory" in his magisterial *A Theory of Justice*, is generally regarded as the most rigorous articulation of proceduralism. "Pure procedural justice obtains," said Rawls, "when there is no independent criterion for the right result: instead there is a correct or fair procedure such that the outcome is likewise correct or fair, whatever it is, provided that the procedure has been properly followed." Implicit in Rawls's definition of pure procedural justice is a still broader axiom—truly, *the* proceduralist axiom of modernity—namely, that the means sanctify the ends: not only do the right means produce the right ends, and only the right means can be relied upon to produce the right ends, but, most important, the rightness of the means is the only criterion for the rightness of the ends.[41]

Among many instances of this elevation of means above ends, the longest standing, and in modernity the one finding the most categorical affirmation and the most universal assent, is the requirement of truth telling. In its ancient origins truth telling was by no means a categorical requirement; it appears in the Ten Commandments only as the requirement of truth telling in local judicial proceedings. Saint Augustine made a strong case for the absolute prohibition of lying, but nowhere in the classical world was that position widely adopted. Truth telling moved into a central position in the chivalric code of medieval Europe and continued so in the early modern definition of a gentleman. But the duty of truth telling these standards imposed did not extend beyond those social strata, and in practice was little observed even within them. Premodernity—for which Shakespeare may be taken as exemplary—was preoccupied, if not pervaded, by treachery and deceit, and correspondingly ambivalent in its view of truth telling. As Steven Shapin acknowledged, "secrecy might be

[40] For orientation, see Amanda Anderson, *The Way We Argue Now: A Study in the Cultures of Theory* (Princeton, N.J., 2006), 161: "Proceduralism is a normative model for the justification of specific political practices and institutions." In the literature that I have encountered, Michael J. Sandel comes closest to taking proceduralism as a general cultural value; "The Procedural Republic and the Unencumbered Self," *Political Theory* 12 (1984): 81–96; *Democracy's Discontent: America in Search of a Public Philosophy* (Cambridge, Mass., 1996). And yet, although Sandel connects proceduralism with, and at times comes close to deriving it from, the culturally dominant conception of personhood, he remains a philosopher, unable to affirm a culturalist interpretation and unwilling to acknowledge the historical prevalence of a perspective—proceduralism—that he himself, as a romantic-postmodern, rejects.

[41] Rawls, *A Theory of Justice* (Cambridge, Mass., 1971), 73. Rawls viewed his theory "as a procedural interpretation of Kant's conception of autonomy and the categorical imperative" (256); i.e., he understood, as did virtually all Kant's modern interpreters, that he should not be misled by Kant's unfortunate formulations reflecting lingering scholastic prejudices that suggested elevation of ends over means—such as "the kingdom of ends" and the imperative to treat men as ends and not as means. See Kant, *Grundlegung zur Metaphysik der Sitten*, 2nd ed. (1786), translated and analyzed by H. J. Paton as *The Moral Law: Kant's Groundwork of the Metaphysic of Morals* (1948; repr., New York, 2009), 66, 69–71, 82–3. Now, however, in postmodernity, it has become philosophically respectable to take such formulations literally and to maintain that Kant's view was "precisely the same" as Aristotle's; Christine M. Korsgaard, *Creating the Kingdom of Ends* (Cambridge, Mass., 1996), and *Self-Constitution: Agency, Identity, and Integrity* (New York, 2009), 10.

laudable, dissimulation circumstantially recommended." Back then, the end justified the means.[42]

Leaping from the end of the sixteenth century to the end of the eighteenth, we land in a different world. An absolute prohibition of lying is widely accepted in the cultivated classes across Europe and North America. Often associated with Kant, this insistence that the truth be told though the heavens fall is by no means fallout from Kant's philosophy. Much rather, it is indicative of Kant's falling in with his times. Indeed, just then when Kant, out east in Königsberg, was polemicizing in defense of an absolute prohibition of lying, the Mozart-Schikaneder duo down south in Vienna was putting the absolute prohibition of lying at the center of *Die Zauberflöte*, and Parson Weems, way out west in Virginia, was constructing the myth of the juvenile George Washington who "cannot tell a lie," a myth that then served as a staple of the moral education of five generations of juvenile Americans.[43] As the nineteenth century advanced, truth telling was subsumed within modernity's broad proceduralist perspective. The scientist's "fanaticism for veracity" was inseparable from the categorical affirmation of method in the then-nascent scientific and scholarly disciplines.[44] By the middle of the twentieth century spokespersons for disciplinary science could claim—and the claim received wide assent—that it was from science that modernity had learned that "the obligation to tell the truth" is "the cement to hold society together."[45]

Today, in Postmodernity

Although the characterization of contemporary society as an "audit society" is increasingly apt, our resort to externally imposed and policed procedures is hardly evidence of a continuing commitment to proceduralism. Audit procedures today, far

[42] Shapin, *A Social History of Truth: Civility and Science in Seventeenth-Century England* (Chicago, 1994), 106. This is not Shapin's "gentlemen are not supposed to lie" thesis, but it is the truth. Repeatedly, Shapin adduces snippets from Shakespeare in support of that thesis, but nowhere does he confront the plain fact that deceit, deception, disguise, or dissimulation is central to the plot structure of most of Shakespeare's plays. Lionel Trilling (*Sincerity and Authenticity* [Cambridge, Mass., 1972]), in his C. E. Norton Lectures dedicated "to my cousin, I. Bernard Cohen," pointed out that "a multitude of Shakespeare's virtuous characters" engaged in such practices some of the time, and that there is good historical reason for thinking that lying was then common (13–5). Shakespeare affirms that the end justifies the means most clearly in *Measure for Measure* and *Much Ado about Nothing*; Inga-Stina Ewbank, "Shakespeare's Liars," *Proc. Brit. Acad.* 69 (1984): 137–68; see 141–3, 164–5. Thus Shakespeare's corpus comports well with the findings of Perez Zagorin (*Ways of Lying: Dissimulation, Persecution, and Conformity in Early Modern Europe* [Cambridge, Mass., 1990], vii, 17, 256) that in the sixteenth and seventeenth centuries "the legitimation and practice of dissimulation were major factors in the lives of religious bodies, intellectuals, philosophers, and men of letters," while "*courtier* became a byword for dishonesty and faithlessness." In particular, "regarding both the literature and politics of Restoration England," Zagorin cited a body of scholarship emphasizing their "exceptional preoccupation with the use of deception and disguise."

[43] On Kant and Weems, see Martin Jay, *The Virtues of Mendacity: On Lying in Politics* (Charlottesville, Va., 2010), 7–8, 65–9. In *Die Zauberflöte* (Emanuel Schikaneder and W. A. Mozart [1791; New York, 1985], 36–7, 47–50), the bird-man Papageno is "tamed" and made a suitable sidekick for the hero Tamino through the imposition of an absolute prohibition of lying. By contrast, early in the eighteenth century Jonathan Swift, satirizing his contemporaries as the untamable Yahoos, created as their antitheses the horse-men Houyhnhnms, for whom lying was inconceivable.

[44] This is T. H. Huxley's frequently quoted phrase, as found in David A. Hollinger, "Inquiry and Uplift: Late Nineteenth-Century American Academics and the Moral Efficacy of Scientific Practice," in *The Authority of Experts: Studies in History and Theory*, ed. Thomas L. Haskell (Bloomington, Ind., 1984), 142–56, on 145.

[45] Jacob Bronowski, *Science and Human Values*, rev. ed. (New York, 1965), 58; the original publication took the form of an entire double issue of *The Nation* at the end of 1957.

from being modernity's self-legitimating one-best-way to an end immanent in the means, are ever-altering, often arbitrary, procedures, the purpose of which is to prevent service providers from substituting their ends for our ends. Thus it is the primacy of ends, not of means, that underlies postmodernity's audit societies.[46]

Nowhere is this primacy of ends clearer than in the revival of the premodern legitimacy of lying. Turning again to the Library of Congress catalog and scanning the titles containing the word *lying* and those classified in the subject "truthfulness and falsehood," one finds that through the first nine decades of the twentieth century there were few scholarly books addressing themselves to the act of lying, and among those all but a very few presented the act as simply wrong.[47] Suddenly, however, in 1990 there opened a gusher of scholarly publications with titles chosen to make clear that their frame of reference turned modernity's upside down. Instead of taking truth telling as the norm and the rule, and departures from it as deplorable exceptions to the rule, here, now, lying is presented as a usual, acceptable, even commendable practice: *Ways of Lying* (1990), *The Prevalence of Deceit* (1991), *The Varnished Truth: Truth Telling and Deceiving in Ordinary Life* (1993), *Lying and Deception in Everyday Life* (1993), *A Pack of Lies: Towards a Sociology of Lying* (1994), *By the Grace of Guile: The Role of Deception in Natural History and Human Affairs* (1994), *The Power of Lies: Transgression in Victorian Fiction* (1994), and on to Martin Jay's *Virtues of Mendacity* (2010). In perfect sync with the appearance of these scholarly reconsiderations of the legitimacy of lying arose a popular literature similarly affirming its indispensability and, taking a step farther, taking delight in the transgressive art of lying.[48] In sum, the new legitimacy of lying—now affirmed far less ambivalently than ever it was in premodernity—effectively differentiates antiproceduralist postmodernity from proceduralist modernity.

Disinterestedness

Disinterestedness—the capacity to think and to act, and the practice of thinking and acting, in disregard of one's personal interest—was the most highly respected quality of mind and character in modernity. Earlier regarded as a luxury available only to

[46] Charles Thorpe, "Capitalism, Audit, and the Demise of the Humanistic Academy," *Workplace* 15 (2008): 103–25; see 105, 107, 121 n. 25; Thorpe, "Participation as Post-Fordist Politics: Demos, New Labour, and Science Policy," *Minerva* 48 (2010): 389–411; Forman, "(Re)cognizing Postmodernity" (cit. n. 5), 163–4, 172 n. 36.

[47] The large number of books on lie-detection techniques and technologies have a Library of Congress subject category all their own; as expressions of the intolerability of lying, they are evidence of the presupposition of the obligation to truth telling in modernity. This literature and its cultural context has been addressed in a hyper-Shapinesque mode by Ken Alder, "A Social History of Untruth: Lie Detection and Trust in Twentieth-Century America," *Representations* 80 (2002): 1–33; see 3, 25 n. 7. Alder there makes a brief, cynical, radically antihistorical case for lying as a temporal and cultural invariant; only the accepted means for detecting lying are historically variable. In his book on the subject (for that much-sought-after nonscholarly audience), *The Lie Detectors: The History of an American Obsession* (New York, 2007), Alder devotes just ten (self-contradictory) lines in his last few pages to the cultural foundations of his subject (see 270).

[48] Typical are Philip Kerr, ed., *The Penguin Book of Lies* (London, 1990), and Gini Graham Scott, *The Truth about Lying: Why We All Do It, How We Do It and Can We Live without It?* (Petaluma, Calif., 1994). Recently the chief judge of the US Court of Appeals for the Ninth Circuit opined that indeed we cannot live without it: "For mortals living means lying" (see the full text of the opinion at http://law2 .umkc.edu/faculty/projects/ftrials/conlaw/usvalvarez2011.html, accessed September 2011). I pointed to the new cultural legitimacy of lying in reviewing Shapin, *Social History of Truth* (cit. n. 42), in *Science* 269 (1995): 707–10.

the gentleman, the man of means, and not expectable of the meaner sort, by the end of the eighteenth century disinterestedness was incumbent on all. Adam Smith had sought to give disinterestedness scope by restricting self-interest to the economic sphere. But Kant took "renunciation of all interest as the specific mark distinguishing a categorical from a hypothetical imperative," and thus as the condition of man's right and claim to moral autonomy. Matthew Arnold famously defined English criticism as "a disinterested endeavor to learn and propagate the best that is known and thought in the world," the which, he said, "may be summed up in one word—*disinterestedness*." This faculty, this virtue, William James saw in its highest degree in those thousands upon thousands who had devoted their "disinterested moral lives" to creating the foundations of "the magnificent edifice of the physical sciences." Freud, meanwhile, "by means of careful investigations (only made possible, indeed, by disinterested self-discipline)," had learned more than he wished were true about the diversity of human sexual perversity.[49] Among the several characteristics commonly alleged as essential to a profession, only disinterestedness was expected in any very high degree from that profession which increased in numbers beyond all others in the course of the nineteenth and early twentieth centuries; namely, the professional civil services. With thirty-one appearances in the concluding chapter of *The Promise of American Life*, *disinterestedness* is about 25 percent more salient than either *discipline* or *method*. Indeed, through disinterestedness these latter two are tightly linked, with proceduralism implying disinterestedness, and disinterestedness implying discipline.[50]

With disinterestedness, as with proceduralism—and, indeed, with disciplinarity and professionality, generally—modernity, characteristically, downplayed the disparity between the ideal and the reality. In practice, this meant that those classes of persons who had a culturally authorized claim to disinterestedness were permitted, indeed encouraged, to represent themselves to the world, as to themselves, as possessing that quality in a preposterously high degree. Of this privilege the natural

[49] On disinterestedness and credibility in the seventeenth century, one can believe Shapin, *Social History of Truth* (cit. n. 42), 237–8. On disinterestedness as the virtue presupposed by late eighteenth-century republicanism, see Gordon S. Wood, *The Radicalism of the American Revolution* (New York, 1992), esp. 103–7, where Wood notes that we can no longer "quite conceive of the characteristic that disinterestedness describes" (103). On Adam Smith as opposing early eighteenth-century interest-based explanations, see Pierre Force, *Self-Interest before Adam Smith: A Genealogy of Economic Science* (New York, 2003), 4, 14, incorporating and extending the findings of Albert O. Hirschman. See also Kant, *Grundlegung* (cit. n. 41), 71, 85; Matthew Arnold, "The Function of Criticism at the Present Time," in *Essays in Criticism* (London, 1865), 1–41, on 18–9, 37–9; George L. Levine, *Dying to Know: Scientific Epistemology and Narrative in Victorian England* (Chicago, 2002); Sigmund Freud, *Introductory Lectures on Psychoanalysis* [1917] (New York, 1977), 377; William James, "The Will to Believe" (1896), as quoted in Forman, "Primacy of Science in Modernity" (cit. n. 7, on 86 n. 95), where the importance of disinterestedness to Marx, Sombart, Veblen, Dewey, and Mumford is also evidenced (16–22, 25–7, 43–5).

[50] The English-language literature generally holds thoughtlessly to a parochial Anglo-Saxon view in which only free professions are counted as professions. So, e.g., Harold L. Wilensky, in "The Professionalization of Everyone?" *Amer. J. Sociol.* 70 (1964): 137–58, listed eighteen professions and wannabe professions, but, without a word of explanation, included no professional civil services among them, whereas Durkheim, in *Professional Ethics and Civic Morals* (New York, 1992), 8, placed civil servants at the top of his list of professions. Robert Adcock, in "The Emergence of Political Science as a Discipline: History and the Study of Politics in America, 1875–1910," *Hist. Pol. Thought* 24 (2003): 481–508, on 499–500, observes that in the 1890s "scholars of history and politics rededicated themselves to the goal of non-partisanship. . . . The goal was seen as requiring a heightened degree of scholarly self-discipline."

scientists availed themselves, but so also, in a still higher degree, did the natural science–emulating students of human affairs in the disciplines of history, sociology, economics, jurisprudence, and political science. For the social sciences, disinterestedness was exactly that which would allow them to attain their ambition of being useful, of being applied sciences; of being, we could well say, professions.[51] Thus when fifteen of their stalwarts drew up the American Association of University Professors' "1915 Declaration of Principles on Academic Freedom and Academic Tenure," their principal underlying consideration was that "to be of use to the legislator or the administrator, he [the university professor] must enjoy their complete confidence in the disinterestedness of his conclusions."[52] Consistent with that consideration, the desideratum most strongly and repeatedly stressed in that document is the university professor's disinterestedness—the appearance of it as well as the reality. Nor should we surmise disingenuousness on the part of the AAUP's Committee of Fifteen. If such a suspicion had then been possible, the declaration would have been too obviously vulnerable to it. Our supercilious certainty that such a claim to disinterestedness can be only a means to the end of influencing action in a preconceived direction is itself a manifestation of postmodernity and its hermeneutics of suspicion.

If for the social scientists, as social scientists, belief in the desirability and attainability of disinterestedness functioned chiefly to justify their claims to authority in matters of social policy, and thus served their professionality at some cost to their disciplinarity, for the natural scientists, as natural scientists, the attachment to disinterestedness functioned oppositely. For natural scientists insistence on disinterestedness had the nominal and unparadoxical function of disengaging their work as scientists from whatever else was going on in their society and in their personal lives. Disinterestedness was thus indispensable to the topological, internal-external, skin-and-innards conception of disciplinary science and was rightly regarded as the keystone in the structure of discipline-sustaining norms.

The most direct and ideologically salient form of disengagement was the decoupling that distinguished disciplines from the professions; namely, the decoupling from just that to which the founders of the social-scientific disciplines so generally aspired: practical applications of their scientifically acquired knowledge. This de-

[51] The centrality for the nascent social sciences of this ambition to be useful in a redirection of society widely agreed to be needful has been emphasized and extensively documented by Henrika Kuklick, "Professional Status and the Moral Order," in *Disciplinarity at the Fin de Siècle*, ed. Amanda Anderson and Joseph Valente (Princeton, N.J., 2002), 126–52. This is also effectively evidenced, though taken largely as a matter of course, by Thomas L. Haskell, *The Emergence of Professional Social Science: The American Social Science Association and the Nineteenth-Century Crisis of Authority* (1977; repr. with additional preface, Baltimore, 2000). To be sure, Weber's stance in "Wissenschaft als Beruf" was for social-scientific purity, but this was, in large measure, a tactic to prevent his political opponents, the holders of the overwhelming majority of German university chairs in history, politics, and economics, from using their lecterns for political speech.

[52] American Association of University Professors, Committee of Fifteen on Academic Freedom and Tenure, report, pt. 1, "General Declaration of Principles" [subsequently commonly referred to as above], *AAUP Bulletin* 1, pt. 1 (1915): 17–39 (and frequently reprinted); as made available on the AAUP website at http://www.aaup.org/AAUP/pubsres/policydocs/contents/1915.htm (accessed April 2010). Thomas L. Haskell, in *Objectivity Is Not Neutrality: Explanatory Schemes in History* (Baltimore, 1998), 174–85, made the important point that in the minds of these fifteen academics the freedom requiring recognition was not that of the individual scholar qua individual, but that of the individual scholar qua member, and authoritative voice, of a discipline. This distinction was lost sight of over the course of the twentieth century with the ever more exclusive understanding of rights as individual, on which see Sandel, *Democracy's Discontent* (cit. n. 40). Therewith one of the most important cultural supports for disciplinarity has gone by the board.

coupling the natural scientists described as purity. With this they stood in a cultural tradition that went back to Greek antiquity. After largely lapsing in the eighteenth century, when Enlightenment utilitarianism had ideological predominance, that cultural tradition was revivified by romanticism's powerful and pervasive antiutilitarian idealism.[53] Although, on the whole, romanticism's role in the creation and sustentation of disciplinarity has been decidedly ambiguous, in this regard romanticism contributed important support to disciplinarity: by exalting the ethical status of pursuing science for its own sake and the ontic status of the products of for-its-own-sake science, and by banishing bound and interested knowledge beyond the borders of the discipline, romanticism helped to maintain a field of knowledge production and validation in which disciplinary controls could more effectively claim to operate exclusively.

Today, in Postmodernity

Who today would even *ask* for evidence that we have lost our modern faith that disinterestedness is, in the main, present and operative there where we would wish it to be? The fact of the legitimacy of lying in postmodernity, evidenced above as a repudiation of proceduralism, is itself evidence that we now take the absence of disinterestedness as fact. Skepticism, cynicism even, about the reality, even the realizability, of disinterestedness has arisen over the past fifty years from so many and such diverse directions that the historian is at a loss to know where to pull on this skein. Yet even cynicism does not alone and of itself amount to a negation of the modern presupposition of the desirability of disinterestedness, is not alone and of itself a negation of the distinctively high value that modernity placed on disinterestedness. Such a positive negation of the desirability of disinterestedness is, however, now widespread in contemporary culture, both popular and academic.[54] Paradoxically, this too is a manifestation of romanticism—of postmodern romanticism, romanticism without its upward-striving, transcendence-seeking dimension. Disdain of disinterestedness appeared first as leftist communitarian romanticism in the 1960s, and then, after remitting in the political hiatus of the 1970s, reappeared as rightist individualistic romanticism in the 1980s and since. Postmodern romanticism's disdain of disinterestedness has not only deprived disciplinarity of this indispensable cultural presupposition, but has demoted the intensely disinterested scientist from the high cultural rank he enjoyed in modernity, elevating into his place the intensely self-interested entrepreneur.[55]

---

[53] Alexander Gode-von Aesch, *Natural Science in German Romanticism* (New York, 1941), 24–31; Perry Miller, "The Romantic Dilemma in American Nationalism and the Concept of Nature" (1955), as reprinted in Miller, *Nature's Nation* (Cambridge, Mass., 1967), 197–207; Cathryn Carson, Alexei Kojevnikov, and Helmuth Trischler, eds., *Weimar Culture and Quantum Mechanics: Selected Papers by Paul Forman and Contemporary Perspectives on the Forman Thesis* (London, 2011), 126–30; Forman, "Primacy of Science in Modernity" (cit. n. 7), 7, 75 n. 24–76 n. 27.

[54] A notable example—the title says it all—came from one of America's most up-and-coming scholars in English literature: David Bromwich, "The Genealogy of Disinterestedness," *Raritan* 1, no. 4 (1982): 62–92. This Nietzsche-invoking, across-the-board assault on Matthew Arnold holds up for admiration in his stead Oscar Wilde, specifically, Wilde's radically romantic, subjectivist nihilism, and his praise of lying.

[55] Forman, "(Re)cognizing Postmodernity" (cit. n. 5), 163, 171 n. 33. Gerald Holton has long emphasized romanticism as the problematic element in contemporary culture; Holton, "The Rise of Postmodernisms and the 'End of Science,'" *J. Hist. Ideas* 61 (2000): 327–41, and earlier publications cited there. Wrongly, I long resisted that understanding.

Autonomy

If one is asked what orientation toward the world is distinctively modern, procedural-ism and disinterestedness are not what first come to mind. What does usually come first to mind is that conception of personhood that Burckhardt, and many follow-ing him, identified as distinctive of the Renaissance, and consequently of modernity; namely, individualism. But the more that individualism is now seeming to be charac-teristic of *post*modern personhood, the less it seems the right way to characterize the presuppositions of modern persons.[56] Moreover, individualism, as such, is incompat-ible with, and could never have led to, disciplinary knowledge production and cura-tion, nor to social-trusteeship professionalism. Given that both were characteristic of modernity, individualism could not have been. Furthermore, individualism's antith-esis, social solidarity, was, as I emphasize below, not merely never absent in moder-nity, but gained strength as a presupposition over the course of modernity. There is, however, a component of the idea of individualism that lies at the heart of discipli-narity; namely, autonomy. As a concept, autonomy is clearer and more general than is individualism, and as a culturally presupposed good it, like solidarity, continually gained breadth and strength through the centuries of modernity.[57]

In contrast with individualism, which is a conception of personhood only, auton-omy is a quality ascribable (or deniable) to corporate entities, not less than to indi-vidual persons—indeed, historically, to corporate entities far more than to individual persons. Only in modernity did the individual's claims to autonomy come to rival, even trump, the claims to autonomy of corporate entities. Only in modernity did au-tonomy become so high and wide a cultural value as to be ascribed not merely to per-sons and formal institutions but even to artifacts of sufficiently exalted cultural labor, as seen, notably, in the modern aesthetic attitude.[58]

While proceduralism invites being likened to a religious doctrine, and disinterest-edness to a religious virtue, personal autonomy literally received its most powerful impetus as a core cultural value from a religious movement, the Protestant Reforma-tion.[59] True, sixteenth-century European culture was still far from ready for complete personal freedom in any realm of thought or society. Hence Luther quickly retreated from according the individual full autonomy in pursuit of a personal relationship with God. Nonetheless, the logic of the Protestant protest against any mediation of the individual's relationship with the Deity was a principal driver of autonomy's con-tinual increase in pervasiveness and prominence as a presupposition of modern cul-ture. However, to be a driver in that direction, the Protestant demand for autonomy had to transcend the specific act of prayerful communion with God and permeate the

---

[56] Steven Lukes, "New Introduction by the Author," in *Individualism* (1973; repr., Colchester, 2006), 1–16; Forman, "(Re)cognizing Postmodernity" (cit. n. 5), 164–5.

[57] Lukes (*Individualism*, cit. n. 56, on 49) quotes Weber, *Die Protestantische Ethik und der Geist des Kapitalismus* (1905): "The term 'individualism' embraces the utmost heterogeneity of meanings." J. G. A. Pocock's *The Machiavellian Moment: Florentine Political Thought and the Atlantic Repub-lican Tradition*, 2nd ed. with a new afterword by the author (Princeton, N.J., 2003), has, to be sure, no critique of the concept of Renaissance individualism, but the word is almost absent in his more than 600 pages. On the contrary, the book can most reasonably be said to be about autonomy, more especially in light of the new afterword.

[58] *HWP*, s.v. "Autonomie," by Rosemarie Pohlmann, citing as exceptional Sophocles's use of the word to describe Antigone's inner state. On the autonomy of the work of art, see n. 67 below.

[59] Lukes (*Individualism*, cit. n. 56, on 55, 58) quotes Luther: "Each and all of us are priests. . . . Why then should we not be entitled to taste or test, and to judge what is right or wrong in the faith?"

practical life of the practicant. We know from Weber's work, and that of the many in-quirers he inspired, that the Protestant ethic and its associated disciplinary personal-ity could not be satisfied by the motto of the monasteries, *laborare est orare* (working is praying). If work was now to count as prayer, that work must be *autonomous* work. Thus here, along with the demand for autonomy, the axiom of proceduralism—the means, and only the means, sanctify the end—demands satisfaction.[60] Without this premium being placed upon autonomous work, professionality and disciplinarity would be unthinkable. Conversely, "the professionalization of everyone" seen by so-ciologists of work in the 1950s and 1960s was in large measure simply the modern universalization of the expectation of, and the demand for, autonomy in work.[61]

Over the course of the eighteenth century the presupposition of personal autonomy gained ideological support both from the diverse Protestant movements emphasizing the availability of a direct, personal relationship with God and from Enlightenment-derived conceptions of the innate rights and capacities of human beings as such. But the support that the presupposition of personal autonomy received from the roman-tic insurgency was more powerful, and would be more enduring, than that received from either, or both, of those two other sources. Kant, who brought all three of these sources together, made autonomy the central structuring principle of his entire phi-losophy—and it is to that emphasis, I venture, that his position as *the* philosopher of modernity is mainly owing.[62] It is thus entirely appropriate that in *A Theory of Jus-tice* (1971), the climactic work of modern, liberal-democratic political philosophy, Rawls "seeks to give specific content to Kant's notion of autonomy," for autonomy is "the supreme human good."[63] Needless to say, one did not need to conceive oneself a Kantian, nor even to have read Kant, in order to give primacy to personal autonomy. Freudian psychoanalysis presumed it, and so also did nearly all other critical psycho-logical and sociological thought in the culminative decades of modernity.[64]

Although romanticism's support for autonomy flowed mainly to personal auton-omy, it was not by any means so restricted. Corporate autonomy—a conception no

---

[60] Autonomy was the sine qua non; it guaranteed nothing. ("Sanctify" is a translation of *heiligen*, the operative verb in the conventional German expression equivalent to "the end justifies the means.")

[61] Bernard Barber, "Some Problems in the Sociology of the Professions," *Daedalus* 92 (1963): 669–88. Wilensky ("Professionalization of Everyone?" cit. n. 50), positioning himself as the tough-minded sociologist, asserted that "this notion of the professionalization of everyone—is a bit of sociological romance" (156). However, Wilensky was himself no less romantic in holding to "the traditional model of professionalism which emphasizes autonomous expertise and the service ideal" (137).

[62] Jerrold Seigel, *The Idea of the Self: Thought and Experience in Western Europe since the Seven-teenth Century* (New York, 2005). Pohlmann ("Autonomie," cit. n. 58, on 707, 709) points out that Kant's more radically romantic followers further radicalized the keystone role of autonomy. Thus Fichte made "the absolute existence and autonomy of the ego" his first and unconditioned postulate. A century later, Hermann Cohen, as a *Neukantianer* freed from the romantic antipathy to method, and finding the real problem to be *das Selbst* in *Selbstverantwortung*, declared that "für diese Aufgabe des Selbst ist die Autonomie die Methode"; Pohlmann, "Autonomie," 712.

[63] The quotations are not of Rawls but of his principal expositor: Samuel Freeman, ed., "Introduc-tion: John Rawls—an Overview," in *The Cambridge Companion to Rawls* (New York, 2003), 1–61, on 26, 34. Google Books finds sixty-one pages in *A Theory of Justice* on which the word *autonomy* appears. Rawls's conception of autonomy being essentially that of Kant, he likewise stresses (251–4, 516–8) the close connection between autonomy and disinterestedness. Thus neither Kant's nor Rawls's is the conception of personal autonomy that we have come to hold today in voluntaristic postmodernity.

[64] See Jamie Cohen-Cole, "The Creative American: Cold War Salons, Social Science, and the Cure for Modern Society," *Isis* 100 (2009): 219–62, whose theme is, equally, the high value placed on au-tonomy, with *autonomy* or *autonomous* appearing more than forty times.

less important for both professionality and disciplinarity—had been, as noted above, the predominant conception of autonomy in all premodern eras and cultures. Early in modernity, however, affinity groups became suspect, even, or especially, when authorized by the state. Romanticism's romanticizing of community created an unprecedentedly persuasive presupposition in favor of corporate autonomy. Thus, while romanticism must bear part of the blame for the ethnoracial nationalisms that blighted the political history of the nineteenth and twentieth centuries, it deserves much of the credit for naturalizing the notion of autonomous associations independent of state authority.[65]

Professions and disciplines profited equally from this communitarian conception of corporate autonomy. But romanticism contributed to the presupposition of autonomy in another way that implicitly discriminated between professions and disciplines: romanticism esteemed, more unequivocally than did any other modern ideology, cultural endeavors conceived to be radically disinterested; that is, for their own sake. Again, admittedly, there are precedents in the Western cultural tradition. Aristotle had presented his natural philosophy as the product of a neutral, disinterested motive, and that packaging had been important to the acceptability of Aristotelianism in Islam and premodern Christendom. But in the early modern period, with Aristotle's overthrow—as both consequence and cause of his overthrow—the ancient and honorable conception of knowledge as truly scientific only if purely contemplative lacked solid ideological support. Just as thinkers otherwise so different as Bacon and Descartes were, remarkably, programmatically at one in giving prime importance to excogitating and expounding a new, post-Aristotelian method for knowledge production, so also were they, even more remarkably, substantively at one in turning to utility as a rationale for the production of scientific knowledge. Utility remained the broadest consensual rationale for the pursuit of science in European societies until the romantic counterrevolution provided a new and deeper foundation for pure, disinterested, for-its-own-sake, autonomous science.[66]

Not that the interests of science were uppermost in the minds, or nearest the hearts, of the romantics. The sciences were, rather, the collateral beneficiaries of the exaltation and reconception of art that emerged with romanticism over the course of the eighteenth century; namely, that art is man's best part, that it is an end in itself, an activity for its own sake, whose products are autonomous objects to be enjoyed contemplatively.[67] Thus, while every profession bears the stigma of utility, of being for the

---

[65] Eric J. Hobsbawm, *Nations and Nationalism since 1780: Programme, Myth, Reality*, 2nd ed. (New York, 1992); John Ehrenberg, *Civil Society: The Critical History of an Idea* (New York, 1999). Although Isaiah Berlin disclaimed the title *historian*, the role of romanticism was better appreciated by him.

[66] As J. L. Heilbron has long emphasized, the Roman Catholic Church was the exception; Heilbron, *The Sun in the Church: Cathedrals as Solar Observatories* (Cambridge, Mass., 1999).

[67] M. H. Abrams, "Kant and the Theology of Art," *Notre Dame English Journal* 13 (1981): 75–106; Abrams, "Art-as-Such: The Sociology of Modern Aesthetics," in *Doing Things with Texts: Essays in Criticism and Critical Theory* (New York, 1989), 135–58, 400–2. Several other essays in this collection recur to that point. Similarly, Martha Woodmansee, *The Author, Art, and the Market: Rereading the History of Aesthetics* (New York, 1994), chap. 1. The concluding chapter of Ernst Cassirer, *The Philosophy of the Enlightenment*, trans. Fritz C. A. Koelln and James P. Pettegrove (Princeton, N.J., 1951), originally published in 1932, is perhaps the first place where the high importance of aesthetics to eighteenth-century philosophy was emphasized. John H. Zammito, in *The Genesis of Kant's Critique of Judgment* (Chicago, 1992), emphasizes the central importance of aesthetics in Kant's intellectual development away from mere philosopher of science. Logically, it is by no means necessary for the artist to be autonomous in order that the work of art be autonomous, but under the presup-

sake of something other than itself, every scholarly or scientific discipline had the attractive option of claiming to be, like art, a for-its-own-sake enterprise. As such, disciplines too were entitled to a higher degree of autonomy. Furthermore, because conceived as engaged in communal production and curation of disinterested knowledge, the scholarly or scientific discipline had a doubly exalted claim to corporate autonomy.[68] Indeed, if one superadds to that doubly exalted claim the claim of the individual scholar or scientist to personal autonomy—well supported by both romanticism and the radical-skeptical Enlightenment—then one has the disciplinary organization of for-its-own-sake knowledge production and curation laced through and through by well-supported claims to autonomy. In short, disciplinarity, more than any other construct of modernity, had autonomy, the hallmark of modernity, stamped all over it.

Today, in Postmodernity

We do not find today with autonomy such a categorical negation of modernity's presuppositions as we found with proceduralism and with disinterestedness—and will find again with solidarity.[69] Postmodernity, on the contrary, has in an important respect brought a great heightening of demands for (and illusions of) autonomy; namely, in our demands for personal autonomy *off the job*.[70] But the claim to autonomy in one's work life, a demand so central, so essential, to modern personhood, is largely renounced in postmodernity.

This renunciation, this cultural cancellation, of a right to personal autonomy in

---

positions regarding autonomy prevailing in modernity, conflation was the only possible conclusion. Conversely, "the anxiety of influence" could hardly arise before the eighteenth century, i.e., before the autonomy of the artist had become a prized presupposition. Harold Bloom, in *The Anxiety of Influence: A Theory of Poetry*, 2nd ed. (New York, 1997), acknowledges that the word *influence* "in our sense—that of *poetic* influence . . . is very late. In English it is not one of Dryden's critical terms, and is never used in our sense by Pope. . . . But," Bloom maintains, "the anxiety had long preceded the usage" (27; emphasis in the original). This contention is contradicted by Bloom's own evidence—to which he himself remains impervious, doubtless because of his belief that absence of influence is the true difference between great and second-rate art, wherefore no serious artist could fail to feel such anxiety.

[68] Thus, as I pointed out in "Primacy of Science in Modernity" (cit. n. 7, on 59–60, 66, 71–2), and again in "(Re)cognizing Postmodernity" (cit. n. 5, on 162, 168 nn. 22–4), a principal basis for the subordinate cultural rank of technology in modernity, and specifically its subordination to science, was technology's lack of autonomy; more exactly, the nonsensicality of a claim of autonomy for technology: just as "without closure there can be no disciplines," so also could the discipline have no closure, if its purposes were technologic. Consequently, the subject-matter fields staked out as scientific disciplines were circumscribed by a boundary within which lay good, true, pure discipline-directed science and beyond which lay applied science. Conversely, the disdain today of pure, basic, curiosity-driven research, and the insistence that research have a technological orientation, is inseparable from the ideological bankruptcy and de facto disintegration of disciplinarity in recent decades.

[69] Associated especially with Foucault, an ostensible rejection of autonomy was generally characteristic of postmodernist theory; Nancy Fraser, "Michel Foucault: A 'Young Conservative'?" *Ethics* 96 (1985): 165–84; John McGowan, *Postmodernism and Its Critics* (Ithaca, N.Y., 1991), 2, 4–5, 20–1, 24–5; Steven Best and Douglas Kellner, *Postmodern Theory: Critical Interrogations* (New York, 1991), esp. 289; Novick, *That Noble Dream* (cit. n. 2), 543, on Barthes. One need not take such theory seriously in order to take that rejection seriously—as a rejection of that which was recognized to be characteristically modern.

[70] This specifically postmodern expansion-cum-restriction of autonomy is aptly expressed by the title of Stanley Fish's *Save the World on Your Own Time* (New York, 2008). Here "save the world" implies no obligation; it is, rather, permission to do as you damn please on, but only on, your own time. On the loss of workplace autonomy in postmodernity, see Freidson, *Professionalism* (cit. n. 27); Elliott A. Krause, *Death of the Guilds: Professions, States, and the Advance of Capitalism, 1930 to the Present* (New Haven, Conn., 1996), 280.

work would alone suffice to transform professionality and disciplinarity from cultural ideals to cultural anomalies. But the situation of professions and disciplines in postmodernity is rendered even more desperate by the withdrawal of their previously presumed right to corporate autonomy. Modernity (apart from its totalitarian statist aberrations) was especially supportive of the autonomy of corporations that made a respectable claim to being disinterested and, as such, serving the public interest. Postmodernity, suspecting all claims to disinterestedness, inverts that perspective, supporting autonomy for private and interested corporations, while withdrawing it from every body pretending to serve the public interest. As Prime Minister Gordon Brown said in 2009, apropos of the British House of Commons' administration of members' expense accounts, "We're in the twenty-first century now. Self-regulation is out the window."[71] Thus also in regard to corporate autonomy, postmodernity brings an inversion of modern presuppositions. Although both inversions are only partial, their very partialness cuts deeply against disciplines and professions.

Solidarity

Those wishing to damn modernity have often held up Soviet Socialist Russia and National Socialist Germany as its measure, pointing especially to the wanton persecution of individuals and the annihilation of entire classes of persons deemed undesirable. Yet such acts were not specifically modern; not more modern than the similarly annihilative aerial bombing campaigns carried out by the United States during the Second World War. There is, however, another, more pertinent respect in which Soviet Socialist Russia and National Socialist Germany were emphatically and specifically modern, while the United States remained not quite: both Russia and Germany were then socialist—or, rather, their governments considered it of prime importance to represent themselves and their national economies as being socialist. In modernity's culminative decades socialism seemed obvious, natural, and inevitable to almost all knowledgeable, reflective observers, worldwide. To be sure, "socialism" did not necessarily mean socialism of the left. Still more indicative of the manner and degree in which socialism was distinctive of modernity were the socialisms of the right. Among these the most important were, of course, the various national socialisms, combining the two most powerful popular ideologies of modernity. But there was also a wide range of religious socialisms on the right (as well as a few religious socialisms on the left). Left or right, wherever one turned in modernity, there was so-

---

[71] The quoted statement is as transcribed by me from the broadcast on WCSP, Washington, D.C., of the prime minister's monthly press conference, May 19, 2009. This statement, as quoted, was not to be found in the official transcript posted at http://www.number10.gov.uk/Page19365 (accessed 20 May 2009). However, that transcript contained a less graphic but more discursive statement by Prime Minister Brown to the same effect: "There has to be transparency, there has to be proper audit, and now I am saying you have to move—like almost every other public organisation has done in the past few decades—from being self-regulated—in other words you make your own rules, you make your own judgements, you make your own discipline—to being independently and statutorily regulated." (With the change in government, that transcript has been removed from the official website of the British Prime Minister's Office. As of November 26, 2011, it appears to be available on the internet only through the DeHaviland "political intelligence" service, at http://www.dehavilland.co.uk/.) Needless to say, if the House of Commons is to be denied self-regulation, no other British public institution can demand it. In this connection, see Thorpe, "Capitalism, Audit, and the Demise of the Humanistic Academy" (cit. n. 46), 113–6.

cialism or pretended socialism. Socialism, in some form or degree, was the cultural consensus of those last decades of modernity.[72]

Viewing the world emerging from World War II, both Friedrich Hayek in Britain and Joseph Schumpeter in the United States saw it just that way, and both deplored what they saw. Hayek, "in no spirit of mockery," dedicated *The Road to Serfdom* (1944) "to the Socialists of All Parties," confident that the Tory socialists and the Anglican socialists would be more receptive than the Labourites to his argument that personal freedom demanded market freedom and a minimal state.[73] Schumpeter, writing *Capitalism, Socialism, and Democracy* early in the war, but reaffirming its perspective in the years following it, asked, "Can capitalism survive?" and answered flatly, "No."

> One may hate socialism . . . and yet foresee its advent. . . .
>   The public mind has by now so thoroughly grown out of humor with it [capitalism] as to make condemnation of capitalism and all its works a foregone conclusion—almost a requirement of the etiquette of discussion. Whatever his political preference, every writer or speaker hastens to conform to this code. . . . Any other attitude is voted not only foolish but anti-social.[74]

What made this so? After all, as Hayek and Schumpeter and their like-minded associates insisted, capitalism realized key modern values, including, in particular, proceduralism and autonomy. As regards autonomy, capitalism was vulnerable to the objection that this universally acknowledged good is vouchsafed to only the few with ample capital. But as regards proceduralism, capitalism had all the logic on its side. Capitalism—no-holds-barred, Schumpeterian capitalism—is, in principle, purely proceduralist: like Darwinian natural selection, capitalism allows the means to determine the (always ad interim) ends. However, such unregulated, unmitigated capitalism offends against so much else implicit in the ideal of proceduralism and demanded by the disciplinary personality. Moreover, capitalism in practice, with its unfair outcomes seemingly locked in, hardly seemed proceduralist to those locked

---

[72] Werner Sombart could ask, "Why is there no socialism in the United States?"—*Warum gibt es in den Vereinigten Staaten keinen Sozialismus?* (Tübingen, 1906). But in fact there was in the Progressive-Era United States lots of support for the sort of socialism to which, underneath, Sombart himself really inclined; Friedrich Lenger, *Werner Sombart, 1863–1941: Eine Biographie* (Munich, 1994). Such quasi-fascistic socialism underlies Croly's *Promise of American Life* (cit. n. 11) and is flagrant in the opinion of Morris L. Cooke, anti-labor-union advocate of scientific management; he could quote Nietzsche in support of the autonomy of the individual worker: "In the Great State production will be made a part of the responsibility of labor." Cooke, "Who Is Boss in Your Shop?" *Ann. Amer. Acad. Polit. Soc. Sci.* 71 (1917): 167–85; see 175, 180. See also Howard Brick, *Transcending Capitalism: Visions of a New Society in Modern American Thought* (Ithaca, N.Y., 2006). Even Keynes, who did his best to save capitalism from itself, thought that in some important respects "anything would be better than the present system"; Roger E. Backhouse and Bradley W. Bateman, "Keynes and Capitalism," *History of Political Economy* 41 (2009): 645–71, on 662.

[73] Hayek, *The Road to Serfdom* [1944]*: Text and Documents*, ed. Bruce Caldwell (Chicago, 2007), 17, 36, 39. All the American commercial publishers approached on Hayek's behalf refused to undertake an American edition on the grounds that Hayek's thesis was very contrary to public sentiment and that sales would therefore be slight.

[74] Schumpeter, *Capitalism, Socialism, and Democracy*, 3rd ed. (New York, 1950), on 61, 63. Schumpeter reaffirmed these views in December 1949, addressing the American Economic Association, on which occasion he referred sarcastically to the Mont Pélerin Society as being completely negligible from the point of view of public opinion (ibid., 417–8).

out. To socialism, by contrast, all attached the proceduralist ideas of order, planning, system, regularity, method, and, as we saw perfected in Rawls, fairness.[75]

While a case in favor of capitalism could be made out in regard to autonomy and proceduralism, in respect of disinterestedness there was simply no contest. As Karl Polanyi said in *The Great Transformation*, his polemical classic—likewise written during the war but from a perspective opposite to that of Hayek and Schumpeter, his two Central European expatriate confreres—"The true criticism of market society is . . . that its economy was based on self interest. Such an organization of economic life is entirely unnatural." Nor was this perspective in any way new when Polanyi articulated it. Attached to Ruskin's name, it had long been widespread in the cultivated classes in Britain and the United States. In France, Durkheim maintained on scientific grounds the unnaturalness of a society based on self-interest, while in Germany, neither prophet nor theorist was needed, for this was simply how everyone thought.[76]

The opposition of disinterestedness to self-interestedness brings us close to socialism's sentimental core, but not yet quite to it. Disinterestedness is a characterological quality of an individual, but makes no reference to any particular group or collectivity. Socialism, however, depends on the individual's affective attachment to a particular collective, which attachment causes the individual to undertake substantially self-denying, even self-sacrificing, behaviors, with the intent of benefiting either that collective generally or particular persons qua members of that collective: "From each according to his abilities, to each according to his needs" was inscribed by Marx, claiming no originality, on the banner of communism as the distilled essence of the socialistic sentiment. There is a word for that specifically socialistic sentiment: *solidarity*. Solidarity was *the* socialist good, declared by Wilhelm Liebknecht "the highest cultural and moral concept," elevated by Eduard Bernstein above "all the other great principles of social rights—whether it is the principle of equality or the principle of liberty."[77]

---

[75] Indicative is that Rawls (*Theory of Justice*, cit. n. 41, on 272, 359–61) was ambivalent about regarding capitalism as proceduralist—partly out of ambivalence toward capitalism, and partly because of the limits of the political theorist's conception of proceduralism. Jürgen Kocka, in "Writing the History of Capitalism," *Bulletin of the German Historical Institute* 47 (2010): 7–24, on 10, surveying the scholarly literature on capitalism since the late nineteenth century, noted that, prior to the late twentieth century, the word and the concept always had, to some extent, "critical, polemical, pejorative connotations."

[76] Karl Polanyi, *The Great Transformation: The Political and Economic Origins of Our Time* (1944; repr., Boston, 2001), 259–60. Note the past tense in the quotation: Polanyi saw the world situation at that time just as did Hayek and Schumpeter, only he applauded the irresistible tide of socialism. On Ruskin, see Roger B. Stein, *John Ruskin and Aesthetic Thought in America, 1840–1900* (Cambridge, 1967); Martin J. Wiener, "Introduction" (2004), in *English Culture and the Decline of the Industrial Spirit, 1850–1980* (1981; repr., New York, 2004), xiii–xviii; Donald Winch, *Wealth and Life: Essays on the Intellectual History of Political Economy in Britain, 1848–1914* (New York, 2009). On Durkheim and his French contemporaries, see J. E. S. Hayward, "'Solidarity' and the Reformist Sociology of Alfred Fouillée, I & II," *American Journal of Economics and Sociology* 22 (1963): 205–22, 303–12; Michael C. Behrent, "The Mystical Body of Society: Religion and Association in Nineteenth-Century French Political Thought," *J. Hist. Ideas* 69 (2008): 219–43. On Germany, see Kocka, "Writing the History of Capitalism" (cit. n. 75); Lenger, *Werner Sombart* (cit. n. 72); Lukes, *Individualism* (cit. n. 56), 33–4.

[77] "From Each According to His Ability, to Each According to His Need," Wikipedia, http://en.wikipedia.org/w/index.php?title=From_each_according_to_his_ability,_to_each_according_to_his_need&oldid=451812464 (accessed 24 September 2011). The ideal collective that Marx envisioned was a universally inclusive, classless, stateless world society—a humanity perfected also in being freed from conflicts between less catholic solidarities; Arthur E. Bestor Jr., "The Evolution of the Socialist Vocabulary," *J. Hist. Ideas* 9 (1948): 259–302, on 273; *HWP*, s.v. "Solidarität," by

If solidarity was elevated so high by the mass movements of that time, it must also underlie scholarly reflection. And, indeed, the centrality of solidarity as a cultural value in modernity is implicit in the prominence given it—how to explain it, how to maintain it—in the social science disciplines emerging at the end of the nineteenth century. Solidarity was the focus of French sociology, and not only, or even first, with Durkheim. If Durkheim, more than any other theorist, provided the foundations for mid-twentieth-century American sociology, that was because in America too there was support from many directions and theorists for his focus on solidarity, and for the close association between *solidarity* and *discipline*.[78] Along with that focus, what, specifically, Durkheim provided was a conceptualization of social evolution as a process of institutional differentiation, which process was conceived as both raising and reconciling the antithetical desiderata of autonomy and of solidarity—autonomy both of the individual and the collective, and solidarity of individuals with the collective and consequently with other members of it. This yang/yin of autonomy and solidarity, which the mid-eighteenth-century protoromantics had discovered to be *the* existential problem of civilized man, and which would emerge in the nineteenth century as more especially *the* existential problem of learned professions and scientific disciplines, retained its centrality so long as modernity lasted, and no longer.[79]

How is it that in the course of the nineteenth century the need for cultivation of solidarity came to be the presupposition of both social-political theory and governmental policy? Why is it that by the last decade of the nineteenth century there was, across Europe and North America, a heightened concern—in some quarters a panicky concern—about the decline of social solidarity? All the accounts with which I am acquainted explain this through a romantic, declensionist framing, alleging that the traditional bases of social solidarity had been destroyed by the development of industrial capitalism and of market society, caused by or causing unrestrained individualism, and resulting in not only high Gini coefficients, but the uprooting of the populace from rural life and its replanting in urban environments, the decline of religious belief and the rise of secularism, and so forth—summed up as "modernity and rationalization whose perverse impulses need to be identified, confronted and resisted."[80]

Andreas Wildt; Wildt, "Solidarity: Its History and Contemporary Definition," in *Solidarity*, ed. Kurt Bayertz (Boston, 1999), 209–20, on 214, where Liebknecht and Bernstein are quoted. See also Steinar Stjernø (*Solidarity in Europe: The History of an Idea* [New York, 2005]), who, however, admits only socialist and social democratic solidarity.

[78] Durkheim, *Professional Ethics and Civic Morals* (cit. n. 50), 10–3; Stjernø, *Solidarity in Europe*, 31, 56, 267, 279; Wildt, "Solidarität," 1007, 1012; Wildt, "Solidarity" (All cit. n. 77), 215, quoting, in translation, Kurt Eisner (1908): "The cold, steely word solidarity . . . provides the whole of society with an iron foundation for the transformation and renewal of all human relations."

[79] Cassirer, *Philosophy of the Enlightenment* (cit. n. 67), 330. On Habermas as envisaging "a self-conscious practice in which solidarity and autonomy can be reconciled," see Peter Dews, "Editor's Introduction" to Jürgen Habermas, *Autonomy and Solidarity: Interviews* (London, 1992), on 29. See also Jerrold Seigel, "Autonomy and Personality in Durkheim: An Essay on Content and Method," *J. Hist. Ideas* 48 (1987): 483–507. Randall Collins, "The Durkheimian Movement in France and in World Sociology," in *The Cambridge Companion to Durkheim*, ed. Jeffrey C. Alexander and Philip Smith (New York, 2005), 101–35, on 108: "Sociology was finding its distinctive turf as the science of social solidarity." Searching this collective volume for *solidarity* via Google Books shows it to appear on 100 pages, or almost one in four. Editor Jeffrey Alexander opines that "people keep reading Durkheim, and arguing about him, to find out whether the determinateness of social structures must involve the sacrifice of autonomy" (136).

[80] Robert H. Wiebe's *The Search for Order, 1877–1920* (New York, 1967) may stand as representative, as it is still the standard reference and is by no means the most obviously romantic-declensionist. The quotation is from Keith Tribe (*Strategies of Economic Order: German Economic Discourse,*

Overlooked, however, when listing such allegedly overpowering forces destructive of social solidarity is the efflorescence in the later decades of the nineteenth century of new forms of social solidarity: consumer and producer cooperative movements, anti-wage-labor movements, political and cultural nationalist movements, internationalist movements of every thinkable sort, and, of course, movements toward the formation of professional associations and scientific and scholarly disciplines.

Neglected, moreover, when affirming that romantic, declensionist framing is the fundamentality and the persistence of the presupposition of the duty of social solidarity throughout modernity. At no time, and nowhere, prior to postmodernity, was simple self-seeking socially approved. From the Renaissance forward, through all the centuries of modernity, even as the premodern limitations on individuation gradually fell away, it was always taken for granted within all the culture-determining classes of those increasingly market-oriented societies that the individual's personal interest could and should ultimately be subordinated to collective needs. This was implicit even in what are conventionally regarded as the most individualistic ideologies produced in modernity—the liberalism and the utilitarianism of nineteenth-century Britain—for they both refused to contemplate a categorical difference between what was good for the individual and what was good for society. "I regard utility as the ultimate appeal on all ethical questions," said J. S. Mill, "but it must be utility in the largest sense, grounded on the permanent interests of man as a progressive being."[81] Thus the creation, the progress, and, by the early decades of the twentieth century, the hegemony of the idea of socialism as a political-economic desideratum reflected not a general failure of social solidarity, but, on the contrary, a revolution of rising expectations of social solidarity.[82]

Thomas Haskell brought together his insightful explorations of these cultural con-

---

1750–1950 [New York, 2007], on xiii), who was seemingly little aware of how largely he was speaking for the body of scholars.

[81] Mill, "On Liberty" (1859), chap. 1, available at http://www.utilitarianism.com/ol/one.html (accessed September 2011). Winch, *Wealth and Life* (cit. n. 76), 6, 10; Atiyah, *Rise and Fall* (cit. n. 14), 234: "It is emphatically *not* true that any influential body of persons ever believed in laissez-faire as a system" (emphasis in the original). Liberalism's ideal, Atiyah points out, was the "man of principle," who, by definition, acted according to rules of universal applicability and ultimate social beneficence.

[82] James T. Kloppenberg's *Uncertain Victory: Social Democracy and Progressivism in European and American Thought, 1870–1920* (New York, 1986) is an admirable exposition of the new and stronger claims of society on the individual asserted by a dozen systematic thinkers, German, French, British, and American, in the decades before the First World War. Although generally Kloppenberg restricted the word and concept *solidarity* to the doctrines of Fouillée and Bourgeois, his sources warrant a much wider use. Thus he quotes (289) University of Wisconsin economist Richard T. Ely putting "social solidarity" front and center, most prominently in *The Social Law of Service* (1896). And had he been willing to cast a wider net he would have quoted Edward Bellamy, *Looking Backward, 2000–1887* (1887), where, toward the end of chap. 12, Dr. Leete, seeking to enlighten Mr. West, who has awaked after a century-long sleep, says, "If I were to give you, in one sentence, a key to what may seem the mysteries of our civilization as compared with that of your age, I should say that it is the fact that the solidarity of the race and the brotherhood of man, which to you were but fine phrases, are, to our thinking and feeling, ties as real and as vital as physical fraternity." (The quotation appears on p. 134 of the Houghton Mifflin Riverside Library edition first published in 1917 and made available at http://www.gutenberg.org/files/25439/25439-h/25439-h.htm [accessed September 2011]. I am indebted to John Burnham for drawing my attention to Bellamy's affirmation of solidarity.) A wide range of turn-of-the-century metaphysical positions, Machian positivism among them, enabled Einstein, lying gravely ill in the winter of 1917/8, to take a step beyond Bellamy and say, "Ich fühle mich so solidarisch mit allem Lebenden, dass es mir einerlei ist, wo der Einzelne anfängt und aufhört"; Carson, Kojevnikov, and Trischler, *Weimar Culture and Quantum Mechanics* (cit. n. 53), 132–3, 237–9.

nections in the question, "What do socialism and professionalism have in common?" The context for this rhetorical question was his discussion of Anglican-Fabian socialist R. H. Tawney's call, in 1920, for a "professionalization" of all the roles through which society carries out its economically productive activities. Haskell's reply to his own question was "Precious little today"—he was writing in the early 1980s—"but everything of consequence to Tawney: Both promised to contain economic individualism and thereby to rescue industrial society from impending moral bankruptcy."[83] In so saying, Haskell spoke well for the romantic Tawney, as indeed for the body of historical scholarship. Overlooked, however, in this paradise-lost perspective is that socialism and professionalism developed in parallel as authentic ideological manifestations of modernity. Neither the one nor the other was primarily a reaction to, or reparation for, or rescue from, modernity. More particularly, both social-trusteeship professionalism, such as Tawney took for granted, and the entrusting of society's knowledge base to scholarly and scientific disciplines, as was then becoming the norm, presumed and required the existence of unprecedentedly high levels of social solidarity. With solidarity, both socialism and professionalism follow naturally if not inevitably; without solidarity, professionalism and disciplinarity, like socialism, are an impossibility.

Today, in Postmodernity

In 1993 Ulrich Beck, bellwether of postmodernity, already famous for "the risk society," published an essay titled "On the Disappearance of Solidarity."[84] Shortly after, Robert Putnam published "Bowling Alone: America's Declining Social Capital." In that article, and in his subsequent similarly titled book, Putnam laid out overwhelming evidence of the pervasive, progressive decline of social solidarity in the United States, beginning in the late 1960s.[85] Putnam's findings are strikingly confirmed, and

[83] Haskell, "Professionalism versus Capitalism: Tawney, Durkheim, and C.S. Peirce on the Disinterestedness of Professional Communities," in *Objectivity Is Not Neutrality* (cit. n. 52), 78–114, on 84. Haskell took the scientific and scholarly disciplines as the model, the template, for the professions. To this he was misled partly by his enthusiasm for Kuhn, and partly by sitting and writing in a period no longer able to credit the idea/ideal of disinterestedness. Thus he conflated self-seeking that is confined within the disciplinary reward system with self-seeking that looks beyond disciplinary boundaries. Consequently he was unable to recognize the predominance of service over competitive self-promotion, which differentiates the professions from the knowledge-producing disciplines, and which is the principal basis for the affinity between professionalism and socialism. Otherwise stated, he failed to appreciate the difference, pointed out above (nn. 51 and 68), between the inner-directed disinterestedness in the natural science disciplines, on the one side, and the other-directed disinterestedness of the applied social sciences and the professions, on the other side.

[84] Beck, "Vom Verschwinden der Solidarität" (1993), translated as "The Withering Away of Solidarity," in *Democracy without Enemies*, trans. Mark Ritter (Cambridge, 1998), 32–8. The fact that *solidarity* appears nowhere except in the title of this essay only augments its suitability to function as a straw in the wind. Daniel T. Rodgers, in *Age of Fracture* (Cambridge, Mass., 2011), describes the continuing fracture of social solidarities in the United States from the sixties through the eighties.

[85] Putnam, "Bowling Alone: America's Declining Social Capital," *Journal of Democracy* 6, no. 1 (1995): 65–78, and *Bowling Alone: The Collapse and Revival of American Community* (New York, 2000). The "revival" is Putnam's wishful thinking; only the collapse is evident. Widespread hopes for revival are unquestionably a large factor in the meteoric uptake of the nonconcept of "social capital" and the translation of Putnam's book into nine languages. For critique, see Margaret R. Somers, "Beware Trojan Horses Bearing Social Capital: How Privatization Turned Solidarity into a Bowling Team," in *The Politics of Method in the Human Sciences*, ed. George Steinmetz (Durham, N.C., 2005), 346–411. Solidarity is a key concept for Sandel (*Democracy's Discontent*, cit. n. 40), even though it is not in his index. He understands it, however, in an early postmodern, now-passé, communitarian

their global reach demonstrated, by *Dilbert*, postmodernity's first hugely successful cartoon strip. Created at the end of the 1980s, within ten years *Dilbert* was appearing in 2,000 newspapers, in 60 countries, with a readership of 200 million. Featuring a hapless, cubicle-dwelling information technology engineer, *Dilbert* provides a sick sort of catharsis for the horrors of that "abandon all claims to autonomy ye who enter here" postindustrial workplace. *Dilbert*'s ostensible theme is that "everyone is an idiot."[86] However, *Dilbert*'s veritable theme is not the idiocy but the cynicism of the postindustrial workplace. Worse, the absence of solidarity of any kind, anywhere— the absence of personal solidarity as well as social solidarity; its absence in the home-place as well as in the workplace. *Dilbert*'s creator, Scott Adams, who himself affects to feel, think, and live that complete lack of solidarity, has grasped the mainspring of postmodernity.[87] Thus, although we may deplore that the members of this or that corporate body—as, for instance, a university faculty—now behave like characters in *Dilbert* strips or in Ayn Rand novels, when push comes to shove, we would grant no corporate body the right to take its stand on solidarity.[88] In postmodernity, solidarity, more thoroughly than autonomy, is out the window.

POSTMODERNITY ENTAILS ANTIDISCIPLINARITY

The fate that has overtaken the ideal and the practice of socialism since the 1980s is well known: its rejection everywhere and in every form in which it had been established in the century from the 1870s to the 1970s, whether that be Russian-style state socialism, or Yugoslav work-site self-management socialism, or Scandinavian social democracy, or British nationalization of key industries and services, or merely U.S.-style progressive extension of redistributive programs. Well into the 1970s Americans were still broadly approving of that ongoing socialization of their society, with radical sentiment in that direction stronger then than it had been at any time since the 1930s. Consequently, conservative Republican Richard Nixon, elected president in 1968, supported far-reaching, left-facing legislation, and continued to do so through the early 1970s, until himself undone by his abuse of his office.[89] Yet had there then been a historian who truly understood just how intimate was the affinity between modernity and socialism, on the one side, and modernity and professionality and disciplinarity, on the other side, that historian, reflecting on the denunciations of the professions and the professional ideal that were then already being pronounced on every side, could have foreseen the fate of socialism.

---

sense, denying categorically that Marx's universal solidarity is solidarity at all. David A. Hollinger, in "From Identity to Solidarity," *Daedalus* 135, no. 4 (2006): 23–31, on 23, though he does not clarify the definitional issue, is nonetheless right that "the problem of solidarity is shaping up as the problem of the twenty-first century."

[86] Scott Adams, *The Dilbert Principle: A Cubicle's-Eye View of Bosses, Meetings, Management Fads and Other Workplace Afflictions* (New York, 1996), 2. This was Adams's twelfth book based on the strip; there would be forty-three more by 2009. For critique, see Norman Solomon, *The Trouble with Dilbert: How Corporate Culture Gets the Last Laugh* (Monroe, Maine, 1997).

[87] To date, the only cartoon strip of postmodernity that in any way rivals the success of *Dilbert* is *Garfield*. *Garfield*'s theme—ostensive and blatant—is a purely domestic version of *Dilbert*'s theme: the multifarious ways in which the title character, a fat and lazy cat, manifests a malicious lack of solidarity with his owner. The contrast with *Peanuts*, modernity's most popular cartoon strip, is total.

[88] Hollinger ("Faculty Governance," cit. n. 3) lays out the ways in which the research university today suffers from that lack of solidarity.

[89] Jacob S. Hacker and Paul Pierson, *Winner-Take-All Politics* (New York, 2010), 96–8.

In 1963 it still seemed a simple fact that "everywhere in American life, the professions are triumphant"—with which declaration cultural historian Kenneth S. Lynn opened an issue of *Daedalus* titled "The Professions."[90] Antiprofessional attitudes appeared abruptly in the mid-1960s and spread widely in the late 1960s, producing striking collisions with continuing expressions by senior figures of untroubled confidence in the professional ideal as the future. The epitome in this regard is Talcott Parsons's article on professions for the *International Encyclopedia of the Social Sciences*, published in 1968. There, in concluding, America's leading sociologist declared that

> the professional complex has already not only come into prominence but has even begun to dominate the contemporary scene in such a way as to render obsolescent the primacy of the old issues of political authoritarianism and capitalistic exploitation.[91]

When writing these words, Parsons, I venture, would have held it inconceivable that, by the time they appeared, younger colleagues would be standing his perspective on its head, applying the term *authoritarianism*, and even *exploitation*, to the professional complex itself. Eliot Freidson, one of those young Turks, looking back from the early 1980s—by then his views had already moderated considerably—described the transformation that had taken place in the regard of professions, both within the culture at large and by the discipline of sociology:

> In the 1960s . . . a shift in both emphasis and interest developed, which has intensified and continued down to our day. The mood shifted from one of approval to one of disapproval, from one which emphasized virtues over failings to one which emphasized failings over virtues. The very idea of profession was attacked, implying—if not often stating—that the world would be better off without professions.[92]

In America this rage against professionalism appears neither to have arisen, nor to have been confined, within any one sector of society. In particular, it was, in its origins, not a generational revolt, not a youthful rebellion, although youth coming of age in the late 1960s became the sector most radicalized by it. Initially, it was, rather, a broadly populist protest, a "revolt of the client." When, in the autumn of 1964,

[90] Lynn, "Introduction," *Daedalus* 92, no. 4, "The Professions" (1963): 649–54. Having, in the spirit of disinterested service, given up his Harvard professorship for one at Federal City College in Washington, D.C., Lynn would face there in the late 1960s attitudes unimaginable anywhere in the early 1960s.

[91] *International Encyclopedia of the Social Sciences*, ed. David Sills and Robert Merton (New York, 1968), s.v. "professions," by Parsons. Virtually identical is the future foreseen by Daniel Bell in *The Coming of Postindustrial Society: A Venture in Social Forecasting* (New York, 1973), whose theses go back to the mid-1960s—on which see my discussion in "Primacy of Science in Modernity" (cit. n. 7), 50–1, 107–9. Similarly, Paul Halmos (*The Personal Service Society* [London, 1970], 25, 117), one of Britain's foremost sociologists, saw professionalism remaking society in yet other, more humanistic, directions.

[92] Freidson, "Are Professions Necessary?" in Haskell, *Authority of Experts* (cit. n. 44), 3–27, on 4. Gerald L. Geison ("Introduction," in *Professions and Professional Ideologies in America*, ed. Geison [Chapel Hill, N.C., 1983], 3–11 and 111–2, on 7) noted "the current fashion for dismissing professional claims as mere self-serving verbiage—as deliberately deceitful smoke screens behind which professional groups can comfortably pursue their monopolistic goals." A fully developed case for social reform through abolition of the professions is made by Randall Collins (*The Credential Society: An Historical Sociology of Education and Stratification* [New York, 1979]), who proposed to prohibit an employer's imposing any formal qualification as a condition of employment.

Berkeley students became the first to revolt, it was chiefly as clients of the University of California. Quickly, however, the revolt of the American client became a global phenomenon led by intellectual elites. An initial upsurge of journalistic indictments was followed by a swell of scholarly works similarly cynical about the purposes and practices of the professions.[93] As the sixties advanced a more radical thrust was superadded to the revolt of the client: a romantic rebellion against discipline in all its forms, not yet present in the mid-1960s, had become prominent by the end of the decade. The world revolution of '68 showed that, to a degree previously unimagined, a common affective culture united a very wide range of politically and linguistically separated societies—showed them united, specifically, in decrying discipline.[94]

Although the sixties' revolt of the client was, in the first instance, directed against the professions, the absence of a clear cultural differentiation between the professions and the scholarly disciplines, together with the upwelling of romantic antipathy to the very notion of discipline, made it inevitable that also the scholarly disciplines would be cried down—in the first instance by their own recruits. Sociology, as said, being marked in especially high degree by the disciplinary triumphalism characteristic of the two postwar decades, was one of the first to experience polemics against its disciplinary identity and ambitions.[95] The "self-loathing" of the discipline of English literature was of equally early onset and has more often been declared devastating, although there are historians who would award that blasted palm to history.[96] Indeed, there is not a humanistic or social-scientific discipline that has not been damned by some secessionist sector of its practitioners or declared to be in crisis by its loyalists.[97]

Thus, beginning in the late 1960s, the positive connotations that had been associated with the concept of a scholarly discipline throughout the first half of the twentieth century, connotations that had continued to grow ever more positive right into the early 1960s, died away in most minds concerned in any way with the creation, curation, or utilization of scientific or scholarly knowledge. University professors and university administrators; officers of funding organizations, both private and public; government officials, both legislative and executive—all have increasingly disparaged disciplines in thought and word. "Disciplines carry the connotation of . . . being static, rigid, conservative, and adverse to innovation," said Peter Weingart in 2000,

---

[93] Marie R. Haug and Marvin B. Sussman, "Professional Autonomy and the Revolt of the Client," *Social Problems* 17 (1969): 153–61. The authors appear to lay claim to the phrase. (Note the feature of professionality that the authors conceive to be jeopardized by that revolt.)

[94] Julie Stephens, in *Anti-disciplinary Protest: Sixties Radicalism and Postmodernism* (Cambridge, 1998), rescues the sixties from political failure by stressing hippie and yippie hostility to every form of discipline and arguing that both postmodernism and the antipathetic attitude toward politics and government emerging in the eighties are its authentic heritors. A somewhat similar thesis, without, however, highlighting discipline, is advanced by David Kaiser, *How the Hippies Saved Physics: Science, Counterculture, and the Quantum Revival* (New York, 2011). Jeremi Suri et al. ("AHR Forum: The International 1968," *Amer. Hist. Rev.* 114 [2009]: 42–96) proceed from the world revolution perspective, perhaps first strongly emphasized by Immanuel Wallerstein.

[95] Terence Halliday, "Introduction: Sociology's Fragile Professionalism," in *Sociology and Its Publics: The Forms and Fates of Disciplinary Organization*, ed. Halliday and Morris Janowitz (Chicago, 1992), 3–42; see 4.

[96] Stanley Fish, "Profession Despise Thyself: Fear and Self-Loathing in Literary Studies," *Crit. Inq.* 10 (1983): 349–64; Novick, *That Noble Dream* (cit. n. 2), 578.

[97] Townsend ("Making History," cit. n. 21, on 21) cites disciplinary histories of anthropology, history, literature, political science, and sociology—all structured by crisis. The history of science is, admittedly, something of an exception, and hence especially in need of this exposition. See n. 106 below.

summarizing his appraisal of the literature. Expressed opinions were often far stronger—as, say, "Disciplines tend toward sclerotic self-satisfaction."[98]

In the natural sciences direct repudiations of disciplinarity are rare, but there are many indications that in the minds of natural scientists too the associations, both affective and formal, with their nominal disciplines, as well as with disciplined thought and action, have become almost monotonically less positive. Returning to the histogram (figure 1) presenting the changing frequency of appearance of different forms of the word *discipline* in *Science*, one sees, as pointed out above, in the first half of the twentieth century both a roughly constant frequency of appearance of *disciplined*, representing the use of the word in a generally characterological sense, and a rapid rise in the frequency of appearance of *disciplines*, representing the use of the word to refer to a field of scientific inquiry. Since the 1960s, however, both usages of the word have decreased in frequency of appearance in the pages of *Science*, more and more from decade to decade. A year-by-year analysis of the three postwar decades yields, surprisingly exactly, what the cultural history of that period would lead one to expect: a rapid rise of the frequency of appearance of *disciplines* to a historic peak in the late 1950s, followed by a slump to half that value in the late 1960s, and then, after a bit of a rebound in the early 1970s, a continuing fall. Insofar as *Science* speaks for them, disciplines and disciplinarity have been in almost unbroken decline for the past fifty years.

Our Antidisciplinarity

Among scholars, by far the most famous expression of this inverted cultural valuation of discipline—the academic emblem of the new animus, the object of countless emblematic citations—is Foucault's *Discipline and Punish*, the title under which *Surveiller et punir* (1975) was published in English in January 1978. The change in title for the English translation is indicative, I suppose, of Foucault's recognition that its prospects lay not in his proprietary concept of *surveillance*, but in the post-'68 consensus on the hatefulness of discipline. Ironically, before the year 1978 was out Foucault would announce to his auditors at the College de France that he "was wrong" in giving central significance to discipline.[99] This recantation, which remained as unknown as it was unimaginable to the great mass of Foucauldians, makes it a bit more difficult to regard Foucault as the prophet who revealed to the intellectual world the true nature of disciplinarity. More to the point of this article, however: as *discipline* in the sense of scholarly or scientific discipline had no prominence in French before

---

[98] Weingart, "Interdisciplinarity: The Paradoxical Discourse," in *Practising Interdisciplinarity*, ed. Weingart and Nico Stehr (Toronto, 2000), 25–41, on 29; Allan Megill, "Recounting the Past: 'Description,' Explanation, and Narrative in Historiography," *Amer. Hist. Rev.* 94 (1989): 627–53, on 631. *Sclerotic* has lost no currency as an epithet attached to disciplinarity; Anthony Grafton [qua president of the AHA], "History under Attack," *Perspectives on History* (2011): 5–7, on 5: "Critics inside and outside the academy have leveled a rich and seemingly plausible indictment at us. . . . We professors are imprisoned in sclerotic disciplines." Some further indications of the breadth of the hostility to disciplinarity are in Paul Forman, "In the Era of the Earmark: The Recent Pejoration of Meritocracy—and of Peer Review," *Recent Science Newsletter* 2, no. 3 (2001): 1, 10–2.

[99] Foucault, *Surveiller et punir: Naissance de la prison* (Paris, 1975), translated as *Discipline and Punish: The Birth of the Prison*, trans. Alan Sheridan (New York, 1978). Sheridan says in his translator's note that "in the end Foucault himself suggested Discipline and Punish." Michael C. Behrent, "Accidents Happen: François Ewald, the 'Antirevolutionary' Foucault, and the Intellectual Politics of the French Welfare State," *J. Mod. Hist.* 82 (2010): 585–634, on 599.

the twentieth century, it could hardly figure in Foucault's sources. Nor did he himself make more than a gesture in that direction. Rather, it was Foucault's scholarly readers, certain that that sense "lurks beneath the surface of the text," who chose to give disciplinarity special importance among the deprecated forms of discipline.[100]

That we are dealing here with an affect and animus springing from the depths of the postmodern *Weltgefühl* becomes evident when we find it expressed even by a scholar still attached to the ideal of disciplinary knowledge production and worried by the growing hegemony of "post-academic science." Unhappy though John Ziman was about what postmodernity had done to the scientific life, he could not refrain, in addressing *Real Science* (2000) to the general reader, from deriding as "academically correct" the practice of citation and discussion of the work of other scholars. Worse, spurning the judgments those other scholars might render on his work, Ziman took the characteristically antidisciplinary postmodern position that acceptance by the general reader is "the real test" of the adequacy of his analysis and description.[101]

But who are those disdained peers? They are not the physicists who so applauded Ziman's contributions in the 1950s and 1960s as to secure him election to the Royal Society at the early age of 42. They are to be found in that self-consciously antidisciplinary knowledge-production enterprise variously known as science studies, social studies of science, or STS. This enterprise Ziman himself helped to create and sustain, first with a 1960 talk on BBC radio, "Science Is Social," and then with a steady stream of books describing with admirable accuracy the successively emerging stages in the progress of our cultural presuppositions, from exalting academic science in the early 1960s to exalting antiacademic science since the 1990s. Yet, his deep reservations about the dedisciplinarization of science notwithstanding, Ziman himself succumbed to that very process. If not he, who, indeed, will remain unmoved in the midst of so powerful and pervasive a Zeitgeist?[102]

To be sure, STS, in every sense a product of postmodernity, offered Ziman no haven in this cultural storm—which, again, began with *The Structure of Scientific Revolutions*. For all that Kuhn presumed disciplinary science to be the true and final form of knowledge production and curation, he bore an unacknowledged animus against it. Thus, while in Kuhn's scheme the only truly disciplinary science was paradigm-

[100] Jan Goldstein, "Foucault among the Sociologists: The 'Disciplines' and the History of the Professions," *Hist. & Theory* 23 (1984): 170–92, on 176–8. Regarding *discipline* in French before 1900, see n. 16 above.

[101] Ziman, *Real Science* (cit. n. 2), xi, xii, 11. Muller ("Discontent in the Historical Profession," cit. n. 12, on 13) found that "among the most distressing trends in academic history is the tendency for scholars to publish and publicize their work in contexts where it is unlikely to be exposed to skeptical scrutiny and intense criticism." But that was more than ten years ago; meanwhile it has ceased to be found distressing—and still less a matter to be treated jokingly, as did David Philip Miller, "The 'Sobel Effect': The Amazing Tale of How Multitudes of Popular Writers Pinched All the Best Stories in the History of Science and Became Rich and Famous while Historians Languished in Accustomed Poverty and Obscurity, and How This Transformed the World; A Reflection on a Publishing Phenomenon," *Metascience* 11 (2002): 185–200. Today we all want to go that way. So, e.g., Shapin, "Hyperprofessionalism" (cit. n. 12), which in the six years since publication has received numerous affirmative citations by leading historians.

[102] Ziman, "Science Is Social," *The Listener*, August 18, 1960, as cited by Ziman, *Public Knowledge: An Essay Concerning the Social Dimension of Science* (New York, 1968), x. Between this volume and *Real Science* (cit. n. 2), Ziman published half a dozen others; John Enderby, "John Michael Ziman," *Phys. Today* 58 (November 2005): 74; Michael Berry and John F. Nye, "John Michael Ziman: 16 May 1925–2 January 2005," *Biogr. Mem. Fellows Royal Soc.* 52 (2006): 479–91. I emphasized the inescapability of postmodernity in "Assailing the Seasons," *Science* 276 (1997): 750–3, a review of P. R. Gross, N. Levitt, and M. W. Lewis, eds., *The Flight from Science and Reason.*

guided "normal science," this he denigrated as consisting largely of "mopping-up operations."[103] For Kuhn, all the truly important and admirable action occurred when a science fell into one of its revolutionary phases. Such phases, however, he portrayed as in every sense antidisciplinary. There paradigm and method are out the window, and external influences fly in. There the cumulativity fundamental to disciplinary science goes by the board. There the discipline in question lapses temporarily into a predisciplinary state. Although Kuhn was not able to recognize his own antidisciplinary animus, the STSers did not mistake his implicit message. Still, for them *SSR* was only a halfway house.

By the 1980s Kuhn's representation of science, anchored to, even if not wholly in, a disparaged disciplinarity, seemed intolerably restrictive to the increasingly antidisciplinary field of science studies. "The new orthodoxy" in the sociology of science, and in the field of science studies more generally, closely associated with the work of Bruno Latour in the late eighties, is manifestly, if not always explicitly, antidisciplinary.[104] Yet the stripping of knowledge production of its disciplinary character is only one side, one consequence, of a far broader, more fundamental rejection of those cultural presuppositions that made social institutions possible and disciplinarity thinkable. This new orthodoxy has generally been described, and describes itself, as "constructivist" or "constructionist," a revealing choice of a label that is not a sociologic term, but an epistemic or ontologic one. Described in sociologic terms, the new orthodoxy is, as Terry Shinn has repeatedly pointed out, antidifferentiationist—radically, categorically, antidifferentiationist.[105]

[103] *SSR*, 24. The text remained unaltered here in the subsequent editions of 1970 and 1996. It is somewhat unfair to Kuhn to apply, as is often done, his more depreciative characterization—"hack work" (30)—to normal science generally. He applied that term only to such work as could "be relegated to engineers and technicians." However, his poorly chosen examples of work that could be so relegated reveal the same bias.

[104] "The new orthodoxy" is Terry Shinn's apt phrase; Shinn and Bernward Joerges, "The Transverse Science and Technology Culture: Dynamics and Roles of Research-Technology," *Soc. Sci. Inform.* 41 (2002): 207–51, on 237–8; Shinn and Pascal Ragouet, *Controverses sur la science: Pour une sociologie transversaliste de l'activité scientifique* (Paris, 2005), 130–4. Andrew Pickering is the most prominent among those giving prominence to antidisciplinarity; Pickering, "Anti-discipline, or Narratives of Illusion," in *Knowledges: Historical and Critical Studies in Disciplinarity*, ed. E. Messer-Davidow, D. R. Shumway, and D. J. Sylvan (Charlottesville, Va., 1993), 103–22; Pickering, *The Mangle of Practice: Time, Agency, and Science* (Chicago, 1995), 214–7 (a section titled "Antidisciplinarity: A New Synthesis") and 224; Pickering, "Culture: Science Studies and Technoscience," in *The Sage Handbook of Cultural Analysis*, ed. Tony Bennett and John Frow (London, 2008), 291–310, on 294. If Latour was not very explicit about this, Pickering (*Mangle of Practice*, 217) was: "I think that antidisciplinarity is what is being recommended in Callon and Latour's frequent attacks on 'sociology' and the 'social sciences.'" No explicit notice of this programmatic antidisciplinarity is taken by John Zammito in *A Nice Derangement of Epistemes: Post-positivism in the Study of Science from Quine to Latour* (Chicago, 2004). Perhaps he finds it too horrifying: he gives, in discussing normal science, a categorical affirmation of disciplinarity as "the soundest vehicle for cultural creativity to which humans have attained" (295 n. 43).

[105] Shinn, "The Triple Helix and New Production of Knowledge: Prepackaged Thinking on Science and Technology," *Soc. Stud. Sci.* 32 (2002): 599–614, on 604; see also Shinn's publications cited in the previous note. Shinn, drawing attention to the antisociological character of the new orthodoxy, declines to admit dedifferentiation thinkable as a social reality, regarding it rather as a characteristic conceptual pathology of contemporary sociology. I have offered an appreciation of Shinn's endeavors to rescue disciplinarity in postmodernity in a foreword to Shinn, *Research-Technology and Cultural Change: Instrumentation, Genericity, Transversality* (Oxford, 2008), vii–xiii. Early on, Gieryn ("Distancing Science from Religion," cit. n. 28, on 588) underscored "the gulf that separates constructivism from the postulate of institutional differentiation," but, there avowing his conversion to constructivism, he took no notice of the skewness of that label from a sociologic perspective.

The postulate of institutional differentiation, on which was built Durkheimian sociology generally, and Mertonian sociology of science specifically, and which rendered the formation and perfection of professions and disciplines natural and inevitable consequences of natural and inevitable social progress, now seems bizarre and groundless. In part this is for the good reason that the implicitly teleological evolutionary assumptions underlying that postulate no longer underlie our best thinking. But more important and more widely consequential is the nominalistic-individualistic ontology implicitly underlying postmodernity. The rejection of the postulate of institutional differentiation is the consequence of the rejection of the reality of social institutions, indeed of society—the rejection of the notion that any collectivity may in principle, or does in practice, impose "an absolute constraint on the manner and means" that an individual "employs in achieving his ends." Thus the antidifferentiationism of contemporary sociology is just the theoretical expression of postmodernity's radical voluntarism.[106]

How then does it stand with our discipline, the history of science? Daston has recently alleged that we are different—different than the "undisciplined" discipline of science studies. We have no disciplinarity problem. Indeed, by Daston's account it is not so much we as they that are different. None of the natural scientific disciplines is disciplinarily challenged, she implies, while among the humanistic disciplines history stands as a perfect model of disciplinarity. In Daston's telling, in the last twenty years, but only in the last twenty years, "historians of science have become self-consciously disciplined, and the discipline to which they have submitted themselves is history." The result is "a less eclectic, more classically disciplined history of science, closely modeled on history."[107] Yet if one sought for a single datum that fully expressed how essentially antidisciplinary our discipline has become in the past twenty years, one need not look farther than this, Daston's dream—this fabulation put into the scholarly literature in the guise of a veridical account of an emergent communal reality.

As today hers is a very model of scholarship, certainly we can have no right to join Sarton in "silencing of dilettante scientists who write their histories 'with a complete lack of scholarly integrity.'"[108] Nor have we any longer the wish to do so. Sarton's ambition to enclose our field and eject all undisciplined practitioners was a taken-for-granted desideratum in the postwar decades of disciplinary consolidation. Two generations on, the one theme common to all the contributions to the December 2007 *Isis* "Focus" symposium on popularized and fictionalized history of science, "History of

---

[106] Gary A. Abraham, "Misunderstanding the Merton Thesis: A Boundary Dispute between History and Sociology," *Isis* 74 (1983): 368–87, on 383, stating positively, affirmatively, sociology's concept of social institutions. Regarding postmodernity's "radically unsociological sociology," Barry Barnes ("Thomas Kuhn and the Problem of Social Order in Science," in *Thomas Kuhn*, ed. Thomas Nickles [New York, 2003], 122–41, on 134–5) observed that since the late 1980s—"the reception of the work of Bruno Latour will serve as mark and symbol"—sociologists have "looked to fantasies of individual agency," for "anything that detracts from an imagined individual autonomy is found embarrassing." My essay "From the Social to the Moral to the Spiritual: The Postmodern Exaltation of the History of Science," in *Positioning the History of Science* [Festschrift for S. S. Schweber], Boston Studies in the Philosophy of Science 248, ed. Jürgen Renn and Kostas Gavroglu (New York, 2007), 49–55, offers bibliometric evidence of our flight from the social.

[107] Daston, "Science Studies" (cit. n. 31), 808, 811. Peter Dear and Sheila Jasanoff, in "Dismantling Boundaries in Science and Technology Studies," *Isis* 101 (2010): 759–74, take Daston to task for disrespecting STS. Nonetheless, they are right in step with her affirmation of the vitality of disciplinarity—while, postmodernly yet impossibly, attributing that vitality to the dismantling of boundaries.

[108] Heilbron, lecture (cit. n. 13), 189, quoting Sarton's inaugural lecture as professor of the history of science at Harvard.

Science and Historical Novels," is that we who identify ourselves with the discipline of the history of science have lost control over the production and curation of knowledge in that field, our field. The fact that, quite sensibly, none of the historians of science contributing to that symposium proposed that we assert the cognitive authority that we as a discipline presumptively possess—the fact that, on the contrary, all saw emulation of the style, and aspiration to the audience, of the trespassing journalists and novelists as the obviously desirable response—shows that we too have abandoned the ideology of disciplinarity.[109] Although we continue habitually to talk of our "discipline," our talk, like all such talk, is increasingly empty, reflecting neither our reality nor our aspiration.

[109] Kathryn M. Olesko, ed., "Focus Section: History of Science and Historical Novels," *Isis* 98 (2007): 755–95. See also nn. 36, 47, and 100 above. Cf. Horace Engdahl, permanent secretary of the Swedish Academy, the body that awards the Nobel Prize for literature, as quoted by Christopher Brown-Humes in "The Nobel Century," *Financial Times*, September 29–30, 2001, Weekend sec., on 1: "The borderline between the literature of fact and the literature of fiction will gradually weaken."

HORIZONS

Moving About and Finding Things Out:

Economies and Sciences in the Period of the Scientific Revolution

*by Harold J. Cook**

ABSTRACT

One of the most common arguments about science is that it leads to economic de-velopment; it is also commonly argued that the rise of science was a critical factor in the rise of the modern economy. This article explores that theme from the viewpoint of the history of northwestern Europe in the early modern period, arguing that rather than either "economy" or "science" producing the other, they were coproduced. In-stitutional forms of organization employed by the urban elite to manage their affairs came to place a high value on descriptive matters of fact, which became the chief matters of exchange in their efforts toward both material betterment and reliable knowledge. In giving pride of place to matters of fact in their knowledge systems, which moved relatively easily across cultural boundaries, it also became possible for urban leaders to imagine a universal form of knowledge, which we often call *science*.

One of the chief generalist visions of the history of science has long fallen outside the mainstream of the field, but is virtually a constant in the public forum: the rise of science has caused modern economic development. The positive contributions of science to economic growth have been assumed and discussed not only by many jour-nalists and others involved with political and economic decision making but also by academics beyond the history and philosophy of science. For many good reasons those of us in that field have generally been wary of entering into these discussions. Yet the enduring view that the modern economy owes a great deal to modern science suggests that however mythological some of these statements might seem, it would be worth examining the reasons why this theme has been kept alive over so many genera-tions. Moreover, if the relationships between "economy" and "science" are taken seri-ously even as they are reformulated, they lead to further questions about the processes of historical change. But in doing so, such investigations require that the history of science be brought into a closer relationship with the history of medicine and tech-nology, and with economic history, than was common in the late twentieth century.

* History Department, Brown University, Box N, 79 Brown St., Providence, RI 02912; Harold _Cook@brown.edu.

I would like to thank the other participants in the conference at the Huntington Library for their comments on an earlier version of this article; those in the Medieval and Early Modern History Semi-nar at Brown who responded with criticism and comment; Lissa Roberts, whose conversations and e-mails have helped to focus some of my thoughts for this article; and most especially, of course, the thoughtful editors, Robert Kohler and Kathy Olesko, and the anonymous referees.

For the purposes of this article, I explore such questions for the period and region with which I am most familiar, the early modern period in Europe and some of the far-flung European enterprises of the era, a time that witnessed the emergence of both modern science and capitalism.[1] The world of commerce and early capitalism not only privileged the mobilization of monetary value in the service of wealth, power, and pleasure but also advantaged certain forms of social interaction and exploitation that placed a very high value on accurate and cumulative information about the material world and its underlying materialistic elements. In taking up this line of analysis, it is necessary to explore not only recent literature in the history of science, medicine, and technology (STM), but some of the new economic history, which in turn offers analytical frameworks that can help us to see old problems in the history of science in new ways. It is likely—as this article will argue—that the modern economy and modern science were not in the kind of relationship in which one produced the other, but that they were coproduced and interdependent phenomena.[2] A further general hypothesis is that the usual statement that science is universal because it is true can be turned on its head, in order to suggest that the ability of certain kinds of information ("matters of fact") to be readily communicated across cultural and linguistic boundaries made it possible to imagine a universal kind of knowledge. This holds out the possibility (again) of showing that while abstractions about nature might be constructed, they are not merely relativistic points of view: there is something other than "culture" at stake, something that includes material life, from which science is made.[3] In any case, the sorts of practices and values that supported both a particular kind of science and a particular kind of economy are worth further consideration.

Such views might be generalizable to other periods and places, but only if the terms *economy* and *science* are pluralized in order to press otherwise essentialist abstractions toward a set of diverse and interactive processes located in time and place, which in turn might invite further historical inquiry. In other words, the relationships between economic activity and scientific activity will change according to the circumstances that produce them both. The manner in which they are interconnected in early modern Europe may not apply to, say, the nineteenth-century United States of America, or twentieth-century East Asia, or other places and times. In considering once again the possible relationships between natural knowledge and material betterment, then, it is very apparent that economic history is as important as intellectual and cultural history for an understanding of STM.

THE SCIENCE AND DEVELOPMENT PARADIGM

Currently, the dominant generalist view about development remains associated with the presumption that the modern economic system originated in Britain and continues to be led by the countries that also went through an industrial revolution in the nineteenth century, although a number of other nations are recognized as gaining

---

[1] For the moment, I use these general terms to capture the spirit of the discussion; I will be more precise later on.

[2] *Coproduced* is an idiom associated with Sheila Jasanoff; Jasanoff, *States of Knowledge: The Coproduction of Science and the Social Order*, International Library of Sociology (London, 2004), 1–6.

[3] For related thoughts, see, e.g., Bruno Latour, "Why Has Critique Run Out of Steam? From Matters of Fact to Matters of Concern," *Crit. Inq.* 30 (2004): 225–48.

through the adoption of similar economic structures. As we will see, the basic premise is that "the West" came to dominate the globe because of certain kinds of rationality coupled with political liberty, which allowed innovators to reap the rewards of their efforts and so led to human betterment as measured by per capita income and life expectancy. I will refer to this version of development, which has material improvement as its end and science as a critical contributor, as the science and development paradigm (SDP).[4] A second version of SDP (SDP_2) more explicitly associates development with moral improvement as well: good governments are instituted for the benefit of the governed, which means that material benefits and human rights are interdependent, being the two fundamental pillars of the hopeful aspects of humanity's collective journey through history. In SDP_2, science also plays a critical role, contributing not only to material improvement but to an investigative and reality-checking discourse that encompasses the human sciences as well, demanding attention to the potential of human capital. SDP_2 more openly asserts that scientific modernity brings political goods as well as material wealth.

At the moment, the chief historical rival of SDP for a generalist vision is the theme of competition between civilizations, currently much discussed in terms of the "clashes" of civilizations, with science sometimes being invoked as one of the chief markers of Western civilization. Because of the moral arguments often attached to SDP_2, ideas about development and ideas about the virtues of Western civilization are sometimes intermingled, resulting in occasional confusion about which of these major paradigms is being discussed. But given that the simple version of SDP—about material improvement—is the most elementary of these three generalist views, and so is shared among them, we can simplify our review by considering its relation to the history of science and (for the moment) setting the others aside.

Versions of SDP have been with Europeans since before the age of Francis Bacon, and they have remained close to the center of political discussion throughout the world since the nineteenth century. To illustrate the purchase that SPD still has on those concerned about current economic development one may turn to Jeffrey Sachs. Because he is one of the most visible figures among those who have argued for reforms that will lead to material improvement for all people—having, for example, served as director of the UN Millennium Project—he can be cited for what might be called current wisdom. The historical remarks he makes in his 2005 book on the possibilities of ending world poverty are therefore meant not to be original but to establish a basis for consensus, and can therefore be considered to be widely shared. Sachs begins his historical discussion by quoting from a 1930 essay by John Maynard Keynes to the effect that all economies were stagnant until the Industrial Revolution in Britain in the mid-eighteenth century. Keynes was among those who thought that all economies bumped up against ceilings to growth (often because of Malthusian limits) until new forms of technology, first invented in Britain, unleashed a new kind of economy, to which a word like *development* can be properly applied. Or as Sachs

---

[4] I am fully aware of the problems in using *science* as a general term, but will continue to do so as a vernacular shorthand—as in "the history of science"—for a collection of activities that included not only natural philosophy, mathematics, and mechanics, but laboratories, natural history, forms of museology, and many aspects of medicine and technology. See also the historical justification in Deborah E. Harkness, *The Jewel House: Elizabethan London and the Scientific Revolution* (New Haven, Conn., 2007), xv–xviii.

himself puts it, slightly more expansively, "The combination of new industrial technologies, coal power, and market forces created the Industrial Revolution. The Industrial Revolution, in turn, led to the most revolutionary economic events in human history since the start of agriculture ten thousand years earlier." It gave rise to urbanization and social mobility, transforming gender roles and family structures, even creating the division of labor. Politically, the major consequence was that "the British Empire became the global political manifestation of the Industrial Revolution."[5] By implication, modern history is the process by which the rest of the world is being caught up in the developmental transformation of material betterment and personal improvement begun in Britain.

Striving for persuasive clarity, Sachs goes on to set out his causal assumptions simply. Why the origin in Britain rather than in China or other "centers of power" in Europe or Asia? Sachs gives six reasons, which include not only coal and geography, but social and political liberty along with, "critically," the prior development of science.[6] Expanding on this last point, Sachs writes that

> after centuries in which Europe was mainly the importer of scientific ideas from Asia, European science made pivotal advances beginning in the Renaissance. . . . The decisive breakthrough came with Isaac Newton's *Principia Mathematica* in 1687, one of the most important books ever written. By showing that physical phenomena could be described by mathematical laws, and by providing the tools of calculus to discover those laws, Newton set the stage for hundreds of years of scientific and technological discovery, and for the Industrial Revolution that would follow the scientific revolution.[7]

Sachs goes on to reemphasize the importance of science when he writes, a few pages later, "I believe that the single most important reason why prosperity spread, and why it continues to spread, is the transmission of technologies and the ideas underlying them."[8] The general message is therefore that while the modern economic system is the product of the British Industrial Revolution, it could not have occurred without the prior Scientific Revolution; technology depends upon science. The simple lessons learned from this history continue to hold out the promise of development for any people, anywhere, even where it has been absent. I will take this to be the general model for contemporary SDP.

While most historians of science have shied away from such claims, Margaret C. Jacob and Larry Stewart have argued a similar position to Sachs's with vigor, underlining the importance of "Newtonianism" in further deepening the relationship between scientific concepts and economic innovation in the lead-up to the Industrial Revolution. "Newton's science in the service of industry and empire" is how they put it in the subtitle of their coauthored book of 2004. There they write, succinctly, that "Newton's followers" and "most of his explicators" focused on how he mathematically described "the mechanics of earthly bodies," which provided "the foundations for the study of fluids in motion and at rest," while also seeing his work as an example of the importance of "experimental evidence." Newton therefore appealed to engineers, and to many entrepreneurs. Such people consequently made the *Principia*

---

[5] Sachs, *The End of Poverty: Economic Possibilities for Our Time* (New York, 2005), 35, 36–7, 33; the quotation from Keynes is on 32.
[6] Ibid., 33–5.
[7] Ibid., 34–5.
[8] Ibid., 41.

"the cornerstone of Western economic development."[9] This comes close to saying, à la Sachs, that without the *Principia* there would have been no Industrial Revolution, which, for reasons developed further below, seems to me implausible.

On the other hand, in also writing about what they term the "culture of Newtonianism" Jacob and Stewart offer a more expansive account about how Newton's *Principia* served as an inspiration for a wide range of people investigating the world and making things from it, rather than a set of particular mathematical problems and solutions per se. That is, it was not the particular conceptual details articulated in the *Principia* but the efforts of those who liked what it stood for that did most of the work in creating change. Similar arguments about how a special kind of scientific and technological "culture" lay behind the Industrial Revolution and subsequent world economic development have become common. In a work suggested for further reading by Sachs, for instance, David Landes adopts the phrase "invention of invention" to describe what he sees as the "pleasure in new and better" that characterized European technological innovation.[10] Although his *Wealth and Poverty of Nations* places the origin of "development and modernity" in Europe in the later medieval period, and stresses the importance of institutional changes rather than political liberty as such—both points to which we will return below—he thinks that European expansion and empire, like the Industrial Revolution of Britain, flowed from the culture of invention.

Similarly, a recent account of the history of technology emphasizes the "culture of improvement" as the driving force for change and something unique to the West,[11] while an eminent member of the American historical profession, Joyce Appleby, has recently declared that England gave rise not only to the Industrial Revolution but to capitalism itself, with its "secret spring" being "innovation," which in turn grew from a new cultural formation that followed the Scientific Revolution.[12] At least one economic historian has fully adopted the cultural argument in explaining that only when capitalistic values are widely shared does economic growth occur.[13] Other grand syntheses connect the culture of innovation tightly to its institutional expressions, as in *The Most Powerful Idea in the World,* where William Rosen states that "human character (or at least behavior) was changed, and changed forever, by seventeenth-century Britain's insistence that ideas were a kind of property," allowing the democratization of the nature of invention through the patent system.[14] In yet another powerful synthesis, Ian Morris points to the "accumulation of technology," which

[9] Jacob and Stewart, *Practical Matter: Newton's Science in the Service of Industry and Empire, 1687–1851* (Cambridge, Mass., 2004), 15. See also Stewart, *The Rise of Public Science: Rhetoric, Technology, and Natural Philosophy in Newtonian Britain, 1660–1750* (Cambridge, 1992), and Jacob, *Scientific Culture and the Making of the Industrial West* (New York, 1997).

[10] Landes, *The Wealth and Poverty of Nations: Why Some Are So Rich and Some So Poor* (New York, 1999). The quoted phrases are from the title of chap. 4 and p. 58.

[11] Robert D. Friedel, *A Culture of Improvement: Technology and the Western Millennium* (Cambridge, Mass., 2007).

[12] Appleby, *The Relentless Revolution: A History of Capitalism* (New York, 2010), on 155; also note her comment on 156 that the true innovators were not mere tinkerers but "genuine geniuses." For her views of the Scientific Revolution, see 141–5.

[13] Deirdre N. McCloskey, *Bourgeois Dignity: Why Economics Can't Explain the Modern World* (Chicago, 2010).

[14] Rosen, *The Most Powerful Idea in the World: A Story of Steam, Industry, and Invention* (New York, 2010), xxiii. For a counterargument to the central importance of patents, see Karel Davids, *The Rise and Decline of Dutch Technological Leadership: Technology, Economy and Culture in the Netherlands, 1350–1800,* 2 vols. (Leiden, 2008).

was in turn the result of an "Atlantic economy that could generate higher wages and new challenges, stimulating the whole package of scientific thought, mechanical tinkering, and cheap power."[15]

But resorting to cultural values is not the only way of framing the argument. In another widely cited study of the relationship between science and economy, Joel Mokyr gives his attention not to "science" or even "knowledge" but to particular forms of them. In *The Gifts of Athena,* for instance, Mokyr argues that the Industrial Revolution was due to a "knowledge revolution."[16] While he gives credit to Jacob for inspiring some of his views, he at the same time criticizes her emphasis on Newtonian ideas as the origin of the kind of "understanding" that led to mechanization.[17] He instead tries to move beyond the usual categories of science and technology to write of both as containing "useful knowledge," which in turn made industrialization possible, while also arguing that useful knowledge could be divided into two kinds: "propositional" knowledge (or *episteme*, about what exists), and "prescriptive" knowledge (or *techne*, about how things work). One can find both propositional and prescriptive knowledge in both science and technology, making the common differentiation between the two pointless. For the moment, we need not go further into Mokyr's definitions, only observe that while he argues for feedback between propositional and prescriptive knowledge, he believes the modern economy to be based mainly on knowledge that is about things rather than reasons, about the intelligence of "how" rather than "why."[18] He stresses that scientific knowledge did not have to be "true" in what are commonly called its "theoretical" claims; instead, the power of the new knowledge practices lay in their descriptive credibility and clarity about the phenomena, which arose from experimental practice, and in the developing custom of making such informational knowledge public. Like Sachs and Jacob, then, he thinks that the origins of the Industrial Revolution in Britain were determined by "the scientific revolution of the seventeenth century and the Enlightenment movement of the eighteenth century."[19] And he urges other economic historians to "re-examine the epistemic roots of the Industrial Revolution."[20] But it is the power of what he calls prescriptive knowledge rather than either Newtonian science or a culture of innovation that is the chief focus of Mokyr's attention.[21]

Something like prescriptive knowledge has of course been of importance to historians of STM for many decades, and that form of knowledge is currently also of much interest to other economic historians who seek to explain the European take-off toward self-sustained growth. For instance, Ian Inkster has long been studying the forms of knowledge and technique that were incorporated into the Industrial

[15] Morris, *Why the West Rules—for Now: The Patterns of History, and What They Reveal about the Future* (New York, 2010), on 499–500 and 502.

[16] Mokyr, *The Gifts of Athena: Historical Origins of the Knowledge Economy* (Princeton, N.J., 2002), 31, 56–76.

[17] Ibid., 30.

[18] Ibid., 24–5, 5, 10, 20.

[19] Ibid., 33.

[20] Ibid., 29.

[21] More recently, however, he has moved more toward culture as an explanation: "Economic change in all periods depends, more than most economists think, on what people believe." Mokyr, *The Enlightened Economy: An Economic History of Britain, 1700–1850*, New Economic History of Britain (New Haven, Conn., 2009), 1.

Revolution and their forms of movement.[22] To articulate the focus of his subject, he adopts the phrase "useful and reliable knowledge" from the economic historian Patrick O'Brien.[23] O'Brien in turn currently has a major comparative project underway funded by the European Research Council, titled "Useful and Reliable Knowledge in Global Histories of Material Progress in the East and the West from the Accession of the Ming Dynasty (1368) to the First Industrial Revolution (1756–1846)."[24] The goal of the researchers is to study the history of several Eurasian regions in order to compare the possibilities for the development of useful and reliable knowledge in Europe and elsewhere as a fundamental aspect of the European takeoff. In a different fashion, but with similar questions in the background, Wolfgang Kaiser, director of studies at l'École des hautes études en sciences sociales and professor of modern history at l'Université Paris I Panthéon-Sorbonne, has a working group studying knowledge and knowledge transfers as critical aspects of economic history.[25] A recent and explicitly anti-Eurocentric work titled *The Eurasian Miracle* by the historically well-informed anthropologist Jack Goody also gives these kinds of knowledge exchanges pride of place in its explanations of historical change.[26] Such examples could be multiplied many times over.

In short, there is a very widespread and important discussion underway that is seeking much greater precision about how particular kinds of investigations into material nature produced the modern economy. They seek to cut through both older formulations about the contributions of "science" and more recent ones about a "culture" of innovation, favoring terms such as "prescriptive knowledge" and "information." Arguments within the history of STM, too, are pointing toward ways in which the so-called Scientific Revolution also placed a very high value on reliable information about material structures and processes. Something like the formation of an information economy might be said to have been at work. I know of no larger set of problems to which historians of STM can contribute.

THE LATE MEDIEVAL AND EARLY MODERN INSTITUTIONAL REVOLUTION

To further clarify the questions now being asked, let us begin with the recent "revolt" among many medieval and early modern historians who have argued that the beginnings of European economic exceptionalism came before the period of the Industrial Revolution. They are arguing that its origins lie in the sixteenth and seventeenth centuries, or even earlier, in the late medieval period, which happens to be precisely the period usually examined for the origins of the Scientific Revolution. The fact that the

---

[22] E.g., Inkster, "Mental Capital: Transfers of Knowledge and Technique in Eighteenth Century Europe," *J. Eur. Econ. Hist.* 19 (1990): 403–41; Inkster, "Cultural Engineering and the Industrialisation of Japan circa 1868–1912," in *Reconceptualizing the Industrial Revolution*, ed. Jeff Horn, Leonard N. Rosenband, and Merritt Roe Smith (Cambridge, Mass., 2010), 291–308.

[23] Inkster, "Potentially Global: 'Useful and Reliable Knowledge' and Material Progress in Europe, 1474–1914," *International History Review* 28 (2006): 237–86; see 260 n. 2 for his credit to O'Brien for using the phrase at a conference in Leiden in 2004.

[24] Described on the website of the Department of Economic History, London School of Economics and Politics, www2.lse.ac.uk/economicHistory/Research/URKEW/aboutUrkew.aspx (accessed 25 October 2011).

[25] Described on the Centre de Recherches Historiques website, crh.ehess.fr/document.php?id=440 (accessed 25 October 2011).

[26] Goody, *The Eurasian Miracle* (Cambridge, Mass., 2010).

rise of global seaborne commerce carried on European ships was contemporary with the rise of what we recognize as science was no coincidence, I believe, but rather a case of codependent phenomena.[27] Is it possible to go further and identify some of the structural changes that might have produced them both?

The importance of commerce to the development of early modern Europe is plain.[28] To be simpleminded about it, in the transition from feudalism to capitalism the taxable monetary wealth derived from urban commerce made new forms of gunpowder warfare possible and gradually undercut older methods of raising armies and navies based on personal loyalties, so that new forms of wealth and power gave rise to new kinds of polities that in turn encouraged commerce as their lifeblood. The late medieval city-states of northern Italy are well-known examples of the process, but many of the free cities of the Holy Roman Empire, the Hanseatic League of the Baltic and North Seas, the heavily urbanized areas of southeast England and the Low Countries, and the commercial cities of France and Iberia shared in a general process of acquiring political rights and privileges in return for monetary payments, making the voices of the merchants noticeable even at the most haughty aristocratic courts. How influential the values of the merchants were of course varied from place to place, but even where a prince's ideological orientation remained focused on controlling the beliefs and behaviors of his subjects, attention had to be given to the question of money and the arrangements necessary for its generation and acquisition.

Behind the development of such particular kinds of political economy lay important changes in behavior and belief. For, as historians of science are well aware, in any exchange between humans the question of trust arises.[29] It is often resolved by acquaintance with the other people involved, being especially strong among families and clans; where the parties are personally unacquainted, other markers such as ethnicity can establish the social bonds necessary for sufficient confidence to allow transactions.[30] In later medieval Europe, new formal methods came into being that allowed nonfamilial groups of merchant-traders to work together. These methods were partly based on new legal forms such as corporations, partly on the political negotiations that gave those corporations the credibility to enforce contracts. Such arrangements allowed a certain amount of confidence to flow from family to family, group to group, and city to city, lowering what the new institutional economics calls "transaction costs": the nonprice costs of conducting business, such as negotiation, the drawing up of contracts, inspections, and the resolution of disputes.[31] And indeed,

[27] Harold J. Cook, *Matters of Exchange: Commerce, Medicine and Science in the Dutch Golden Age* (New Haven, Conn., 2007).

[28] A masterwork is Fernand Braudel, *Civilization and Capitalism, 15th–18th Century*, trans. Siân Reynolds, 3 vols. (New York, 1979).

[29] To be very brief, Bruno Latour and Steve Woolgar made the question of the "credibility" of scientific facts an issue, and Steven Shapin and Simon Schaffer identified a key part of that as being a matter of "trust"; Latour and Woolgar, *Laboratory Life: The Social Construction of Scientific Facts* (Beverly Hills, Calif., 1979), and Shapin and Schaffer, *Leviathan and the Air-Pump: Hobbes, Boyle, and the Experimental Life* (Princeton, N.J., 1986); Shapin, *A Social History of Truth: Civility and Science in Seventeenth-Century England* (Chicago, 1994). Shapin has also brought the analysis of trust to a study of the modern system of capital formation and science in *The Scientific Life: A Moral History of a Late Modern Vocation* (Chicago, 2008).

[30] I am here indebted especially to Janet T. Landa, *Trust, Ethnicity, and Identity: Beyond the New Institutional Economics of Ethnic Trading Networks, Contract Law, and Gift-Exchange* (Ann Arbor, Mich., 1994), and to conversations with David Harris Sacks.

[31] A pioneering and still useful work in the field of the new institutional economics is Douglass C. North and Robert Paul Thomas, *The Rise of the Western World: A New Economic History* (Cambridge,

when one measure of transactions costs is mapped—the cost of money, which is an indicator of the level of confidence in the borrower to pay back what is owed—it appears that interest rates in Western Europe declined significantly during the late medieval period, in the fifteenth century reaching something like a modern rate, a low of 5 to 6 percent.[32]

In other words, while neoclassical economics focused on the role of price as a coordinating mechanism in economic behavior, a new band of historians have investigated other issues of coordinating exchange, which might generally be described as the social relationships behind economic forms. Exploring such relationships has taken them into questions of noncooperative behaviors such as coerced exchange (stealing, defrauding, and breach of contract), as well as what enabled those involved in economic exchanges to overcome such barriers. Such economic historians have also departed from neoclassical views in arguing that methods of establishing the formal mechanisms of trust can sometimes take place alongside the mechanisms of established governance—indeed, some have gone on to argue that state formation was a result of the kinds of formal arrangements for negotiation put in place by the merchants.[33] (One might say that these historians are Lockeans rather than Hobbesians.) In other words, it is not technology and the development of markets that came first but certain kinds of social and political relationships, which in turn gave rise to institutional structures (including "property rights") that underpinned the "modern" economy. Perhaps what we recognize as modern science grew from the same processes.

To grasp a few of the details, it is key to explore further the views of an important group of historians who have been arguing that the Industrial Revolution was only one outcome—if a fundamental one—of an economic transformation being wrought in Europe from the late Middle Ages forward. One of the foremost among these historians, Jan Luiten van Zanden, has called the new view a "revolt of the early modernists,"[34] which might equally be called a revolt of the Northwest Europeanists. It was most visibly marked by the appearance in 1995 of a study of the early modern Dutch economy by Jan De Vries and Ad van der Woude that was translated two years later under a bold title: *The First Modern Economy*.[35] If the appearance of "the market" rather than "markets" is the best sign of capitalism, then the modern European economy did not emerge until later in the sixteenth century; but most parts of the Low Countries had authentic markets in goods, labor, land, and capital by the beginning of the sixteenth century, along with measurable economic growth (which was not necessarily a "success story," however, since the consequent proletarianization

1973); see also North, *Institutions, Institutional Change, and Economic Performance* (Cambridge, 1990), and Joel Mokyr, "The Institutional Origins of the Industrial Revolution," in *Institutions and Economic Performance*, ed. Elhanan Helpman (Cambridge, Mass., 2008), 64–119.

[32] See esp. chap. 1 in J. L. van Zanden, *The Long Road to the Industrial Revolution: The European Economy in a Global Persepective, 1000–1800* (Leiden, 2009); for the figures on interest rates, 22–3.

[33] E.g., Landa, *Trust, Ethnicity, and Identity* (cit. n. 30), 49–65; Avner Greif, *Institutions and the Path to the Modern Economy: Lessons from Medieval Trade* (New York, 2006).

[34] Van Zanden, "The 'Revolt of the Early Modernists' and the 'First Modern Economy': An Assessment," *Econ. Hist. Rev.* 55 (2002): 619–41.

[35] De Vries and Van der Woude, *Nederland, 1500–1815: De eerste ronde van moderne economische groei* (Amsterdam, 1995), translated as *The First Modern Economy: Success, Failure, and Perseverance of the Dutch Economy, 1500–1815* (Cambridge, 1997). The title took up an argument De Vries advanced as early as 1973: De Vries, "On the Modernity of the Dutch Republic," *J. Econ. Hist.* 33 (1973): 191–202.

of labor was only slightly beneficial for the population as a whole, while social welfare overall declined).[36] Van Zanden's own recent analysis, titled *The Long Road to
the Industrial Revolution*, does not quarrel with the phenomenon of an economic
takeoff from about 1800 associated with the Industrial Revolution, but he demonstrates the rise of market institutions and real economic growth from the late medieval period, which provided the "long runway" that made the modern takeoff possible. From modeling real wages and productivity, moreover, he concludes that "the
transition towards modern economic growth" occurred in both the Dutch and English
economies not in the nineteenth century but "at some point between the 1590s and
1620s" (with Dutch productivity growth declining after 1670 and starting up again
after 1820). The "Little Divergence" of these parts of northwest Europe was, he argues, in turn founded on unique processes of family formation that emerged in the
fourteenth and fifteenth centuries (which made for fewer children but higher investment in human capital) and on consequent methods for making nonfamilial institutions credible and trustworthy.[37]

Historians of other regions of Europe, most notably northern Italy, might wish to
quarrel with some of the conclusions of the Northwesterners, but for the moment
we can take away two propositions: new schools of economic and historical thought
have been signaling that late medieval and early modern social and institutional
changes were fundamental to the development of the modern economy, and the Low
Countries were a region where this new economy was clearly emergent. One can also
call the new economy capitalistic by at least the seventeenth century, when financial
institutions existed that allowed investors to have confidence that they could make
money from money without engaging in trade themselves. In 1621, the word *capitalist* was even used in Holland as a legal term (for those who owned more than 2,000
gilders' worth of movable property).[38] While some recent work shows that not all
markets in Europe and the Middle East were convergent in terms of price—a sign
of "the market" at work—by the seventeenth century Dutch markets were among
the most integrated of all.[39] In exploring the possible connections between emergent
capitalism and the Scientific Revolution, then, this is a period and place worth our
attention.

In general terms, moreover, one can say that through processes of the Dutch Revolt
the capitalists had become sovereign in the United Provinces. The practical sover-

[36] Martha C. Howell, *Commerce before Capitalism in Europe, 1300–1600* (New York, 2010); Bas
van Bavel, "The Medieval Origins of Capitalism in the Netherlands," *Bijdragen en mededelingen
betreffende de geschiedenis der Nederlanden—Low Countries Historical Review* 125 (2010): 45–79.

[37] Van Zanden, *Long Road* (cit. n. 32), quotations on 253–4; see also his summary on 264–5. For
similar kinds of arguments, although differing in emphasis and detail, see Michael Mitterauer, *Why
Europe? The Medieval Origins of Its Special Path*, trans. Gerald Chapple (Chicago, 2010), and Bas
van Bavel, *Manors and Markets: Economy and Society in the Low Countries, 500–1600* (New York,
2010).

[38] Marjolein C. 't Hart, *The Making of a Bourgeois State: War, Politics and Finance during the Dutch
Revolt* (Manchester, 1993), 122–3.

[39] Larry Neal, "The Integration and Efficiency of the London and Amsterdam Stock Markets in the
Eighteenth Century," *J. Econ. Hist.* 47 (1987): 97–115; Jonathan I. Israel, "The Amsterdam Stock
Exchange and the English Revolution," *Tijdschr. Gesch.* 103 (1990): 412–40; Süleyman Özmucur
and Pamuk Şevket, "Did European Commodity Prices Converge during 1500–1800?" in *The New
Comparative Economic History: Essays in Honor of Jeffrey G. Williamson*, ed. T. J. Hatton, Kevin H.
O'Rourke, and Alan M. Taylor (Cambridge, Mass., 2007), 59–85.

eignty of urban merchant-magistrates—soon called the *regenten*, or regents—meeting in formal committees to seek consensus and pound out agreements, created a large enough network of personal acquaintance and credit to finance a defensive war.[40] The polity that emerged, internationally recognized as a sovereign state in the Treaty of Westphalia of 1648, was often referred to as the Dutch Republic. It was a true republic during the period from 1650 to 1672 when, in most of the provinces, no *stadholder*, or commander of the armed forces, also held high political office, but the details of governance—which even Dutch politicians themselves were sometimes at a loss to explain—need not delay us long. Simply put, the republic was a place where powerful merchants ruled. Revolt against Habsburg centralization in the Low Countries, which included resistance to the imposition of the Inquisition and new taxes, was framed not as a revolution advocating new political rights but as a defense of the liberties of the nobles, cities, and provinces of the region.[41] After Philip II sent Spanish troops to restore his authority in the later 1560s, bitter warfare left the seven United Provinces of the north de facto independent and the Spanish Netherlands of the south, including Flanders and Brabant, under Habsburg dominion. For various reasons, Amsterdam profited from the decline of Antwerp, becoming the most powerful city in the most powerful province, Holland, but the republic contained many other important cities, such as Middleburg, Rotterdam, Dordrecht, Utrecht, Deventer, Delft, Leiden, Haarlem, Alkmaar, Enkhuizen, Leeuwarden, Groningen, and so on.

The result was a polity in which the regents had the upper hand in making their voices heard in provincial assemblies, while the provinces in turn retained most sovereign rights, giving way to collective "national" decision making only for the sake of warfare. In matters of law and taxation the provinces were virtually independent. Even after the restoration of the stadholder (William III) in 1672, and his purging of local governments to ensure that his supporters held power, it was regents who governed the cities and who in turn dominated the provinces. They were effectively collective sovereigns who ruled via often difficult negotiations with one another and, at times, with the stadholder. They even privatized much of their long-distance warfare by contracting it out to some of themselves via the East India Company (which was granted sovereign powers east of the Cape of Good Hope) and the West India Company. Unsurprisingly, the cultural values of the merchants also dominated the republic, so that the kinds of knowledge they valued most highly also took precedence.[42]

The institutional processes that were producing something resembling a modern economy and polity were therefore also producing something like modern technoscience. The forms of knowledge that Dutch early modern merchants authored and valued have not attracted as much attention as those of the British during the

---

[40] For the fiscal mechanisms enabling this, see James D. Tracy, *The Founding of the Dutch Republic: War, Finance and Politics in Holland, 1572–1588* (Oxford, 2008); 't Hart, *Making of a Bourgeois State* (cit. n. 38).

[41] For a masterful survey that shows the medieval roots of the Dutch Republic, see Marten Prak, *The Dutch Republic in the Seventeenth Century: The Golden Age*, trans. Diane Webb (Cambridge, 2008).

[42] It is my impression that the cultural influence of the court of Orange had its greatest effect when the princes were working with the regents rather than against them, as in the period of Frederick Henry, but this deserves further investigation. For an example, see Marika Keblusek and Jori Zijlmans, eds., *Princely Display: The Court of Frederick Hendrik of Orange and Amalia Van Solms* (The Hague, 1997).

Industrial Revolution, no doubt because they are not as obviously related to technology as the much-memorialized experimental tinkering of steam engines and spinning jennys. But it was a period and place of enormous technical innovation. A recent and important study by Karel Davids has investigated changes in land use and water management, fishing and shipping, the movement of goods, energy production, institutional regulations (from guilds and markets to patents and municipal underwriting), manufacturing industries (building, clothing, brewing, shipbuilding, salt and soap boiling, metalworking—including the making of armaments and the minting of coins—textile and book printing, and the manufacture of gilt leather, ceramics, glass, tobacco pipes, and paper), and processing industries (oil pressing, barley hulling, hemp crushing, timber sawing, tanning, sugar refining, diamond cutting, tobacco processing, distilling, chemical manufacturing and pharmaceutical preparation, and, after 1770, the use of steam engines).[43] The patterns of change in each of these sectors were subject to different constraints of supply, demand, capital, organizational and manufacturing methods, and expertise. But Davids finds that from about 1580 to 1700 the overall rate and scope of technological innovation was pronounced in the Dutch Republic, although after 1700 changes became much more uneven according to sector, with technological leadership being retained in only a few areas (such as armaments and minting).

While economic factors such as demand, wage rates, and capital costs help to explain Dutch technological innovation, they cannot do it alone: the effects of nonmarket institutions were also very important (although not religion, since Davids finds that the positive developments came despite Calvinism rather than because of it). As many other recent studies do, Davids also concludes that the many microinnovations that improved technologies already in place, which were in turn often the result of the introduction of knowledge and skills by way of immigration, were more important than breakthrough discoveries. More generally, a relatively open society that encouraged the intermingling of discourses among a variety of social groups, and political decentralization, together with targeted encouragement for innovation from the regents, were critical. Borrowing from Mokyr's distinction between propositional and prescriptive forms of useful knowledge, Davids also finds that science made a difference once the methods of representing the latter (the how-it-is form of knowledge) began to be written down and supplemented with numbers, pictures, and models in ways that made this knowledge somewhat independent of the persons who created or employed it. The values of the regents were therefore fundamental to technological development. Their management of institutional incentives for innovation was probably the most important factor of all. Decline set in as they moved from being an entrepreneurial to a rentier class: instead of close involvement with the material of economic production and the exchange of goods, the development of financial instruments shifted the most powerful regents into dependence on legal abstractions such as investments in land, bonds, and shares, and so led them to take less interest in the details of the means of production, while at the same time their increasingly comfortable political hold on power increased their reliance on secrecy, protectionism, and patriotism, which interfered with the exchange of propositional and prescriptive knowledge.

[43] Davids, *Rise and Decline* (cit. n. 14).

MATTERS OF FACT AND COMMERCIAL ECONOMIES

To understand the development of the Scientific Revolution in the republic,[44] then, we need to understand more about the kind of knowledge of nature valued by the Dutch regents. The basis of their information economy required attentiveness to matters of fact, and so their science was framed by the same attentiveness to the objects of nature (i.e., they valued "objectivity" highly). In recent decades, especially following the work of Barbara Shapiro, many historians of early modern science have placed this concern for matters of fact at the center of their accounts.[45] Sir Francis Bacon was deeply involved in such reforms, and it is therefore no accident that those writing in English have often described this kind of science as "Baconian."[46] But the valuing of facts was well established before Bacon's time and in other places, and it did not require his endorsement.[47] Matters of fact may be either propositional or prescriptive (in Mokyr's language) or constitute reliable information (in O'Brien's terms). While some kinds of facts are constructed through the elaborate procedures of modern laboratories, there are others that refer to recognizable natural kinds, such as gold, that are not constructed in the same way. Even in the case of complex constructions (such as the identification and naming of trace compounds in the body), their material manifestations are not constructions in the same way that linguistic expressions about them might well be.[48] In the early modern period, at least, it was the matters of fact known from the senses and experience that gained increased attention as the basis for a new kind of philosophy.

For various reasons, the history of science long gave its chief attention to theories and concepts, and more recently to the linguistic expressions of constructed knowledge. But early modern matters of fact are different from concepts in many ways.

[44] For recent overviews, see K. van Berkel, "The Dutch Republic: Laboratory of the Scientific Revolution," *Bijdragen en mededelingen betreffende de geschiedenis der Nederlanden* 125 (2010): 81–105; Eric Jorink, *"Het boeck der natuere": Nederlandse geleerden en de wonderen van Gods schepping, 1575–1715* (Leiden, 2007), recently translated as *Reading the Book of Nature in the Dutch Golden Age, 1575–1715*, trans. Peter Mason (Leiden, 2010).

[45] Shapiro, *Probability and Certainty in Seventeenth-Century England: A Study of the Relationship between Natural Science, Religion, History, Law, and Literature* (Princeton, N.J., 1983); Shapiro, *"Beyond Reasonable Doubt" and "Probable Cause": Historical Perspectives on the Anglo-American Law of Evidence* (Berkeley and Los Angeles, 1991); Shapiro, *A Culture of Fact: England, 1550–1720* (Ithaca, N.Y., 2000). See also Simon Schaffer, "Making Certain," *Soc. Stud. Sci.* 14 (1984): 137–52; Lorraine Daston, "The Factual Sensibility," *Isis* 79 (1988): 452–67; Peter Dear, *"Totius in Verba*: Rhetoric and Authority in the Early Royal Society," *Isis* 76 (1985): 145–61; William Eamon, *Science and the Secrets of Nature: Books of Secrets in Medieval and Early Modern Culture* (Princeton, N.J., 1994); Pamela O. Long, *Openness, Secrecy, Authorship: Technical Arts and the Culture of Knowledge from Antiquity to the Renaissance* (Baltimore, 2001). My thoughts are especially indebted to David S. Lux, *Patronage and Royal Science in Seventeenth-Century France: The Academie de Physique in Caen* (Ithaca, N.Y., 1989).

[46] E.g., Shapiro, "Law and Science in Seventeenth-Century England," *Stanford Law Review* 21 (1968): 727–66; Shapiro, "Law Reform in Seventeenth-Century England," *American Journal of Legal History* 19 (1975): 280–312; Shapiro, "Sir Francis Bacon and the Mid-Seventeenth-Century Movement for Law Reform," *American Journal of Legal History* 24 (1980): 331–62.

[47] E.g., Harold J. Cook, *The Decline of the Old Medical Regime in Stuart London* (Ithaca, N.Y., 1986); Paula Findlen, *Possessing Nature: Museums, Collecting and Scientific Culture in Early Modern Italy* (Berkeley and Los Angeles, 1994); Lorraine Daston and Katharine Park, *Wonders and the Order of Nature, 1150–1750* (New York, 1998); Harkness, *Jewel House* (cit. n. 4).

[48] Ian Hacking, "The Participant Irrealist at Large in the Laboratory," *Brit. J. Phil. Sci.* 39 (1988): 277–94, responding to Latour and Woolgar, *Laboratory Life* (cit. n. 29).

Concepts can be explained, for instance. If they are associated with a proof, it can be demonstrated. In either case, while argument may ensue, in principle agreement can be reached through a process of "reasoning." Facts, however, are propositions about truth founded on what can be known through the senses and so are subject to the vagaries and certainties of sensory experience. Many modern European languages other than English have words to indicate this kind of knowledge, words like German and Dutch *kennen*, or French *connaître* and *connaissance*, all of which indicate knowing by acquaintance rather than by reasoning. Put another way, what one knows from the senses as filtered by taste and experience has to do with familiarity rather than universal propositions. Hence doubts about the validity of testimony via the potentially misleading senses were vigorously attacked by many other early modern figures, including René Descartes.[49]

As has been much noticed, however, public acknowledgment and conveying of facts require the witnessing of intermediaries in the form of trusted persons and institutions. Individually reported facts are not only subject to doubt about the state of an observer's senses, but often involve particular instances rather than repeatable events. It is significant, then, that as Shapiro pointed out, the English word derives from the legal term *factum*, meaning an event that was agreed to have happened. In other words, a fact is what a consensus of eyewitnesses agree took place. It is reminiscent of the point made by Martin Heidegger about the word *thing*: it, too, derives from legal language. As his translators comment, *das Ding* originally designated "the tribunal, or assembly of free men. The *thing* was a cause one negotiated or reconciled in the assembly of judges. Heidegger in a later work refers to this in setting forth the notion of *thing* as what *assembles* a world."[50] Put another way, the facts of a matter are those elements of a consensus about an object or event that are agreed upon by the participants in a meeting before they move on to debate questions about what decisions to make in light of the facts. As Steven Shapin and Simon Schaffer argued eloquently, the problem for early modern science was therefore to establish the credibility of descriptions of objects or happenings, often by assembling witnesses for demonstrations and writing reports that allowed the reader to participate vicariously as a "virtual witness."[51]

Similar processes were at work in the methods of keeping track of business affairs on paper. Miles Ogborn, for instance, has argued that the English East India Company was constituted by an assemblage of writing practices to which the parties had to conform. By showing how information and interests were embedded in routines of formal recording and communication, he was able to uncover some of the mechanisms of conveying trust from authorities in one place to their deputies far away, even when those faraway persons had many motivations to act against the best interests of

---

[49] See Michele de Montaigne's famous "Apologie for Sebon," in *The Complete Works of Montaigne*, trans. Donald M. Frame (Stanford, Calif., 1989), esp. 443–57, refuted in Descartes's sixth meditation. See also the still-worthwhile Richard H. Popkin, *The History of Scepticism from Erasmus to Descartes* (New York, 1964).

[50] Heidegger, *What Is a Thing?* trans. W. B. Barton Jr. and Vera Deutsch (South Bend, Ind., 1967), 5; see also the translators' n. 3, pp. 8–9.

[51] Shapin and Schaffer, *Leviathan* (cit. n. 29). See also the employment of the concept by others, as in Michael Aaron Dennis, "Graphic Understanding: Instruments and Interpretation in Robert Hooke's *Micrographia*," *Sci. Context* 3 (1989): 309–64, and Rob Iliffe, "Material Doubts: Hooke, Artisan Culture and the Exchange of Information in 1670s London," *Brit. J. Hist. Sci.* 28 (1995): 285–318.

those in London.[52] The ability to establish agreement about goods and persons, and what happened and when, was the groundwork of the activities that allowed people to calculate probable risks and benefits and to plan for the future. The Dutch East and West India Companies were, if anything, even more bureaucratic than the English. In other words, the factual sensibility encouraged attentiveness to descriptions of material substances, which, like other goods in which merchants had interests, became trustworthy nuggets of information that could be moved easily from place to place despite local cultural barriers.

The centerpiece of the Dutch market was the Amsterdam *Beurs,* or "exchange," where merchants could speculate on the differences in price between different qualities and quantities of goods at various places and on potential future value given a variety of circumstances, and even purchase shares in businesses. Such markets allowed strangers to buy and sell under publicly acknowledged conditions, so that financial affairs no longer needed to be transacted by clans or family firms or others who were personally bound to one another: trust could be diffused through more impersonal institutions. All things for sale (and we have noted that this now included land and labor as well as goods and even money itself) could then acquire a value measured in a common currency—could become commensurable—allowing a kind of lowest common denominator that existed as a figure written on paper to substitute for the material thing itself. This kind of abstract exchange was in turn based on the flow of information. No market information was completely transparent, and people and firms held back important information when they could and when it was to their advantage. But to make an exchange on the Beurs, some modicum of information about the facts at hand needed to be accepted by both parties. Hence the beginnings of newssheets and later newspapers, which conveyed information about the goods themselves, the price other people had been willing to pay for the same commodities at different times and places, the people involved in the transactions, the firms and markets from which they came, the availability and quality of transport and shipping, the state of war and peace, and so forth. The exchange was, then, first and foremost a meeting place for the exchange of information, which was in itself highly valued and which supported the flow of credit (from the Latin *credo,* "I believe").[53]

Neoclassical economists long ago emphasized the importance of information in any economy, although usually assuming that access to information is a transparent process. Questions of secrecy aside, it was from placing a high value on factual knowledge and the means by which its truths were assessed that commercial exchanges might happen; on such kinds of values the new science was founded, too.[54] More important, finding out the facts was a process to which people had become accustomed through their daily activities, a process in which they commonly invested some of their resources (as transaction costs), so that acting by the familiar rules of that game outside of business hours, so to speak, might bring delight as well as

[52] Ogborn, *Indian Ink: Script and Print in the Making of the English East India Company* (Chicago, 2007); the book draws not only on the work of Shapin and Schaffer but on Adrian Johns, *The Nature of the Book: Print and Knowledge in the Making* (Chicago, 1998), among other works of the history of science.

[53] Cook, *Matters of Exchange* (cit. n. 27), 49–53.

[54] Long, *Openness, Secrecy, Authorship* (cit. n. 45); William R. Newman and Anthony Grafton, eds., *Secrets of Nature: Astrology and Alchemy in Early Modern Europe* (Cambridge, Mass., 2001); William Eamon, *The Professor of Secrets: Mystery, Medicine, and Alchemy in Renaissance Italy* (Washington, D.C., 2010).

material benefit.[55] Factual knowledge of nature did not have to be directly useful or financially profitable (although it might). But attentiveness to descriptive information about nature was part of a more general attention to those matters of fact that allowed a wide variety of transactions to take place.

Moreover, among the shared values in both commercial exchanges and scientific communication was clarity of speech, at the time called "plain speech." As the Dutch physician Cornelis Bontekoe put it, "I am accustomed to pay more attention to the subject, and to the truth of what I say than to the fair choice of words and eloquence of style: all the more since I believe I am eloquent enough if I can make myself understood, since the only standard of speaking and writing is that of being understood."[56] Similar statements can be found among early proponents of the Royal Society of London. In the work of historians of science of decades ago, plain speech was therefore seen as one of the chief characteristics of science, although it was often associated with "puritan" preaching.[57] But it may have even deeper roots in the kinds of exchange of accurate and precise information and promises necessary for business. Almost from the beginning of printing, publishers traded on the provision of information in order to supplement the manuscript newssheets on which merchants relied.[58]

As an example of how the values common to both business and science were evidenced in one life, we can turn to a Dutch regent-savant, Nicolaes Witsen, born in 1641.[59] The son of an Amsterdam patrician and Muscovy merchant, Witsen rose to prominent positions in business and politics as well as becoming a well-known advocate for the new science and a patron of learning-as-information. As a young man he traveled with embassies to London and Moscow before embarking on a grand tour of France and Italy. He later gained election to the city council of Amsterdam—serving as burgomaster thirteen times—and to the governing board of the Dutch East India Company (the VOC), represented his city as an ambassador extraordinary in England after the Glorious Revolution, and hosted the visit of Tsar Peter to Amsterdam in 1697–8. He took a keen interest in land and sea routes, and maps, languages, and peoples, as well as new information about nature, collecting an impressive cabinet of *artificialia* and *naturalia* from throughout the Dutch East and West Indies. In 1671 he published a book on shipbuilding (*Scheepsbouw en Bestier*; revised and expanded as *Architectura Navalis* in 1690) and printed a map of what are now called Central and East Asia, which became the basis for a book on the subject (*Noord en Oost Tartarije*, 1692; expanded in an edition dated 1705 on the copperplate frontispiece but not published until many years after his death). Witsen also promoted expeditions to Namaqualand in southern Africa and to the maritime "southland" (western Australia, or Nieuw Holland); was a patron of many men of learning and the arts, including many informants about Asia; ventured into the republic of letters; became a fellow

---

[55] See also Anne Goldgar, *Tulipmania: Money, Honor, and Knowledge in the Dutch Golden Age* (Chicago, 2007).

[56] Bontekoe, *Tractaat van het excellenste kruyd thee* [Treatise about the most excellent herb tea], vol. 14, *Opuscula Selecta Neerlandicorum de Arte Medica* (Amsterdam, 1937), 127.

[57] Richard F. Jones, *Ancients and Moderns: A Study of the Rise of the Scientific Movement in Seventeenth-Century England* (1936; repr., St. Louis, 1961).

[58] Andrew Pettegree, *The Book in the Renaissance* (New Haven, Conn., 2010), 130–50.

[59] I rely on Marion Peters, *De wijze koopman: Het wereldwijde onderzoek van Nicolaes Witsen (1641–1717), burgemeester en VOC-bewindhebber van Amsterdam* (Amsterdam, 2010), and Bruno Naarden, "Witsen's Studies of Inner Asia," in *The Dutch Trading Companies as Knowledge Networks*, ed. Siegfried Huigen, Jan L. de Jong, and Elmer Kolfin (Leiden, 2010), 211–39.

of the Royal Society of London; and much else. His personal library was sizable but not outstanding, being for his use rather than display and aimed at assembling information rather than model literature. Moreover, in his attempts to discover the truth of things in the midst of misinformation, rumor, and legend, he not only assessed his agents and compared accounts, but sought pencil sketches of people, places, and things, clearly placing a high value on picturability. Things with shape and solidity, which could be described carefully and acknowledged as real by the eyes, vouched for the nature of truth.[60]

Or take another example, Witsen's cousin (and fellow burgomaster) Johannes Hudde. He is best known to historians of science as someone who contributed to the development of the mathematics of probability, but he was deeply involved in many kinds of political, mercantile, and scientific projects. The Huddes and Witsens of the world were necessarily highly numerate, and Hudde had a further excellent education in the mathematical school at Leiden, while in office he gave his attention to concerns such as raising funds for public institutions via lotteries, which in turn caused him to take up the problems of calculating probabilities.[61] His mathematical explorations were in keeping with his experimental interests generally: for instance, he taught both Johannes Swammerdam and Antoni van Leeuwenhoek how to make the single-lens microscopes that made their own reputations, and he proved his personal abilities as a virtuoso in other ways, too.[62] In other words, like Witsen, Hudde placed a very high value on discovering new and accurate information about the natural world, information that was available through the senses—sometimes with the aid of instruments—and equally encouraged the use of mathematics in support of both accuracy (as in Witsen's mapmaking) and generalization (as in Hudde's own calculations). One could multiply these examples by pointing to other urban patricians and improving landowners and princes throughout early modern Europe.[63] Their attentiveness to material detail fits well with what Shapiro terms a factual sensibility, Mokyr calls useful knowledge, and O'Brien characterizes as useful and reliable knowledge.

Of course, less wealthy and powerful people also took a major part in describing and analyzing the world of natural facts that the leaders of commerce valued so highly. Isaac Beeckman exemplifies many of these well-educated but middling sorts.[64] Best known as the person who introduced Descartes to the methods of mathematical physics, Beeckman was, among other things, a craftsman, businessman, physician, and schoolmaster. Intended by his father to be a minister, Beeckman had a fine education at a Latin school, also picking up practical mathematics from a teacher in Rotterdam, Jan van den Broecke, before continuing to read philosophy and mathematics during 1607–8 on the advice of Leiden professor Rudolph Snell.

[60] This may give further support to an older idea that associates the new science with the outlook of Petrus Ramus, although it does not reduce it to that; Walter J. Ong, *Ramus: Method, and the Decay of Dialogue; From the Art of Discourse to the Art of Reason* (1958; repr., Cambridge, Mass., 1983). See also Klaas van Berkel, *Isaac Beeckman (1588–1637) en de mechanisering van het wereldbeeld* (Amsterdam, 1983), revised and translated edition forthcoming.

[61] Ian Hacking, *The Emergence of Probability: A Philosophical Study of Early Ideas about Probability, Induction and Statistical Inference* (Cambridge, 1975), 114–8.

[62] Marian Fournier, "Jan Swammerdam en de 17e eeuwse microscopie," *Tijdschrift voor de Geschiedenis der Geneeskunde, Natuurwetenschappen, Wiskunde en Techniek* 4 (1981): 75–6; J. van Zuylen, "The Microscopes of Antoni Van Leeuwenhoek," *Journal of Microscopy* 121 (1981): 309–28; see 310.

[63] For an earlier example of the argument, see Pamela Smith, *The Business of Alchemy: Science and Culture in the Holy Roman Empire* (Princeton, N.J., 1994).

[64] For this account I rely on Van Berkel, *Isaac Beeckman* (cit. n. 60).

Returning from the university without a degree, he took up the candle-making trade of his father, then tried again for the ministry under the influence of the Pietist movement, while in business he developed an expertise in the piping of water for breweries, fountains, and other uses. He also began studying medicine and apparently became convinced of the truth of atomism after finding the arguments against it offered by Galen, Fernel, and others to be insufficient (he took an MD from Caen in 1618).

Two years later, in Breda, the city of his future bride, whom he was visiting, Beeckman met the young cavalier Descartes at the *rederijkerskamer* Het Vreuchdendal (one of the urban literary and debating societies common in the region). By then he had a new position as a schoolmaster in one of the most distinguished Latin schools in Utrecht, but he also acted as a technical consultant, advising on the dredging of the harbor of Dordrecht, for example. At the end of 1620 he moved to a position as rector of an illustrious school in Rotterdam, where he became a member of various social circles of regents, merchants, craftsmen, and tradesmen. He was frequently consulted on many industrial and infrastructural projects in the rapidly developing city and was among the eight founders of a *collegium mechanicum* there. He moved again in 1627, to the somewhat grander city of Dordrecht, as the rector of its distinguished Latin school, where he gave an inaugural oration on what he termed *fysisch-mathematische wijsbegeerte* (physico-mathematical philosophy) and where he built a tower full of instruments for the study of physical phenomena. He was connected to people who knew William Harvey, Giordano Bruno, and Galileo Galilei as well as to libertines and Rosicrucians, and hosted Pierre Gassendi and Marin Mersenne on their visits to the republic, although he quarreled bitterly with his former friend Descartes when he returned to the Netherlands in 1629. After returning to Zeeland in 1635, and before his death in 1637, he investigated methods for grinding lenses. He was a truly important figure for his ideas as well as for his investigative energies and technical abilities. But he also represents a group of very skilled and able students of natural phenomena who moved easily between the worlds of formal education, commerce, and governance, between medicine and natural philosophy, and between personal information networks and the distribution of information and ideas through impersonal books and devices. In cases like Beeckman's, it makes little sense to try to decide whether his abilities in craftsmanship and sales, engineering, politics, or learning, or even religion, led the way—they were of a piece, held together by the formal and informal institutions of the day.

A philosophical formulation of such values was given in 1632, on the occasion of the founding of an athenaeum in Amsterdam. The city's merchant princes had wanted a university, but were prevented by the states of Holland and Zeeland, who supported the supremacy of the university in Leiden, so they had to settle for an advanced school that could not award degrees. As the first professors in their new institution they appointed two of the most eminent philosophers in the country: Caspar Barlaeus and Gerhard Joannes Vossius. For his inauguration, Barlaeus delivered an address titled *Mercator Sapiens* ("The Wise Merchant")—afterward printed in both Latin and Dutch versions.[65] For his theme, he gave a modern twist to Martianus Cappella's late classical *Marriage of Mercury and Philology*, which would have been well known to

---

[65] For a recent study and translation into French, see Catherine Secretan, *"Le marchand philosophe" de Caspar Barlaeus: Un éloge du commerce dans Hollande du siècle d'or; Étude, texte et traduction du "Mercator Sapiens"* (Paris, 2002).

his distinguished audience. He turned the god Mercury into *Mercatura* (trade), and Philology into *Sapientia* (wisdom), and showed how their union in Amsterdam had created the most powerful and knowledgeable of polities.[66]

Barlaeus could almost take it for granted that the merchants cared deeply about formal education, not only because they were establishing the athenaeum but because from the later Middle Ages they had promoted one of the most extensive systems of schooling in Europe.[67] Apprenticeship usually required initial literacy and numeracy, and further training in these skills was often stipulated by the guilds.[68] Cities in the Low Countries—as in the Breda of Beeckman and Descartes—also fostered rederijkerskamers, in which many adults could continue to pursue their enthusiasm for learning. Literacy in the republic was probably the highest in Europe, the rate around 1500 being approximately 30 to 40 percent of men and 20 to 30 percent of women, rising to about 85 and 64 percent, respectively, by 1800, and was even higher in the countryside than in the cities.[69] (Even rural workers might be involved in proto-industrial occupations, since imported grain from the Baltic freed much labor from the simple and heavy tasks of agricultural routine.) Many Dutch cities offered formal schooling in navigation, surveying, and other technical mathematical subjects, while private schoolmasters and tutors advertised their expert abilities to further educate pupils in Latin and other languages, mathematics, and other subjects.[70] Some families invested in advanced tutoring for their female children, too, with a few Dutch women gaining widespread reputations for their learning, such as Anna Maria Schurman, or Descartes's interlocutor, the princess Elizabeth.[71]

In Amsterdam, moreover, something like 7 percent of boys went on to universities, which was considered essential for any responsible position in business, law, politics, or religion.[72] While Philip II had developed plans to establish a new university in Deventer, William the Silent and the provinces of Holland and Zeeland had founded a university in Leiden in 1575.[73] The province of Friesland followed that example by establishing a university in Franeker in 1585, and Groningen did the same in 1614. Many other cities also established athenaea or "illustrious schools," which gave local boys the equivalent of an introduction to university education: Harderwijk (in Gelderland) in 1600, Deventer in 1630, Utrecht in 1634, both Dordrecht and Den Bosch in 1636, Breda in 1646, Middelburg in 1650, Nijmegen in 1655, Rotterdam in

[66] For additional thoughts on Barlaeus's address, see Cook, *Matters of Exchange* (cit. n. 27), 68–73, and K. van Berkel, "Rediscovering Clusius: How Dutch Commerce Contributed to the Emergence of Modern Science," *Bijdragen en mededelingen betreffende de geschiedenis der Nederlanden* 123 (2008): 227–36.

[67] Engelina Petronella de Booy, *De Weldaet der scholen: Het plattelandsonderwijs in de provincie Utrecht van 1580 tot het begin der 19de eeuw*, Stichtse historische reeks 3 (Haarlem, 1977).

[68] Stephan R. Epstein and Maarten Roy Prak, eds., *Guilds, Innovation, and the European Economy, 1400–1800* (Cambridge, 2008).

[69] Van Zanden, *Long Road* (cit. n. 32), 190–5.

[70] A project on the culture of early modern Dutch mathematics is coming to a conclusion at the University of Twente, led by Fokko Jan Dijksterhuis and including Tim Nicolaije and Arjen Dijkstra.

[71] Mirjam de Baar et al., eds., *Choosing the Better Part: Anna Maria Van Schurman (1607–1678)*, trans. Lynne Richards (Dordrecht, 1996).

[72] Willem Frijhoff, "Het Amsterdamse Athenaeum in het academische landschap van de zeventiende eeuw," in *Athenaeum Illustre: Elf studies over de Amsterdamse Doorluchtige School, 1632–1877*, ed. E. O. G. Mulier Haitsma et al. (Amsterdam, 1997), 37–65; see 40–1.

[73] Frijhoff, "Deventer en zijn gemiste universiteit: Het Athenaeum in de sociaal-culturele geschiedenis van Overijssel," *Overijsselse historische bijdragen* 97 (1982): 45–79; M. W. Jurriaanse, *The Founding of Leyden University*, trans. J. Brotherhood (Leiden, 1965).

1681, Maastricht in 1683, and Zutphen in 1686. In Utrecht, the athenaeum was converted into a university proper in 1636 (i.e., it gained the legal right to grant degrees), followed by Harderwijk in 1648. Prince Maurits, who used the latest technology to excellent effect in his military campaigns, also insisted on setting up a formal engineering school at the university in Leiden under the leadership of an adviser and tutor of his, Simon Stevin: the curriculum was in Dutch, being called the *Nederduytsche Mathematique*, or "Dutch mathematics."[74] While the governance structures of the advanced schools and universities varied, they usually contained representatives of the academic body together with a majority of members from the town council and provincial estates, who funded the schools partly from their own pockets via taxation. Even at this level of education the interests of the ruling merchants were dominant.

In his address to the Amsterdam regenten, then, Barlaeus could spend most of his words not explaining why merchants should be educated but countering the arguments of those who feared that engagement with worldly activity would undercut academic excellence. Thus, he explained why giving attention to the creation of wealth increased "ruminations of the mind," too. Some of his arguments drew on the views of the famous jurist Hugo Grotius, French advocates of amour propre, and others who were showing why the pursuit of self-interest was natural and good.[75] But in looking around at the magnificent buildings of the expanding city, the canals and locks, docks and warehouses, and great ships, which brought bustling commerce in precious things from all the world over, Barlaeus not only praised the activities that created the city's wealth and might but also demonstrated that its worldly interests were founded on virtue and learning. Both the merchant and the sage needed to cultivate honest conduct, and to value all things that helped them to discover what they sought after; moreover, a kind of love of worldly learning produced knowledge necessary both for trade and for learning. In going on to explain what he called "speculative philosophy" Barlaeus pointed not to abstract topics but to tangible subjects such as geography, natural history, astronomy, languages, and the study of other peoples' dress, habits, and so forth.

In turning the attention of people to the "philosophical" implications of descriptive and factual subjects by revisiting ancient myths in light of this new world, Barlaeus was contributing to a Europe-wide discourse. His more famous English contemporary, Sir Francis Bacon, had of course been making similar claims as to how the conjunction of power and majesty depended on cultivating a knowledge of the world rather than of words; in the later sixteenth century, the French royal cosmographer Louis Le Roy had similarly shown not only that all great nations had united both power and wisdom, but that their cycle of rise and fall could be broken if the learned turned their attention to the preservation of the arts and sciences and all things necessary for life.[76] For Barlaeus's contemporaries, there was even a new name for the kind of person who loved knowledge of the world virtuously: *liefhebber*. (At about

---

[74] Willem Otterspeer, *Groepsportret met dame: Het bolwerk van de vrijheid, de Leidse Universiteit, 1575–1672* (Amsterdam, 2000), 200–2.

[75] A good basic introduction to these themes remains Arthur O. Lovejoy, *Reflections on Human Nature* (Baltimore, 1961), 129–51.

[76] Le Roy, *Of the Interchangeable Course or Variety of Things in the Whole World, and the Concurrence of Armes and Learning thorough the First and Famousest Nations, from the Beginning of Civility, and Memory of Man, to This Present*, trans. R A (London, 1594); for one of the many suggestive works on Bacon, see Julie Robin Solomon, *Objectivity in the Making: Francis Bacon and the Politics of Inquiry* (Baltimore, 1998). See also Adrien Delmas, "Writing History in the Age of Discovery,

the same time, the English adopted the Italian *virtuoso*; the French would later use the word *amateur*, from *amare*, to indicate this kind of lover.)[77] Liefhebbers became particularly known for their collections of paintings, prints, and objects, both artificialia and naturalia.

It would certainly be possible at this point to start listing examples of the many other people in the Dutch Republic who would have shared the values placed on this kind of factual information as observed in Witsen, Hudde, and Beeckman and as articulated by Barlaeus. At almost the same moment that Barlaeus was speaking, Dr. Tulp commissioned from Rembrandt the famous painting of his anatomy lesson, a painting rich in symbolism about how it is possible to know the wonders of God's most magnificent creation, humanity, from the careful study of our physical bodies. It would not be long before people like Beeckman would be carefully considering William Harvey's new idea about the circulation of the blood, which was based on careful experimentation, or Descartes would be proposing a physiological account of human behavior in his work on the passions. Plans were also then being discussed to create a medical faculty in Utrecht, and an anatomy theater would soon be constructed in Amsterdam, where the apothecary Jan Swammerdam senior was only the best known of the many collectors of naturalia. All of them, despite their different backgrounds and various religious preferences, agreed on one thing: the first duty of anyone who wished to find out about natural things was to find out the matters of fact, in all their fine detail, before going on to speculate about their causes.

"EAST AND WEST" IN EARLY MODERN MEDICINE

Looked at from the home country, then, the northern Dutch world had become an entrepôt for scientific and technical information and ideas as much as an entrepôt for goods and finance. But at the same time, the Dutch had become powerful partly because they were not limited to activities in their home country: they had become sometimes-brutal merchants to the world, interacting not only with their European neighbors but with peoples of North America and Brazil; with the Caribbean, Angola and the Slave Coast of Africa, the South African Cape, the Malabar Coast of South Asia, Ceylon, Sumatra and Java, Taiwan, and Japan; and with other VOC cities and trading posts strung along the Indian Ocean from Basra and Mocca to what is today Thailand. In the home metropolis manuscripts and collections were accumulated, housed, and preserved, inventories were taken and sometimes published, and redistribution of the value-added information and objects was initiated. The cabinets of curiosities and gardens of men like Witsen were filling with specimens from the East and West Indies, and maps, engravings, travel writings, and books chock-full of information were pouring from the presses. Matters of fact from all over the world were uncovered and assessed, and at the same time that Dutch soldiers and sailors, and mercenaries, were to all intents and purposes destroying some indigenous peoples in order to retain a monopoly position in the spice trade, respect was offered to other people whose knowledge of the uses of nature was admired. Some of the indigenous

---

According to La Popelinière, 16th–17th Centuries," in Huigen et al., *Dutch Trading Companies* (cit. n. 59), 297–318.

[77] Jan de Jong et al., *Virtus: Virtuositeit en kunstliefhebbers in de Nederlanden*, Nederlands kunsthistorisch jaarboek 2003, vol. 54 (Zwolle, 2004); Craig Ashley Hanson, *The English Virtuoso: Art, Medicine, and Antiquarianism in the Age of Empiricism* (Chicago, 2009).

peoples, such as the Chinese and Japanese, even seemed to possess a kind of scientific knowledge that equaled or bettered that of the Dutch, at least in some areas. The rise of science in the republic (as well as elsewhere) therefore demands some account of how it interacted with, and compared to, the information and concepts of other places.

A couple of examples might help to make the problem clear. The first is the physician Jacobus Bontius—son of the first professor of medicine at Leiden University—who took ship for the capital of the Dutch East Indies, Batavia (now Jakarta), in 1627. He survived for four years, during which time he endured a punishing schedule of administrative and medical duties and two sieges of the city by the Matamarese in which his own health suffered terribly; yet he also gathered together a great deal of information from local informants, from which he composed five major studies of the medicine and natural history of the region. He did so in the years shortly after the VOC virtually exterminated the population on some of the Banda Islands in order to insure its monopoly over the production and distribution of nutmeg. But Bontius himself grew to think so well of many of his local acquaintances that he considered them superior in knowledge and skill to any European medical practitioner. He then turned the strange and exotic customs he encountered into parcels of useful and accurate information, information that could be easily transported from Java to literal-minded readers elsewhere.

Bontius's efforts have been explored at length recently.[78] For the moment it is enough to note a few points. One is that he had the governors of the VOC in mind when he composed his works, although the manuscripts were sent to his brother (a lawyer in Leiden) and published many years after his death, when increased interest in the Dutch enterprises overseas became manifest. Second, despite the difficulties he encountered, he managed to collect a very large body of information. In doing so, he explained, he was an eyewitness to the truths of the Indies that none of his European medical forebears had experienced, yet in retrospect it is clear that he was heavily dependent on local informants for the details he gathered about the use of plants, animals, and minerals to treat diseases. While he does not explain his methods directly, from various hints it would appear that most of his informants were women, and some of them slaves. In interacting with them, it would appear that he came to admire them, frequently defending them against usages of his contemporaries that included calling them "barbarians." Third, when he wrote down his findings, he left out some topics. While he loudly praised the God-given medicinal effects of local plants and the uses made of them by local people, at the same time he dismissed out of hand the local meanings attached to the plants and practices. That is, his concern for descriptive information—matters of fact—gave him a blind spot when it came to what we would call the intellectual system of, or cultural assumptions behind, local medical practices. He says nothing about any learned medical traditions he might have encountered, nor does he reveal anything substantial about the practices that

---

[78] Harold J. Cook, "Global Economies and Local Knowledge in the East Indies: Jacobus Bontius Learns the Facts of Nature," in *Colonial Botany: Science, Commerce, and Politics in the Early Modern World*, ed. Londa Schiebinger and Claudia Swan (Philadelphia, 2005), 100–18, and *Matters of Exchange* (cit. n. 27), 191–209. For his works, see Jacobus Bontius, *De Medicina Indorum* (Leiden, 1642); William Piso, *De Indiæ Utriusque Re Naturali et Medica Libri Quatuordecim* (Amsterdam, 1658); modern translations in *Opuscula Selecta Neerlandicorum de Arte Medica*, 19 vols. (Amsterdam, 1907–48), vol. 10 (1931).

were deeply embedded in much of local medical ritual, although he makes occasional dismissive remarks about "superstition" and "idolatry." In other words, Bontius did not simply encounter and speak with various sorts of people; he was boiling things down to their lowest common denominator, information units that could be circulated in just about any context. He (re)produced knowledge, accumulated it, and handed it on in a form that made his information about nature commensurable with that produced in any other place.

Or take the example of another physician in the East Indies, Willem ten Rhijne.[79] He is well known as the first European university-educated physician to live in Japan, which he did for two years in the 1670s. He was there because the Japanese government had ordered the VOC to send such a person. Many well-placed Japanese scholar-physicians were then seeking to modify the principles of classical Chinese medicine into a Japanese approach and were intrigued by the possibility of using Western medical knowledge as part of their reforms. Many of the Japanese translators who worked with the Dutch were scholars in their own right, and because of the close link between classical learning and medicine, many were also physicians, or at least knowledgeable about medicine. By the later 1660s, someone close to the shogun wanted access to more than Dutch surgeons, which had led to the orders that caused the VOC to seek someone like Ten Rhijne.

To be brief, during the two years he spent on the tiny island of Deshima, in the harbor of Nagasaki—to which the Dutch were confined for the purposes of trade—Ten Rhijne was kept busy answering questions and even visiting patients and pharmacies. But in the process he also managed to turn the personal relationships with his Japanese acquaintances into opportunities to ask questions of them in turn. The difficulties of finding a vocabulary for the exchanges were clearly evident from some of Ten Rhijne's complaints, for not all of the official translators understood him easily. But there had implicitly been much preparation on the part of the translators before they started to work, and indeed some of them became well versed in medicine. Most of the questions they asked concerned the composition of Western medicines, and their use, but some were of a more general nature, such as "Why do you touch only the left radial pulse?" and "How do you discriminate the yang-type and yin-type carbuncles?" He replied with explanations based on European physiology, such as that the circulation of the blood meant that there was no difference between the left and right pulses, and that the yang and yin referred to hot and cold, which were qualities from an older medical system that no longer had meaningful diagnostic significance. Other kinds of questions, such as those regarding the uses of certain kinds of botanicals, he could answer with less concern for conceptual incompatibilities. He also helped the translators learn more about the latest anatomical findings from Europe. In turn, Ten Rhijne learned from the Japanese.

In the late 1670s and early 1680s, living on the west coast of Sumatra, he had a chance to put many of his medical thoughts in order and wrote a letter to Henry Oldenburg, the secretary of the Royal Society of London, asking whether the society

---

[79] I have also written about him at greater length: Harold J. Cook, "Medical Communication in the First Global Age: Willem Ten Rhijne in Japan, 1674–1676," *Disquisitions on the Past and Present* 11 (2004): 16–36, and *Matters of Exchange* (cit. n. 27), 339–77. See also Wolfgang Michel, "Willem Ten Rhijne und die japanische Medizin (I)," *Studien zur deutschen und französischen Literatur* 39 (1989): 75–125; Michel, "Willem Ten Rhijne und die japanische Medizin (II), " *Studien zur deutschen und französischen Literatur* 40 (1990): 57–103.

would be interested in seeing his work. One part of his manuscript contained the first detailed report by a European on the practice of acupuncture, as well as further information on moxibustion. The letter caused considerable discussion at a meeting of the Royal Society, which quickly agreed to publish it—it appeared in 1683, helping to create a brief flourishing of popularity for acupuncture in Europe.[80] But it is noticeable that Ten Rhijne's attempts to summarize Chinese medical concepts, as filtered through his Japanese acquaintances who had studied in China, were hopelessly garbled, so that even in our own day the proper names of Chinese authors to whom he referred remain unrecognizable.[81]

SCIENCE AND "OBJECTIVITY"

One could elaborate much further on these and other examples of early modern East-West medical exchanges.[82] But the general point is that some matters were successfully exchanged while other matters were not. Information about medicines and practices—descriptions about tangible things available to the senses—moved as readily between places as other goods, while meanings and explanations, and some kinds of concepts, presented a great deal of linguistic and philosophical difficulty, leading to elision, speculation, and (sometimes creative) misunderstanding on each side. In other words, to invoke a distinction from decades ago between knowledge claims that refer to sensations and those that refer to words: "Observation sentences peel nicely; their meanings, stimulus meanings, emerge absolute and free of all residual verbal taint. Theoretical sentences such as 'Neutrinos lack mass,' or the law of entropy, or the constancy of the speed of light, are at the other extreme. For such sentences no hint of the stimulatory conditions of assent or dissent can be dreamed of that does not include verbal stimulation from within the language."[83] Some Europeans, Ten Rhijne among them, made considerable efforts to understand not only the medical practices of East Asians, but their terminology and even their medical views, just as some Japanese tried to understand aspects of European medical practices, anatomical findings, and basic principles; in the process, some fundamental concepts, such as *yin* and *yang,* resisted attempts at translation. But other aspects of their work moved easily, because they were more descriptive: more objective. This is to turn upside down the common claim that scientific ideas are universal, saying instead that those words and formulae that can be conveyed from place to place are by definition "scientific" (in the German sense of *wissenschaftlich*).

Such episodes did not lead to any major revolutions in medical practice on either side, but they make it clear that contemporary commerce enabled personal networks to convey a great deal of information about the natural world around the globe. The information was communicated in more than one direction (although like others who do not know Asian languages, my own knowledge is about how a host of locales con-

[80] Ten Rhijne, *Dissertatio de Arthritide* (London, 1683).

[81] For example, see the struggles to interpret his views by Gwei-Djen Lu and Joseph Needham, *Celestial Lancets: A History and Rationale of Acupuncture and Moxa,* new ed., ed. Vivienne Lo (London, 2002), 271–6.

[82] Harold J. Cook, "Conveying Chinese Medicine to Seventeenth-Century Europe," in *Science between Europe and Asia: Historical Studies on the Transmission, Adoption and Adaptation of Knowledge,* ed. Feza Günergun and Dhruv Raina (Heidelberg, 2011), 209–32.

[83] Willard Van Orman Quine, "Meaning and Translation," in *On Translation,* ed. Reuben Brower (Oxford, 1966), 148–72, quotation on 171.

tributed to the accumulation of information in European entrepôts). Ideas and information flowed from place to place in the minds or on the notebook pages of travelers often enough to make it possible to speak of wider intellectual "movements." While recent work in the history and sociology of science has stressed how knowledge was produced in close-knit groups of local investigators, local knowledge became part of more general developments, giving rise to the Renaissance, the Scientific Revolution, the Enlightenment, and so on.[84] But something we often term "culture"—rooted in ways of life and language—made some aspects of early modern knowledge systems very difficult to move about.

The transmission of information and, sometimes, ideas from West to East and vice versa is therefore a particularly illustrative example of a more general problem: how did the knowledge claims of people in the commercial parts of the Dutch Republic interact with those in other regions? The interactions might be at far distances, or among regions or nations that bordered one another, or between a commercial capital and its rural hinterland—even between one person and another. But while there are exciting debates occurring about the economic exchanges between "the West and the rest," we have a more rudimentary understanding of the exchanges of knowledge.

A particularly important and long-standing argument for our present purposes is the Needham thesis, which has set the stage for so many of the comparisons between Europe and China. Joseph Needham was both a practicing scientist and a historian of embryology who fell in love with a woman from China and spent time there during the Second World War while beginning to document the country's history of science in order to show that it was responsible for many of the material innovations most important for humankind. The many volumes of the huge project for which he is responsible began to appear in 1954, and as of 2000 they had been published in twenty-one parts.[85] Although he changed some of his eclectic views over time as his research accumulated, his impressive studies immediately posed a challenge to standard accounts of the history of science. During the period of the European Middle Ages, the Chinese could be shown to have been more innovative and possessed of more technically superior methods than any other people on earth; but it was the Europeans rather than the Chinese who had the Scientific Revolution. Why? It is a question that has continued to set a framework for the history of science in China—and indeed, in many other non-European contexts, too[86]—despite attempts by many of its practitioners to move beyond it.[87]

[84] Also, Harold J. Cook and David S. Lux, "Closed Circles or Open Networks? Communicating at a Distance during the Scientific Revolution," *Hist. Sci.* 36 (1998): 179–211.

[85] Needham, *Science and Civilisation in China*, 7 vols. (21 parts) (Cambridge, 1954). See also Colin A. Ronan, ed., *The Shorter "Science and Civilization in China": An Abridgement of Joseph Needham's Original Text*, 2 vols. (Cambridge, 1978–81). For an introduction to Needham's biography and views, see H. Floris Cohen, *The Scientific Revolution: A Historiographical Inquiry* (Chicago, 1994), 418–29; Alain Arrault and Catherine Jami, eds., *Science and Technology in East Asia: The Legacy of Joseph Needham* (Turnhout, 2001); Jack Goody, *The Theft of History* (Cambridge, 2006), 125–53; and Simon Winchester, *The Man Who Loved China: The Fantastic Story of the Eccentric Scientist Who Unlocked the Mysteries of the Middle Kingdom* (New York, 2008).

[86] Dhruv Raina, "Cognitive Homologies in the Studies of Science in Indian Antiquity: A Historiographic Axis of the Indian Journal of History of Science," in Arrault and Jami, *Science and Technology in East Asia* (cit. n. 85).

[87] E.g., see Mark Elvin, *The Retreat of the Elephants: An Environmental History of China* (New Haven, Conn., 2004); Benjamin A. Elman, *On Their Own Terms: Science in China, 1550–1900* (Cambridge, Mass., 2005); Francesca Bray, *Technology and Society in Ming China (1368–1644)*

It may be, however, that Needham's is a *question mal posée*. While on the one hand his question elevated science in China to world significance, on the other it also made the Scientific Revolution an even more singular instance of Europe's exceptionalism. At the moment, many of the historians of European science who have taken Needham's question to heart have used it mainly to explore the special qualities of Europeans. An example is Floris Cohen's impressive studies.[88] In a recent book on the Scientific Revolution written for a Dutch audience, when pondering the question, why Europe? Cohen ends up supplying an answer based on what he takes to be special qualities of European culture: "greater openness, more intensive curiosity, stronger energy, more powerful individualism and judgment—and a stronger seeking of everything in an active earthly existence."[89] In his more recent analysis for a more scholarly audience, he explains these differences as being due to "Europe's singular dynamism," which was in turn the result of "how European civilization aimed in action *and* in thought to intervene in the world and manipulate it for human ends."[90] He is certainly not alone in offering such explanations. Purported special cultural attributes are often proposed for European exceptionalism, or even superiority. To mention just two more examples: In a recent stimulating account on globalization in the early modern period, Geoffrey Gunn explains that "the 'divergence' was as much intellectual as it was material. Eventually, European science and technology triumphed" because of a unique attribute that he characterizes with the word *curiosity*, which he thinks the Europeans possessed in abundance but which was lacking elsewhere.[91] In another recent work promising a global perspective on the history of science Toby Huff agrees. He finds evidence "of a wide and deep embedding of an infectious *ethos of scientific curiosity* across Europe that remained unmatched outside Europe during the seventeenth and eighteenth centuries."[92]

If we do not think that Europeans exhibited characteristics such as a drive for domination or curiosity because of some inherent trait—and I certainly do not—then this sort of explanation must carry weight only when the authors who invoke it show what elements in Europeans' form of life caused them to be more curious than other people. Otherwise we fall into the vexed problem of comparing civilizations. One of the chief advocates for including civilizations additional to the Western in the post–World War II American curriculum, Marshall Hodgson, defined a civilization as "what is carried in the literature of a single language, or of a single group of culturally related languages." He broadened the definition further to include cultural formations committed to "major lettered traditions," which in turn made for continuity "in social and economic institutions generally. All cultural traditions [therefore] tend to be

---

(Washington, D.C., 2000); and Vivienne Lo's introduction to Lu and Needham, *Celestial Lancets* (cit. n. 81).

[88] See esp. Cohen, "Joseph Needham's Grand Question, and How to Make It Productive for Our Understanding of the Scientific Revolution," in Arrault and Jami, *Science and Technology in East Asia*, 21–31, and his earlier considerations in Cohen, *Scientific Revolution*, 418–29 (Both cit. n. 85).

[89] Cohen, *De Herschepping van de Wereld: Het onstaan van de moderne natuurwetenschap verklaard* (Amsterdam, 2008), 267; similar sentiments are expressed on 149–50.

[90] Cohen, *How Modern Science Came Into the World: Four Civilizations, One 17th-Century Breakthrough* (Amsterdam, 2010), 138, emphasis in the original; see also 47, where Cohen highlights his debt to Landes's idea of "latent developmental potential" to explain the differences.

[91] Gunn, *First Globalization: The Eurasian Exchange, 1500–1800* (Lanham, Md., 2003), quotation on 7–8, and see esp. 4 for his summary of his views about curiosity among Europeans.

[92] Huff, *Intellectual Curiosity and the Scientific Revolution: A Global Perspective* (Cambridge, 2011), 299; emphasis mine.

closely interdependent."[93] In other words, civilizations are defined by their core texts and the cultural values that are embodied in them and flow from them, which give rise to unities of outlook and behavior. A recent popular reformulation of this argument sees contemporary events in terms of the "clash of civilizations," one influential statement of which, by Samuel Huntington, gives a six-part definition that amounts to saying that civilizations are ideological formations that underlie politics but are not political systems per se; he also adds that "religion is a central defining characteristic of civilizations."[94] Ironically, perhaps, in light of the arguments above, he goes on to write that while "the causes for the rise of the West are multiple" they rely on technological means of coercion. "The West won the world not by the superiority of its ideas or values or religion (to which few members of the other civilizations were converted) but rather by its superiority in applying organized violence. Westerners often forget this fact; non-Westerners never do."[95]

But one of the most helpful aspects of recent versions of world and global history is that they are led by questions about economies, which challenge us to rethink national, cultural, ideological, and even civilizational units in order to examine levels of human organization that are closer to ordinary life. In the past two decades, several powerful arguments along such lines have offered a fundamental revision of previous assumptions about comparisons between European and other places. Immanuel Wallerstein and Fernand Braudel have both taken the view that European economic domination of the world's trading system, which began with the upswing in urban commerce in the later medieval period and was consolidated under the Spanish and Portuguese overseas commercial empires of the sixteenth century (and the later successful interloping of the Dutch and English), created a commercial-financial engine of growth.[96] They were stimulated in part by earlier work on the commercial system of the Indian Ocean, which in the early modern period came to be dominated by the European trading companies, implicitly showing how much wealth was transferred from Asia to Europe.[97] Other historians, such as Jonathan Israel, attribute much economic growth in Europe (in his case, the Dutch Republic) to the wealth acquired from overseas commerce.[98] Asian goods were clearly highly sought after and brought tremendous wealth to the European merchants controlling the trade.[99]

But a powerful counterblast has come from economic historians who are keenly

[93] Hodgson, *Rethinking World History: Essays on Europe, Islam, and World History*, ed. Edmund Burke III (Cambridge, 1993), on 82, 84.

[94] Huntington, *The Clash of Civilizations and the Remaking of the World* (New York, 1996); for the definition see 41–5; for the quotation, 47.

[95] Ibid., 51. Anthony Pagden gives more weight to the Scientific Revolution, though not because of useful "science" in the usual, technological, sense, but because it led to a better understanding of humankind in the Enlightenment (i.e., he incorporates SDP_2 in his account); Pagden, *Worlds at War: The 2,500-Year Struggle between East and West* (New York, 2008), esp. 252–61.

[96] Wallerstein, *The Modern World-System*, vol. 1, *Capitalist Agriculture and the Origins of the European World-Economy in the Sixteenth Century*, and vol. 2, *Mercantilism and the Consolidation of the European World-Economy, 1600–1750* (New York, 1974 and 1980); Braudel, *Civilization and Capitalism* (cit. n. 28).

[97] Kristof Glamann, *Dutch-Asiatic Trade, 1620–1740* (Copenhagen and The Hague, 1958); Niels Steensgaard, *The Asian Trade Revolution of the Seventeenth Century: The East India Companies and the Decline of the Caravan Trade* (Chicago, 1974).

[98] Israel, *Dutch Primacy in World Trade, 1585–1740* (Oxford, 1989).

[99] Jan De Vries, "Connecting Europe and Asia: A Quantitative Analysis of the Cape-Route Trade, 1497–1795," in *Global Connections and Monetary History, 1470–1800*, ed. Dennis O. Flynn, Arturo Giráldez, and Richard von Glahn (Aldershot, 2003), 35–106.

interested in the Eurasian story from the viewpoint of East Asia. Three of them argue that throughout the early modern period the Chinese economy rather than the European one dominated global commerce. Andre Gunder Frank's *ReOrient*, for instance, takes the view that there was nothing in European commercial methods that the Chinese did not also have; in fact, the Europeans were latecomers to the Asia trade. According to Frank, then, the "Rise of the West" came only in the nineteenth century and has caused our historical and social models, based on the supremacy of Europe and mainly articulated in the period of European dominance, to be fundamentally mistaken. Most important in terms of the common SDP arguments, his summary of the reasons why Europe rather than China had an Industrial Revolution makes no case for political or intellectual advantages, only for a conjuncture of ecological/economic incentives and demographical/economic structures.[100] R. Bin Wong similarly argues that Chinese history cannot be interpreted according to the widely used formulations of Marx, Weber, and others, and that the importance of the early Industrial Revolution in Europe has been overrated.[101] The third of this group, Kenneth Pomeranz, offers a somewhat more familiar view of the Industrial Revolution as separating parts of Europe from the path of labor-intensive growth previously shared with regions of China and the rest, making for a profound historical difference, however temporary in the long term. To do so, however, Pomeranz compares and contrasts the economic development of "core areas" in Europe and East Asia rather than nations, and without holding one to be the norm, finds that the Industrial Revolution emerged from new sources of capital made possible by the appropriation of New World resources through force and slavery, assisted by the peculiar institutions of armed long-distance trading companies. For Pomeranz, too, then, what has made for economic advantage has nothing to do with science or technology per se, let alone representative polities or special ideologies.[102]

Since these works of the late 1990s, the question of why the Industrial Revolution occurred in Europe rather than China has been hotly debated, gaining the attention of scholars of the new comparative economic history, for instance.[103] Members of that group have made the coercive power of gunpowder and naval technology even more important in the shifting story of the world economy.[104] For our purposes, we need only observe that among the conclusions of such debates is that while areas of northwest Europe were becoming part of an integrated market, elsewhere there were many regional and local markets. Long-distance firms like the Dutch and English East India Companies therefore made a great deal of their profit from something like

---

[100] Frank, *ReOrient: Global Economy in the Asian Age* (Berkeley and Los Angeles, 1998); see esp. 312–4.

[101] Wong, *China Transformed: Historical Change and the Limits of European Experience* (Ithaca, N.Y., 1997). See also the critique of mainline Western social science principles in Goody, *Theft of History* (cit. n. 85).

[102] Pomeranz, *The Great Divergence: China, Europe, and the Making of the Modern World Economy* (Princeton, N.J., 2000); for a summary of his views, see 8–21.

[103] T. J. Hatton, Kevin H. O'Rourke, and Alan M. Taylor, eds., *The New Comparative Economic History: Essays in Honor of Jeffrey G. Williamson* (Cambridge, Mass., 2007), 1–8.

[104] Ronald Findlay and Kevin H. O'Rourke, *Power and Plenty: Trade, War, and the World Economy in the Second Millennium* (Princeton, N.J., 2007). Another emerging alternative is to emphasize the peculiarities of European urbanization, which forced industry behind city walls because of high levels of warfare; Jean-Laurent Rosenthal and R. Bin Wong, *Before and Beyond Divergence: The Politics of Economic Change in China and Europe* (Cambridge, Mass., 2011).

arbitrage, exploiting the price differences among the markets by moving goods from one place to another.[105]

If the early modern world continued to exhibit several different forms of economy, were there perhaps also several forms of science? It would appear so if we pay attention to recent works in the history of Chinese science, for instance. Mark Elvin's investigations bring us back again to terms such as *useful and reliable knowledge*, suggesting that the critical difference between Europe and China was the lack of fact-finding in the latter, or what we might better say was a lack of suitable institutional arrangements to produce facts. Elvin takes the example of an early modern scholar writing an encyclopedic work that contains a section on dragons. Why does he not explain that dragons are legendary rather than real? "Xie has no procedure that will systematically tend to establish facts and to discredit falsehoods. Hence it is possible for him to show a relatively high level of logic in his arguments, but to make no useful progress, because often what he takes to be facts are not." He elaborates,

> The fact was not known in premodern China. By a "fact" in this sense, I mean a publicly recorded and accessible statement about an observable aspect of the world, set in the context of a systematic evaluation of the evidence that yields an approximate probability of its being true, and subject to a continuing public scrutiny and re-evaluation, with the results and the evidence being publicly recorded and accessible. China came quite close; what was mainly lacking was the continuity of public scrutiny and the circulation of the results. In other words, the feedback process.[106]

In yet other words, China did not have the institutional arrangements for assessing natural facts that had become so familiar in early modern Europe because of the extent and power of commercial arrangements.

CONCLUSIONS

If we wish to understand the values of European merchants thinking about the connections between distant places, we need to consider not only how they valued their material goods, but the high value they placed on factual descriptions of a variety of people, places, objects, and conditions at many places. Together, they probed for the similarities and differences to find common denominators. Their institutional solutions for reducing transaction costs depended on being able to exchange matters of fact among themselves, which in turn allowed them to exert long-distance control over their representatives, employees, and laborers while at the same time negotiating with colleagues, rivals, law courts, and the abstract reckoning of markets. In such circumstances, commercial materialism happily went hand-in-hand with the new science, which also offered careful descriptions of physical phenomena and explanations couched in terms of philosophical materialism. To people like the merchants and those around them who held similar values, focusing on inquiries and explanations that invoked nothing more than material facts might provide comfort as well as offer openings toward the productive management of such phenomena, and the familiarity of such matters could sometimes become joyful indulgence or fond admiration for unusual information or elegant explanations. But it was not everyone's

---

[105] See esp. the thoughts of De Vries, "Connecting Europe and Asia" (cit. n. 99), 94–7.
[106] Elvin, *Retreat of the Elephants* (cit. n. 87), 388.

cup of tea. Clerics often objected to the limitations and conclusions of such lines of inquiry—the Calvinist ones could be as obstreperous as the Catholic ones about Copernicanism, for instance—while they were probably of little interest to those of other frames of mind.[107] Elsewhere, such forms of knowledge could provoke both admiration for their exactness and frustration at their moral emptiness. In Japan, for instance, one neo-Confucian scholar admitted that the descriptive power of European astronomy was very great, but held that the attention given to the appearances of things was vulgar.[108] So were merchants in his eyes, no doubt.

We sometimes forget that Copernicus wrote on monetary policy, or that Robert Boyle held a place on the board of the English East India Company and held shares in the Hudson's Bay Company, or that Newton—the master of the mint who ferreted out and bound over for execution a large number of counterfeiters—lost his fortune in the South Sea bubble (although Sir Hans Sloane managed to get out in time).[109] Such scientists moved easily from commerce to science and back again, since both were founded on the factual sensibility. The shape of the concerns of, say, a director of the VOC such as Witsen would have been quite familiar to his fellow virtuosi in seventeenth-century London, and in Hamburg, Augsburg, Paris, Marseilles, and Venice; and it was probably familiar enough to those in Osaka, Canton, Cochin, Basra, and many other trading capitals. But it may be that those outside Europe did not have durable institutional arrangements for organizing their activities according to something like lawful regularities among people who were of different clans and ethnicities and so could not draw on as many allies when in need.[110] Moreover, merchants had political authority—even sovereign power—almost nowhere else beyond a few parts of Europe, those in the rest of the world usually being as despised by their rulers as their counterparts had been by European kings of distant generations. That is the explanation once offered by Needham for the differences between Chinese and European science: despite having all the seeming prerequisites, China did not give rise to the Scientific Revolution because it was ruled by the mandarins, with merchants being placed considerably lower on the social scale. The rise of European commercial power and science at the same time was no accident. Both appeared on the back of new forms of social life that are described as formal and informal institutions and that made information exchange sovereign.[111]

What I am suggesting is that science and the ways of life that are often termed economic share many sources, so that the relationships between them might illuminate

[107] Rienk Vermij, *The Calvinist Copernicans: The Reception of the New Astronomy in the Dutch Republic, 1575–1750* (Amsterdam, 2002).

[108] Shigeru Nakayama, *A History of Japanese Astronomy: Chinese Background and Western Impact* (Cambridge, Mass., 1969), 88–98.

[109] J. Taylor, "Notes and Documents: Copernicus on the Evils of Inflation and the Establishment of a Sound Currency," *J. Hist. Ideas* 16 (1955): 540–7; Timothy J. Reiss and Roger H. Hinderliter, "Money and Value in the Sixteenth Century: The Monete Cudende Ratio of Nicholas Copernicus," *J. Hist. Ideas* 40 (1979): 293–313; Sarah Irving, *Natural Science and the Origins of the British Empire* (London, 2008); Carl Wennerlind, *Casualties of Credit: The English Financial Revolution, 1620–1720* (Cambridge, Mass., 2011).

[110] Timur Kuran, *The Long Divergence: How Islamic Law Held Back the Middle East* (Princeton, N.J., 2011).

[111] On the importance of information exchange to early modern polities, see Filippo De Vivo, *Information and Communication in Venice: Rethinking Early Modern Politics* (Oxford, 2007), and Jacob Soll, *The Information Master: Jean-Baptiste Colbert's Secret State Intelligence System* (Ann Arbor, Mich., 2009).

some of their commonalities in other circumstances as well. The means of production powered by coal may have shifted attention to concepts of energy, the telegraph and other forms of communication to theories of signaling, the German chemical dye industry to cell theory, and so on, all of which involve ways of understanding nature that could not have been imagined in the early modern period. Such a strategy may help us to penetrate the sometimes obscure formulations of the "culture" of science by allowing us to examine the material relationships entangled in it rather than treating it as a more encompassing form of intellectual history. Newton's science may not have created the Industrial Revolution, but it was certainly a careful articulation of a massive amount of information about the physical universe compiled using the most precise methods available, as well as calculations and proposals about how the details all fit together. He therefore became deservedly famous among the businessmen and entrepreneurs, as well as the virtuosi, of his age. It may be that numerate and critical merchants could understand what he was proposing far better than their aristocratic contemporaries. And it may be that such a form of science, based on exchangeable units of reliable information, remains fundamental to our lives, like commerce, despite the fact that the methods of both have been enormously altered since his time. We need not think that Newton's scientific or economic concerns are the same as ours, although there are familiar elements in them. But we should remind ourselves of the reason why development economists regularly associate modern science and modern economies: they are coproduced.

Calling for attention to the connections between economic history and intellectual history is no longer a revolutionary move. A new generation of work in the history of science is attentive to what one recent book title calls the "mindful hand"; this work can offer thoughtful assistance to the many who are concerned with SDP.[112] The recent study by Lissa Roberts on steam machines and entrepreneurs in the Dutch Republic, France, and Britain in the eighteenth century is an excellent example of how very different the picture of the Industrial Revolution looks if one moves away from the standard figures of the British story and considers knowledge and economy as interactive rather than distinct processes.[113] For a somewhat earlier period, examples include the works of William Eamon, Pamela Long, and Deborah Harkness on practical knowledge and science; Paula Findlen, Lorraine Daston, and Katherine Park on collections; Pamela Smith on artisanal knowledge of nature; Karel Davids on technological development and the knowledge economy of the urban Dutch Republic; and Mario Biagioli on Galileo's "instruments of credit," as well as a continuing interest in the medical marketplace and collected volumes on natural history, commerce, and empire.[114] Other work, like that of Ursula Klein, has emphasized the material culture

[112] Lissa Roberts, Simon Schaffer, and Peter Dear, eds., *The Mindful Hand: Inquiry and Invention from the Late Renaissance to Early Industrialisation* (Amsterdam, 2007).

[113] Roberts, "Full Steam Ahead: Entrepreneurial Engineers as Go-betweens during the Late Eighteenth Century," in *The Brokered World: Go-betweens and Global Intelligence, 1770–1820*, ed. Simon Schaffer et al. (Sagamore Beach, Mass., 2009), chap. 5.

[114] Eamon, *Science and the Secrets of Nature*; Long, *Openness, Secrecy, Authorship* (Both cit. n. 45); Harkness, *Jewel House* (cit. n. 4); Findlen, *Possessing Nature*; Daston and Park, *Wonders* (Both cit. n. 47); Smith, *The Body of the Artisan: Art and Experience in the Scientific Revolution* (Chicago, 2004); Davids, *Rise and Decline* (cit. n. 14); Biagioli, *Galileo's Instruments of Credit: Telescopes, Images, Secrecy* (Chicago, 2006); Schiebinger and Swan, *Colonial Botany* (cit. n. 78); Timothy Walker, "Acquisition and the Circulation of Medical Knowledge within the Early Modern Portuguese Colonial Empire," in *Science in the Spanish and Portuguese Empires, 1500–1800*, ed. Daniela Bleichmar et al.

of scientific practice.[115] Moreover, prompted in part by some of the literature on early modern global interconnectedness, and in part by Bruno Latour's imagining of networks of practice, the boundary crossers and intermediaries who made the exchange of knowledge among people of different languages and cultures possible and profitable are now receiving the kind of attention they deserve.[116] Since the authors of such work often explore the experiences of Europeans abroad that challenge set narratives of colonialism and empire, they offer much evidence of the kind of interactions that produced knowledge, which should interest a new generation of global historians. Investigations such as these, at the intersections of the grand themes of the history of science and economy, have the potential to open up new vistas on historical contingency and structure, exploring how small and seemingly unrelated events can have larger consequences at far distances.

A world in swift motion and transformation, full of potential dangers as well as benefits, pictured against a steadying backdrop of facts that remained constant: by the sixteenth century, this was becoming an increasingly acceptable way to know the things constituting nature in the academies of Europe. It was worldly knowledge, gained from the bubbling up of countless encounters with the material world rather than a trickling down of appreciation for eternal verities. It was the knowledge gained in the first instance not by the professors, but by the people who moved about, working with things and other people, gaining confidence from their acquisition of experience and knowledge of precise details, and sharing at least some of it. It may not have been the big picture, but in the midst of change one might take comfort in the stability of small things, through which, sometimes, one might glimpse the larger interconnections of all things that are.

Maybe the moral threat of materialism to the established politico-religious order was real. Horrors like the events in the Banda Islands certainly were. But at the same time, seaborne commerce provided new means of exchange, exchange that necessarily brought people around the globe into conversation with one another, creating the framework for a kind of discourse about what everyone could agree was true, which we used to call objective science. In some cases, at least, those conversations encouraged not only finding utilitarian benefits for human life but respect for one another. Attention to material things gave no hint of personal salvation, but it fostered the building of information networks that stretched across the globe. For Grotius, Spinoza, and many other philosophers of the early modern period, developing forms of sociability were now the key to a kind of secular hope of betterment in both knowledge and material life. Perhaps, then, the values of the exchange economy were not simply a religion of the damned, but gave glimpses of new worlds as well.

(Stanford, Calif., 2009); Mark S. R. Jenner and Patrick Wallis, eds., *Medicine and the Market in England and Its Colonies, c. 1450–c. 1850* (Houndsmills, 2007).

[115] E.g., Klein and Wolfgang Lefèvre, *Materials in Eighteenth-Century Science: A Historical Ontology* (Cambridge, Mass., 2007); Klein and E. C. Spary, eds., *Materials and Expertise in Early Modern Europe: Between Market and Laboratory* (Chicago, 2010).

[116] E.g., Pamela H. Smith and Paula Findlen, eds., *Merchants and Marvels: Commerce, Science, and Art in Early Modern Europe* (New York, 2002); Kapil Raj, *Relocating Modern Science: Circulation and the Construction of Scientific Knowledge in South Asia and Europe, 17th–19th Centuries* (Delhi, 2006); Schaffer et al., *Brokered World* (cit. n. 113); Steven J. Harris, "Mapping Jesuit Science: The Role of Travel in the Geography of Knowledge," in *The Jesuits: Cultures, Sciences, and the Arts, 1540–1773*, ed. John W. O'Malley et al. (Toronto, 1999), 212–40; Latour, *Reassembling the Social: An Introduction to Actor-Network-Theory* (New York, 2005).

Reflections of a
Troglodyte Historian of Science

by Edward Grant[*]

ABSTRACT

For most of the twentieth century, historians of science pursued their discipline largely from the standpoint of the history of ideas (internalism), and only toward the end of the century did they expand its horizons to the social context (externalism), which now dominates the discipline. I explain how I came to accept the importance of the social context approach and then reject the claim that the history of science has been largely Eurocentric. Natural philosophy flourished in the late Middle Ages largely because the Church came to support it, rather than oppose it, and also because it became a fundamental part of the university curriculum. In the last part of my article, I consider the relations between science and religion, focusing on counterfactuals, when natural philosophers sought to show that what Aristotle regarded as naturally impossible was supernaturally possible by God's absolute power. These discussions helped lay the foundations of early modern science.

In the mid-twentieth century, when the history of science emerged as a significant new discipline in the United States, ideas about it differed radically from what they have become today. It would be accurate to characterize the approach to the history of science at that time as one based on ideas—the history of ideas. It was an almost exclusively ideas-oriented discipline. The ideas were derived from scientific texts or documents that discussed those texts. The context within which science existed played only a peripheral role, if even that.

The major exception to this generalization is the approach of Marxist historians of science who believe—and have believed—that science was wholly shaped by workers, artisans, farmers, and, in general, largely illiterate individuals, rather than by the intellectual labors of the likes of Galileo and Isaac Newton, who were the ungrateful beneficiaries of their social inferiors. Even if this were true—and it most certainly is not—how would science have emerged if Aristotle, Archimedes, Alhazen (Ibn al Haytham), Fibonacci, Galileo, Kepler, Newton, and a host of others had not been there to integrate the various bits and pieces of information provided by the laboring classes about the workings of machines and instruments? All the knowledge that may have been derived about machines, instruments, agriculture, navigation, and other similar activities would have been of little value if it had not been integrated into a larger view of the world that we may call "science." Indeed, Clifford D. Conner, the author of a recent significant defense of the Marxist interpretation of the history of

[*] Department of History and Philosophy of Science, 130 Goodbody Hall, Indiana University, Bloomington, IN 47405; grant@indiana.edu.

science, focuses his book "on *empirical* as opposed to *theoretical* processes," because he believes "that the foundations of scientific knowledge owe far more to experiment and 'hands-on' trial-and-error procedures than to abstract thought."[1] Without theory, empirical observations would by themselves be unrelated bits and pieces of knowledge that would be of limited value. Apart from this odd book, the Marxist interpretation of the history of science has virtually disappeared.

THE HISTORY OF SCIENCE AS THE HISTORY OF IDEAS

Few, I believe, will deny the claim that in the period from approximately 1940 to 1970, the history of science was studied largely from the standpoint of intellectual history, or the history of ideas. The first lecture of a series called "The Voice of America Forum Lectures," delivered by Professor Duane H. D. Roller of the University of Oklahoma, was titled "The History of Science and Its Study in the United States" (undated, but probably written around 1962–3). He explains that it was in the 1950s that the history of science emerged "as an academic and intellectual discipline in its own right." As evidence of this, he mentions that "some 70 colleges and universities in the United States offer courses of lectures in the history of science." And then, in a telling statement, Roller declares: "Most historians of science in the United States have come to regard the history of science as being a branch of intellectual history. That is, they find themselves most interested in and concerned with *ideas* and with the origin, development, transmission and influence of these ideas."[2] This description would embrace most of the eminent historians of science who flourished in the period 1950 to 1980, such as Alexandre Koyré, Marshall Clagett, Stillman Drake, Alistair Crombie, E. J. Dijksterhuis, Charles C. Gillispie, Marie Boas Hall, A. Rupert Hall, and many others. Indeed, I exemplified Roller's description when I dedicated a book (*Much Ado about Nothing: Theories of Space and Vacuum from the Middle Ages to the Scientific Revolution* [1981]) to

> Marshall Clagett
> Pierre Duhem
> Alexandre Koyré
> Anneliese Maier
>
> who taught me the meaning of conceptual history

"Conceptual history," of course, has virtually the same signification as "the history of ideas."

An extraordinary emphasis on the analysis of scientific texts largely explains why the history of ideas held sway in the history of science during the period of its emergence in the 1950s and 1960s. Many historians of science believed that in order to know how science had fared over the past three thousand years, it was essential to read the texts of treatises that were regarded as relevant to the development of science and natural philosophy. Most of these were either in unedited manuscripts or, with

---

[1] See Conner, *A People's History of Science: Miners, Midwives, and "Low Mechanicks"* (New York, 2005), 11; emphasis in the original.

[2] Sol Tax et al., eds., *The Voice of America Forum Lectures* (Washington, D.C., 1964), on 4, 6; emphasis in the original.

the advent of printing, in books that had not yet been carefully studied. In this seminal period, numerous texts in manuscript form were edited and translated from one language to another, many from Greek, Arabic, and Latin to various European vernacular languages. Under Clagett's expert guidance at the University of Wisconsin's Department of History of Science, many of his students—myself included—wrote PhD dissertations based on editing, translating, and commenting upon a medieval Latin text that was previously little known, or virtually unknown. Whatever science may signify, or however it is defined, its ultimate source is a text, published or unpublished. This judgment encompasses science from its origins to the present. Thus, we may appropriately declare that the formative period of the history of science emphasized the study of scientific texts—or what scholars judged to be scientific texts. This included discussions of Aristotelian natural philosophy in a variety of formats as well as treatises on the exact sciences, which included mathematics, optics, astronomy, mechanics, and medicine.

Historians of science who did not edit or translate texts usually focused their scholarly efforts on comparing the texts of different authors on a given theme. For example, they might compare changes in knowledge and approach with regard to problems of motion, or optics, between a thirteenth-century author and a fifteenth-century author; or they might compare changes in astronomical or cosmological interpretations between various medieval authors and some of their sixteenth- and seventeenth-century successors. In virtually all of these endeavors, the comparisons are based on knowledge of relevant texts.

In the emergent period of the history of science, the wider context of science—its societal setting—was not wholly ignored, but it received only meager attention, largely because it was regarded as having little relevance for the understanding of scientific development. By the 1980s, however, societal concerns began to displace the emphasis on texts, and by the 1990s, the larger contexts of science became the overriding concern for historians of science. By the 1990s, many, if not most, historians of science were investigating topics that could be appropriately classified as belonging to the sociology of science. Numerous others had moved away from the physical sciences, which had been previously dominant, and focused their attention on the occult sciences, medicine, and natural history. In general, it was the impact of science on society, and the impact of society on science, that came to hold sway and shape the history of science. These are now the most significant motivations for historians of science.

This shift of emphasis from ideas in scientific texts to the social relations of science is not completely the result of the broadening research interests of historians of science. It is, I would suggest, also a consequence of the diminishing knowledge of science that prevents, or seriously hinders, historians of science from writing about the content of modern sciences such as physics, astronomy, chemistry, biology, medicine, and others. That task must be left to those scientists who have a desire to present the modern history of their subject to modern readers.[3] The simple fact is that fewer and fewer historians of science have the requisite scientific knowledge to study

---

[3] An excellent example is Roger G. Newton's *From Clockwork to Crapshoot: A History of Physics* (Cambridge, Mass., 2007). Newton is distinguished professor emeritus of physics at Indiana University, Bloomington. He concludes his book with themes from modern physics that would be beyond the ability of historians of science who lacked training in modern physics.

historical texts that are technically scientific. But that does not explain the shift from the history of science as the history of ideas to the history of science as the history of the impact of society on science and of science on society.

THE SOCIAL CONTEXT OF SCIENCE

By the 1990s, the relatively narrow focus of the history-of-ideas approach to the history of science became increasingly unappealing to students of the discipline, who were drawn more and more to the ways in which science affected and altered our society and, therefore, our lives. They were equally attracted to an examination of the ways in which society affected the development of the different sciences. Indeed, after many years as a historian of science who sought to interpret his discipline within the format of the history of ideas, I came to realize how important institutional and societal phenomena were for the evolution of science. In the preface to my book *The Foundations of Modern Science in the Middle Ages: Their Religious, Institutional, and Intellectual Contexts* (1996), I expressed my new sentiments about the role played by the Middle Ages in the Scientific Revolution (I am aware that this expression has become controversial, but I use it here for convenience). My *Foundations* book was originally intended as a modest revision of *Physical Science in the Middle Ages* (1971), but became a major overhaul, largely because of a particular idea that had occurred to me a few years before I began the revision, an idea that I describe in the preface of *Foundations*. If we assume that early modern science took shape in the Scientific Revolution of the seventeenth century, I inquire whether that revolution would have been possible if, by the seventeenth century, natural philosophy and the exact sciences had been at the approximate level they had attained in the first half of the twelfth century, when Greco-Islamic science had not yet been translated into Latin, the universities had not yet been established, and the Church had not yet confronted the problems that would arise from Aristotle's natural philosophy. The reply to this question is unequivocally *no*.

The explanation is simple: without the translations of Greco-Arabic science into the Latin language in the twelfth and thirteenth centuries, Western Europeans could not have acquired the level of scientific knowledge that was an essential prerequisite if the likes of Kepler, Galileo, Newton, and numerous others were to produce a scientific revolution. Without the translations, there would have been no natural philosophy to speak of, and without a well-developed natural philosophy there would have been little advance in the exact sciences. The popularity of Aristotle's natural philosophy, which literally opened up the physical world to medieval scholastics, made it feasible and desirable to establish centers to study and discuss what Aristotle had said about the world and its operations. And so it was that by around 1200, not long after most of the translations had been made from Arabic into Latin (most of the translations from Greek into Latin occurred in 1269), the first three great medieval European universities were in existence: the Universities of Paris, Oxford, and Bologna. By 1500, there were approximately sixty-five universities in Europe, extending as far east as Poland. The basic curriculum in the majority of these universities was taught in the arts faculty and included natural philosophy, logic, arithmetic, geometry, music, and astronomy. A degree in the arts faculty was required before a student could enter one of the three higher faculties of theology, medicine, or law. Thus, the basic and common curriculum of a medieval university was essentially scientific, emphasizing reason and analysis. This went on for more than four hundred years, until Aristotelian natural philosophy was rejected and abandoned.

What role did the universities play in preparing the way for early modern science? Over the centuries the universities taught thousands of students a curriculum based on Aristotle's natural philosophy, thus disseminating a common cosmological view and worldview across Europe. For the first time in history, a relatively large intellectual class came into being that was well educated in natural philosophy and science, a class that extended from the clergy to the broader governmental and courtly reaches of medieval society. An examination of the sorts of questions medieval scholastics posed in their Aristotelian commentaries quickly reveals that they had developed a pervasive and deep-seated spirit of inquiry that was guided by an extraordinary emphasis on reason. As I expressed it elsewhere, scholastics were strongly motivated by a spirit of "probing and poking around."[4]

In addition to the translations and the establishment of universities, a very important third factor also played a vital role in establishing the foundations of early modern science. This concerns the attitude of the Church and its theologians toward the new Aristotelian secular learning, which was initially viewed as potentially dangerous, as was evident in 1210, when the provincial synod of Sens forbade, at the risk of excommunication, the public or private reading of Aristotle's natural philosophy. This decree failed to prevent the study of those works. In 1231, Pope Gregory IX tried a new tactic—instead of banning Aristotle's works, the pope ordered them to be purged of errors. This was apparently never carried out. In 1255, all of Aristotle's natural philosophical works were listed as textbooks at the University of Paris. One other effort was made to curb students of Aristotle's natural philosophy. In 1270, the bishop of Paris condemned thirteen articles drawn from the works of Aristotle, and in 1277, he condemned 219 articles. The penalty for defending any one of them was excommunication. Although some of the condemned articles had an interesting history, the Condemnation of 1277 had virtually no adverse effect on the development of natural philosophy. Indeed, I believe it stimulated natural philosophers to probe into unusual cosmological themes.

What is the significance of all this? It is simply that theologians were favorably disposed toward natural philosophy, and virtually all of them had been trained in the discipline when they were undergraduate students in arts faculties. Indeed, many medieval theologians may be appropriately categorized as theologian–natural philosophers, because they wrote important commentaries on the works of Aristotle and also inserted a great deal of natural philosophy into their theological treatises. Among such theologians we might mention Albertus Magnus, Peter John Olivi, Thomas Aquinas, Thomas Bradwardine, William of Ockham, Gregory of Rimini, Nicole Oresme, and numerous others who wrote important treatises on natural philosophy and, for the most part, did not intrude theology into their natural philosophy, but rather used natural philosophy to cope with many theological problems. This positive attitude toward Aristotelian natural philosophy is of great importance because if the theologians had found natural philosophy dangerous to the faith, it could never have taken root in the universities and become a commonly studied discipline all across Europe. The Church would have banned it. But with their stamp of approval, natural

---

[4] See Edward Grant, *A History of Natural Philosophy from the Ancient World to the Nineteenth Century* (Cambridge, 2007), 325–6.

philosophy flourished in the late Middle Ages. Indeed, the background that theologians acquired in natural philosophy and logico-mathematical techniques when they were students and teachers in the arts faculties "enabled them to transform theology into a rationalistic, analytic discipline during the thirteenth and fourteenth centuries. Theology was often more natural philosophy than it was theology." It is an unavoidable conclusion "that while natural philosophy was virtually independent of theology, theology was utterly dependent on natural philosophy."[5] Had events led the Church in the Middle Ages to view natural philosophy as dangerous to the faith and as a challenge to its authority, it would have stymied and stunted the development of natural philosophy. The Scientific Revolution would not have occurred in the seventeenth century, and perhaps might not yet have occurred. Fortunately for Western Europe, this was not the case, and the Scientific Revolution emerged to transform the world.

What I have presented here does not depend in any way on the level of scientific achievement in the late Middle Ages, a subject that has been controversial since Pierre Duhem exaggerated the influence of medieval science on the Scientific Revolution. Although I believe that some medieval scientific concepts and ideas—most notably in the physics of motion and cosmology—influenced Copernicus, Galileo, and others in the sixteenth and seventeenth centuries, the truth or falsity of such claims is irrelevant to my claim that the factors I have described above for the late Middle Ages were instrumental in laying the foundations of modern science.

The crucial events that I have described would not have occurred without a conducive cultural atmosphere. Western European culture, which had been influenced by a long tradition of Greek and Roman thought that antedated Christianity, possessed the essential prerequisites for the development of societal activities and institutions that made the emergence of a significant natural philosophy and science possible. In my judgment, the most relevant societal activities and institutions were the translations, the universities, and the Church with its theologian–natural philosophers. These institutions and activities encouraged, or simply permitted, the doing of science and natural philosophy.

An intellectual and societal atmosphere that encourages the doing of natural philosophy and science will not, however, guarantee that those disciplines will advance and flourish. To ensure this, genuine contributions to science and natural philosophy must be made, or the conducive societal conditions will be of no avail. Moreover, all of the numerous, individual contributions must be constantly integrated into a larger whole to produce a bigger picture and thus expand and deepen the science to which those contributions are relevant. But who furnishes the various specific contributions that will lead to the further development of this or that science?

HAS THE STUDY OF PRE-SEVENTEENTH-CENTURY
HISTORY OF SCIENCE BEEN EUROCENTRIC?

According to Arun Bala, in his recent book *The Dialogue of Civilizations in the Birth of Modern Science*, most of those contributions were made by other civilizations—especially China, India, and Islam. Indeed, without those contributions the Scientific Revolution could not have occurred. Bala argues that the history of science as written

---

[5] Ibid., 273.

in the West is Eurocentric. By Eurocentric history of science, Bala means "any account of the birth and growth of modern science that appeals solely to intellectual, social, and cultural influences, causes, and ideas within Europe, and that marginalizes the importance of contributions, if any, of cultures beyond Europe to the birth and growth of modern science."[6] Simply because modern science developed in Europe does not mean that the roots of that science are to be found only in Europe. Bala insists, "This is wrong. The roots of modern science are 'dialogical'—that is, the result of a long-running *dialogue* between ideas that came to Europe from a wide variety of cultures through complex historical and geographical routes."[7] Although Bala is convinced that most of the scientific ideas and concepts that were instrumental in producing a scientific revolution in Europe were imported from other civilizations, he is also convinced that

> a multicultural history of modern science acknowledging non-Western contributions can be written without denying that there was a revolutionary break involved in the emergence of modern science. Even if modern science was forged out of ideas, methods, and technologies developed in multicultural contexts, it is a sufficiently radical and unique achievement to be rightly described as revolutionary. Consequently, it is hardly ethnocentric to claim that a scientific revolution occurred in Europe—unless one denies the plausible claim that modern science first emerged in Europe. What would be ethnocentric is to deny the dialogical contributions that made the revolutionary break possible.[8]

I fully agree with Bala that there was a scientific revolution in Europe and that it would not have occurred without dialogical contributions from other civilizations. I have argued that "the modern science that emerged in the seventeenth century in Western Europe was the legacy of a scientific tradition that began in Ancient Greek and Hellenistic civilization, was further nurtured and advanced in the far-flung civilization of Islam, and was brought to fruition in the civilization of Western Europe, beginning in the late twelfth century." I thought it appropriate to describe the science that emerged as "Greco-Arabic-Latin" science, because I believe that

> the collective achievement of these three civilizations, despite their significant linguistic, religious, and cultural differences, stands as one of the greatest examples of multiculturalism in recorded history. . . . It was possible only because scholars in one civilization recognized the need to learn from scholars of another civilization.[9]

I also fully agree with Bala that all scientific influences on the development of European science from other civilizations should be properly accredited to the originating civilizations. But most of the numerous claims Bala makes are devoid of any relevant historical evidence. This, however, does not deter Bala, who simply rejects the need for historical evidence to support his claims. In taking issue with H. Floris Cohen, Bala rejects Cohen's "strong criterion that establishing transmission requires us to present direct, independent evidence—a condition so stringent that it biases history toward a Eurocentric direction."[10] As we shall see, Bala either ignores evidence

---

[6] Bala, *The Dialogue of Civilizations in the Birth of Modern Science* (New York, 2006), 21.
[7] Ibid., 1; emphasis in the original.
[8] Ibid., 38.
[9] Edward Grant, *The Foundations of Modern Science in the Middle Ages: Their Religious, Institutional, and Intellectual Contexts* (Cambridge, 1996), 205–6.
[10] Bala, *Dialogue of Civilizations* (cit. n. 6), 48.

or is ignorant of it. For the most part, he makes it up to fit the situation he wishes to claim as a dialogical influence. As a substitute for evidence, he conjures up circumstances that will support his ideas. In the following passage, we see how Bala twists history to serve his purposes:

> Clearly we cannot decide if an influence occurred by simply asking whether certain ideas in modern science can be found in the premodern ideas of any particular culture. What is crucial is to decide when an idea from outside Europe can be said to have influenced European thinkers even when the same idea may have been proposed earlier by a thinker within Europe. If the idea had not been taken seriously in Europe until its significance was shown by contact with the culture outside, we have reason for presuming influence. If an idea remained a recessive theme in Europe before the emergence of modern science, and its significance was not perceived until Europeans confronted it as a dominant theme in another culture, which then led them to articulate it into a major theme in modern science, then we have to concede influence. It is reasonable to assume that the rise of the conceptual theme in modern science was the result of the influence of the non-European culture.[11]

Bala finds the major source of support for his ideas in Chinese science, especially astronomy and cosmology. He sees the beginning of Chinese influence in the sixteenth century, "and given that views were to develop in Europe that paralleled Chinese views in the seventeenth century, is it reasonable to suppose that European astronomers were not influenced by these Chinese ideas? Such a conclusion would hardly be credible."[12] Many of these ideas were allegedly transmitted to the West by the Jesuits in China, who arrived in China in the sixteenth century. Of the differences between the Chinese and Western cosmological systems, the most significant "was the Chinese belief, associated with the *xuan ye* theory, that space is empty and infinite. This was later to be seen as one of the most important changes brought about by the new cosmology of modern science." By contrast, Bala declares that "the notion of an infinite empty universe was not even possible to entertain within the Aristotelian worldview of the Jesuits, in which infinity could only be an attribute of God, and nature was held to abhor the vacuum."[13]

Bala asserts that Koyré identified Nicholas of Cusa as the one who introduced an infinite empty space into modern astronomy.[14] Although Bala accepts this, it is in fact false: Nicholas of Cusa did not assume an infinite, empty space.[15] Bala acknowledges that the idea of infinite space goes back to the Greek atomists Diogenes Laertius and Epicurus (he mistakenly includes Lucretius, a Roman author, with the Greek atomists). But did the Greek atomists play the instrumental role in causing infinite space to be accepted in the seventeenth century? Bala emphatically rejects such a suggestion, because he assumes their influence at that time was marginal. It was Chinese astronomical influence, with its emphasis on infinite space, that drew attention to that idea. "Is it not possible," Bala conjectures, that once this idea, and others, "had come to be seriously entertained European scientists foraged within their own tradition for precedents to give them historical legitimacy and authority?" The answers to such

---

[11] Ibid., 47.
[12] Ibid., 138.
[13] Ibid., 137.
[14] See Alexandre Koyré, *From the Closed World to the Infinite Universe* (Baltimore, 1957), 19–20.
[15] See Edward Grant, *Much Ado about Nothing: Theories of Space and Vacuum from the Middle Ages to the Scientific Revolution* (Cambridge, 1981), 139; for the Latin text on space, see 348 n. 113.

questions, Bala insists, will determine how we pursue the history of science. "If we treat the ideas in Europe as a case of independent discoveries that can be traced back to European antecedents we would be reinscribing a Eurocentric history. However, if we see them as products of the influence of Chinese ideas we open the door to a dialogical perspective."[16]

In all of this, Bala is wholly in error. When one describes the concept of infinite, void space that was adopted in the West, it becomes immediately apparent that it could not have been imported from China. This is so because the concept of infinite, void space, which was first adopted in the West by Thomas Bradwardine and Nicole Oresme in the fourteenth century, was, until the early eighteenth century, assumed to be God's infinite immensity. It was also the view of infinite space held by Newton.[17] God played no role in the Chinese view of infinite space, and therefore the Chinese view of space could have played no role in the conception of infinite space that developed in the West from late medieval sources. Nor indeed, as Bala suggests (see the passage from 139 cited above), did European scientists forage within their own traditions for precedents to give scientific ideas imported from China, or other non-European civilizations, historical legitimacy and authority. If they had done this, it would mean that early modern European natural philosophers and scientists regarded medieval scholastic natural philosophers of the thirteenth and fourteenth centuries as the source and origin of various scientific ideas. This is contrary to everything we know about how Renaissance, or early modern, scholars viewed their medieval predecessors, for whom they usually had contempt. Under no circumstances would they have sought to detect scientific ideas in the writings of their scholastic predecessors. Bala's claim has no historical support whatever.

The invention of the telescope has always been regarded as having occurred exclusively in the West, but Bala seeks to claim it for a non-Western civilization: in this instance, Islam. He argues that the Islamic scientist Alhazen (956–1039) made the invention possible by his optical theory, which was influential in the West until Kepler's optical theory replaced it in the seventeenth century. Therefore, Bala insists,

> should we not consider Alhazen's theory to have played an important role in the design of the telescope, and thereby, on the whole history of the discoveries in the heavens it made possible? Indeed it is conceivable that without the theory and Alhazen's study of the behavior of magnifying glasses, the emergence of the modern telescope may have been unlikely. If this is the case then the theoretical discoveries of Arabic science can be considered to have laid the basis for some of the most important instrumental innovations in modern science, such as the telescope and microscope—both of which opened up new dimensions of the universe with far-reaching consequences for the future history of science.[18]

---

[16] Bala, *Dialogue of Civilizations* (cit. n. 6), 139.

[17] For a summary account of the relations between God and void space, see Grant, *Foundations of Modern Science* (cit. n. 9), 122–6. For a quite detailed discussion of medieval scholastic views of infinite void space, see Grant, *Much Ado about Nothing* (cit. n. 15), pt. 2, "Infinite Void Space beyond the World." The chapters in part 2 are as follows: chap. 5, "The Historical Roots of the Medieval Concept of an Infinite, Extracosmic Void Space"; chap. 6, "Late Medieval Conceptions of Extracosmic ('Imaginary') Void Space"; chap. 7, "Extracosmic, Infinite Void Space in Sixteenth- and Seventeenth-Century Scholastic Thought"; chap. 8, "Infinite Space in Nonscholastic Thought during the Sixteenth and Seventeenth Centuries." To gain an overview of medieval and early modern views of infinite space, see pt. 3, "Summary and Reflections."

[18] Bala, *Dialogue of Civilizations* (cit. n. 6), 165.

This is a completely false and misguided interpretation of the telescope's invention. Spectacles were invented in Italy toward the end of the thirteenth century, and thereafter Italy became a center for the manufacture of lenses. In 1608, Hans Lippershey (or Lipperhey; 1570–1619), who was born in Germany but became a citizen of the Netherlands, may have invented the first telescope, although two other contemporary residents of the Netherlands (Sacharias Janssen and Jacob Metius) have also been regarded as claimants to that honor.[19] These Dutchmen were lens makers and spectacle makers. Albert Van Helden explains that

> by about 1600 we can point to a steadily improving glass-making and lens-grinding industry, and to the presence of both convex and concave lenses in the shops of spectacle makers. There was also a well-established belief in the possibility that some combination of lenses would produce miraculous magnifying effects, and there was a growing demand for convex lenses with longer focal lengths. . . .
>   When all is said and done, we are still left with the fact that the earliest undeniable mention of a telescope is to be found in the letter of 25 September, 1608, which Lipperhey carried to The Hague and that Lipperhey was the first to request a patent on the telescope.[20]

But Van Helden does not regard this as sufficient evidence to award Lipperhey the honor of inventing the telescope. All that is fairly certain is that it was invented around 1608 in the Netherlands.

We may, furthermore, be certain that Alhazen's optical theory played no role in the invention of the telescope. The men who are candidates for the honor were craftsmen—spectacle and lens makers. They knew nothing about optical theory, and it is certain that they did not construct a telescope on the basis of Alhazen's optical theory. Indeed, as Van Helden explains, "only after the invention of the telescope does the study of lenses and lens systems become an important part of formal optics."[21]

Bala includes numerous other claims for dialogical scientific influences on European science. But all are devoid of evidence, although this is of little concern to Bala. Indeed, he does not even entertain the possibility of independent scientific discoveries. John North offers a possible example of independent discoveries when he declares, "Early in the fourth century AD, the [Chinese] astronomer Yü Hsi (who was active between 307 and 338) discovered the changing longitudes of the stars, our 'precession of the equinoxes,' seemingly independently of western knowledge of it, as first found by Hipparchus."[22] Presumably, this example would be of little interest to Bala, because it is dialogical in the wrong direction: from West to East, rather than East to West. But North's example might be far more commonplace than one might suppose. Similar ideas about nature and its operations can occur to individuals in different civilizations. There will also be differences. It is the historian's task to sort them out.

In his incisive and profound review of Bala's book, Peter Sobol begins with this

---

[19] For the history of the invention of the telescope, I rely on the excellent monograph by Albert Van Helden, "The Invention of the Telescope," *Transactions of the American Philosophical Society*, n.s., 67, no. 4 (1977): 1–67. For a discussion of the three Dutch candidates as inventors of the first genuine telescope, see sec. 4, "Lipperhey, Metius and Janssen."

[20] Ibid., 16; 25, col. 2.

[21] Ibid., 12, col. 2.

[22] North, *Cosmos: An Illustrated History of Astronomy and Cosmology* (Chicago, 2008), 140.

question: "Do you believe that seventeenth-century Europeans invented modern science with no more inspiration than what they received from ancient Greece?"[23] In the final paragraph of his review, Sobol answers his question: "No historian of science would answer the question posed at the top of this review in the affirmative." But equally, "neither would most agree that particular cases of influence are established by the mere similarity of ideas between two entities, even when they communicate. We want evidence." To illustrate his significant point, Sobol offers this example: "Bala may be right that Europe must have received an empowering dose of Chinese astronomy and technology. But that is a working hypothesis, not a conclusion." Without evidence, Bala's claims will remain mere hypotheses, and historians of Western science will have no good reason to accept most of the unsupported claims for extra-European influences. As a consequence of his refusal to be guided by evidence, we may conclude that although Bala is a scientist and philosopher, he does not appear to be a historian.

CONTEXT AND TEXTUAL CONTENT

For better or worse, my concern, as well as that of all who will read this, has been overwhelmingly with the history of science in Europe. I shall now inquire as to how science and natural philosophy were related to the societies in which they were produced. I am convinced that knowing the societal and cultural background within which contributions were made to natural philosophy and the exact sciences will not reveal much, if anything, about the actual content of the resulting treatises. In other words, the societal context will not shed much light on, or reveal the meaning of, the text to which it is related. To become familiar with scientific and natural philosophical treatises, it is essential to study them and also become familiar with the historical background that underlies all such works. That is, to maximize one's possibilities of properly understanding the text of a given treatise, it is essential to pursue its most *relevant context*. For a treatise in mathematics, for example, one must try to determine what other mathematical treatises the author may have known and used and whether or not he used them wisely. Knowing all this will put the research scholar in a reasonable position to judge whether the author contributed anything significant or novel to the history of mathematics. The same may be said for the other subject areas. One must study the literature of the given subject in order to determine the significance, if any, of the treatise under investigation. Thus, the context is not societal, or generally external, but is internal, because it is largely confined to an examination of treatises treating, for the most part, much the same subject.

But what about the life and times of authors who wrote scientific and natural philosophical texts? Are they not instrumental in comprehending what an author wrote in any particular treatise, or treatises, under investigation? They are important for many reasons. We may learn much from knowledge of an author's life and times. We may learn where the author acquired knowledge about the subject in question; we may learn about the kind of intellectual environment in which the author lived; whether the author was in regular contact with individuals who had a similar interest and with whom he or she could discuss the subject in question; and so on. Knowledge of this

---

[23] For Sobol's review, see *Isis* 98 (2007): 829–30.

kind will, however, not be of much use in resolving questions about the text itself, as, for example, What did the author mean in a particular passage; or, Who may have influenced him or her in arriving at a particular approach or a particular result?

The so-called internalist-externalist controversy is basically a problem between a text and its context. In an important article ("Historical Contextualism: The New Historicism?"), Preston King explains the relationship that normally obtains between a text and its context. To begin with, every text unavoidably has a context of some kind. The pertinent question is "whether this idea or that has been placed in the context that is *most apt* for one's purposes—literary, economic, religious, historical, social or whatever."[24] Assuming one has the most appropriate context for a text, what has one achieved? One has added details to the text that are not found in the text. But this has come at a price, for the context "must in equal measure, and at the same time, sacrifice detail." King offers this significant example: "To view a house down below in the context of the valley, from the height of the mountain above, is never to apprehend it in the detail ready to hand and eye from a seat on its veranda." King further explains that the context one makes for a text "only amounts to additional data—to supplementary claims and arguments." If every text needs a context, then the context itself needs a context, and so on indefinitely. As King explains,

> If I am *always* to contextualise, then the context itself (which is substantively reducible to one or more additional "texts") must in turn be contextualised. Let us say that I place text t_1 in context by juxtaposing to it text t_2. I place this in context in turn by adding to it text t_3. But to contextualise this I must append . . . text t_n. I cannot of course continue endlessly in this way. Am I being enjoined to do what I cannot do? If a necessary condition of my securing a reliable grasp of a statement, of a text, is that I be able to place it in context, then I shall never be able to accomplish this. For at the moment that I create the context, I only do so at the expense of leaving the latter out of context.[25]

If King's interpretation of texts and contexts is sound, it seems that historians of medieval science have, over these many years, interpreted medieval texts in an appropriate manner. The contexts within which most have operated in establishing and interpreting scientific and natural philosophical texts were largely confined to manuscripts and other treatises that were historically related to the focal text. They did not explain the texts and their influences by expanding their contextual field to the farther reaches of society.

THE RELATIONSHIP OF THEOLOGY AND RELIGION
WITH NATURAL PHILOSOPHY IN THE LATE MIDDLE AGES

Theology and religion were perhaps the most significant contexts within which natural philosophy functioned. Although I have mentioned the exact sciences, they will play almost no role in what follows. Theology, and religion generally, had virtually no influence on the exact sciences of mathematics, astronomy, optics, and mechanics. There were no significant issues in the exact sciences that posed problems for the Christian religion and its theologians. I am not aware of any ideas or concepts in the exact sciences that medieval Christians regarded as offensive to the faith.

---

[24] The article appears in *Hist. Europ. Ideas* 21 (1995): 209–33; quotation on 210 (emphasis in the original).
[25] Ibid., 224.

When one speaks of the relationship between science and religion in the Middle Ages, it is almost exclusively natural philosophy that is intended. More specifically it is Aristotle's natural philosophy, which formed the solid core of medieval natural philosophy.

When Aristotle's natural philosophy was translated into Latin in the late twelfth century and was quickly disseminated to the university cities of Paris and Oxford, it entered a society in which the Christian religion had been dominant for nearly a millennium. The numerous treatises that constituted Aristotle's natural philosophy, which ranged over virtually all types of natural phenomena, were essentially devoid of religious sentiments. Aristotle viewed all natural phenomena as the natural effects of natural causes. Although Aristotle believed in the existence of a God, the role he accorded to the deity virtually divorced it from the physical cosmos. He assumed that both God and the universe were eternal, uncreated entities. In his philosophy, God is an immaterial, immobile being who not only did not create the universe, but has nothing whatever to do with it. Indeed, its only activity is thought, but it thinks only about itself because it is the only object worthy of its thought. And yet the world is in continuous motion, because the intelligences that move the celestial spheres do so from their love of God, giving rise to the expression "'tis love that makes the world go round." Because the celestial intelligences are driven to move their orbs by virtue of their love of God, who is not even aware of their existence, the latter came to be known as "the Unmoved Mover."

Religion thus plays virtually no role in Aristotle's natural philosophy. He sought to understand and describe the operations of nature by reasoned analysis and natural causes and effects. He was the quintessential analytic thinker of the Western world. Thus, when his works became available in Latin translation, they presented a picture of the world that was truly new and dramatically different from anything known before. Although it had enormous immediate appeal to students in Paris and Oxford, it also aroused opposition in the thirteenth century, when it survived various efforts to wholly reject it, and then to censor it (discussed in detail below). By the end of the thirteenth century, however, Aristotle's natural philosophy was accepted and admired by virtually all scholastic natural philosophers and theologians. Moreover, it was fully accepted by the Church, almost all of whose theologians had studied natural philosophy in some European university—many of them at the Universities of Paris and Oxford. Indeed, medieval theologians of the Catholic Church were usually as much natural philosophers as they were theologians. In light of this, it is hardly surprising that Aristotelian scholastic natural philosophy flourished in Europe for more than four hundred years.

When we inquire about the relationship between natural philosophy and theology (or religion), we are immediately involved in a two-way relationship comprising (1) the influence of natural philosophy on theology or religion and (2) the influence of theology or religion on natural philosophy. The first theme is one that I shall merely mention, before focusing my attention on the second topic. The influence of natural philosophy on theology during the late Middle Ages was extensive and profound.[26] "Logic and natural philosophy were applied to the deepest mysteries of the Christian faith: the Trinity and Eucharist." This occurred because medieval theologians

---

[26] For a detailed discussion of that influence, see Grant, *History of Natural Philosophy* (cit. n. 4), 262–73. The section is titled "Did Natural Philosophy Influence Medieval Theology?"

learned logico-mathematical techniques as students in the arts faculties before they began the study of theology. This "enabled them to transform theology into a ratio-nalistic, analytic discipline during the thirteenth and fourteenth centuries. Theology was often more natural philosophy than it was theology."[27] However, because this is an article on the history of science, it is far more important to determine what influ-ence, if any, theology or religion may have had on natural philosophy, which I shall now discuss.

We may distinguish two major categories in which mentions of God or religious ideas may be classified. The first embraces religious influences on natural philosophy that involved mentions of God, or the inclusion of religious sentiments, within a dis-cussion of some relevant question or problem in natural philosophy. The second cate-gory concerns theological reactions to Aristotle's assertions that were in conflict with the Christian faith. These were important, because all natural philosophers—both theologians and nontheologians—were compelled to take cognizance of them, by either explicitly or implicitly rejecting Aristotle's claims. As we shall see, the second category affected natural philosophy far more significantly than the first, which had almost no detectable impact.

The first category includes all mentions of God and religious entities that appeared in the broad range of natural philosophy treatises from approximately 1200 to 1600. Natural philosophers found many occasions to insert religious sentiments into their treatises. In virtually all such instances, however, the sentiments did not affect the substantive content of the basic theme or discussion. Where Aristotle himself had occasion to mention something about the divine, his medieval commentators would usually respond with some statement drawn from religion, a statement that often included God. For example, in *On the Heavens* (*De caelo*), Aristotle asserted that "we recognize habitually a special right to the name 'heaven' in the extremity or upper region, which we take to be the seat of all that is divine." In commenting on this passage in his *Commentary on "De caelo,"* Thomas Aquinas declared that "'up' [*sursum*] is customarily said to be the place of all divine things . . . : for it was said above that all men attribute to God the place that is 'up.'"[28] In this instance, Aquinas is simply agreeing with Aristotle and offers no new or qualified argument.

Albertus Magnus also cited God, and synonyms for God, in direct response to Aristotle's mention of gods or God. This is especially true in Albertus's *Commentary on Aristotle's "Physics."* In the eighth book of that treatise, where Aristotle refers to God by various synonyms, Albertus speaks about God more frequently than in any other of the eight books of his commentary: "Thus, of the 64 occurrences of *primus motor,* that is, first mover, or God, 55 occur in book 8; of the 69 occurrences of *causa prima,* that is, first cause, or God, 37 occur in book 8; and of the 78 occurrences of *deus* (God), 40 occur in the eighth book."[29] Many similar instances could be cited. But of great significance is the fact that none of these instances had a substantive im-pact on the discussions in which they occurred.

The same may be said for a second kind of insertion of religious terms and con-

[27] Ibid., 273.

[28] My translation from Aquinas's *Commentary on "De caelo,"* bk. 1, lectio 20, par. 199, in *S. Thomae Aquinatis in Aristotelis Libros "De caelo et mundo," "De generatione et corruptione," "Meteorologi-corum" Expositio,* ed. Raymundi M. Spiazzi (Turin, 1952), 98, cols. 1–2. See also Grant, *History of Natural Philosophy* (cit. n. 4), 259.

[29] Edward Grant, *God and Reason in the Middle Ages* (Cambridge, 2001), 194.

cepts. Medieval natural philosophers frequently found it useful to strengthen an argument by invoking God analogically. In his *Questions on "De anima"* (bk. 3, question 2), Nicole Oresme presented this analogy: "Some power makes this or that operation anew without changing itself, just as is obvious with God who continuously produces new effects without any change in Himself." In a similar fashion, Themon Judaeus, in his *Questions on the "Meteorology"* (bk. 4, question 5), used God analogically when he declared that "a pure element is understood [to be] simple, but not simple absolutely, as is God, or an intelligence."[30] The term for God (*deus*), and its numerous synonyms cited above, occur in natural philosophy treatises, but they do not affect the discussions in any substantive manner, nor do they appear frequently.

It is the second category of mentions of God or religious ideas that had the greatest impact. Aristotle held a number of positions about the physical world that were contrary to the Christian faith and could not be left unchallenged. Undoubtedly, the most important of Aristotle's contrary-to-faith ideas was his firm conviction that the world is eternal—it has no beginning and will never end—an idea that contradicted the account in Genesis that God created the world from nothing and would eventually destroy it. Aristotle also argued that all accidents, or qualities, without exception, must inhere in a substance. This ran counter to the Eucharist, or Mass, which assumes that God transforms the bread and wine of the Mass into the body and blood of Christ, but that the accidents of the bread and wine continue to exist and be visible although they do not inhere in any substance. Another point of contradiction arose from Aristotle's belief that only the rational part of the human soul is immortal and that its other parts perish with the death of the body, a view that was in direct conflict with the strong Christian belief that the soul exists through eternity following the death of the body. Other ideas of Aristotle were also deemed offensive to the Christian faith and the Catholic Church.

During the thirteenth century, the Church tried initially to ban the natural philosophical works of Aristotle (in 1210). When that proved unsuccessful, there was an effort (in 1231) to delete the offensive parts of Aristotle's philosophy by censorship, but this was never carried out. Finally, in 1277, the bishop of Paris issued a condemnation of 219 articles, many of them drawn from the works of Aristotle and his great Islamic commentator, Averroes (Ibn Rushd). Anyone who held, or defended, even one article was subject to excommunication.[31] Many, if not most, of the condemned articles were drawn from Aristotle's natural philosophy. Some twenty-seven articles condemned different versions of the eternity of the world.

But the most significant aspect of the Condemnation of 1277 was not the condemnation of Aristotle's offensive ideas about Christian doctrine, but rather the condemnation of his ideas that seemed to limit God's absolute power to do whatever he pleased short of a logical contradiction. The following list of condemned articles immediately reveals the ways in which Church authorities reacted to Aristotle's various claims that this or that phenomenon was "naturally impossible." They firmly believed that for God Aristotle's natural impossibilities were all "supernaturally possible."

---

[30] Both examples appear in my article "God, Science, and Natural Philosophy in the Late Middle Ages," in *Between Demonstration and Imagination: Essays in the History of Science and Philosophy Presented to John D. North*, ed. Lodi Nauta and Arjo Vanderjagt (Leiden, 1999), 243–68, on 260.

[31] For the relations between Aristotle's natural philosophy and the Church in the thirteenth century, see Edward Grant, *Science and Religion, 400 B.C. to A.D. 1550: From Aristotle to Copernicus* (Westport, Conn., 2004), 176–84.

34. That the first cause [that is, God] could not make several worlds.

35. That without a proper agent, as a father and a man, a man could not be made by God [alone].

38. That God could not have made prime matter without the mediation of a celestial body.

48. That God cannot be the cause of a new act [or thing], nor can he produce something anew.

49. That God could not move the heavens [or world] with a rectilinear motion; and the reason is that a vacuum would remain.

63. That God cannot produce the effect of a secondary cause without the secondary cause itself.

141. That God cannot make an accident exist without a subject nor make several dimensions exist simultaneously.

147. That the absolutely impossible cannot be done by God or another agent—An error, if impossible is understood according to nature.[32]

Article 147, the final condemned article cited here, seems to serve as a generalization of all the articles just cited, as well as numerous others among the 219 condemned.

The most important of these natural impossibilities was probably Aristotle's firm denial of the possible existence of other worlds, for which he provided supporting arguments. From the 1230s to the early 1270s, there were already reactions to this natural impossibility. Scholastic authors, such as Michael Scot, William of Auvergne, Thomas Aquinas, and Roger Bacon, argued that God could create other worlds if he wished, but they all found one or more arguments to convince their readers that God really had no good reason to create other worlds. Scot, for example, believed that nature was incapable of receiving more worlds and, therefore, God, knowing this, would have no reason to create any. Aquinas believed it was best for God to make just a single, perfect world rather than diffuse perfection over many worlds, none of which would be as perfect as our world. Others proposed similar arguments.[33]

After the Condemnation of 1277, scholastic natural philosophers might have followed the same path. That is, they could have simply acknowledged that God could easily make realities out of Aristotle's natural impossibilities and then found reasons to explain why God would not do so. But most scholastic natural philosophers did not follow that path. After 1277, they not only chose to imagine that all of Aristotle's natural impossibilities were possible, as well as others that he had never considered, but they assumed, hypothetically, that God had actually performed them. This was a momentous occurrence, because it led scholastic natural philosophers into a realm of hypothetical problems—or counterfactuals—about our world and what, if anything, might lie beyond it. They entered the realm of "let's pretend" and began to discuss topics that were literally out of this world, which stirred their imaginations in remarkable ways. They sometimes identified such arguments as *secundum imaginationem,* the Latin phrase for "according to the imagination." Ironically, a major result was that these natural philosophers often used Aristotle's concepts and interpretations as the basis for investigating the behavior of phenomena under conditions that Aristotle regarded as impossible—but that they considered supernaturally possible, by God's

---

[32] Quoted ibid., 183. All articles, except article 63, were originally translated in Edward Grant, *A Source Book in Medieval Science* (Cambridge, Mass., 1974), 48–50. Article 63 appears in Ralph Lerner and Muhsin Mahdi, eds., *Medieval Political Philosophy: A Sourcebook* (Ithaca, N.Y., 1963), 343.

[33] For the arguments proposed by these authors, see Edward Grant, "The Condemnation of 1277, God's Absolute Power, and Physical Thought in the Late Middle Ages," *Viator* 10 (1979): 217–9.

absolute power to do anything short of a logical contradiction. The major consequence of all this was that hypothetical and counterfactual discussions became a vital
part of medieval natural philosophy. The medieval imagination was free to "probe
and poke around." A few of their achievements were destined to influence some of
the major figures of the Scientific Revolution. Thus, some of the most important medieval achievements in the history of science were ultimately motivated by religious
concepts about God's attributes.

THE WORLD OF COUNTERFACTUALS

Of the eight condemned articles cited earlier, numbers 34 (that God could not make
other worlds) and 49 (that God could not move the world rectilinearly, because a
vacuum would be left in the space vacated by the world) were probably the most significant for the history of science.

The Possible Existence of Other Worlds

Although he firmly believed that nothing whatever could exist beyond our world—
neither void spaces nor other worlds—Aristotle acknowledged that those who believed that an infinite magnitude and an infinite number of worlds existed beyond our
world did so because the human mind is always capable of imagining more things beyond any limit. This is so because these things "never give out in our thought."[34] Most
scholastic authors agreed with Aristotle in denying the existence of other worlds as
well as any kind of extracosmic void space, but they could imagine such things. A
plurality of worlds could be imagined as either successive or simultaneous. Successive worlds were not often discussed, although all agreed that God could destroy one
world and replace it with another, and he could do this endlessly. It was simultaneous
worlds that captured their attention.

  One of the most imaginative interpreters of possible simultaneous worlds was Nicole Oresme (ca. 1320–82), who considered two kinds of simultaneous worlds—
concentric or eccentric—and separately existing worlds. He focused attention on
simultaneous concentric worlds, which had already been discussed and rejected by
William of Auvergne and Roger Bacon in the thirteenth century. Oresme's presentation bears no resemblance to those of his predecessors. Among a number of illustrations he gave, Oresme imagined that another world lies concentrically within
ours—that is, it lies at the center of our Earth—and another world lies concentrically
wrapped around our world, and another world lies concentrically around that world.
Oresme did not take concentric worlds seriously. He regarded them as "another speculation . . . which I should like to toy with as a mental exercise." The world imagined
inside our world would be very small, and the worlds wrapped around our world
would be very much larger than our world and become ever larger as we conceive of
more and more concentric worlds further and further removed from our world. Would
these differences in size signify that people and things would be very small in worlds
within our Earth and very large in worlds concentrically around ours? In response to
this problem, Oresme explained that "*large* and *small* are relative, and not absolute,
terms used in comparisons." He "imagines that if between now and tomorrow, our

---

[34] Aristotle, *Physics* 3.4.203b.26.

world were made one hundred or one thousand times larger or smaller than it is now, with 'all its parts being enlarged or diminished proportionally, everything would appear tomorrow exactly as now, just as though nothing had been changed.'"[35] Oresme regarded the existence of concentric worlds as improbable, but not impossible, "because," as he put it, "the contrary cannot be proved by reason nor by evidence from experience, but also I submit that there is no proof from reason or experience or otherwise that such worlds do exist."[36]

In rejecting the existence of other worlds, Aristotle had in mind worlds that were identical to ours, but that existed separately and simultaneously. His major reason for regarding the existence of other worlds as impossible was that he believed our spherical world could have only one circumference and one center. But if other worlds existed simultaneously with ours, there would be as many independent centers and circumferences as there were worlds. Since all these worlds would be identically equal, the element earth of another world would try to reach the center of our world. To do this, it would have to rise up in its own world—contrary to its natural inclinations—enter our world, and descend naturally toward the center. Thus, it would have two natural motions: one up, and one down. Aristotle regarded this as absurd and concluded that it was impossible for other worlds to exist.

The response of numerous scholastics to Aristotle was simple and direct. The elements of one world would not leave their world to move to another. Each world is self-contained and independent. Thus it was not impossible, as Aristotle had argued, for a plurality of worlds, each with its own center and circumference, to exist independently of one another. Although this was seemingly a major departure from Aristotle, and would have subverted his physics and cosmology, it was merely an exercise of reasoned imagination and had little impact, because no one really believed in the existence of other worlds. It was not until 1506 that John Major (1467/8–1550), a scholastic theologian and Aristotelian commentator at the University of Paris, proposed the actual existence of a plurality of worlds—indeed, an infinity of worlds. In this Major followed Democritus, whom he cited. "Naturally speaking," said Major, "there are infinite worlds, [and] no argument can convince one of the opposite."[37]

Although there were widespread discussions about the existence of other worlds, the possibility of life elsewhere in the cosmos or in other worlds was not discussed until the fifteenth century, when Nicholas of Cusa (1401–64), in his treatise *On Learned Ignorance,* conjectured that the same life forms found on Earth might also be found on other planets and stars and that the living beings on other planetary worlds would be no more noble or perfect than the inhabitants of our Earth. Indeed, their worlds would be subject to the same kind of corruption and generation as is our world. Cusa's conception of the world represents a total rejection of Aristotle's cosmos.

Also in the fifteenth century, William Vorilong (d. 1463) conjectured, in a theo-

---

[35] Based on Grant, *History of Natural Philosophy* (cit. n. 4), 227–8. The direct quotations are from Nicole Oresme, *Le livre du ciel et du monde,* ed. Albert D. Menut and Alexander J. Denomy, trans. Menut (Madison, Wis., 1968), 169.

[36] Grant, *History of Natural Philosophy* (cit. n. 4), 228; Oresme, *Livre du ciel et du monde* (cit. n. 35), 171.

[37] Quoted in Edward Grant, *Planets, Stars, and Orbs: The Medieval Cosmos, 1200–1687* (Cambridge, 1994), 167.

logical treatise known as the *Sentence Commentary* (bk. 1, distinction 44), that there might be life on other worlds and that the species on those worlds might differ from those of our world. As for the existence of human life on other worlds, William explained that if men did exist in these worlds, they would not exist in sin, because they did not spring from Adam. In answer to the question whether, by dying, Christ also redeemed the inhabitants of these other worlds, William replied in the affirmative. Christ's unique death in our world was sufficient to save the inhabitants of all worlds.[38]

God Moves the Cosmos Even Though a Vacuum Is Left Behind

We saw that article 49, one of those condemned in 1277, denied that God could move the world, because a vacuum would be left in the place formerly occupied by the world. If God moved the spherical cosmos, a vacuum would indeed remain where the cosmos formerly rested. By this action, Aristotle's ideas about vacuum, place, and motion would be subverted. Aristotle had assumed that the place of a body is the innermost, immobile surface of the body that surrounds it. In a world that is a plenum, every material body is surrounded by another body. Motion in our plenistic world goes from place to place. But no body surrounds our cosmos, and yet God moves it rectilinearly through a void space, even though the cosmos cannot be said, in Aristotelian terms, to be moving from one place to another. Oresme considered the rectilinear motion of the cosmos to be an absolute motion, because the motion was not relatable to any other body. Given the assumed conditions, Oresme accepted the motion of the cosmos as possible and plausible.[39]

If the world were moved out of its original place, into what would it move? It had to move into other void places; otherwise, as Jean de Ripa argued around 1344, God would be unable to move it. Therefore, there had to be imaginary void places outside of the material world in which bodies and angels could be received. If there were void spaces outside of our world, there seemed no good reason to assume that there was any end to such spaces. This led some scholastics to assume the actual existence of an infinite void space. In these numerous discussions, article 49 was frequently mentioned.

The idea of God moving the world with a rectilinear motion was introduced into seventeenth-century discussions about space, void, and the infinite. In his famous controversy with Gottfried Leibniz (1646–1716), Samuel Clarke, who was Newton's spokesman, invoked this idea. Pierre Gassendi (1592–1655), in a posthumously published work, also assumed that God moved the world through an endless void space.[40]

Infinite Space beyond Our World

Up to this point, my discussion of scholastic natural philosophy in the thirteenth and fourteenth centuries has been concerned only with counterfactual situations. But a few theologians assumed the actual existence of an infinite space beyond our

---

[38] On Cusa and Vorilong, see ibid., 168.

[39] For further details, see Grant, *History of Natural Philosophy* (cit. n. 4), 205–6.

[40] Gassendi, *Syntagma Philosophicum* (Lyon, 1658). See Grant, *Much Ado about Nothing* (cit. n. 15), 209.

world. In the fourteenth century, Thomas Bradwardine (ca. 1290–1349) and Oresme adopted a spatial concept that was destined to play a significant role in seventeenth-century cosmology, strongly influencing Newton. In this departure from Aristotle, God was the central character. Around 1344, Bradwardine composed a theological treatise titled *In Defense of God against the Pelagians.* In a chapter titled "That God Is Not Mutable in Any Way," Bradwardine presented five corollaries in which he proclaimed that "God is necessarily everywhere in the world and all its parts" and that he is "also beyond the real world in a place, or in an imaginary infinite void."[41] God must therefore be immense and unlimited, from which Bradwardine inferred that although "a void can exist without body," it cannot exist without God's presence. Why is void space infinite? Because God could have created the world in any empty space whatever, and therefore must have had an infinite number of choices. But the void spaces in which God might have created the world cannot have existed from eternity, for then they would constitute an uncreated entity co-eternal with God; nor can they be independent of God. The conclusion for Bradwardine, and a few other scholastics, was simply that infinite void space must be God's infinite immensity. However, because God is immaterial and without extension, and infinite void space is his immensity, it followed that infinite void space is extensionless.

Some years later, in his *Livre du ciel et du monde* (1377), a French translation and commentary on Aristotle's *On the Heavens,* Oresme adopted much the same position as Bradwardine. Almost as if following Aristotle's idea that the human mind always seems to imagine something beyond a given limit, Oresme was convinced that something must exist beyond our finite world. "The human mind," he declared, "consents naturally . . . to the idea that beyond the heavens and outside the world, which is not infinite, there exists some space, whatever it may be, and we cannot easily conceive the contrary." This space differs from any other kind of corporeal, or dimensional, space. Oresme declared that although "we cannot comprehend nor conceive this incorporeal space which exists beyond the heavens," we know it is out there, because "reason and truth . . . inform us that it exists." What is this space? Oresme explained that it "is infinite and indivisible, and is the immensity of God and God Himself, just as the duration of God called eternity is infinite, indivisible, and God Himself."[42]

Although he did not mention it, it is highly likely that Oresme agreed with Bradwardine that infinite void space is dimensionless. In the Middle Ages, God would never have been regarded as a corporeal being, because corporeal beings are divisible, and God could not be regarded as divisible in any way. In the seventeenth century, however, Newton, following Henry More (1614–87), came to regard infinite void space as three-dimensional and equated it with God's infinite immensity. For Newton, then, three-dimensional infinite void space was an attribute of God—indeed, it was his immensity—and yet Newton thought of God as incorporeal! It seems that Bradwardine was more consistent than Newton. Bradwardine regarded God as extensionless and therefore regarded infinite void space as extensionless, because it was God's immensity. Newton also thought of God as extensionless, but he identified God's immensity with a three-dimensional infinite space. It seems, therefore, that Newton made God a three-dimensional being and therefore, presumably, a divisible

---

[41] Cited from my translation in Edward Grant, *A Sourcebook in Medieval Science* (Boston, 1974), 556–7.

[42] Oresme, *Livre du ciel et du monde* (cit. n. 35), 177.

being. However, one cannot imagine that Newton would have thought of God as a divisible being. Whether Newton was aware of these issues, or confronted them, I do not know.

Did Theology Influence the Substantive Content of Natural Philosophy?

We have now seen that theology and religion played a significant role in shaping medieval natural philosophy. They had the authority to encourage the pursuit of certain themes that might not otherwise have been pursued. If theologians had not denounced Aristotle's natural impossibilities and countered them with God's absolute power to render them possible, pursuit of the kinds of counterfactuals described in the preceding sections would very likely not have materialized. Important speculations about infinite void space and the existence of various versions of a plurality of worlds would probably not have occurred. As we have seen, some of these discussions were of sufficient importance to play a role in the seventeenth century, after the abandonment of Aristotle's physical cosmos. Indeed, the medieval interest in an infinite God undoubtedly incited discussions about infinites in various contexts.

If religious concepts and beliefs spurred numerous discussions, they could also discourage inquiry. For example, medieval natural philosophers could not proclaim belief in an eternal world, although they could assume it hypothetically, provided they left no doubt that they believed in a created world. They generally sought to avoid defending positions that could be construed as contrary to the faith. In that sense, of course, religious beliefs could function as obstacles to scientific inquiry, although almost any belief could be assumed hypothetically, for the sake of an argument.

In general, I believe that religious ideas and beliefs cannot directly influence genuine discussions in natural philosophy. If any natural phenomenon is explained by invoking God as its cause, or attributing it to a miracle or angelic intervention, we have shifted from natural philosophy to supernatural philosophy. I believe many medieval scholastic natural philosophers were well aware of this, although few expressed their opinions.[43] For the most part, they regarded matters of faith and religion as distinct from natural philosophy. One could, however, make reference to biblical passages and religious ideas, as long as they were not presented as explanations of natural phenomena. All medieval natural philosophers believed that God had created the world and assigned it the physical laws that govern its cause-and-effect relationships, and that he could intervene at any time and alter as many of those relationships as he pleased. But they also believed that God rarely did such things and that it was the task of the natural philosopher to discover the causes of all natural effects. John Buridan (ca. 1300–ca. 1360) spoke for virtually all medieval natural philosophers when he declared, "In natural philosophy, we ought to accept actions and dependencies as if they always proceed in a natural way."[44] In another treatise, Buridan asserted that truth is possible in natural science if "a common course of nature (*communis cursus*

---

[43] Among those who did were Albertus, Aquinas, John Buridan, and Oresme. On the first three, see Grant, *History of Natural Philosophy* (cit. n. 4), 251–7; on Oresme, see Grant, *God and Reason in the Middle Ages* (cit. n. 29), 202–3, and Grant, *Foundations of Modern Science* (cit. n. 9), 84.

[44] See Grant, *Foundations of Modern Science* (cit. n. 9), 145, and 211 n. 12. The translation is mine from Buridan's *Questions on "De caelo,"* bk. 2, question 9, in the edition by Ernest A. Moody, *Ioannis Buridani Quaestiones super Libris Quattuor "De caelo et mundo"* (Cambridge, Mass., 1942), 164.

naturae) obtains in natural things, and in this way it is evident to us that fire is warm and that the heaven moves, although the contrary is possible by God's power."[45]

Buridan was not a theologian and, as he clearly indicated, sought to avoid theological issues that might arouse the ire of professional theologians. Indeed, he was well aware that nontheologians were not permitted to discuss theological questions, but if it was unavoidable, they were to resolve the issue in favor of the faith. Although theologians could introduce theology into their natural philosophical discussions, they rarely tried to explain natural phenomena by appeals to the Bible or miracles, or anything religious. What they usually did, however, was to apply natural philosophy, and logico-mathematical techniques, to theological problems. This they did almost exclusively in theological treatises, primarily in the *Commentaries on the Sentences of Peter Lombard,* the basic theological textbook from around 1200 to 1600. One notable exception to this is Oresme, who injected much theological material into a few of his natural philosophical treatises. Oresme was many things: he was not only a distinguished theologian and defender of the faith, but was probably the most brilliant scientific thinker of the Middle Ages, both as a natural philosopher and as a mathematician.

Because of his significant mathematical contributions on irrational ratios and celestial incommensurability, which he applied to celestial velocities, Oresme became convinced that our knowledge of causation in nature could only be approximate and that we could not have precise knowledge of nature.[46] However, he believed that approximate knowledge was sufficient for human purposes. He therefore concluded that our knowledge of nature and natural causes was no more intelligible than our knowledge of the articles of faith. Reason was no more suitable to aid natural philosophy than it was to interpret the articles of faith. Thus did Oresme weaken confidence in natural philosophy.

And yet, he did not intermingle faith and natural philosophy and did not attempt to Christianize science or natural philosophy. One can say this despite the fact that in a few of his treatises, Oresme cited numerous biblical passages. In his *Configurations of Qualities and Motions,* he included approximately fifty citations to twenty-three different books of the Bible, but "he did so only by way of example or for additional support, but in no sense to demonstrate an argument." In *Le livre du ciel et du monde,* Oresme made many references to scripture, frequently introducing biblical passages by way of example and as appeals to faith. He even devoted the last chapter of his commentary to "the body of Jesus Christ."[47] But none of these religious references were employed to explain any natural phenomenon. Oresme strongly opposed those who explained natural phenomena by appeals to magic and the supernatural and was very hostile to astrology and astrologers. In the prologue to a later treatise, *On the Causes of Marvels (De causis mirabilium,* also known as *Quodlibeta),* written around 1370, and therefore one of his later works, Oresme revealed his genuine attitude:

---

[45] Grant, *Foundations of Modern Science* (cit. n. 9), 145. The sentence occurs in Buridan's *Questions on the "Metaphysics,"* bk. 2, question 1; for the full reference, see *Foundations of Modern Science,* 211 n. 8.

[46] For Oresme's major treatises on these themes, see Oresme, *"De proportionibus proportionum" and "Ad pauca respicientes,"* ed. and trans. Edward Grant (Madison, Wis., 1966), and Grant, ed. and trans., *Nicole Oresme and the Kinematics of Circular Motion: "Tractatus de commensurabilitate vel incommensurabilitate motuum celi"* (Madison, Wis., 1971).

[47] See Oresme, *Livre du ciel et du monde* (cit. n. 35), bk. 4, chap. 12.

In order to set people's minds at rest to some extent, I propose here, although it goes beyond what was intended, to show the causes of some effects which seem to be marvels and to show that the effects occur naturally, as do the others at which we commonly do not marvel. There is no reason to take recourse to the heavens, the last refuge of the weak, or demons, or to our glorious God as if He would produce these effects directly, more so than those effects whose causes we believe are well known to us.[48]

* * *

The four medieval scholastics I have mentioned in this section—Albertus, Aquinas, Buridan, and Oresme (see n. 43 above)—would not have explained natural phenomena by appeals to God, or miracles, or any supernatural actions. They may be taken as generally representative of the whole range of scholastic natural philosophers in the period from approximately 1200 to 1600. This is rather amazing when one realizes that most medieval natural philosophers were theologians. Although the supernatural was their business, they did not intrude it into their discussions about nature and natural phenomena. It was because of their attitude that natural philosophy remained an essentially secular discipline during the late Middle Ages. And it was because natural philosophy remained a secular discipline that it could become, to use Francis Bacon's apt description, the "Great Mother of the Sciences" in the seventeenth century, when it was finally merged with the exact sciences by Johannes Kepler and others. Kepler played the major role when, in his *Epitome of Copernican Astronomy* (1618), he discussed "what is astronomy" and concluded,

> It is a part of physics [i.e., natural philosophy], because it seeks the causes of things and natural occurrences, because the motion of the heavenly bodies is amongst its subjects, and because one of its purposes is to inquire into the form of the structure of the universe and its parts.

A few lines later, Kepler declares that physics, or natural philosophy,

> is popularly deemed unnecessary for the astronomer, but truly it is in the highest degree relevant to the purpose of this branch of philosophy, and cannot, indeed, be dispensed with by the astronomer. For astronomers should not have absolute freedom to think up anything they please without reason; on the contrary you should be able to give *causas probabiles* [i.e., probable causes] for your hypotheses which you propose as the true causes of the appearances, and thus establish in advance the principles of your astronomy in a higher science, namely physics or metaphysics.[49]

---

[48] Bert Hansen, ed. and trans., *Nicole Oresme and the Marvels of Nature: A Study of His "De causis mirabilium"* (Toronto, 1985), 137.

[49] The translation is by N. Jardine. I have added the bracketed phrases. For a lengthy discussion of the relationship between natural philosophy and the exact sciences, see Grant, *History of Natural Philosophy* (cit. n. 4), 305–13. For the full title of Jardine's work, see 312 n. 106. The passage quoted from Kepler appears on 312.

The Sciences of the Archive

by Lorraine Daston*

ABSTRACT

Since the mid-nineteenth century, classifications of knowledge have opposed the bookish, history-conscious humanities to the empirical, amnesiac sciences. Yet in the sciences of the archive, the library stands alongside the laboratory, the observatory, and the field as an important site of research. The sciences of the archive depend on data and specimens preserved by past observers and project the needs of future scientists in the creation of present collections. Starting in the early modern period, distinctive practices of weaving together the data of the archives and of present investigation have created a hybrid hermeneutics of reading and seeing.

PEOPLES OF THE BOOK

Since the mid-nineteenth century, it has been a melancholy academic commonplace that whereas the humanities are the guardians of memory, the sciences cultivate amnesia.[1] The story runs something like this. Humanists lovingly preserve their texts in libraries and reanimate them through the arts of exegesis, commentary, and interpretation; scientists, in contrast, ignore or (worse) discard any publication more than twenty years old, and the range of their citations rarely reaches back farther than five years. Scientists may pay homage to their great forebearers—the pantheon of Copernicus, Galileo, Newton, Darwin, Einstein—but they seldom read them and almost never cite them. In contrast, humanists are still in imagined dialogue with Plato and Dante, Shakespeare and Kant, Manu and Montaigne. Scientists notoriously take up the history of their field only in their dotage; if active scientists turn a stray thought to history, then it is usually to deride the errors of their predecessors. But the humanists seem to be engaged in a perpetual séance, raising the illustrious dead in their libraries.

This opposition of the bookish humanities, guardians of memory in the library, and the hands-on sciences, discoverers of timeless truths in the laboratory and the observatory, has its roots in mid-nineteenth-century classifications of the disciplines—first in bellwether German universities, with later echoes worldwide as other coun-

---

* Max Planck Institute for the History of Science, Boltzmannstr. 22, 14195 Berlin, Germany; ldaston @mpiwg-berlin.mpg.de. I would like to thank Robert Kohler and Kathryn Olesko for editorial comments on an earlier draft of this article and my colleagues at the Max Planck Institute for the History of Science for a bracing discussion and fruitful suggestions. All translations are my own unless otherwise specified.

[1] Thomas S. Kuhn drew a similar contrast between science and art (and mathematics): "For reasons and in ways that remain obscure to me, the sciences destroy their past more thoroughly than do mathematics or the arts." Kuhn, "The Halt and the Blind: Philosophy and History of Science," *Brit. J. Phil. Sci.* 31 (1980): 181–92, on 190 n. I thank Skuli Sigurdsson for drawing my attention to this passage.

tries instituted their own versions of the Humboldtian model of teaching united with research. In an influential speech delivered at the University of Heidelberg in 1862, for example, the physicist and physiologist Hermann von Helmholtz admitted that whereas the philologist and the historian, the jurist and the theologian considered the historical dimension essential to their disciplines, the scientist (*Naturforscher*) was "strikingly indifferent to literary treasures, perhaps even to the history of his own discipline." The humanist's memory must be richly provisioned with examples and cases in order to support the "artistic brand of induction" characteristic of these disciplines, but the regularity and generality of natural laws made the cultivation of this faculty largely otiose in the sciences.[2] On the side of the humanities, philosophers and historians of the stature of Wilhelm Windelband and Wilhelm Dilthey drove home the message that one of the essential distinctions between the *Geisteswissenschaften* and the *Naturwissenschaften* was the deeply historical sensibility of the former and the timeless perspective of the latter.[3]

Yet at about the same time that Helmholtz, Windelband, and Dilthey were drawing bold lines between the sciences of memory in libraries and the sciences of natural laws in laboratories and observatories, the most advanced scientific research institutions in Europe and North America were not only purchasing state-of-the-art instruments and building new temples to science; they were locating libraries at the heart of observatories and laboratories and busily stocking them. Here is the American astronomer Benjamin Gould rhapsodizing over the Russian Observatory at Pulkova, "the El dorado of astronomers," in 1849—the "magnificent" telescope with its Frauenhofer lens; the "exquisite execution" of the Repsold great meridian circle; and, directly opposite the main entrance, the library, replete with old as well as new books: "Of the library we will merely say, that no book of value on any department of astronomy is wanting which has been obtainable since the observatory was established, and that agents are employed all over Europe, ready to avail themselves of the first opportunity of acquiring rare books as they may happen to be in the market."[4] Or consider the Institute for Physiology at the University of Berlin (established 1882), which housed, in addition to dissection rooms, workshops, lecture halls, and lodgings for the staff, a library as large as one of the main laboratories (fig. 1). Or for that matter, the 1974 plans for the renovated Cavendish Laboratory at Cambridge University, which locate a library in the Bragg Building (fig. 2). All of these establishments were considered state-of-the-art when erected, powerhouses of the most advanced scientific research. And all of them contain libraries. Scientists too are people of books, and not just brand-new ones.

---

[2] Helmholtz, "Ueber das Verhältnis der Naturwissenschaften zur Gesammtheit der Wissenschaft" (Akademische Festrede gehalten zu Heidelberg beim Antritt des Prorektorats, 1862), in Helmholtz, *Vorträge und Reden*, 5th ed. (Braunschweig, 1903): 158–85, on 166, 171, 175–8.

[3] Windelband, "Geschichte und Naturwissenschaft" (Rektoratsrede, Universität Strassburg, 1894), in *Strassburg Universität: Gelegenheitsschriften, 1892–96* (Strasbourg, 1896): 15–41; Dilthey, "Der Aufbau der geschichtlichen Welt in den Geisteswissenschaften," in *Abhandlungen der Preußischen Akademie der Wissenschaften, Philosophisch-Historische Klasse* (Berlin, 1910): 1–123.

[4] Gould, "The Observatory at Pulkowa," *North American Review* 69 (1849): 143–62, on 143, 153, 155, 156. Among the rare books acquired by the Pulkova library were Johannes Hevelius's *Machina Coelestis* (1673) and unpublished manuscripts of Kepler, hardly the newest literature in the field circa 1850. See also Simon Werrett, "The Astronomical Capital of the World: Pulkova Observatory in the Russia of Tsar Nicholas," in *The Heavens on Earth: Observatories and Astronomy in Nineteenth-Century Science and Culture*, ed. David Aubin, Charlotte Bigg, and H. Otto Sibum (Durham, N.C., 2010), 33–57.

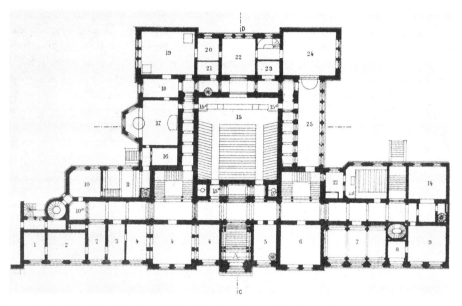

Figure 1. *Institut Physiologique de Berlin, plan du rez-de-chaussée. Nr. 7 is the library and reading room; nr. 17 is a physiological laboratory. Adolphe Wurtz,* Les hautes études pratiques dans les universités d'Allemagne et d'Autriche-Hongrie (Paris, 1882), *vol. 2, planche IX. Reprinted by permission of the Max Planck Institute for the History of Science, Berlin.*

But what do they do with them? Books alone do not historians make. What is the historical consciousness of the sciences? It is hardly news to historians of science that the sciences have their own histories, and of several sorts. There are the intrinsically historical sciences, such as geology and evolutionary biology, which study natural processes fully as sensitive to context and contingency as any in human history across epochs and eons of billions of years.[5] There are the well-funded jubilees, most recently the Darwin Year, for which scientific societies solemnly gather to commemorate the life and works of some past titan.[6] There are the anecdotes, biographies, autobiographies, and lore that make up the long-lived mythology of scientific disciplines, praising heroes and blaming villains and more generally exemplifying core values like perseverance in adversity (as emblematized by Marie Curie stirring those vats of pitchblende) or studied disdain for social convention (as celebrated by almost all Einstein lore).[7] Yet these are not the genres that fill the shelves of working

[5] Martin J. S. Rudwick's *Bursting the Limits of Time: The Reconstruction of Geohistory in the Age of Revolution* (Chicago, 2005) provides the most comprehensive overview of how these sciences became historicized.

[6] On the history of such rituals of public commemoration and their spread from religious to secular occasions, see Winnfried Müller, ed., *Das historische Jubiläum: Genese, Ordnung und Inszenierungsgeschichte eines institutionellen Mechanismus* (Münster, 2004), and Paul Münch, ed., *Jubiläum, Jubiläum: Zur Geschichte öffentlicher und privater Erinnerung* (Essen, 2005). To my knowledge, there is no monograph devoted exclusively to the history of such anniversary commemorations in science (as opposed to eulogies and obituaries).

[7] This genre ultimately derives from Diogenes Laertius's *Lives of the Philosophers*; see Robert Goulet, *Études sur les vies des philosophes dans l'antiquité tardive* (Paris, 2001). For the history of

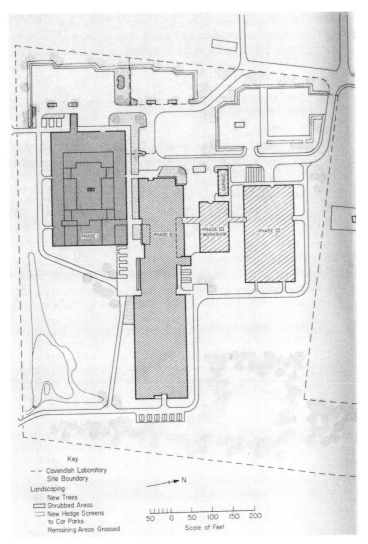

Figure 2. *Site plan of the New Cavendish. The Bragg Building (Phase II) is the middle structure. J. G. Crowther,* The Cavendish Laboratory, 1874–1974 *(New York, 1974), 424. Reprinted by permission of the University of Sussex Library.*

scientific libraries. There is another kind of historical consciousness—and historical practice—at work in at least some of the sciences. The way in which it works is the subject of this article: history *in* science.

By no means all sciences look to the past (or the future) with keen attentiveness. Moreover, the historicity of those that do is of different kinds. For some sciences,

---

science specifically, see Michael Shortland and Richard Yeo, eds., *Telling Lives in Science: Essays on Scientific Biography* (Cambridge, 1996), and Thomas Söderqvist, ed., *The History and Poetics of Scientific Biography* (Aldershot, 2007).

such as astronomy, geology, demography, and meteorology, the superhuman time-scale of the phenomena under investigation and the extreme difficulty of discerning subtle correlations dictate the careful preservation and consultation of past observations, from ancient Babylonian star catalogs to medieval weather diaries to parish church records of births and deaths. Without the treasures stored up in libraries and archives, there would be no way to discern long-term trends in, say, climate or human mortality.

For other sciences, such as botany or zoology, the stability of the objects of inquiry depends crucially on a long disciplinary memory. The theoretical basis of the classifications of plants and animals has changed dramatically and repeatedly in the past three centuries, from morphology to phylogeny to cladistics to genetics. For just that reason, naturalists since the eighteenth century have attempted to standardize nomenclature and weld names of species permanently to the same objects (and since the early twentieth century, even to the same, unique "type specimens"[8]), no matter how a species may be defined.[9] The annotations made upon the herbarium sheet by past botanists who have studied a particular specimen are scrupulously preserved, along with the object itself (fig. 3); botanists conduct historical research to find out who collected which specimen where and whether the plant flattened on the herbarium sheet is identical with the one held in the hand of whoever made the initial classification. Linnaeus's herbarium, repository of many type specimens avant la lettre, is still a research tool in active use by botanical systematists.[10] Finally, almost any science can suddenly if briefly turn historical when confronted by a novel or anomalous phenomenon, which immediately prompts the question, Has this ever happened before, and if so, when, where, and how? History—more specifically, archiving data and specimens[11]—is integral to these sciences, which might collectively be called the sciences of the archive.

The historical consciousness of the sciences of the archive itself has a history. How long a history depends on the archiving practices of the various sciences: millennia for astronomy; only a few centuries for demography and botany. The intrinsically historical sciences, such as evolutionary biology and geology, do not necessarily qualify as sciences of the archives. Although fossils have been collected for centuries, and for the most diverse ends—to hang from church ceilings, to adorn *Wunderkammern*, to display nature's plastic powers, to study now-extinct species—it was only when they began to be collected and preserved systematically, with an eye toward future users,

---

[8] Lorraine Daston, "Type Specimens and Scientific Memory," *Crit. Inq.* 31 (2004): 153–82.

[9] Geoffrey C. Bowker, "The Game of the Name: Nomenclatural Instability in the History of Botanical Informatics," in *Proceedings of the 1998 Conference on the History and Heritage of Science Information Systems*, ed. Mary Ellen Bowden, Trudi Bellardo Hahn, and Robert V. Williams (Medford, N.J., 1999), 74–83. For the latest developments see Rebecca Ellis, "Rethinking the Value of Biological Specimens: Laboratories, Museums and the Barcoding of Life Initiative," *Museum and Society* 6 (2008): 172–91.

[10] Charlie Jarvis, "A Concise History of the Linnean Society's Linnaean Herbarium, with Some Notes on the Dating of the Specimens It Contains," in "The Linnaean Collections," special issue no. 7, ed. B. Gardiner and M. Morris, *Linnean* 23 (2007): 5–18. The Linnaean Herbarium, already preserved in a strong room in London, is now being digitized, a further step in archiving its contents, as described on the website of the Linnean Society, http://www.linnean.org/index.php?id=326 (accessed 30 August 2011).

[11] Although the focus of this article will be the storage and retrieval of data, the practices of collecting objects in the geo- and biosciences constitute an important parallel; see Bruno Strasser, "Laboratories, Museums, and the Comparative Perspective: Alan A. Boyden's Seriological Taxonomy, 1925–1962," *Hist. Stud. Nat. Sci.* 40 (2010): 149–82, and Strasser's article in this volume.

Figure 3. Carduus sp. *(LINN 966.34). Digitalized specimen from Linnaeus's herbarium, including not only the plant itself but also annotations by Johan Leche, a professor of medicine and botanist who sent Linnaeus observations and plant specimens. The Linnean Society of London Collections Online, http://www.linnean-online.org/9833/ (accessed 1 September 2011). Copyright Linnean Society of London. Reprinted by permission of the Linnean Society of London.*

that paleontology became a science of the archives. The idea of taking and storing core samples from various locations on the Earth as a reference collection for future geologists is another (and still more recent) archival practice.[12] It is not just a deep time dimension in the phenomena investigated that is the hallmark of the sciences of the archives, but rather the practice of storing up materials for future investigators. One historical precondition for such archival practices is the sense of a community of inquirers—not necessarily a discipline, much less an institutionalized one—that extends into the future as well as the past. And this sense in turn often goes hand-in-hand with an awareness of the enormity of the investigative task, which demands a legion of inquirers to do it justice.

Like all stereotypes, the one of humanists worshiping their ancestors, faces turned reverently toward the past, and scientists committing disciplinary patricide, gaze riveted on the future, contains elements of truth. And like all stereotypes, it distorts and occludes. It freeze-frames a nineteenth-century contrast of disciplines steeped in history versus those oblivious or even hostile toward it, of the curators of the library and

---

[12] National Research Council, *Geoscience Data and Collections: National Resources in Peril* (Washington, D.C., 2002).

the museum versus those of the laboratory and the observatory. My aim here is to undermine these habitual oppositions—and in so doing, to challenge reigning classifications of knowledge that still mold the ways in which the histories of science and scholarship are written.

COLLECTIVE EMPIRICISM: ENDLESS DATA, INFINITE LABOR

Pace the stereotypes, the library has never ceased to be a site of scientific knowledge, alongside the laboratory and the observatory—often literally alongside, as architectural plans of research institutions from the seventeenth century to the present reveal.[13] Its enduring and indispensable presence at the heart of scientific endeavor bears witness to the deep historicity of certain sciences. The library in these sciences is literal, not metaphorical. It is a repository of not only books but also manuscripts, registers, notebooks, and correspondence—an archive. It is the product of human labor, not natural processes (in contrast to, e.g., fossils metaphorically understood as a "record" of past life forms: "the many volumes of botany representing in the same quarry the oldest library in the world," as the eighteenth-century French naturalist Antoine de Jussieu described his discovery of plant imprints in the stones of the Lyonnais[14]). What distinguishes the sciences of the archive from other sciences is not just a historical dimension of the phenomena they study nor even the practice of taking, making, and keeping data. Rather, it is practices of collection, collation, and preservation conceived as an intrinsically collective undertaking—and one that extends into both past and future. The sciences of the archive are either too grand in scale or simply too much work for an individual or even a generation.

The history of collective empiricism in the sciences varies by discipline: astronomy pioneered its practices in the ancient world; medicine organized itself into an exchange network in the early modern period; chemistry was a relative latecomer. René Descartes was perhaps the last major figure to fantasize about science conducted in splendid solitude, alone in a bare study without books, in search of "truths that can be deduced from things known and ordinary," as opposed to the sort of knowledge that "required, first of all, to have researched all the plants and stones that come from the Indies, to have seen the phoenix, and in short to overlook nothing of all that is strangest in nature."[15] But even Descartes recanted in his *Discours de la*

[13] Anthony Grafton has described the importance of libraries for early modern science, but there are very few studies about their role thereafter (in contrast to the large literature on scientific museums or on the history of libraries more generally); Grafton, "Libraries and Lecture Halls," in *Early Modern Science*, ed. Katharine Park and Lorraine Daston, vol. 3 of *The Cambridge History of Science* (New York, 2006), 238–50. A notable exception is the excellent introductory essay on the eighteenth-century scientific library in Marco Beretta, *Bibliotheca Lavoisieriana: The Catalogue of the Library of Antoine Laurent Lavoisier* (Florence, 1995), 13–58. Geoffrey C. Bowker's *Memory Practices in the Sciences* (Cambridge, Mass., 2005) is an essential introduction to metaphors and practices in selected modern sciences and highly suggestive concerning the ways in which scientists imagine data and the past.

[14] Jussieu, "Examen des causes des impressions des plantes marquées sur certaines pierres des environs de Saint-Chaumont dans le Lionnais," *Mémoires de l'Académie Royale des Sciences* (Paris, 1718), 287–97, on 289.

[15] Descartes, "La recherche de la verité par la lumière naturelle" (comp. 1628/9?; pub. in Dutch 1684, Latin 1701), in *Oeuvres de Descartes*, ed. Charles Adam and Paul Tannery (Paris, 1966), 10:459–532, on 503. The dating of this unfinished manuscript is uncertain, but the most recent critical edition argues for a date before 1637; Ettore Lojanco with Erik Jan Bos, Franco A. Meschini, and Francesco Saita, eds., *"La recherche de la verité par la lumière naturelle" de René Descartes* (Milan, 2002).

méthode (1637), admitting that his grand project for a new natural philosophy would require "many experiments," and explained that one of his principal motivations in publishing his preliminary discoveries was "to urge men of ability to continue the work by contributing, each one according to his inclinations and abilities, to the experiments that must be made."[16] In a private letter to King James I accompanying the presentation copy of the *Novum Organum* (1620), Francis Bacon similarly hoped that even premature publication would attract recruits to "help in one intended part of this work, namely, the compiling of a natural and experimental history, which must be the main foundation of a true and active philosophy."[17] Bacon failed to win royal support, and his own ambitious project for 130 natural histories of everything from comets to sleep and dreams faltered for lack of time, money, and manpower.[18]

Nothing daunted, the fledgling northern European scientific academies founded in the mid-seventeenth century reiterated calls for volunteers far and wide to submit observations and experiments to be published in their journals: the *Miscellanea Curiosa* of the Academia Naturae Curiosorum (est. 1652, later known as the Leopoldina), the *Philosophical Transactions* of the Royal Society of London (est. 1660), and *Histoire et mémoires de l'Académie Royale des Sciences* in Paris (est. 1666). The challenge of recruiting, training, and coordinating an army of observers and experimenters, both paid and volunteer, has preoccupied the sciences ever since, as research-grant proposals from Bacon's requests to James I to the Internet *levée en masse* for the Encyclopedia of Life project bear witness.[19] Because nature is vast and labyrinthine, empiricism in the natural sciences is hugely labor intensive—and therefore collective.[20]

Of necessity collective empiricism spans continents and generations. The phenomena to be investigated are myriad and global (if not cosmic); merely to catalog them all would be the work of centuries (if not millennia). Whatever the rhetoric of firsthand observation (*sola autopsia* echoing the Protestant *sola scriptura* since the sixteenth century), the reality is that the modern empirical sciences have always and essentially depended on testimony.[21] Like historians, scientists must gather, weigh, and amalgamate the testimony of witnesses who are more or less sagacious, more or less thorough, more or less trustworthy. Depending on whether the dimension of space or time dominates in the collective empiricism practiced by a particular science, the witnesses may be contemporaries in other places or observers from other epochs. Because the bulk of scientific testimony has accumulated over decades and

---

[16] Descartes, "Discours de la méthode" (1637), in Adam and Tannery, *Oeuvres de Descartes* (cit. n. 15), 4:130–47. On the interplay of deduction and experiment in Descartes's natural philosophy, see Daniel Garber, *Descartes Embodied: Reading Cartesian Philosophy through Cartesian Science* (Cambridge, 2001), 85–110.

[17] Bacon to James I, 12 October 1620, quoted in Lisa Jardine and Alan Stewart, *Hostage to Fortune: The Troubled Life of Francis Bacon* (New York, 1998), on 438. Bacon reiterated his request more pointedly to James himself: "This comfortable beginning makes me hope further, that your Majesty will be aiding me in setting men to work for the collecting of a natural and experimental history; which is the *basis totius negotii*" (439). See also Bacon, "Epistola Dedicatoria," in *Novum Organum*, vol. 9 of *The Works of Francis Bacon*, ed. Basil Montagu (London, 1828), on 150.

[18] See the "Catalogus Historiarum Particularum" appended to the "Parasceve ad Historiam Naturalem, et Experimentalem," in Montagu, *Works of Francis Bacon* (cit. n. 17), 11:427–36.

[19] See "Help Build EOL" on the Encyclopedia of Life website, http://www.eol.org/content/page /help_build_eol (accessed 29 August 2010).

[20] Lorraine Daston and Peter Galison, *Objectivity* (New York, 2007), 19–27, 367–8.

[21] Steven Shapin, *A Social History of Truth: Civility and Science in Seventeenth-Century England* (Chicago, 1994); see also Martin Kusch and Peter Lipton, "Testimony: A Primer," *Stud. Hist. Phil. Sci.* 33 (2002): 209–17.

often centuries, collective empiricism must always be in part historical, albeit to different degrees: the data reside in the library as well as in the laboratory and the observatory. There is nothing new about data mining per se.

The word *data* is, however, a slippery one; "the givens" (originally, those things that are given in a Euclidean geometric proof or construction) are indeed given, but by whom, and to what end? Historians have evolved a princess-and-the-pea sensibility concerning what their own archives give them: what was selected for preservation (and what not) and why; how data were organized both physically and conceptually; who had access to them; what purposes they were meant to serve.[22] Scientists also cross-examine their current data: the circumstances under which they were collected; the instruments used; the reliability of the reporter; the possible sources and size of error; the robustness of the phenomenon. Do they also query past data? To do so is immediately to plunge into the same quandaries that confront historians and to run the same risks of anachronism. For example, both the social historian and the historical demographer must probe the local circumstances under which parish records were kept and determine where systematic and strategic omissions are likely.[23] To use past data is willy-nilly to become a historian.

Unlike historians, however, scientists occasionally construe their own present data as the past data for future scientists: they become archivists. They self-consciously create the archives for an imagined community of disciplinary descendants, just as they embrace past observers in an imagined disciplinary lineage.[24] Just how far this sense of a community of data inherited and passed on stretches is variable, according to the timescale of the science. In astronomy, for example, it has extended across millennia in both directions, from ancient Mesopotamian star watchers in 1200 BCE to the astronomers of the year 3000 CE. This Janus-faced perspective of the scientific archive, reaching back into the past and forward into the future, sets it apart from the traditional historical archive.[25]

How does this vision of a century-spanning imagined community inform the creation and use of scientific archives of data? What are the ways in which present and past data, first- and secondhand experience, come to be spliced together? If the experiment is the reigning practice of the laboratory and the observation that of the

[22] For the history of science specifically, see Michael Hunter, ed., *Archives of the Scientific Revolution: The Formation and Exchange of Ideas in Seventeenth-Century Europe* (Woodbridge, 1998), and Ann Blair and Jennifer Mulligan, eds., "Toward a Cultural History of the Archives," special issue, *Archival Science* 7, no. 4 (2007).

[23] E.g., in the case of the frequency of infanticide; for discussions of the impact of shifting definitions and sanctions on records, see Richard van Dülmen, *Frauen vor Gericht: Kindermord in der frühen Neuzeit* (Frankfurt, 1991); Adriano Prosperi, *Die Gabe der Seele: Geschichte eines Kindermordes*, trans. from Italian by Joachim Schulte (Frankfurt, 2007); Keith Wrightson, "Infanticide in European History," *Criminal Justice History* 3 (1982): 1–20.

[24] I borrow the evocative phrase "imagined community" from Benedict Anderson, *Imagined Communities: Reflections on the Origin and Spread of Nationalism* (London, 1983). Anderson's prime example is the modern nation, but the term might be applied with equal justice to the republic of letters or the scientific community, both of which embrace members who may never meet face-to-face and who may even remain anonymous to one another. The imaginary dimension is if anything even more pronounced in these conjured confraternities of the learned, since they lack even the territorial concreteness of the nation.

[25] This may be changing, though, as archivists overwhelmed by the sheer volume of materials try to divine what future historians will be interested in so as to develop criteria for what should and should not be preserved—or simply abandon the process to random sampling; K. J. Smith, "Sampling and Selection: Current Policies," in *The Records of the Nation: The Public Record Office, 1838–1988; the British Record Society, 1888–1988*, ed. G. H. Martin and Peter Spufford (Woodbridge, 1990), 49–59.

observatory, what goes on in the scientific library? The answers to these questions deserve volumes. What I offer here are some examples, drawn from several disciplines and time periods, that suggest the richness of the topic—and the magnitude of its implications for how we understand science and its histories. The next section describes how early modern genres of empirical scientific inquiry, such as *historiae* and *observationes*, tended to blur the boundary between words and things by applying similar practices to book learning and the study of nature. I then examine in detail how reading and observing practices merged in early modern natural history, creating a hybrid hermeneutics of first- and secondhand experience. The article concludes with some reflections on the conceptions of history and imagined community that govern the taking, making, and keeping of data in the scientific archive.

GRANARIES AND TREASURIES OF KNOWLEDGE

The seventeenth-century *Kunstkammer* of the Regensburg iron-dealer family Dimpel exhibits the usual profusion of *naturalia* and *artificialia*: cannons (paying homage to the family business), exotic shells, globes and instruments, Chinese porcelain, paintings—and books, stacked on the table and arrayed on the shelf (fig. 4). The combination of books with other collectibles in the same space was not exceptional in the early modern period,[26] even in the large princely libraries that later became the core of nineteenth-century national libraries. Sometimes the proximity of books and specimens was even closer, alarmingly so: the Paris Académie Royale des Sciences, housed in the Bibliothèque du Roi until 1699, conducted dissections and vivisections in the library, "fitting a table with straps to restrain live subjects."[27]

The promiscuous mixture of words and things in these early modern sites of science is almost as disconcerting to modern eyes as the studied miscellany of art and nature in the Wunderkammern. Nor would the incongruity be much softened by paying attention to early modern pronouncements on the immiscibility of book learning and the reformed study of nature. Many sixteenth- and seventeenth-century manifestos polemically opposed the knowledge of words to that of things: from Paracelsus's railings against learned doctors to Royal Society apologetics, from Francis Bacon's critique of the Idols of the Marketplace to Gottfried Wilhelm Leibniz's fascination with Chinese pictograms as thing-like words—the acute sense of a chasm yawning between words and the things they purportedly represented was ubiquitous. On the side of words, schemes for artificial languages, like those of Amos Comenius or John Wilkins, and on the side of things, new forms of disciplined experience, such as collecting and experimenting, were devised to bridge this gap. Yet what was explicitly conceived as a challenge of representation—how best to mirror the world of things faithfully in both word and image—was implicitly also one of practice: how were bookish scholars trained to handle, appraise, and coin words to learn to manipulate, assay, and order things?

As the architectural juxtaposition of books, natural history collections, and even

---

[26] For other examples, see Beretta, *Bibliotheca Lavoisieriana* (cit. n. 13), 20–7. For a sense of the early modern connections between research in books and on things, see Anthony Grafton, "Where Was Salomon's House? Ecclesiastical History and the Intellectual Origins of Bacon's *New Atlantis*," in *Worlds Made by Words* (Cambridge, Mass., 2009), 98–113.

[27] Alice Stroup, *A Company of Scientists: Botany, Patronage, and Community at the Seventeenth-Century Parisian Royal Academy of Sciences* (Berkeley and Los Angeles, 1990), 39.

Figure 4. *Joseph Arnold,* Cabinet of Art and Rarities of the Regensburg Iron Dealer and Mining Family Dimpel *(1668). Copyright Ulmer Museum, Ulm, Germany; photo: Oleg Kuchar, Ulm. Reprinted by permission of the Ulmer Museum.*

dissecting rooms and chemical laboratories suggests, the reformed natural philoso-phers of the early modern period retooled reading into observing. Thanks to the work of Ann Blair, Anthony Grafton, Gianna Pomata, Nancy Siraisi, Brian Ogilvie, Adrian Johns, and other historians of early modern science and scholarship, we know a great deal about how Renaissance humanist techniques like the making of compendia and the keeping of commonplace books merged with the collection of observationes, the assembling of historiae, and an eagle eye for minute differences—whether in collat-ing manuscripts of Cicero or classifying plant species.[28] Two examples, both central

[28] There is a large literature on early modern reading practices, but still seminal is Roger Chartier, *Lectures et lecteurs dans la France d'ancien régime* (Paris, 1987). On learned reading, see Guglielmo Cavallo and Chartier, eds., *A History of Reading in the West* (1997), trans. Lydia G. Cochrane (Cambridge, 1999), especially the essays by Jacqueline Hamesse, "The Scholastic Mode of Reading," 103–19, and Grafton, "The Humanist as Reader," 179–212; Bernhard Fabian, "Der Gelehrte als Leser," in *Buch und Leser,* ed. Herbert G. Göpfert, Wolfenbüttler Arbeitskreis für Geschichte des Buchwesens (Hamburg, 1977), 48–88; Blair, "Humanist Methods in Natural Philosophy: The Commonplace Book," *J. Hist. Ideas* 53 (1992): 541–51; Blair, *The Theater of Nature: Jean Bodin and Renaissance Science* (Princeton, N.J., 1997); Johns, "Reading and Experiment in the Early Royal Society," in *Reading, Society and Politics in Early Modern England,* ed. Kevin Sharpe and Steven Zwicker (Cambridge, 2003), 244–71; Blair, "Note-Taking as an Art of Transmission," *Crit. Inq.* 31 (2004): 85–107; Blair, "Scientific Reading: An Early Modernist's Perspective," *Isis* 95 (2004): 64–74; Blair, *Too Much to Know: Managing Scholarly Information before the Modern Age* (New Haven, Conn., 2010). On the use of such techniques in the compilation of observationes, see Pomata, "Observation Rising: Birth

to early modern learned empiricism, show how fluid the boundary between the study of books in a library and the study of the book of nature could be during this period: the compilation of a historia and the making of an observatio.

The historia, from the Greek work for enquiry, was traditionally opposed and subordinated to *philosophia* (or poetry) following Aristotle's contrast of the two genres in the *Poetic*s: "Hence poetry is something more philosophic and of graver import than history, since its statements are rather of the nature of universals, whereas those of history are singulars."[29] Historia in the ancient, medieval, and early modern sense did not necessarily involve a temporal dimension (a usage still fossilized in the term *natural history*). As Aristotle's derogatory comparison indicates, the defining characteristic of historia was that it dealt in particulars—and was therefore at best a means to the end of universal generalizations. In the course of the sixteenth century, however, the epistemic prestige of both civil and natural history was on the rise, and the sturdy particulars of historia were increasingly regarded as more reliable than the airy universals of philosophia.[30] By the early seventeenth century, natural history seemed to some the foundation upon which natural philosophy must be rebuilt. In his programmatic *Distributio Operis* (1620), Bacon saw "no hope therefore of greater advancement and progress unless by some restoration of the sciences. But this must commence entirely with natural history."[31] The historia was now understood to be an omnium-gatherum of all that was known on a subject, following the example of Pliny's encyclopedic *Historia naturalis*, which boasted in its preface that it included 20,000 items:[32] the winds, life and death, quicksilver, smell and odors. Bacon's ambitious "Catalogue of Particular Histories" was to be executed by "the joint application of others," but he began several model histories on his own, including the unfinished *Sylva Sylvarum*, in order to recruit and guide collaborators.[33] He was probably drawing upon models of historiae compiled in the sixteenth century, mostly by doctors, who had already recognized that only a collective could accomplish such works.[34]

These collectives were largely virtual in nature, composed of correspondents and,

---

of an Epistemic Genre, 1500–1650," in *Histories of Scientific Observation*, ed. Lorraine Daston and Elizabeth Lunbeck (Chicago, 2011), 45–80; and on the overlap between Renaissance philological and botanical practices, see Ogilvie, *The Science of Describing: Natural History in Renaissance Europe* (Chicago, 2006).

[29] Aristotle, *Poetics*, 1451b5–7, in *The Complete Works of Aristotle*, 2 vols., ed. Jonathan Barnes (Princeton, N.J., 1984), 2:232. Still seminal for understanding the convoluted meanings of historia in early modern Europe is Arno Seifert's *Cognitio Historica: Die Geschichte als Namengeberin der frühneuzeitlichen Empirie* (Berlin, 1976); for the sense of historia discussed in this article, see esp. 116–89.

[30] Paula Findlen, "Natural History," in Park and Daston, *Early Modern Science* (cit. n. 13), 435–68; Gianna Pomata and Nancy G. Siraisi, eds., "Introduction" and "The Ascending Fortunes of Historia," in *Historia: Empiricism and Erudition in Early Modern Europe* (Cambridge, Mass., 2005), 1–40.

[31] Bacon, "The Distribution of the Work," in Montagu, *Works of Francis Bacon* (cit. n. 17), 14:14–24, on 20.

[32] Pliny the Elder, "Preface," in *Natural History*, 10 vols., trans. H. Rackham, Loeb Classical Library (Cambridge, Mass., 1989), 17. On the many early modern editions and reception of Pliny, see Charles G. Nauert Jr., "Humanists, Scientists, and Pliny: Changing Approaches to a Classical Author," *Amer. Hist. Rev.* 84 (1979): 72–85.

[33] Bacon, "A Preparation for a Natural and Experimental History" (1620), in Montagu, *Works of Francis Bacon* (cit. n. 17), 14:213–6, on 215. Of the 130 historiae listed in Bacon's catalog, over thirty were on medical topics.

[34] Gianna Pomata, "*Praxis Historialis*: The Uses of Historia in Early Modern Medicine," in Pomata and Siraisi, *Historia* (cit. n. 30), 105–46.

above all, authors of books, both ancient and modern. Despite a great deal of speech-ifying about the superiority of firsthand over secondhand experience and the dangers of bowing to ancient authorities, the lion's share of the work of putting together a his-toria, that bulwark of early modern empiricism, was done in a well-stocked library. Against the background of a rhetoric that trumpeted the virtues of things over words, experience over erudition, this is perhaps shocking, but it should not be surprising. The ambitions of early modern historia, which aimed to embrace all the phenomena of the universe, could hardly be realized by a single investigator or even by a legion of them, no matter how long and diligently they labored. Nothing less than the collec-tive experience of all of humanity would be adequate to the task. And the repository of that *longue durée* experience was the library, not the observatory, the laboratory, or even the museum.

Take the case of Bacon's own historia, the *Sylva Sylvarum*. Of the thousand items (ten "centuries" of one hundred items each) assembled therein, at least a third were taken from other sources, ranging from Aristotle and Pliny to Giambattista della Porta and Girolamo Cardano—a proportion so large that a modern editor of Bacon's manuscripts felt obliged to defend him against charges of plagiarism.[35] Bacon's sur-viving manuscript notes for the project demonstrate how particulars derived from books and from experiment were literally interleaved: the marginalia that annotate the "experiments" to be performed sometimes reference a source (e.g., "Aristotle"), sometimes are reported as testimony ("It was reported by a sober Man, that an Artifi-cial Spring may be made thus"), and sometimes as personal observation ("Done Oc-tob. 10 put a green Apple into Hay, and leave another of the same Apple to compare with it, and see how much sooner the one will sweeten and ripen than the other").[36] Such intermingling of reading notes and hearsay with experiments and observations does not imply that Bacon was indiscriminate or a traitor to his own empiricist prin-ciples; rather, it suggests that reading and observing were activities almost always pursued in tandem—and together with others, both the quick and the dead.[37]

Other examples of seventeenth-century historiae bear out this claim. The Aca-demia Naturae Curiosorum, established in Schweinfurt in 1652 as an academy of learned doctors (mostly municipal physicians practicing in the Holy Roman Empire), resolved in its early statutes that each member would earn his stripes (signified by a ring and a special nickname) and further the advance of knowledge by writing a historia on the medicinal properties of some natural substance: "To be researched is the name of the substance under investigation, the synonyms, the development, place of origin, the differences, the species, the selection, the effects of the whole as well as its parts, the usual and chemical medications that can be prepared from it, both simple and mixed. . . . To this end, he [the author] will draw upon recognized authors,

---

[35] Graham Rees, "An Unpublished Manuscript of Francis Bacon: *Sylva Sylvarum* Drafts and Other Working Notes," *Ann. Sci.* 38 (1981): 377–412, on 388–9.

[36] [Francis Bacon/William Rawley], British Library St. Pancras Additional MSS 38, 693, fols. 31r, 32r, 43v. On the provenance and identification of the manuscript, see Rees, "Unpublished Manu-script" (cit. n. 35), 378–81. See also Richard Yeo, "Between Memory and Paperbooks: Baconianism and Natural History in Seventeenth-Century England," *Hist. Sci.* 45 (2007): 1–46.

[37] On the influence on the early Royal Society of Bacon's "granary" model of archiving scientific observations, see Mordechai Feingold, "Of Records and Grandeur: The Archive of the Royal Society," in Hunter, *Archives of the Scientific Revolution* (cit. n. 22), 171–84.

his own observations, and credible reports and perceptions of others."[38] A few works were completed and published along these lines, offering copious erudition and experience intermingled.[39] The prerequisites for fulfilling this requirement were leisure and a substantial personal library, like that of the first president, Johannes Laurentius Bausch.[40] Neither was readily available to the intended membership of *Stadtphysici*; fourteen of the twenty-three candidates initially invited to join declined.[41]

The expedient adopted by the Academia Naturae Curiosorum to overcome these impediments to recruitment might at first seem a decisive shift from bibliographic to empirical research. Instead of composing a weighty *historia*, physicians and other learned persons might instead send in short *observationes* on medical and natural historical topics, which would then be collected and published in the Academia's journal *Ephemerides* (sometimes titled *Miscellanea Curiosa*), with full credit given to the correspondents who submitted the observations. The call was issued to Europe at large, promising that correspondents would be treated "in a courteous, honorable, and friendly manner" and that they would enjoy fame and the gratitude of mankind without having to take the time and trouble to write a big book.[42] As academician Philipp Jakob Sachs von Lewenhaimb explained in his letters to Chemnitz town physician Christian Friedrich Garmann, even those devotees of things "physico-medical" not in the possession of "medical libraries" could participate.[43]

Yet although many of the short, numbered *observationes* were firsthand reports by named contributors, they were interspersed with items excerpted from other accounts: for example, "Observatio XXXIX," sent in by a Dr. Joachim Elsner, quotes an Italian account of male and female conjoint twins, modestly separated by a placenta, along with a Latin commentary that the cohabitation of brother and sister is contrary to natural law (in the legal sense: *jus naturae*, not *lex naturalis*) (fig. 5).[44] Throughout the volume are strewn scholia relating this or that isolated observation to similar phenomena reported in ancient and modern sources.[45] Whatever the

---

[38] [Academia Naturae Curiosorum], *Leges* (III, IV) [1652], quoted in Uwe Müller, "Die Leopoldina unter den Präsidenten Bausch, Fehr und Volckamer, 1652–1693," in *350 Jahre Leopoldina: Anspruch und Wirklichkeit*, ed. Benno Parthier and Dietrich von Engelhardt (Halle, 2002), 45–93, on 50. The complete text of the 1662 statutes is reproduced in Andreas Büchner, *Academiae Sacri Romani Imperii Leopoldino-Carolinae Naturae Curiosorum Historia* (Halle/Magdeburg, 1755), 187–97.

[39] E.g., Philipp Jacob Sachs von Lewenhaimb's *ΑΜΠΕΛΟΓΡΑΦΙΑ, sive Vitis Viniferae* (Leipzig, 1661), which runs to 671 pages and includes a "catalogus authorum" of works cited.

[40] On Bausch's large library of some six hundred books, see Uwe Müller, Claudia Michael, Michael Bucher, and Ute Grad, *Die Bausch-Bibliothek in Schweinfurt*, Acta Historica Leopoldina 32 (Halle, 2004).

[41] Wieland Berg, "Die frühen Schriften der Leopoldina—Spiegel zeitgenössischer Medizin," *NTM—Schriftenreihe Geschichte der Naturwissenschaften, Technik, Medizin* 22 (1985): 67–76, on 69.

[42] "Epistola Invitatoria ad Celeberrimos Medicos Europae," *Miscellanea Curiosa Medico-Physica Academiae Naturae Curiosorum* 1 (1670): n.p. The observationes (160 in toto) are prefaced with a "syllabus" listing the names of the thirty-six contributors (only ten of whom were members of the Academia Naturae Curiosorum).

[43] Sachs von Lewenhaimb to Garmann, 4 July 1670, quoted in Müller, "Die Leopoldina" (cit. n. 38), on 62. The correspondence makes clear that Sachs von Lewenhaimb modeled the *Ephemerides* on the *Philosophical Transactions* of the Royal Society of London and the French *Journal des savants*.

[44] Joach[im] Georg Elsner, "Observatio XXXIX," *Miscellanea Curiosa* 1 (1670): 119–27. The Italian report is quoted on 127: "La sagicissima Natura in una particular membrana separata dal Maschio conserva la femmina."

[45] Most but not all of the observationes are followed by a scholion that relates them to other observations and texts and sometimes attempts an explanation of the reported phenomenon. The learned scholion was already a fixture of the medical *curationes* and observationes literature by the latter

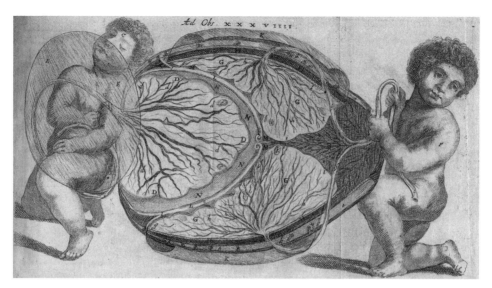

Figure 5. *Male and female twins modestly holding a placenta between them. Elsner, "Observatio XXXIX" (cit. n. 44), 127. Reprinted by permission of the Max Planck Institute for the History of Science, Berlin.*

bibliographic resources of the contributors, the editors saw it as their role to knit the firsthand observations submitted to them into a web of literature stretching from the latest number of the *Philosophical Transactions* all the way back to Galen and Hippocrates.[46]

And for the most Hippocratic of reasons: ars longa, vita brevis. As Bausch explained in the founding document of the Academia Naturae Curiosorum, the works of God in the vegetable, mineral, and animal realms were "innumerable, and to such an extent, that the lifetime of a single person does not suffice to investigate and know them precisely no matter how ardent the desire for knowledge; [but] this failing, which lies in the fact that the lifespan of an individual is too short to do justice to the innumerable natural phenomena to be researched, can perhaps be compensated for by several people banding together to work with shared dedication."[47] For so enormous a task no academy—not even the entire republic of letters—was adequate. Observations and experiments, especially of rare phenomena like comets or monstrous births, were too precious to ignore, however long ago and far away and therefore suspect their origins might be. Early modern compendia of these valuable items were often referred to as *thesauri*, "treasuries."[48] This was a fortiori the case for astronomical observations, which were difficult and expensive to make, and often tracked pe-

half of the sixteenth century; see Pomata, "Observation Rising" (cit. n. 28), 56. The early *Histoires* published by the Paris Académie Royale des Sciences served much the same contextualizing purpose for the *Mémoires*.

[46] There are obvious parallels between these compilation techniques and those used by humanists composing florilegia, commonplace books, and other collections of texts (complete with complaints about the endless labor involved); Blair, *Too Much to Know* (cit. n. 28), 173–229.

[47] Johann Laurentius Bausch, "Epistola Invitatoria" (1652), quoted in Müller, "Die Leopoldina" (cit. n. 38), 49–50.

[48] Pomata, "Observation Rising," on 56; see also Blair, *Too Much to Know*, on 113–6 (Both cit. n. 28).

riodic phenomena with cycles that were decades, centuries, or even millennia long. When, for example, Parisian astronomer and academician Gian Domenico Cassini I observed a strange light in the constellations of the zodiac on March 18, 1683, his first impulse was to survey the literature from Anaxagoras in antiquity to his own observations of 1668 to find references to similar phenomena.[49] To a surprising degree, early modern empirical inquiry was an archival science.

But not in the usual early modern sense of the word *archive*: medieval and early modern archives were usually institutional (pertaining to, e.g., the church, state, or municipality) or familial and largely legal in character. They preserved edicts, genealogies, court records, and, above all, titles and deeds to property, to be produced should dynastic legitimacy, ownership, tax revenues, or other rights be challenged.[50] When, for example, Leibniz foraged in the archives in the service of the House of Braunschweig-Lüneburg, it was in order to establish its credentials as the legitimate successor of the Guelf dynasty.[51] Early modern archives were bastions of authenticity, places of proofs and pedigrees. In contrast, the language used to describe what would now be called scientific archives was that of stored-up riches: "granaries," "warehouses," "treasuries." Collections of historiae and observationes were precious repositories (another favored early modern word[52]) to be drawn upon to make comparisons, extend inductions over past similar cases, and support generalizations. Like the diplomatic archives plumbed by nineteenth-century historians of the Ranke school, the sixteenth- and seventeenth-century scientific archives were sites of discovery and serendipity.

They were also provisions laid up for future inquirers. Astronomers and meteorologists, whose dependence on observational records stretched back into earliest antiquity,[53] were particularly conscious of the value of such legacies (which were often considered family property, handed down father to son, as in the case of the Cassini dynasty at the Paris Observatory). Tycho Brahe allowed only a privileged few access to his hoard of astronomical observations; the legal disputes concerning their ownership after his death were notorious.[54] The French astronomer Joseph-Nicholas Delisle traveled all over Europe buying up the books, observation journals, manuscripts, and correspondence of defunct astronomers, which he sold to the French crown in exchange for handsome annuities for himself, his secretary, and his assistant Charles Messier.[55] Even rough, badly made observations were considered valuable enough to be preserved for posterity. Jacques-Dominique Cassini IV pronounced

---

[49] Cassini, "Découverte de la Lumiere Celeste qui paroist dans le Zodiaque," *Mémoires de l'Académie Royale des Sciences, 1666–99*, 11 vols. in 14 (Paris, 1733), 8:179–278.

[50] For an overview, see Patrizia Angelucci, *Breve storia degli archivi e dell'archivistica* (Perugia, 2008), 29–58.

[51] Maria Rosa Antognazza, *Leibniz: An Intellectual Biography* (Cambridge, 2009), 230–3.

[52] On the etymology and early modern associations of *repertorium*, see Blair, *Too Much to Know* (cit. n. 28), 120.

[53] Daryn Lehoux, *Astronomy, Weather, and Calendars in the Ancient World* (Cambridge, 2007), 56–65, 101–13; on medieval astrometeorological observations, see Katharine Park, "Observation in the Margins, 500–1500," in Daston and Lunbeck, *Histories of Scientific Observation* (cit. n. 28), 15–44.

[54] Brahe, *Tycho Brahe's Description of His Instruments and Scientific Work* [*Astronomiae Instauratae Mechanica*, 1598], trans. and ed. Hans Raeder, Elis Strömgren, and Bengt Strömgren (Copenhagen, 1946), 108–10; John Robert Christianson, *On Tycho's Island: Tycho Brahe, Science, and Culture in the Sixteenth Century* (Cambridge, 2000), 302–3.

[55] E. Doublet, *Correspondance échangée de 1720 à 1739 entre l'astronome J.-N. Delisle et M. de Navarre* (Bordeaux, 1910), 7. The riches of Delisle's collection have since been dispersed across several archives, but some sense of their contents can be derived from the partial inventory still held by

some loose sheets of astronomical and meteorological observations made by the academician Sédileau in 1691–3 "uncertain" and "in rather bad order" but nonetheless urged that the Académie Royale des Sciences keep them: "However, original manuscripts of observations are always precious, they can provide clarification when compared with others, [and] consequently I believe that one would do well to preserve these."[56] In this spirit, the big, leather-bound observation journals begun by Cassini I on September 14, 1671, at the Paris Observatory were not only continued but also carefully annotated and corrected by his successors.[57] Who knew when an apparently trivial or even sloppy observation might turn out to be invaluable?

Overwhelmed by the vasty vastness of the universe of particulars awaiting investigation and sharply conscious of the time, labor, and expense involved in observing even a portion thereof, early modern inquirers often became possessive to the point of miserliness when it came to preserving even slipshod jottings, like those of Sédileau. Two classes of observations were particularly precious: daily astronomical and meteorological records, and firsthand descriptions of rare phenomena (e.g., an aurora borealis, a monstrous birth, or even cyclic events with long periods, such as the transits of Venus). These extremes of the mundane and the marvelous were alike in that they could never or only seldom be repeated. The English astronomer Edmond Halley was bitterly disappointed to have missed the first part of an aurora borealis on March 6, 1716, exceptionally visible in the London sky, and therefore to be forced to report the not always reliable observations of others: "Thus far I have attempted to describe what was seen, and am heartily sorry I can say no more as to the first and most surprizing Part thereof, which however frightful and amazing it might seem to the vulgar Beholder, would have been to me a most agreeable and wish'd for Spectacle; for I then should have contemplated *propriis oculis* all the several Sorts of Meteors I remember to have hitherto heard or read of."[58]

Yet utterly unspectacular daily weather readings were even more fleeting, as ephemeral as today and never to return. Under these circumstances, the old motto *carpe diem*, "seize the day," could powerfully motivate observers. When the events of the French Revolution threatened to interrupt the continuous series of weather observations conducted at the Paris Observatory since 1671, the astronomer Joseph-Jerôme de Lalande wrote a terse, peremptory, and, given how recently the Académie Royale des Sciences had been dissolved and Lavoisier guillotined by revolutionary decrees, bold letter to the Directory: "The Bureau of Longitudes, as charged by the law of the Observatory, must continue meteorological observations in the cellar of the Observatory; we request that you have the seals [blocking access] removed immediately."[59] These were the kinds of irreplaceable observations, quotidian or

---

the Paris Observatory: [Delisle], "Inventaire des livres d'astronomie," Bibliothèque de l'Observatoire de Paris, B5.14.

[56] Folder "Sédileau, Observations astronomiques depuis le 1er Novembre 1691 jusqu'au 4 Septembre 1693," in Pochette de séance 1693, Archives de l'Académie des Sciences, Paris. The annotations are signed "J. D. Cassini."

[57] See, e.g., the corrections (apparently by Jacques Cassini II, usually in red ink) inserted in the margins of the observational journals kept for 1671–4; Gian Domenico Cassini I, "Journal des observations faites à l'Observatoire, 1671–1674," Bibliothèque de l'Observatoire, Paris, AD 1–16.

[58] Halley, "An Account of the late surprizing Appearance of the *Lights* seen in the *Air*, on the sixth of March last, with an attempt to explain the Principal *Phaenomena* thereof," *Phil. Trans. Royal Soc. London* 29 (1714–6): 406–28, on 416.

[59] Lalande to the Directoire, 4 Prairial [there is no year, but the Directory held power 1795–9], Dossier Joseph-Jerôme de Lalande, Archives de l'Académie des Sciences, Paris.

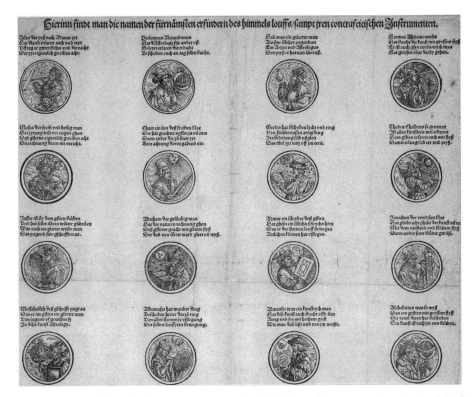

Figure 6. *A genealogy of famous astronomers. Heinrich Vogtherr the Elder and Jacob Ruf,* Hierinn findt man die Namen der fürnamsten Erfindern des Himmels Louffs *(1546). Copyright Staatliche Museum zu Berlin, Kupferstichkabinett, Inv. 140–1889. Reprinted by permission of the Staatliche Museum zu Berlin, Kupferstichkabinett.*

once in a blue moon, that were hoarded by families (and sometimes sold by heirs for a tidy sum to collectors like Delisle) and by institutions such as observatories and academies for perusal as future occasions might require.

The time dimension of these scientific archives avant la lettre extended into the past as well as into the future. The collaborators enlisted in the writing of early modern historiae and observationes included not only coeval correspondents and fellow academicians but also authors who survived only in books and manuscripts. Just as humanists lined their studies with portraits of illustrious predecessors since Plato, astronomers constructed genealogies of observers linking past and present (fig. 6).[60] Historians of science have noted the affective bonds that often united correspondents who exchanged specimens, images, and observations.[61] Joined in a common

[60] Anthony Grafton, "From Apotheosis to Analysis: Some Late Renaissance Histories of Classical Astronomy," in *History and the Disciplines: The Reclassification of Knowledge in Early Modern Europe*, ed. Donald Kelley (Rochester, N.Y., 1997), 261–76. See also Grafton, *Bring Out Your Dead: The Past as Revelation* (Cambridge, Mass., 2001), for parallels with Renaissance humanist construction of learned genealogies.

[61] For the early modern period, see, e.g., Florike Egmond, "Clusius and Friends: Cultures of Exchange in the Circles of European Naturalists," in *Carolus Clusius: Towards a Cultural History of a Renaissance Naturalist*, ed. Edmond, Paul Hoftijzer, and Robert Visser (Amsterdam, 2007), 9–48, and Pomata, "Observation Rising" (cit. n. 28).

undertaking and in a common enthusiasm for their subject matter, the members of these epistolary communities rang the changes on the register of friendship and linked-arm dedication to a noble goal. But behind the ranks of the living citizens of the republic of letters stood the dead, no longer regarded with the reverence due to demigods but rather with the fellow feeling (and at times, the asperity and rivalry) due to colleagues. We have already seen how early modern astronomical and medical observations could be treated either as private property with cash value or as a common good to be recompensed only by gratitude and glory.[62] Then as now, the republic of letters was located in a twilight zone between the realms of honor and lucre. In certain ways, the relationship with the dead was more intimate than that with the living.[63] Learned pen pals separated by mountains or oceans or warring religions might never meet. The dead, in contrast, were always at hand, as close as the books of one's own library, patiently awaiting a consultation—and unable to revenge a plundering.

THE HYBRID HERMENEUTICS OF READING AND SEEING

The resemblance between these two pieces of early modern learned furniture, the one designed for natural and medical specimens and the other for reading excerpts (figs. 7 and 8), is striking. Drawers further subdivided into compartments sorted mobile objects—rocks, scraps of paper—into orders that could be expanded and revised as further acquisitions dictated. Reading and seeing, collecting words and collecting things, were closely intertwined practices in the study of nature during this period. Observations recorded in words and images had to be intercalated with autopsia, and despite much rhetoric to the contrary, it was not always the case that the firsthand experience trumped the secondhand testimony delivered in books and records. A much more complex process of compilation, comparison, correction, and calibration brought together past descriptions and present experience. Sometimes the calibration was straightforward, if arduous, as when French astronomer Jean Picard traveled to the ruins of Tycho Brahe's Uraniborg to determine its coordinates in order to compare Tycho's observations with those made at the Paris Observatory.[64] Sometimes it was subtle, because interpretive, as when Cassini I mined ancient works of astronomy, meteorology, natural history, and history to find precedents for the strange celestial light he observed in the Parisian skies in the spring of 1683.[65] In contrast to observations preserved because they were unusual or unrepeatable, multiple observations of what were arguably the same or similar objects posed challenges of

---

[62] Once again, there are strong analogies with humanist collections of reading notes and excerpts; Blair, *Too Much to Know* (cit. n. 28), 63, 188–202.

[63] Niccolò Machiavelli famously wrote (in a 1513 letter to Francesco Vettori) about how he donned his finest robes before entering his library in order to be worthy of the company of the dead dignitaries assembled on his shelf. On more ambivalent attitudes of humanists to ancient authors, see Anthony Grafton, *Defenders of the Text: The Traditions of Scholarship in an Age of Science, 1450–1800* (Cambridge, Mass., 1991), 23–46.

[64] Kurt Møller Pedersen, "Une mission astronomique de Jean Picard: Le voyage d'Uraniborg," in *Jean Picard et les débuts de l'astronomie de précision au XVIIIe siècle*, ed. Guy Picolet (Paris, 1987), 175–203.

[65] Gian Domenico Cassini I, "Nouveau phenomene rare et singulier, d'une lumiere Celeste, qui a paru au commencement du Printemps de cette Année 1683," *Mémoires, 1666–99* (cit. n. 49), 8: 179–278.

Figure 7. Cabinet for mineral specimens. Johannes Kentmann, Nomenclatura Rerum Fossilium *(Zurich, 1565), sig. a. Reprinted by permission of the Staatsbibliothek zu Berlin, Preussischer Kulturbesitz, Abteilung Historische Drucke in sign Mq 211: P.16.*

surfeit rather than dearth: how to combine many, sometimes divergent perspectives into one.[66]

Early modern reading practices reinforced the exigencies of so little time, so much to observe in order to preserve the library alongside the field, the observatory, and the laboratory as a research site. Past observations were not just physically contained in books and manuscripts; they were *read* as books and manuscripts were. It is not enough for the sciences of the archive to store information; they must also invent ways to use it. The ways of reading cultivated by early modern savants were as distinctive as today's computer search algorithms and just as consequential for

[66] These problems became acute when continuous and repeated observation became the norm, first in astronomy in the sixteenth century and in almost all other sciences by the early eighteenth century; Lorraine Daston, "The Empire of Observation, 1600–1800," in Daston and Lunbeck, *Histories of Scientific Observation* (cit. n. 28), 81–113, on 91–5.

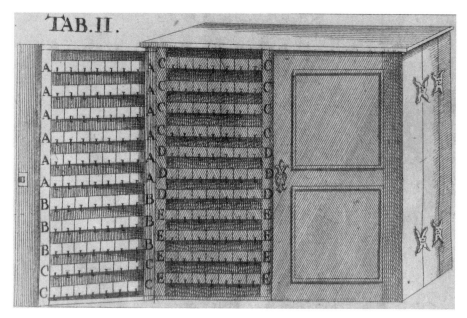

Figure 8. *Cabinet for reading excerpts. Vincent Placcius,* De Arte Excerpendi: Vom gelahr-
ten Buchhalten *(Hamburg, 1689), tab. II. Reprinted by permission of the Staatsbibliothek
zu Berlin, Preussischer Kulturbesitz, Abteilung Historische Drucke in sign A.1998:R.*

what counted as knowing and knowledge. Many, though not all, early modern naturalists had been trained in applying humanist techniques to texts: making excerpts,
taking notes, assembling commonplace books and other collections, compiling indexes.[67] Some of these methods for reading books were retooled for reading the book
of nature.

The habit of keeping a commonplace book in which to record excerpts from reading could, for example, blend seamlessly with that of noting down observations. The
keeping of commonplace books of especially moral adages culled from the reading
of classical authors was a pillar of early modern education in rhetoric. Schoolboys
were enjoined to fill notebooks specially designated for this purpose with choice
morsels from their reading, as an aide-mémoire and source of apposite quotations
and sententiae in set themes and speeches.[68] In contrast to the medieval florilegium,
which was recommended but not mandatory, the Renaissance commonplace book,
after the advent of printing made many more texts available, became what Ann Moss
has called the "interpretative grid" for schoolboy reading and classroom dictation,
"a paradigm for reading analysis" throughout learned Europe in the sixteenth century.[69] During the early modern period, numerous instructions for the keeping of
well-ordered commonplace books were published, recommending various systems

[67] See the excellent account of these techniques of "scholarly information management" in Blair,
Too Much to Know (cit. n. 28).
[68] Sister Joan Marie Lechner, *Renaissance Concepts of the Commonplaces* (Westport, Conn., 1962),
153–99.
[69] Moss, *Printed Commonplace Books and the Structuring of Renaissance Thought* (Oxford,
1996), 136.

of indexes, headings, and cross-references, entrenching the practice still further.[70] Blair has shown in the case of Jean Bodin how these methods could be applied to the composition of a work in natural philosophy; Johns has described their role in the experiments performed and recounted at the early Royal Society.[71]

A few examples have come to light that also point to an intimate connection between the reading practices of the commonplace book and those of observing nature; more research on the *Nachlässe* of early modern naturalists will probably yield more. Ogilvie has described how the Zurich humanist and naturalist Conrad Gessner kept commonplace books both for his botanical observations (including dried specimens) and for his reading and correspondence. He filed all letters he received according to topic, sometimes cutting them up and distributing the clippings topically.[72] John Locke kept his daily meteorological observations (arranged in tabular form) in the same bound commonplace book, entitled "Adversaria Physica" and dated 1660, as his handwritten excerpts from his reading in Latin, French, and English. Scattered amid excerpts from the works of Isaac Newton, Robert Boyle, Christiaan Huygens, and numerous other natural philosophers are recipes, both medical and culinary (e.g., for "Pudding. You've your Pann of ye bignesse you'l have it . . ."), useful information (e.g., a list of the best French pears, in descending order of deliciousness), and Locke's own observations, initialed "JL" (e.g., on thunder and lightning). Locke's weather observations for the period 1666–1703 were recorded at the back of this volume; there is a hint (the outline of a pressed plant) that he also used the "Adversaria Physica" as a makeshift herbarium.[73] For Gessner and Locke, there seemed to be little distinction between observations they had conducted themselves and excerpts they had made from written sources: all were extracted and filed, bits of books and experience cut out and reassembled in new patterns.

Both reading and observing were first and foremost exercises in seeing. Intermediate between the word and the thing was the image, or even, in the case of the botanical herbarium, the dried plant flattened on a page and bound into a book. Recent work by historians of art and science on early modern collections of images hint at the fungibility of specimens and images, which were gathered, classified, and exchanged in similar ways.[74] Woodcuts and engravings could mediate between verbal descriptions and the immediate and sometimes overwhelming experience of raw nature: the cornucopia of novel flora and fauna of the tropics, the gore and mess of the dissected

[70] Ibid., 186–214. Among the most famous of these instructions was [John Locke], "Méthode nouvelle de dresser des recueils," *Bibliothèque universelle et historique* 2 (1686): 315–28; see Richard Yeo, "John Locke's 'New Method' of Commonplacing: Managing Memory and Information," *Eighteenth Century Thought* 2 (2004): 1–38.

[71] Ann Blair, "Annotating and Indexing Natural Philosophy," in *Books and the Sciences in History*, ed. Marina Fresca-Spada and Nicholas Jardine (Cambridge, 2000), 69–89; Johns, "Reading and Experiment" (cit. n. 28).

[72] Ogilvie, *Science of Describing* (cit. n. 28), 180–1.

[73] John Locke, "Adversaria Physica," MS Locke d.9, Bodleian Library, Oxford, on 240, 236, 42. The table of weather observations (running backward and unpaginated) is at the back of the volume; the imprint of the plant is on 109.

[74] See, e.g., David Freedberg, *The Eye of the Lynx: Galileo, His Friends, and the Beginnings of Natural History* (Chicago, 2002); Alessandro Alessandrini and Alessandro Ceregato, *Natura Picta: Ulisse Aldrovandi* (Bologna, 2007); Daniela Bleichmar, "The Geography of Observation: Distance and Visibility in Eighteenth-Century Botanical Travel," in Daston and Lunbeck, *Histories of Scientific Observation* (cit. n. 28), 373–95; Susan Dackerman, ed., *Early Modern Prints and the Pursuit of Knowledge* (Cambridge, Mass./New Haven, Conn., 2011).

human corpse in the anatomy theater, the delicate and fleeting effects of phosphors. Sachiko Kusukawa has described how the students who attended Andreas Vesalius's anatomy lectures at the University of Padua in the mid-sixteenth century triangulated between the corpse being demonstrated in front of them, the much more distinct image of the same organs in the textbook they simultaneously held in their hands, and the oral and written descriptions of the object.[75] By reading both book and book of nature side by side, the novice gradually developed an expert eye for the essential, the normal, and the typical. Halfway between reading and observing, the study of printed images linked the skills that made texts legible and nature intelligible.[76]

Something of how this process of visual calibration worked can be gleaned from the working notes of the huge, expensive, and never-completed project for a *Histoire des plantes* launched by the Paris Académie Royale des Sciences in 1667.[77] This project attempted to calibrate the senses of botanists and artists: by teaching them how to read, see, taste, and smell in unison, multiple observers would, it was hoped, synchronize themselves—and not only with each other, but also with the generations of botanists who had preceded them. Their working tools were live plants, dried herbarium specimens, drawings and engravings, and books. To describe a plant properly involved looking (as well as tasting and smelling), drawing, and also reading all previous published descriptions, textual and visual. In contrast to the humanist techniques for managing texts (excerpting, compiling, glossing), the Parisian academicians sought a hybrid hermeneutics that merged the testimony of botanists since Theophrastus and Dioscorides with that of their own senses.

Instilling such a collective hermeneutics could only have been contemplated within a framework of close coordination and supervision. The first line of botanist Denis Dodart's 1676 publication of the preliminary results of the project underscored its collective nature: "This book is the work of the entire Academy."[78] But the ambitions of the *Histoire des plantes* overflowed the bounds of what the academicians could contribute in the way of their own observations and experiments. As Dodart explained, he and his fellow academicians aimed to provide descriptions and figures of all plants identified by both ancient and modern botanists, in order to correct errors, clarify conflated identifications, and distinguish not only all known species unambiguously from one another but also from as-yet-unknown species that might be discovered in the future.[79] Moreover, plants would be described on the basis not only of the external appearance of all their parts (flower, leaf, bud, seed, root, etc.)

[75] Kusukawa, *Picturing the Book of Nature: Image, Text, and Argument in Sixteenth-Century Human Anatomy and Medical Botany* (Chicago, 2012), 198–227; Kusukawa, "The Uses of Pictures in the Formation of Learned Knowledge: The Cases of Leonhard Fuchs and Andreas Vesalius," in *Transmitting Knowledge: Words, Images, and Instruments in Early Modern Europe*, ed. Kusukawa and Ian Maclean (Oxford, 2006), 73–96.

[76] On the role of scientific atlases in calibrating ways of seeing, see Daston and Galison, *Objectivity* (cit. n. 20), 19–26.

[77] Claude Perrault sketched out the project at meetings of the Académie Royale des Sciences in January 1667: "Projet pour la Botanique," *Procès-Verbaux* 1 (22 décembre 1666–avril 1668): 30–8; Perrault's original manuscript, which diverges in some places from the fair-hand minutes of the meeting, is preserved in the Pochette de séance for January 15, 1667. Records of the chemical analyses performed on plants are preserved in the Fonds Bordelin, Archives de l'Académie des Sciences, Paris. The fortunes of the project are described in Stroup, *Company of Scientists* (cit. n. 27).

[78] Dodart, "Avertissement," in *Mémoires pour servir à l'histoire des plantes* (Paris, 1676), n.p.

[79] Dodart, *Mémoires* (cit. n. 78), 3.

but also of their internal structures, examined microscopically, and of their chemical composition.

This was an undertaking that required the constant consultation of all extant botanical literature and further invited "savants and other persons expert in these matters to communicate their thoughts to us," with the promise that all those who had "contributed something to the perfection of this work" would be thanked by name in future publications.[80] And these were only the named participants in the project. At least as crucial to the project's success were the efforts of its artists, especially Abraham Bosse, Nicolas Robert, and Louis de Chastillon, whose names appeared only in small print in the lower corner of the plates. Yet the *Histoire des plantes* was in part inspired by Robert's opulent illustrations of flowering plants for Gaston d'Orléans, and aside from Dodart's 1676 *Mémoires*, the sole publication ever to issue forth from the mammoth project was of the 319 plates, produced at a cost of (by 1700) more than 25,000 livres and published without text in part in 1719 and in full in 1788.[81] Most of the surviving manuscripts concerning the project appear to have been Dodart's minute instructions to the artists, of whom he sometimes despaired, and other anonymous assistants employed to make observations, write descriptions, and draw plants.[82]

These unpaginated loose sheets, each headed with the name of a plant, take the form of numbered queries, corrections, and instructions, usually in a fair hand (presumably dictated), with occasional replies in another hand. So, for example, under "Grande Absinthe" (fig. 9; replies, in a different hand, are here rendered in italics):

I.   The seed? by the Microscope.
II. Of how many florets is the flower composed? *around 50 or 60.*
III. The leaf and the bud are they truly furry? look at under the magnifying glass. *they are covered with cotton which renders them white-ish.*
IV. The root is it really bitter? *the bitterness is not extreme but it is piquant.*
V.   Distilled water.
VI. What Absinthes are in the Royal Garden?[83]

The style of seeing and describing enforced by these queries and instructions was unrelentingly comparative: artists and assistants were sternly reminded to consult all previous descriptions and figures in previous botanical works before beginning their own; the single most frequent exhortation was to repeat an observation, either to

---

[80] Ibid., sig. Ar.

[81] Perrault mentions the Orléans collection explicitly: "Suivant cette methode on peut travailler à une histoire des plantes generalle et acomplie qu'il sera facile d'enrichir et rendre tout a fait magnifique et Royale en continuant et achevant de travail qui a été commencé par l'ordre de feu Mr. le Duc d'Orleans et faisant graver beaux portraits des plantes qu'il nous a laissez si naivement tiré." Perrault, "Projet pour la Botanique," in Pochette de séance, January 15, 1667 (cit. n. 77), n.p. On the subsequent publication history of the plates, see *Recueil des plantes, gravées par ordre de Louis XIV* (Paris, 1788), 1:v; and on the vellums, see Yves Laissus, "Les plantes du roi: Note sur un grand ouvrage de botanique préparé au XVIIe siècle par l'Académie Royale des Sciences," *Rev. Hist. Sci.* 22 (1969): 193–236.

[82] Camille Frémontier and Alice Stroup, "Les sources iconographiques de l'histoire de l'Académie Royale des Sciences," in *Histoire et mémoire de l'Académie des Sciences*, ed. Eric Brian and Christiane Demeulenaere-Douyère (Paris, 1996), 379–93. Some of Dodart's manuscript annotations refer to the observations of "J. B."—possibly the academician Jacques Borel (also sometimes known as Borelly), who was also involved in the *Histoire des plantes*.

[83] "Grande Absinthe," in "Notes sur l'*Histoire des plantes*," MS 450, Bibliothèque du Muséum national d'histoire naturelle, Paris.

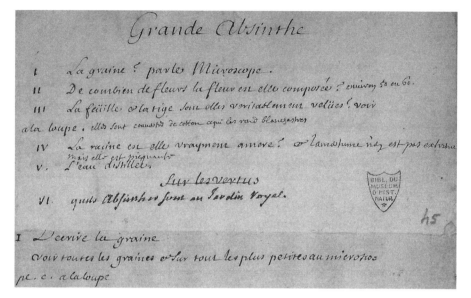

Figure 9. *"Grande Absinthe" (cit. n. 83) with numbered queries and replies. Reprinted by permission of the Bibliothèque centrale du Muséum national d'histoire naturelle, Paris, MS 450.*

check that of a predecessor or to sharpen one's own. At the bottom of the queries to "Aconitum hyemale luteum bulbosum" followed a firm injunction:

> I.   Do not make any description of a Plant already described without having read the best description, or at least [not] without confronting it with the one [already] made.
> II.   Try to read all the descriptions in order to verify them from nature [*sur la nature*], and to compare them with ours.[84]

The copious and exacting instructions, paired with the laconic and sometimes dissenting replies, bear witness to a struggle between ways of seeing, touching, smelling, and tasting. The syntheses achieved by the *Histoire des plantes* depended on such comparisons, corrections, and repetitions (fig. 10). Observation shuttled between texts, images, and plants and back again until all three coalesced. These were the practices and challenges that anchored even the most militantly empirical naturalists to the library, as well as to the herbarium and museum. There they found armies of collaborators waiting on the shelf to help with the challenge of observing nothing less than all the phenomena in the universe; there they exercised the ways of seeing, parsing, recording, and ordering that made sense of the written page, the printed image, and the prodigious variety of nature.

The hybrid hermeneutics of reading and seeing, and of melding multiple and multimedia observations into a single definitive one, has its own history. Andrew Mendelsohn has shown how the Société Royale de Médecine used refined editing techniques to synthesize hundreds of meteorological and medical observations sent in

---

[84] "Aconitum hyemale luteum bulbosum" (see also "Aconitum pyramidale multiflorum"), in "Notes sur l'*Histoire des plantes*" (cit. n. 83).

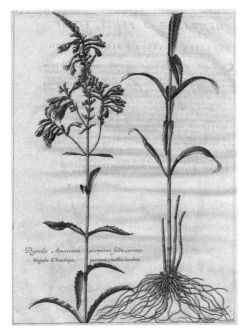

DIGITALIS AMERICANA
PURPUREA, FOLIO SERRATO.

DIGITALE D'AMERIQUE, POURPRE'E, A FEÜILLE DENTELEE.

LA racine de cette Plante est blanche & fibreuse. Elle pousse une seule tige, haute de quatre pieds, quarrée, noüeuse en distances égales d'un pouce & demy, & moüelleuse. Les feuilles sont longues de trois pouces, & larges d'un demy-pouce, fort pointuës, dentelées, lisses, d'un vert-brun, avec une coste blanche. Elles sortent des nœuds de la tige, deux à deux opposées l'une à l'autre, en sorte que celles d'un nœud croisent celles de l'autre. Du haut de la tige naissent des branches opposées deux à deux, les unes croisant les autres, revestuës vers le haut de quantité de cornets gris-de-lin, longs environ d'un pouce, estroits dans leur origine, d'où ils vont s'eslargissant jusques au bout, où ils sont divisez en deux levres. L'inferieure est coupée en trois parties. Celle du milieu est la plus grande, & tachetée de pourpre comme à la Digitale vulgaire. A la levre superieure sont attachez quatre filets couleur de citron, qui naissent du fonds de la fleur, & ne s'en destachent que vers l'extremité. Ils ont chacun un sommet de la mesme couleur. Chaque fleur naist d'un calice divisé en cinq, lequel venant à se grossir, est remply de quatre graines brunes triangulaires.

La racine paroist d'abord insipide. Mais quand on l'a beaucoup maschée, elle fait sentir une acreté considerable, meslée de quelque amertume. Les feuilles aussi sont assez acres, mais on n'y remarque que cette saveur.

Cette Plante est vivace. Elle fleurit en Juillet.

Elle vient également bien à l'ombre & au soleil, mais il luy faut une bonne terre. On la peut semer en Automne en pleine terre, ou sur couche au Printemps.

Figure 10. Description and figure of Digitalis americana. *The text contains a gustatory as well as a visual description of the plant: "The root at first seems insipid, but when much chewed, it exudes considerable acidity, mixed with some bitterness." Dodart,* Mémoires *(cit. n. 78), 79. Reprinted by permission of the Staatsbibliothek zu Berlin, Preussischer Kulturbesitz, Abteilung Historische Drucke in sign Lw2250.*

by provincial doctors; Jed Buchwald has described in detail how discrepant measurements were dealt with before the adoption of probabilistic methods like that of least squares.[85] But we still lack anything like the detailed and textured histories now available for early modern reading practices (or the senses) about how scientific reading and seeing were schooled in unison as new methods of combining past and present observations, text and autopsia, developed. What evidence we do have suggests that this history was dynamic, as rich in innovation and sophistication as the evolving hermeneutics of the nineteenth-century historians who perfected *Quellenkritik*, the twentieth-century informatics specialists who put together databases, and the twenty-first-century computer scientists who devise search algorithms.

One example from the history of taxonomy must suffice here. When the comparative anatomist Georges Cuvier sought to settle the question of whether mammoth remains corresponded to any still-extant species, he examined fossils, dissected elephants—and read books, lots of them: Polybius and Titus Livy, Linnaeus and Buffon, Athanasius Kircher and Felix Platter. Cuvier was critical of his sources, especially the older ones: "As one approaches our epoch, observations become more

---

[85] Mendelsohn, "The World on a Page: Making a General Observation in the Eighteenth Century," in Daston and Lunbeck, *Histories of Scientific Observation* (cit. n. 28), 396–420; Buchwald, "Discrepant Measurements and Experimental Knowledge in the Early Modern Era," *Arch. Hist. Exact Sci.* 60 (2006): 565–649.

exact [*positives*]."[86] Yet he relied upon them to make comparisons, correct errors, sharpen vague reports, and canvass all the locales where fossil pachyderm remains had been discovered to date. His voluminous survey of the literature (which he regretted was not still more thorough) allowed him to conclude that a "prodigious quantity of such bones" had already been discovered with a distribution from Siberia to the Americas, mostly in alluvial basins that also yielded fossils of rhinoceroses, antelopes, horses, and marine animals.[87]

Cuvier's evaluation of the quality of past observations was that of a connoisseur, cautious and discerning. His reading practices were no longer those of the humanist compendium and commonplace book, indiscriminately jumbling together reading excerpts and observations. He drew a bold line between his own observations, those of witnesses whom he trusted (e.g., Johann Friedrich Blumenbach), and those of witnesses he did not (e.g., Peter Simon Pallas). Particularly notable is his reinterpretation of past observations in light of present knowledge: for example, he explained how Platter could have mistaken the bones dug up in Lucerne in 1577 for those of a human giant (which the city then proudly emblazoned on its coat of arms), pointing out that the missing teeth prevented him from identifying the remains as those of a quadruped.[88] But even as he read his sources against the grain, he did read them, carefully, comprehensively, and fruitfully. Along with the museum, the dissecting table, and the field excavation, the library was still an indispensable site of research for the most advanced sciences of the early nineteenth century—and long thereafter (fig. 11).

Although the organization of observer networks and the emergence of new reading and observing practices in the late eighteenth and early nineteenth centuries altered the relationship of scientific inquiry to book learning,[89] the library remained an indispensable site of natural history, supplemented by mushrooming collections of specimens at institutions such as the Paris Muséum d'histoire naturelle, the British Museum, Kew Gardens, and the New York American Museum of Natural History.[90] Especially in systematics, as Robert Kohler has noted, rules of nomenclature "oblige those who would name a new species to actively engage the literature back to the Linnaean big bang."[91] Once the practice of linking species names to individual type

[86] Cuvier, *Recherches sur les ossemens fossiles, où l'on rétablit les caractères de plusieurs animaux dont les révolutions du globe ont détruit les espèces*, 2nd ed. (Paris, 1821–4), 1:104. On the context of Cuvier's investigations of pachyderms, see Martin J. S. Rudwick, *Georges Cuvier, Fossil Bones, and Geological Catastrophes* (Chicago, 1997), 13–41; Claudine Cohen, *The Fate of the Mammoth: Fossils, Myth, and History* (1994), trans. William Rodarmor (Chicago, 2002), 105–24.

[87] Cuvier, *Recherches* (cit. n. 86), 1:159, 201.

[88] Ibid., 1:113.

[89] Mendelsohn, "World on a Page" (cit. n. 85); Anne Secord, "Coming to Attention: A Commonwealth of Observers during the Napoleonic Wars," in Daston and Lunbeck, *Histories of Scientific Observation* (cit. n. 28), 421–44.

[90] On the growth of these collections and their impact on nineteenth-century taxonomy, see (for the British Museum) Gordon McOuat, "Cataloguing Power: Delineating 'Competent Naturalists' and the Meaning of Species in the British Museum," *Brit. J. Hist. Sci.* 34 (2001): 1–28; (for American institutions) Robert E. Kohler, *All Creatures: Naturalists, Collectors, and Biodiversity, 1850–1950* (Princeton, N.J., 2006); (for Kew Gardens) Jim Endersby, *Imperial Nature: Joseph Hooker and the Practices of Victorian Science* (Chicago, 2008); and (for the Muséum d'histoire naturelle) Pierre-Yves Lacour, "La république naturaliste: Les collections françaises d'histoire naturelle sous la Révolution, 1789–1804" (PhD diss., European Univ. Institute, 2010).

[91] Kohler, "Reflections on the History of Systematics," in *Patterns in Nature: Historical and Conceptual Foundations of Systematics*, ed. Andrew Hamilton (Berkeley and Los Angeles, 2012), 16–37. I am grateful to Professor Kohler for allowing me to read his essay in manuscript.

AUTEURS consultés.	DÉTAILS SUR LES DÉFENSES.	LONGUEUR en suivant la grande courbure.	DIAMÈTRE au gros bout.	POIDS.	OBSERVAT.
Daubenton, t. XI.	N° DCDXCVI de Sibérie, tronquée en avant. ...	Long. du tronç. 5' 4'''	6'' et à l'ant. bout 5'' 4'''	89 l. 4°	
	N° DCDXCV de Sibérie, tronquée aux deux bouts.	5'	4'' 8''' et à l'autre bout 4''	4 15°	
	N° DCDXCIV de Sibérie, tronquée aux deux bouts.	3' 4''	3''' 10''' aux deux bouts.	15 5°	
	N° DCDXCII, tronquée aux deux bouts.	3' 3''	2''' 9''' et 1'' 10''' à l'ant. bout.	9 12°	
Faujas, Géol., 293.	Défense des environs de Rome, trouvée par MM. La Rochefoucauld et Desmarets; fort tronquée aux deux bouts et cassée en trois morceaux.	5'	8''	»	On a estimé que si elle était entière elle aurait quatorze pieds de longueur; mais il est difficile de conclure la longueur d'après un tronçon parce que la diminution du diamètre ne se fait pas toujours uniformément.
Fortis, 11, p.	Défense trouvée au Serbaro, près de Vérone, par Fortis et le comte de Gazola, tronquée aux deux bouts, rendue par infiltrations.	7' 6'' de ver.	9'' à 10''' id.	»	
	Défense fossile de Toscane.	8' 6''	»	»	
Camper.	Défense de Sibérie du cabinet de M. Camper.	5' et plus.	»	»	
Zach.	Premier éléph. de Burgtonna.	8'	»	»	
	Deuxième, id.	10'	»	»	
Pallas, Nov. Com Petr., XIII, pag. 473.	La plus grande défense de Sibérie, du cabinet de Pétersbourg, tronquée aux deux bouts	8'	6'' 6''' et à l'autre bout 6'' 4'''	»	
Autenrieth et Jæger.	La plus grande défense de Canstadt, très-courbée, tronquée aux deux bouts.	5' 6''	5'' et à l'aut. bout 3''	»	
	Reisel et Spleiss disent qu'il y en avait de.	10'	»	»	

AUTEURS consultés.	DÉTAILS SUR LES DÉFENSES.	LONGUEUR en suivant la grande courbure.	DIAMÈTRE au gros bout.	POIDS.	OBSERVAT.
Natter.	La plus grande trouvée au même lieu en 1818.	8'	1'	»	
Messerschmidt et Breynius, Triphil., 40, p.	Une défense très-courbée de Sibérie.	13' 6'' 5''' rom.	6''	13,L1*p. d'apoth.	
Tilesius, Mém. de Petersb., v, pl. x.	Les défenses du squelette d'Adams à en juger par le dessin.	12'	7''	»	
Adams.	Une défense observée à Iakoutsk.	15'	8'' 8'''	7 pouds ou 234 l.	M. Adams dit deux toises et demie, mais ne s'explique pas quelles toises; il n'eût pas non plus s'il a mesuré ou estimé ces dimensions, et s'il ne les rapporte pas seulement de souvenir.
Hermann, Prog.-pecul.	La défense suspendue dans la cathédrale de Strasbourg, très-courbée.	6' 7''	3'' 5'''	»	
Id. Lettres.	Id. de Wendenheim.	4' 10''	5'' 6'''	»	

Figure 11. *Synoptic chart of past observations of elephant fossils. Georges Cuvier,* Recherches sur les ossemens fossiles, *4th ed. (Paris, 1834), 2:194. Reprinted by permission of the Max Planck Institute for the History of Science, Berlin.*

specimens was enshrined in early twentieth-century international codes of botanical and zoological nomenclature, engagement with the literature could, in the absence of a designated holotype, extend to manuscript research in the Nachlass of the first naturalist to identify a new species in order to establish at least a lectotype.[92] Keeping words firmly attached to things—in this case, names to species—required feats of collective scientific memory that stretched centuries back into the past.

THE ARCHIVES OF THE FUTURE

Naturalists routinely embraced their predecessors but rarely their successors in their imagined community of collaborators. Astronomers were also avid curators (and sharp critics) of past observations, often with an undertone of sympathy even for efforts they judged to be flawed or obsolete. When Picard journeyed from Paris to the Danish island of Hven in 1671, he was firmly convinced that the accuracy of the Paris Observatory had surpassed that achieved by Tycho at Uraniborg a century earlier. Nevertheless, he was shocked to find the once-celebrated observatory in ruins: "I could not without indignation view this famous place, which will be talked about for as long as there are astronomers, filled with the old carcasses of animals like some foul rubbage-heap."[93]

---

[92] See W. Greuter et al., eds., *The International Code of Botanical Nomenclature* (Saint Louis Code, 1999), chap. 2, sec. 2, art. 9, recommendation 9A, available on the website of the Botanischer Garten und Botanisches Museum Berlin-Dahlem, http://www.bgbm.org/iapt/nomenclature/code/saintlouis /0000st.luistitle.htm (accessed 3 January 2011); Daston, "Type Specimens" (cit. n. 8), 170–7.

[93] Quoted in Pedersen, "Mission astronomique de Jean Picard" (cit. n. 64), on 179.

For as long as there are astronomers: how long would that be? The superhuman scale of celestial phenomena demanded communities of observers that were more ancient and more enduring than any human civilization in history. If early modern astronomers had projected their virtual communities deep into the past, their nineteenth-century counterparts vaulted into the far future. Not content simply to preserve the archive inherited from ancient observers in Mesopotamia, India, and China, they also aimed to create an archive for the astronomers thousands of years hence. The most ambitious of these archives for the future was the Carte du ciel or Great Star Map, an international collaboration of observatories launched in 1887 and ended (though not completed) by the International Astronomical Union in 1970, which aimed to compile a complete photographic map of the sky with all stars down to fourteenth magnitude (and a catalog down to eleventh magnitude, not completed until 1964).[94] The popular scientific press described the undertaking in the awed tones reserved for the Egyptian pyramids or the greatest medieval cathedrals: "Thus the science of the nineteenth century will bequeath to posterity an unimpeachable and imperishable record of the starry sky, which, in future centuries, will serve as a certain foundation for the solution of the grand problem of the constitution of the universe."[95] The astronomers themselves were hardly less modest and a great deal more practical: "This result [a new method for encasing the astrophotographs in glass] would have an enormous importance for the project, since it would assure the perfect conservation of images that we want to will to the astronomers of the year 3000 at least, inasmuch as today we do not know whether the chemical coating deposited on the glass will remain eternally unalterable."[96] Civilizations might rise and fall, but at least until the year 3000, there would be astronomers who would appreciate the archives prepared by their predecessors.

An imagined community that transcends time is one of the preconditions for the sciences of the archives. Another is the conviction that information about empirical particulars is intrinsically valuable and worth saving: either because the observations are part of a time series, like those of the daily weather or planetary positions, or because the phenomena are rare, like a supernova or an omega-minus particle in a bubble chamber, or just because there is not world enough and time to observe all the universe contains. Under these circumstances, observations will be not only collected but also preserved. The material and institutional underpinnings of transmission vary by historical epoch, from baked clay cuneiform tablets to optical discs made of silver and aluminum alloys, and from hereditary scribal classes to great national and university libraries. Some media are more durable than others—incunabula printed on parchment are still legible after over four centuries, whereas the estimated lifetime of a compact disc is thirty years—but all records depend on cultural will for their preservation and accessibility, as well as for the investment of labor and resources

---

[94] On the Carte du ciel see Suzanne Débarat, J. A. Eddy, H. K. Eichhorn, and A. R. Upgren, eds., *Mapping the Sky: Past Heritage and Future Directions* (Dordrecht, 1988); Théo Weimer, *Brève histoire de la Carte du ciel en France* (Paris, 1987); and Jérôme Lamy, ed., *La Carte du ciel: Histoire et actualité d'un projet scientifique international* (Paris, 2008). Correspondence relating to the project is available in Ileana Chinnici, *La Carte du ciel: Correspondance inédite conservée dans les Archives de l'Observatoire de Paris* (Paris, 1999).

[95] Camille Flammarion, "Le congrès astronomique pour la photographie du ciel," *Revue d'astronomie populaire, de météorologie et de physique du globe* 6 (1887): 161–9, on 169.

[96] E. B. Mouchez to David Gill, 30 April 1887, Observatoire de Paris, MS IV.A, "Comité international de la Carte du Ciel," carton 7.

needed to make them in the first place. Some sciences of the archive may exploit information not originally intended to serve their purposes, as, for example, historical demographers use parish records of births and deaths. Yet the effect of such serendipitous troves of data is usually to strengthen the conviction among scientists that special measures should be taken to preserve them (on microfiche, by digitalization, on backup servers, or even in fireproof vaults), rather than to trust to finder's luck and investigative ingenuity in the future. The sciences of the archive create traditions as durable as those of the Vatican.

But it is not enough to insure material transmission and institutional stability—the library, whether physical or virtual,[97] still nestled at the heart of the laboratory and the observatory. Traditions must be continually reinterpreted in light of present needs and notions in order to survive. In the humanities, this is the role of commentary and exegesis. Do the sciences of the archive cultivate analogous practices? The early modern hermeneutics of reading and seeing described above was one such practice of bringing past and present together but by no means the only one. New ones are constantly being invented—for example, in the construction and mining of databases.[98] In the early modern case, the practices of reading the book of nature depended heavily on older practices of reading books, but by the late eighteenth and early nineteenth centuries, at least some sciences of the archives had developed ways of sifting and synthesizing observations that were specific to discipline and sensitive to historical dynamics: witness Picard's and Cuvier's view that the accuracy of observation in their respective disciplines had advanced and that older observations must therefore be used with care. Yet the overwhelming sense of historicity that engulfed the humanities in the nineteenth and twentieth centuries, and that transformed the ancient methods of commentary and exegesis into exercises in historical contextualization, seem to have left the sciences of the archives largely untouched. They pursue history in science without history of science.

History *in* science differs from most other kinds of history by its curious indifference to the contours of time. Historians are the sculptors of time. They divide dynasties and mold the contours of periods: antiquity, the Ming, modernity. They calibrate the chronologies that connect and sound the depths of ruptures that separate, from the neolithic revolution to the computer revolution. The shape of time defines the physiognomy of history, its most prominent features and overall coherence, and various schools of history have drawn revealingly different physiognomies: even within the compact province of European history, the German *Barockzeitalter*, the Italian *rinascimento*, the French *âge classique*, and the Anglo-American "early modern period" (not to mention the history of science's "Scientific Revolution") overlap in the time periods covered but diverge sharply in themes and implications. Chronologies, timelines, and tables are to history what atlases are to geography: a map by

[97] Even if the library is located on the Web rather than in a building, the challenges of material durability (e.g., maintaining servers and translating old formats like floppy disks and microfilm into new ones) and archiving are at least as daunting; see the Web Archiving Service, http://webarchives.cdlib.org/ (accessed 30 August 2011). This may be one reason why large, architecturally ambitious libraries are still being built on a grand scale; Markus Eisen, "Zur architektonischen Typologie von Bibliotheken," in *Die Weisheit Baut sich ein Haus: Architektur und Geschichte von Bibliotheken*, ed. Winfried Nerdinger (Munich, 2011), 261–306.

[98] Of course, practices in the humanities also evolve; see Andrew Abbott, "Library Research Infrastructure for Humanistic and Social Scientific Scholarship in the Twentieth Century," in *Social Knowledge in the Making*, ed. Michèle Lamont, Charles Comic, and Neil Gross (Chicago, 2011), 43–88.

which to navigate unmarked territory.[99] To carve up time is the first, essential step to making sense of history—and therefore perpetually controversial.

The same might be said of at least some of the sciences, on a still-grander scale. If the time of history dwarfs the human lifetime, the time of the sciences of earth and sky beggars history. Geologists and evolutionary biologists layer their epochs like the strata of rock and embedded fossils they name: the lower Cretaceous, the upper Cretaceous. Epochs are swallowed up by periods and periods in turn by eras, each at least 150 million years long. Astronomers track the birth and death of stars, indeed of the entire universe—beside which the period of the precession of the equinoxes, a mere twenty-six thousand years, shrinks to a pinpoint. To contemplate these oceans of time stokes both poetic sublimity and scientific explanation: the vertiginous possibilities of Kant's infinity or Darwin's natural selection. Yet even the mind-boggling expanses of geological and astronomical time must be mapped. In the historical sciences, as in history, the precondition for knowledge is a cartography of time.

So it is striking that the practices of the sciences of the archives, whether oriented toward the past or the future, are so indifferent to those continents of time called "periods" or "epochs" that make oceans of time navigable. The early modern naturalists who combed libraries for past observations of plants or volcanoes or seasonal winds and the astronomers at the turn of the twentieth century who laid down an archive for the turn of the thirtieth century largely believed, like Cuvier, in a vague sort of progress in the quality of observation: modern, firsthand, authored observations were generally superior to the ancient, secondhand, anonymous sort; detailed descriptions and above all precise measurements were preferable to nebulousness; the accuracy and reliability of instruments had improved steadily in the past and could be counted upon to continue to do so in the future. But this overall vector of amelioration did not lead them to abandon either libraries or archives in disgust (declaring that past observations are useless) or despair (fearing that future observers would find their observations to be useless), much less to a periodization of the ages of observation. There were no iron, silver, and golden ages; not even ages of good record keeping and bad. There were just data strewn across the flat expanse of time.

As in the case of Cuvier sifting observations of pachyderms since classical antiquity, users of past scientific data developed their own highly refined form of Quellenkritik, poking and prodding past accounts for accuracy and reliability. But in contrast to the historian's practice of that wary art of cross-examining one's sources, there was little awareness of the systematic distortions of anachronism. Historians acquire a period ear, analogous to the period eye of historians of art. Beliefs, turns of phrase, orthography, even punctuation bear the stamp of a time and place, and even tiny anomalies can jar the period ear and inflame suspicions of forgery. For the scientists, in contrast, there were only good observers and bad observers, and both kinds could appear anywhere, in a dusty manuscript centuries old or in the latest journal. This leveling view of history was and remains the precondition for the centrality of the library in the sciences of the archive.

For historians, nothing could be more puzzling, shocking even, than this tone deafness for anachronism (however exquisite the sensibility for other aspects of past data). If the cardinal sin of anthropologists is ethnocentrism, that of historians is

---

[99] Daniel Rosenberg and Anthony Grafton's *Cartographies of Time: A History of the Timeline* (New York, 2010) offers a remarkable collection of such ways of visualizing the epochs of history.

anachronism—in both cases, the criminally naive assumption that the inhabitants of other times and places are just like one's next-door neighbors, albeit oddly attired and quaintly spoken. But this neighborly assumption was the very basis for collective empiricism: a conception of a community that of necessity transcended space and time. The word *community*, with its cozy, egalitarian associations, is perhaps not le mot juste here and in any case does not appear to have been used systematically in a scientific context much before the late nineteenth century. Scientific utopias, from Bacon's House of Salomon to the Saint-Simonian phalanxes, could be rigidly hierarchical; the division of labor could resemble that of a Victorian factory. What remained constant were the voracious demand for scientific labor and the challenge of organizing it in ways unprecedented except perhaps in organized religion. This may be why utopias recur so frequently in modern science, visions of alternative polities that sprawl over continents and centuries, in the teeth of the realities of local institutions and the human life span. Only in the library and the archive could their time-defying aspirations be realized.

The Semblance of Transparency:
Expertise as a Social Good and an Ideology in Enlightened Societies

*by Thomas Broman**

ABSTRACT

Our modern idea of scientific expertise is compounded from two historically distinct elements: occupational expertise and the expertise claimed by scientists as privileged knowers of truths about the world. This article examines the historical roots of these two kinds of expertise in terms of the conceptual separation of theory and practice and the social status of scholars. In the Enlightenment, the value of theoretical knowledge began to be measured against its ability to guide action, and scientific knowledge acquired a public role as the ideological justification for the reform of society. The consequences of these changes are illustrated by Condorcet's writings on education during the French Revolution and discussions of science and the public in the 1940s and 1950s.

INTRODUCTION: EXPERTISE BETWEEN STATE AND SOCIETY

No one who is even minimally acquainted with current affairs will deny that ours is an era caught up in debates over how—and how much—the fruits of science and medicine are to be relied upon in matters of public interest. Whether the issue is global climate change, the use of genetically modified organisms in medicine and food production, or the regulation of stem-cell research, deep divisions between different political and social factions make the possibility of finding real consensus remote indeed. One of the most striking aspects of these controversies is the extent to which scientists' ability to articulate basic truths about nature or society, and to have those statements of fact accepted as such by all parties to the debate, has become compromised and an object of open contestation.[1] And not surprisingly, when the results and uses of scientific research become controversial, the nature of expertise and its relationship with politics attract critical examination by scholars.[2] Harry Collins

* Department of History of Science, University of Wisconsin–Madison, 1225 Linden Ave., Madison, WI 53706; thbroman@wisc.edu.

This article has benefited from discussions with a great many colleagues and friends, particularly including John Carson, Theodore Porter, Volker Hess, and Pamela Long. Robert Kohler and Kathryn Olesko merit special thanks for their encouragement and their help in shaping the final product.

[1] As striking as the contemporary debates are, it is worth noting that they are not unprecedented. For an examination of similar debates in the 1960s, see Jon Agar, "What Happened in the Sixties?" *Brit. J. Hist. Sci.* 41 (2008): 567–600.

[2] For a sample of this literature, see Stephen P. Turner, *Liberal Democracy 3.0: Civil Society in an Age of Experts* (London, 2003); Sheila Jasanoff, *Designs on Nature: Science and Democracy in*

and Robert Evans, for example, point up what they label a "problem of legitimacy" in the resolution of scientific disputes with broader implications for political policy, by which they mean the question of which actors have the right to claim a knowledgeable contribution to the formulation of policies. Conspicuously, Collins and Evans take it as given that a generation's work in science and technology studies (STS) has effectively deconstructed scientists' claims that their research produces objective knowledge. They urge that such STS-bred skepticism has gone too far in denying the epistemological claims of science, and they affirm that scientists do possess knowledge and practical skills that can be brought to bear on current problems.[3] Sheila Jasanoff too assumes that STS has eliminated once and for all the idea that science can stand aloof from politics, or indeed that it ever did. In place of an ideal of pure science, Jasanoff cites her own idea that the "coproduction" of the natural and social orders is the cornerstone of the modern view of science as thoroughly embedded in the socially and culturally contingent processes of its creation.[4]

There can be no denying that more than a generation's work in the Sociology of Scientific Knowledge has revolutionized our understanding of how scientific knowledge is created and how it is embedded in its various contexts—social, cultural, and institutional. Yet in speaking of how "our" understanding has been revolutionized, the claim refers to an exceedingly small group of people. What nearly all scholars in STS take as resolved and obvious would, one imagines, still shock most of the public even today, more than a decade after the so-called science wars flickered into oblivion.[5] The apotheosis of science as the most reliable and objective knowledge available to us remains largely unshaken. It is incessantly reiterated in the press and in the choices that both individuals and governments make in managing their affairs. And yet there is also no denying the paradoxical situation concerning the contentiousness of science in debates over stem cells and global warming. The public's faith in *science* might remain firm, but its willingness to trust *scientists'* representations of science seems to have diminished considerably.[6]

How can we account for this bifurcation in the public's view of science? One explanation invokes what Jasanoff describes as the "instrumentalization" of science by the modern state. To an unprecedented degree, she argues (using biotechnology to exemplify science more generally), governments have harnessed the fruits of scientific research as potential engines of economic growth. Universities have been handed mandates to develop deeper partnerships with industry, a move that promises to reshape universities' traditional status as institutions devoted largely to teaching

---

Europe and the United States (Princeton, N.J., 2005); Philip Kitcher, *Science, Truth, and Democracy* (Oxford, 2001); Harry Collins and Robert Evans, *Rethinking Expertise* (Chicago, 2007).

[3] Collins and Evans, *Rethinking Expertise* (cit. n. 2), 2. See also Collins and Evans, "The Third Wave of Science Studies: Studies of Expertise and Experience," *Soc. Stud. Sci.* 32 (2002): 235–96.

[4] Jasanoff, *Designs on Nature* (cit. n. 2), 22–3. Jasanoff attributes the origin of the coproduction thesis to Steven Shapin and Simon Schaffer, *Leviathan and the Air-Pump: Hobbes, Boyle, and the Experimental Life* (Chicago, 1985).

[5] Philosophers of science, it should be noted, have been conspicuously rare in this consensus, with many among them advocating a mitigated form of scientific realism, such as Philip Kitcher in *Science, Truth, and Democracy* (cit. n. 2).

[6] An important qualification must be inserted here, which is that this article will not attempt to sort out the precise limits of terms such as *scientist, expertise,* and *authority.* No doubt some of its arguments would be more persuasive were such an effort to be made, but if the reader will grant that such terms can be meaningfully applied to different individuals and occupational groups, then I think we can proceed.

and the pursuit of pure science. Demands for more intensive cooperation between university-based scientists and industry have been made all the more attractive by legislation and supportive judicial decisions that have widened the ability of scientists to claim their discoveries as intellectual property. These too have contributed to making science look increasingly like just another form of profit-driven enterprise. At the same time, Jasanoff and Theodore Porter have written of how the instrumentalization of science by the state has also resulted in the enlistment of various species of scientists by governments as consultants in the formulation of public policy.[7] Small wonder, then, that when members of the public see scientists chasing patents and profits on the one hand and fronting as expert pitchmen for controversial government policies on the other, those citizens become skeptical of the claims being made on behalf of science.

This explanation has much to recommend it; not least the fact that it accounts for developments that nearly everyone working in the modern university environment can observe firsthand on an almost daily basis. Yet in my opinion the understandable attention paid to scientific expertise in the service of governmental and intergovernmental or nongovernmental policy matters can be potentially misleading about the nature of scientific expertise for at least two reasons. First, the mobilization of scientific resources by the state and other public entities is scarcely a twenty-first-century novelty, a point made in the recent *Osiris* volume on expertise and the early modern state, among numerous such examples.[8] In light of this longer history, therefore, it would be helpful to know more clearly what distinguishes the contemporary situation from its historical precedents. As I hope to show below, pursuing this question can yield a substantial reward in the understanding we can bring to the contemporary situation.

Second, the attention paid to expertise in the service of policy can create the misleading idea that the authority of scientific expertise is based exclusively or primarily on the legitimacy it derives from the state. Expertise has both a broader and a deeper basis than this implies. It is broader in the sense that the cultural and social authority of experts is an ideal widely shared by the public, and not just imposed by government.[9] It is also deeper in that it is compounded out of two distinct elements that can trace ancient roots. The first element is expertise as a form of productive knowledge arising from the division of labor in society. This is not a profound or obscure notion. To the contrary, expertise of this kind exists in any society that recognizes different kinds of skilled labor: the labor of weavers, midwives, witch doctors, architects, and civil engineers, among many others. Whatever else it may also be, I claim that scientific expertise is specialized labor of precisely this kind. As such, it can be bought and

---

[7] Jasanoff, *The Fifth Branch: Science Advisors as Policymakers* (Cambridge, Mass., 1990); Porter, "Speaking Precision to Power: The Modern Political Role of Social Science," *Soc. Res.* 73 (2006): 1273–94.

[8] Eric H. Ash, ed., *Expertise and the Early Modern State*, vol. 25 of *Osiris* (2010). This theme has also been dominant in the abundant historiography of travel and colonialism in the early modern period, as evidenced in Daniela Bleichmar et al., eds., *Science in the Spanish and Portuguese Empires, 1500–1800* (Stanford, Calif., 2009), and James Delbourgo and Nicholas Dew, eds., *Science and Empire in the Atlantic World* (New York, 2008). Finally, for nineteenth-century antecedents, see Chandra Mukerji's *A Fragile Power: Scientists and the State* (Princeton, N.J., 1989), which identifies clear origins for the mobilization of scientific resources by states in that century.

[9] Indeed, if scientific experts and their practices were understood to be wholly creatures of the state, I believe they would possess even less authority than they currently seem to have.

sold, it can orient itself toward consumer demand, it can be mobilized in the service of political authority, and entire scientific disciplines can flourish or decay as the expertise they offer passes into or out of the public's favor.

The second element of scientists' expertise consists of their role in representing what our society knows about nature and society. Put another way, scientists' expertise consists of guarding, increasing, and organizing the stockpile of what once might have unproblematically been called "theory" or "theoretical knowledge." To be sure, to invoke a distinction between theory and practice, as I am doing here, appears to run against the grain of contemporary STS, which routinely denies that anything called "theory" exists separate from the practices of its generation and application. As an analytical point made by historians and sociologists about the contexts of scientific knowledge, this has proven fruitful. But here, too, what those in STS take to be an obvious truth flies in the face of a distinction between science and technology or pure and applied knowledge that is equally self-evident to most of the public.[10] Kitcher has addressed this issue by invoking the image of a chemist who studies a particular molecule from a purely theoretical point of view, perhaps in terms of its distinctive structure or the kinetics of its reactions. He then proposes we try telling this researcher that there is no meaningful distinction between pure and applied research, or between science and technology, and watch the response such a claim would elicit. Now multiply this chemist's sensibility by millions of similarly inclined people (i.e., nearly everyone who would have an opinion at all), and one begins to appreciate how much the standard STS view of science departs from the public's.[11]

In order to uncover and evaluate the roots of modern scientific expertise, therefore, it is crucial that we restore theory and practice as categories that have been constitutive of the study of nature virtually since it can be found in history as an organized activity.[12] At first glance it may seem that theory and practice represent the two elements of modern expertise I identified above. But scientific expertise cannot be packaged as neatly as this suggests. One part of scientists' expertise, their expert practices, consists of these practices' being one kind of skilled labor in an exchange economy. This kind of expertise is strongly demarcated socially. By contrast, the theoretical element of expertise does not operate in a defined social milieu of exchange. Instead of directing a technology or the manufacture of a product, or guiding an action in the political arena, scientists' theoretical knowledge exists simply to frame our understanding of reality. Moreover, this function is essentially ideological, as will be shown below, and therefore it has no clear social demarcations. As a consequence the two parts of expertise function in strikingly different ways. Whereas occupational expertise exists because some people can be recognized as superior in performing certain skilled tasks, the theoretical element of expertise is not meant to be part of a system of exchange, and therefore the demonstration of the relevant expertise is not based in a product or service.

In order to understand why theory operates so differently in constituting expertise,

---

[10] On this point, see Peter Dear's article in this volume.

[11] Kitcher, *Science, Truth, and Democracy* (cit. n. 2), 89–90. In fact, this example is invoked by Kitcher in a chapter titled "The Myth of Purity." But while Kitcher argues against applying the distinction between pure and applied science too facilely and broadly, his point is not to deny that such a distinction is meaningful in principle. His aim is rather to show how it can be misused.

[12] See G. E. R. Lloyd, *The Ambitions of Curiosity: Understanding the World in Ancient Greece and China* (Cambridge, 2002).

we need to know something about the history of concepts of theory and practice. This will be discussed in greater detail further on, but for the moment let us simply recall that theory, defined as contemplation of the nature of things, can be identified as far back as Plato and Aristotle, when such knowledge was clearly and unequivocally separated from knowledge that had more direct practical uses. Only in the Enlightenment did the connection between theory and practice receive its modern formulation when, in addition to designating a set of truths about nature, theory also became a social good, a bedrock of knowledge on which enlightened society could engineer its own progress. That is, whereas previously the possession of theory or learnedness had been an ascriptive quality of the educated person, in the Enlightenment theory assumed the guise of truths of direct use to society, and the means of advancing theoretical knowledge and making such knowledge available to society became issues of supreme importance. Most historians would take it as conventional wisdom that the Enlightenment marked the first time when a public and widely reiterated mandate was placed on scientific knowledge to make itself useful for social improvement. In effect, the claim that a rationally organized society necessarily depended on science for its definition projected scientific knowledge as an ideology, and this same ideology still guides a great deal of public policy making today. This point is fundamental for the story being told here, for this ideology removed theoretical knowledge from its former role as the possession of a certain class of individuals and transformed it into a pillar of the Enlightenment as a whole.

This program for the rationalization of society raised two questions that were already recognized during the eighteenth century and have been repeatedly discussed ever since. First, how was such knowledge to be made available to an enlightened society? Second, and related to the first, if the program of spreading enlightenment were to succeed, what social role would scientists play in an enlightened society? Would they continue to be recognized as those individuals whose role (one of whose roles) was to be oracles for what science claims to be true about the world? Or would the public, because it had achieved the necessary enlightenment to understand and give its assent to the claims of science, assume the position formerly occupied by scientists? As we shall see, these questions motivated the thinking of the French mathematician Jean-Antoine-Nicolas de Caritat, Marquis de Condorcet, in the 1790s, at the moment when the French Revolution was about to enter its most radical phase. Condorcet argued that citizens in the new French republic were to be educated in the rational and scientific basis of the new social order, but they were not to have an equal share in the determination of what that rational basis was. That job was to be left to experts—to scientists. Condorcet's formulation of this issue is no mere historical curiosity, however, for a strikingly similar answer was offered by American scientists and educators in the late 1940s and 1950s, when they confronted the question of how to educate the public about natural science. As John Rudolph has shown, American scientists too envisioned education in science as giving citizens the rational tools for assenting to scientists' claims about the natural world.[13]

The argument developed here concerning the ideological character of scientific knowledge in modern societies bears directly on another strand in the STS literature, one that focuses on the relationship between scientific expertise and the liberal-

---

[13] Rudolph, *Scientists in the Classroom: The Cold War Reconstruction of American Science Education* (New York, 2002).

democratic state. This work takes its point of departure from a supposed threat posed by scientists' expertise to the function of liberal-democratic systems that takes one of two possible forms: The first possibility is that such expertise reduces the exercise of democratic choice by the public and its attendant debates to mere theatrics, since the experts have the ultimate say over which alternatives are feasible and how they might be implemented. The second is that the government, acting in the name of the collective good, enlists such experts to formulate and carry out its own policies, thereby sacrificing the neutrality it ought to maintain vis-à-vis the public's right to have a share in determining which among different courses of action should be taken. Either way, the experts' superior grasp of the matter at hand puts them in the position to dictate which choices in given situations are to be made.

Collins and Evans's project to inaugurate the "third wave" of science studies is clearly meant to mitigate these difficulties by examining how scientific expertise can be reconciled with the interests and knowledge of others who may lack the scientists' credentials and professional stature, but who may be in a position to contribute meaningfully to the decision-making process. Other scholars whose interests more directly pertain to political theory, such as Yaron Ezrahi and Stephen Turner, have offered explanations of how expertise functions in liberal-democratic polities that reaffirm the viability of reconciling expertise with democratic politics.[14] I will turn to these approaches in the conclusion, noting for the present that by discounting the ideological function of scientific expertise, Turner and Ezrahi suggest that expertise can be reconciled with liberal democracy in ways that I argue need to be reconsidered. My point is not to revive the criticisms of Max Horkheimer and Theodor Adorno and, following their lead, the early Jürgen Habermas, who attacked the instrumentality of scientific knowledge for reducing democratic function to a mere sham.[15] Yet we need to recognize that because scientific expertise functions ideologically in certain respects, it is not subject to criticism in the same way that scientists can be criticized and engaged with when they function as occupational experts or as advisers to governments in formulating policy choices.

Ultimately, therefore, the scientist's role in society is a complex amalgam consisting of one part occupational expertise and one part ideological or theoretical expertise, with the two joined by the rhetoric linking theory and practice that pervades our post-Enlightenment understanding of science and its social uses. In what follows, I will explore both varieties of expertise in turn. This should not be understood as a historical account in the conventional sense, one that presents a narrative of expertise from its earliest roots to the present. Instead the aim here is more essayistic and analytical in nature: to uncover the roots of expertise in its contemporary guise.

[14] Ezrahi, *The Descent of Icarus: Science and the Transformation of Contemporary Democracy* (Cambridge, Mass., 1990); Turner, "What Is the Problem with Experts?" *Soc. Stud. Sci.* 31 (2001): 123–49. Turner's view of expertise in liberal democracy was expressed somewhat less optimistically in his *Liberal Democracy 3.0* (cit. n. 2). There he saw expertise as corroding the valuation formerly placed on civil society as a cornerstone of liberal democracy.

[15] On the instrumentalization of reason with science as a prime exemplar, see Horkheimer and Adorno, *Dialektik der Aufklärung: Philosophische Fragmente* (Frankfurt, 1988), especially the first essay, "Begriff der Aufklärung." Addressing similar themes and still eminently readable are two essays by Habermas from the 1960s, "Technical Progress and Social Life-World" and "Technology and Science as 'Ideology,'" in Habermas, *Toward a Rational Society: Student Protest, Science, and Politics*, trans. Jeremy J. Shapiro (Boston, 1970), 50–61 and 81–122.

EXPERTISE AS PRODUCTIVE KNOWLEDGE

Experts are everywhere, as are experts on experts. Psychologists and educators often refer to "expertise" when describing the differing talents of children, especially those individuals with unusual abilities for playing chess or the violin.[16] Philosophers employ the term in mounting criticisms of the argument from authority, an old chestnut among the list of fallacies, and an issue relevant for the practices of experts in situations such as courtroom testimony.[17] Whatever their individual approaches, all of these experts on experts emphasize the ubiquity of expertise in our society as a touchstone of modernity. Or, as Thomas Haskell put it in the introduction to one well-known volume of this genre, "All of us defer to the authority of experts."[18]

Yet Haskell appears to assume too much. That experts are everywhere can scarcely be doubted. But how is it that we come to believe that experts have "authority"? As Haskell's discussion unfolds, he throws a bewildering array of expert practices together to exemplify his meaning: weather forecasts shape our day-to-day plans; dentists tell us that a tooth needs to be filled or extracted; mechanics attribute difficulties with an automobile to a faulty voltage regulator; biologists tell us that there is sufficient evidentiary basis for asserting that life has evolved on Earth by a process in keeping with Darwin's description of it; and so on. To be sure, all of these examples involve interacting with others in one way or another. Whether they all involve submitting to "authority," as opposed to merely transacting the purchase of a service, as is clearly the case in a couple of his examples, is another matter.[19] At the very least, a consideration such as this one should prompt us to take a careful look at what authority has to do with expertise, without assuming too much about how the connection is constituted.

We should be wary, too, of any claims that the ubiquity of expertise is unique to modern societies. Eliot Freidson, the late sociologist whose entire career was spent studying the sociology of the professions, appeared to stake a claim for expertise as a strictly modern phenomenon when he argued that to do away with experts, we would have to return to a time just before the Industrial Revolution, and "abolish all but small, largely economically self-sufficient communities and accept a very severe reduction in the number of needs permitted to their inhabitants." In the very same paragraph, however, Freidson traced a far more ancient root for expertise in the basic fact that for any society to develop an acceptable standard of living, it would have to accept some expertise in the division of labor.[20]

---

[16] E.g., Carl Bereiter and Marlene Scardamalia, *Surpassing Ourselves: An Inquiry into the Nature and Implications of Expertise* (Chicago, 1993).

[17] Douglas Watson, *Appeal to Expert Opinion: Arguments from Authority* (University Park, Penn., 1997).

[18] Haskell, ed., "Introduction," in *The Authority of Experts: Studies in History and Theory* (Bloomington, Ind., 1984), ix–xxxix, on ix.

[19] The use of "authority" here is probably an indirect allusion to what Robert Merton described as the "cognitive authority" of scientific experts. For a discussion of this point, see Turner, "What Is the Problem with Experts?" (cit. n. 14), 128.

[20] Freidson, "Are Professions Necessary?" in Haskell, *Authority of Experts* (cit. n. 18), 3–27, on 15. For a more comprehensive picture of professions and the division of labor, see Freidson, *Professionalism: The Third Logic* (Chicago, 2001). The latter book, Freidson's final testamentary work on the sociology of the professions, demonstrates how thoroughly his idea of the professions was linked to a functionalist view of modern societies, and it featured his utopian belief (or hope) in the professions' continued ability to maintain control over their own practices.

If expertise is a universal consequence of the division of labor, as it undoubtedly is, then our task must be to explain how some forms of expertise came to be vested with the authority alluded to by Haskell. After all, our ultimate aim here must be to understand the nature of scientific expertise, and it will not serve our purposes to permit scientists to remain lumped together with plumbers, weavers, and tinsmiths in our inquiry. Yet in premodern society—that is, for the purposes of this article, Europe before 1650 or 1700—this is largely just where astronomers and astrologers, alchemists, apothecaries, and architects belong, as several among a large assortment of occupations.[21] Needless to say, these expert practitioners can be differentiated in any number of ways, including in terms of their place in the social hierarchy. In this regard, someone's claim to the possession of theory or learnedness was a significant marker of social distinction.

In order to understand such claims in the premodern era of history, we can make no more useful beginning than to recall Aristotle's distinction between what he names the "intellectual virtues" in book 6 of his *Nicomachean Ethics*. He begins by distinguishing two kinds of rational knowledge, one that contemplates things that are changeless and universal, and the other that contemplates what is changeable. Aristotle devotes little effort to describing the first form of knowledge, saying only that things known in this way are known by necessity and demonstratively.[22] He cites no examples, but others would later adduce geometry, mathematics, and astronomy as exemplary kinds of theory, and by the Middle Ages this category had become quite capacious, including theology, metaphysics, and even Aristotle's own writings on physics. In terms of knowledge of things that are changeable, Aristotle had in mind two principal kinds: *phronesis*, which can be called practical wisdom or prudence, and *techne*, the knowledge of how to make things. This distinction between phronesis and techne in turn arose from consideration of the difference between action and production, a distinction that Aristotle claimed was well established and therefore one that he need not linger over.[23] Both involve deliberation over means and ends, but the end sought in action, or *praxis*, by means of phronesis is happiness, which makes phronesis the domain of politics and ethics, whereas the ends of techne are a house, a pair of sandals, or the healing of a sick person.

Although occupying only a modest portion of a lengthy treatise on ethics, the division between forms of knowledge introduced by Aristotle would have a long, long history, not only because of the legacy of the *Nicomachean Ethics* as Aristotle's most important writing on ethics, but also because his division resonated with later writers down through many centuries. Kant's distinction in his first two critiques between "pure reason" and "practical reason," to take only one obvious example, shows that

[21] This point might appear similar to that made in Pamela H. Smith, *The Body of the Artisan: Art and Experience in the Scientific Revolution* (Chicago, 2004). However, I am not arguing the question of whether artisans did or did not contribute to the new sciences equally with those individuals more obviously identifiable as "scholars."

[22] Aristotle, *The Nicomachean Ethics*, trans. J. A. K. Thomson, rev. with notes and appendixes Hugh Tredennick (London, 2004), 144–66, esp. 148. For a helpful exposition of issues at work in book 6, see C. D. C. Reeve, "Aristotle on the Virtues of Thought," in *The Blackwell Companion to Aristotle's "Nicomachean Ethics,"* ed. Richard Kraut (Oxford, 2006), 198–217.

[23] Aristotle, *Nicomachean Ethics* (cit. n. 22), 149. The distinction between the two may not have been so self-evident even to Aristotle's audience, for, as Pamela Long has noted, there was language that linked praxis and techne in the Periclean era of Greek history; see Long, *Openness, Secrecy, Authorship: Technical Arts and the Culture of Knowledge from Antiquity to the Renaissance* (Baltimore, 2001).

the Aristotelian distinction was still alive and well as late as the 1780s. But by the later eighteenth century two important developments in the conceptualization of theoretical, practical, and technical knowledge gave a new pattern to their relationship. As Reinhart Koselleck argued more than fifty years ago, in the late eighteenth century the practice of "criticism" began to be touted as a powerful form of action or praxis, a process that aestheticized praxis and decontextualized it from its previous connection with politics and ethics.[24] At the same time, the Enlightenment program for useful knowledge tended to forge a tight link between theory and productive knowledge, whereby theoretical insight was thought to be applicable instrumentally.

The distinction between theoretical knowledge on the one hand and practical or technical knowledge on the other had considerable resonance beyond the writings of philosophers. In the Middle Ages, as Jacques Verger has shown most recently, one key factor in differentiating occupations hierarchically was their attachment to literacy and higher schooling, and especially literacy in Latin, which functioned as every scholar's second language. Latin played such a decisive role in defining scholarly identity, Verger reminds us, because of its centrality in two crucial respects. First, it was the language of scripture (the Vulgate Bible) and of church liturgy. It was thus a quotidian presence in every Christian's life and, because it was unintelligible to the uneducated and largely illiterate population, a clear reminder of who could and could not claim the privilege of its secrets. Second, Latin was the language of antiquity, and even before the emergence of the classicizing cultural movement of humanism in the late fourteenth century, access to the writings of the ancients played a clear role in demarcating a scholarly caste, with universities serving as gatekeepers to membership.[25] For our purposes, a third feature of Latinate literacy is noteworthy, which is that a university education was understood to confer on students the knowledge of theoretical principles on which professional identities depended.

Although higher education in premodern Europe was clearly intended to make its beneficiaries socially useful in a general way, it was not necessarily aimed toward providing a specific set of theoretical tools that would be applied in practice.[26] Pamela Long's article on efforts to improve the water system in sixteenth-century Rome illustrates this point nicely in terms of how the display of literary accomplishment in the humanist style was variously connected with occupational expertise.[27] Long's study covers two specific projects: the efforts to control the flooding of the Tiber River after the devastating floods of 1557 and a slightly later effort to restore the ancient Roman aqueduct, the Acqua Vergine, in 1570. She examines a number of writings authored by men from quite variable occupations, all of whom drew upon ancient sources to buttress their recommendations for how these projects should be carried out. Whatever a particular applicant's occupational position, it seems, all felt

[24] Koselleck, *Critique and Crisis: Enlightenment and the Pathogenesis of Modern Society* (Cambridge, Mass., 1988), 98–123, originally published in 1959. See also Nicholas Lobkowicz, *Theory and Practice: History of a Concept from Aristotle to Marx* (Notre Dame, Ind., 1967), 215–34. Lobkowicz's discussion pertains to the journalism of the left Hegelians in the 1830s and 1840s, before Marx and Engels arrived on the scene.

[25] Verger, *Men of Learning in Europe at the End of the Middle Ages,* trans. Lisa Neal and Steven Randall (Notre Dame, Ind., 2000).

[26] In fact, Verger specifically denies any direct theory-practice linkage in medieval higher education; see ibid., 102–3.

[27] Long, "Hydraulic Engineering and the Study of Antiquity: Rome, 1557–1570," *Renaiss. Quart.* 61 (2008): 1098–138.

it necessary to frame their project proposals and related contributions in the language of humanist reverence for ancient authority.

As Long points out, all of the actors in her story shared the conviction, as she puts it, that "classical antiquity offered essential guidance for the solution of contemporary problems."[28] As a general statement of the aims and methods of humanism, this captures the mentality of the figures in Long's story. Yet it is not clear whether the "essential guidance" provided by the ancients was believed to furnish specific theoretical and practical aids for remedying the Tiber's flooding and restoring the Acqua Vergine, or whether such references were offered as displays of their authors' suitability for papal patronage by virtue of possessing the requisite literary accomplishments. To be sure, there is no need for us to draw a firm line between these two alternatives, and the actors probably did not do so either. But the essential point is that the display of humanistic learning had no effective role in determining who was, or was not, an expert in the particular circumstances, nor was it meant to.

The variable association between theory or learnedness and occupational expertise is evident as well when we turn our attention to healing in the same period.[29] The inclusion of medical teaching as part of university structures, first in Bologna, Montpellier, and Padua, then in northern European universities such as those of Paris and Oxford, established physicians as a distinct class of healers defined by their command of medical theory inherited from translated Arabic-language sources and taught as part of university curricula. What is conspicuous about physicians in this context, in comparison to the other medieval university professions of law and theology, is that they practiced their occupation in a social environment where they faced competition from other healers. Some of these had identifiable occupations—midwives, surgeons, barbers, and apothecaries, for example—but many other individuals undertook healing less formally as a part-time occupation. Such an efflorescence tells us that physicians enjoyed no particular advantages in terms of their expertise at the bedside, except insofar as their learnedness would make them more or less acceptable socially to similarly literate classes.[30]

Yet for all their lack of a clear advantage as healers, physicians' theoretical knowledge evidently did count for something, because surgeons—practitioners of a lower-class occupation because it was a manual craft—made strenuous efforts to advance their own position as Latinate scholars in command of theory. As early as the fourteenth century in Paris, the prominent French surgeon Guy de Chauliac wrote his widely circulated *Chirurgia Magna* to argue specifically that surgery too commanded the same theoretical and ancient grounding as that claimed by physicians.[31] Guy's initiative would be echoed down through the succeeding years to the time of Jacopo Berengario da Carpi and Andreas Vesalius in the sixteenth century, who likewise

[28] Ibid., 1099.

[29] Learnedness and theoretical knowledge obviously were not the same thing. One could be learned without claiming to possess theory, although in the premodern context the converse—claiming to have theory without being learned in the classical heritage—was more difficult to maintain.

[30] For general discussions of medieval physicians, see Nancy G. Siraisi, *Medieval and Early Renaissance Medicine: An Introduction to Knowledge and Practice* (Chicago, 1990), and Vivian Nutton, "Medicine in Medieval Western Europe, 1000–1500," in *The Western Medical Tradition: 800 BC to AD 1800*, ed. Lawrence I. Conrad et al. (Cambridge, 1995), 139–206.

[31] Cornelius O'Boyle, "Surgical Texts and Social Contexts: Physicians and Surgeons in Paris, c. 1270 to 1430," in *Practical Medicine from Salerno to the Black Death*, ed. Luis Garcia-Ballester et al. (Cambridge, 1994), 156–85.

defended—in Latin, needless to say—surgery's learned status by drawing out its theoretical basis in the study of anatomy. The need to reiterate over and over the learned status of surgery strongly suggests that, whatever the technical accomplishments of its most renowned experts and advocates, surgery remained a practice of somewhat precarious social status.[32]

The defense of surgery's occupational status by linking it to anatomy reminds us that in premodern Europe it certainly was possible for theory to be thought of in connection with occupational expertise, even if there was no necessary connection in a larger sense. Another good example of this can be found in the *Disputationes contra Deliramenta Cremonensia* (Disputations against Cremonan Absurdities) by the fifteenth-century astronomer and printer Regiomontanus. In the preface to that work, Regiomontanus took to task those individuals who acquired a superficial knowledge of astronomy by perusing the basic texts of John of Sacrobosco or Gerard of Cremona, and then presumed to advise princes by casting horoscopes. His point was not that astronomy could not provide the theoretical underpinnings for the practice of astrology. After all, no less an ancient authority than Ptolemy himself had sanctioned the connection between them, and Regiomontanus too cast horoscopes. His point was rather that to become a useful astrologer required a deep and detailed knowledge of astronomy, far more than what one could obtain from an elementary text such as Sacrobosco's.[33]

As these several examples suggest, access to theoretical knowledge in different forms could contribute significantly to shaping one's social status in the premodern world. It did not do so because theory was thought to be directly applicable to occupational practices, although as we have seen, it was possible to make such claims. Before concluding this section, two further observations about expertise in the premodern context will help us specify what has and has not changed between the premodern world and our own. First, although the range of occupations in premodern society can be identified in any number of ways—for example, in tax records and wills, by means of trade organizations such as guilds, or in university subjects and faculties—the same cannot be said for experts. That is to say, in the premodern world occupational expertise always became manifest in the delivery of the product itself, and only there. Individuals in premodern society knew what expertise was—even if they did not have a word for it, they had guides for anticipating whether any particular individual commanded the expertise necessary for a given job—but ultimately someone's status as an expert revealed itself in a transaction. A university degree, the grant of a title by a king, or even the patronage of a pope did not make someone an expert.[34]

---

[32] Roger French, *Dissection and Vivisection in the European Renaissance* (Aldershot, 1999); French, "Anatomical Rationality," in *Medicine from the Black Death to the French Disease*, ed. French et al. (Aldershot, 1998), 288–323.

[33] Regiomontanus, *Disputationes contra Deliramenta Cremonensia*, lines 31–60. Thanks to my colleague Michael Shank for sharing with me his unpublished translation of this work. More generally on astrology as an expert practice, see Shank, "Academic Consulting in 15th-Century Vienna: The Case of Astrology," in *Texts and Contexts in Ancient and Medieval Science: Studies on the Occasion of John E. Murdoch's Seventieth Birthday*, ed. Edith Sylla and Michael McVaugh (Leiden, 1997), 245–70, and Darin Hayton, "Expertise ex Stellis," *Osiris* 25 (2010): 27–46.

[34] My interpretation differs in this respect from the excellent introduction to the *Osiris* volume on expertise and the early modern state by Eric Ash, who asks rhetorically when someone "becomes" an expert. This seems to assume a normative or transhistorical definition of the term, whereas my definition attempts to maintain a highly contextualized stance. See Ash, "Introduction: Expertise and the Early Modern State," in *Expertise* (cit. n. 8), 1–24, on 6.

This brings me to my second point about occupational expertise, which is how deeply rooted it was in social relations dependent on trust, precisely the kind of trust described by Steven Shapin in *A Social History of Truth*. Shapin underscores how dependent early modern natural philosophers were on communally shared norms of truthfulness and consistency in their dealings with one another. He characterizes these relationships as ones based on trust, and his interpretation gives trust a strong moral coloration.[35] This placement of social relationships based on trust in a moral context is supported both theoretically, by sociologists such as Georg Simmel, whom Shapin cites, as well as by the basic definitions of action or praxis we encountered earlier in Aristotle's ethics. Trust clearly operates as well in the context of occupational expertise. In advance of actually obtaining the product of someone's putative expertise, the purchaser of the service had to exhibit trust, a trust warranted by a variety of factors, including in some cases evaluations of the person's command of ancient literature or theoretical knowledge. Here again, it should be stressed that command of theory was not necessarily a factor in making someone more *expert* than another person, but instead more *trustworthy*.

Needless to say, a great deal has changed between the period of history described in this section, when transactions with experts largely involved one-on-one negotiations of the kinds featured in the examples I have provided (the hiring of engineers to repair aqueducts, the patronage of astrologers by rulers seeking advice), and today, when experts occupy roles as members of institutionally credentialed and often licensed professions, masters of unintelligible mathematical formulations, and so forth. To try to treat that story would take us far beyond the limits of a single article. Yet I think modern historians too often fail to appreciate how many of the basic institutional mechanisms for bestowing a social and legal imprimatur on occupational expertise had already appeared by the later Middle Ages. Thus, although a rich and useful story could be told about how occupational expertise evolved in its relationship to learnedness after this time, perhaps enough has been said about its basic principles to justify not doing so here.

THE TRANSPARENCY OF TRUTH

When Aristotle, or Regiomontanus, or Vesalius, claimed to have theoretical knowledge of a particular topic, the claim was made in reference to an object of the understanding and served to describe the nature of the belief under consideration. However, in the eighteenth century theoretical knowledge took on a modified significance in three important respects. First, it became routine to insist that theory ought to drive practice, and what justified someone's claims to possess theoretical knowledge was the ability to put such knowledge to work in one way or another. Of course, demands for useful knowledge had already been voiced in the seventeenth century, most prominently in the rhetoric of Francis Bacon, yet only in the eighteenth century did those calls become widespread and insistent. Second, the most prominent feature of the program to link theory and practice was its aim of achieving the reform of society. Society was to be placed on a rational basis, so the argument ran, with the

[35] See Shapin, *A Social History of Truth: Civility and Science in Seventeenth-Century England* (Chicago, 1994), chap. 1, esp. 7–8, where the author restricts his use of *trust* to social interactions and assigns them a moral valence.

sciences providing lessons regarding what constituted the natural order. Third and most important for the purposes of this article, the reformation of society was to be a thoroughgoing one, embracing not only social and political institutions, including schools, churches, and systems of taxation, but also popular attitudes. The public was to be taught to discard its superstitious ways and adjust its habits and attitudes to a more natural—and therefore more authentic—basis.

Thus, although the Enlightenment was partially a reform from above, as seen in the way that various academies of science and other reformist institutions busied themselves with projects of improvement, its advocates also sought to enlist the public's active participation in making things happen. The consequence of this intention was to focus attention on the question of how the public should be taught to use its reason. One widely copied approach was popularized in the early part of the century by Joseph Addison and Richard Steele in their papers the *Spectator* and the *Tatler*, where an anonymous commentator spoke not ex cathedra, as a priest would, but instead as a neighbor and partner in conversation. But Addison's and Steele's papers were concerned with moral philosophy and the arts, not with natural science. It was one thing to teach the public about stoicism and cultivate its taste for Italian opera; it was quite another to educate it about the mechanics of motion. With respect to the latter form of education, public lecture courses on natural philosophy began to be offered throughout Europe. Judging by their numbers and the number of different places one can find such activity, these lecture courses found a sizable market.[36]

Accompanying these new forms of pedagogy there arose an important question: If the public was to be educated into the truths of the new science, what was to be the result of such pedagogy and what sort of relationship between teacher and pupil would produce the most effective outcome? Would there come a time when the public had attained a sufficient level of enlightenment that it no longer needed instruction from its masters? Addison's and Steele's essayistic style suggested that the distance between teacher and pupil was not very wide. Rousseau took a similarly undogmatic line in *Emile*, undoubtedly the most influential treatise on pedagogy in the entire century. Students must be permitted by their tutors to explore the world with a considerable degree of freedom to perform their own trials and draw their own conclusions. No permanent social barrier, no fixed category of occupational expertise, would separate the young Emile from his tutor, for ultimately the world itself was his teacher.[37]

Yet other writers postponed the public's emancipation from tutelage to a more distant future, or simply canceled the graduation ceremonies entirely. The French physiocrats, who engaged with this question as part of their examination of the relationship between social science and pedagogy, were explicit in this respect. Writing in 1767, Pierre Le Mercier de la Rivière argued that the social order must be based on the natural order, and for this to be possible, the public must be educated in science, in order that it be able to appreciate the rational basis of society's laws. It was nei-

---

[36] See esp. Larry Stewart, *The Rise of Public Science: Rhetoric, Technology and Natural Philosophy in Newtonian Britain, 1660–1750* (Cambridge, 1992); Jan Golinski, *Science as Public Culture: Chemistry and Enlightenment in Britain, 1760–1820* (Cambridge, 1992); Geoffrey V. Sutton, *Science for a Polite Society: Gender, Culture, and the Demonstration of Enlightenment* (Boulder, Colo., 1995); Michael R. Lynn, *Popular Science and Public Opinion in Eighteenth-Century France* (Manchester, 2006).

[37] Jean-Jacques Rousseau, *Emile*, trans. Barbara Foxley (Rutland, Vt., 1993), book 2.

ther necessary nor desirable that everyone have an equally thorough knowledge of science, Le Mercier continued, but it was important that everyone have an education sufficient for understanding the enlightened laws that are in place. In this way, reason would displace coercion as the principle of government, becoming thereby "the sole despot of the universe." This same theme of legal despotism was echoed by Pierre-Samuel Dupont de Nemours, who declared that legal despotism was the first principle of rational government, followed closely by public instruction as the necessary second principle.[38]

It is conspicuous that both Le Mercier and Dupont believed that the public had nothing to fear from legal despotism, because enlightened public opinion, instructed in the use of reason by the schools, stood ready to criticize and amend the ruler's and his ministers' erroneous ways. They appeared unaware of any possible asymmetry between the public's rudimentary grasp of social science and the more advanced knowledge possessed by themselves and their colleagues. Presumably, this is because they saw themselves serving as neutral arbiters in any disputes, making their expertise available either to the public or to the government, as needed. Thus, as both teachers and consultants, they can be seen as having had a kind of expertise, but it was a distinctly different sort of expertise from what we saw previously in the case of occupational expertise. There, expertise was related to productive knowledge or techne, and it was manifested in the act of production itself. Here, by contrast, the expertise was represented by a claim to possession of theory, which, crucially, is not subject to validation in a social transaction via production.

Undoubtedly the most vivid illustration of how scientists could serve as ideological guardians of truth for an enlightened society came from Condorcet. Condorcet made his name as a mathematician while a young man in the 1760s, but his involvement in political life during the ministry of Turgot in the mid-1770s turned his thinking toward the social sciences and the application of probability to the determination of outcomes in balloting. By the time the French Revolution began in 1789, Condorcet had become both a leading figure in the French scientific elite, occupying the position of secretary of the Academy of Sciences, as well as an outspoken advocate for the reform of the French political system. Unlike most other aristocrats, even those who generally supported the reformist aims of the early revolution, Condorcet strongly advocated the full leveling of political rights and universal suffrage, including granting the vote to women. Such an equality of political and civil rights, however, was only possible in an environment where the public had been educated to its new responsibilities. Thus, for Condorcet the eventual realization of France's revolutionary aspirations was fundamentally a pedagogical problem and could only find its fulfillment in the context of a thoroughly reformed system of education. Condorcet's opportunity to take a hand in crafting a new educational structure for France came in 1791, when as a member of the National Assembly he was elected to the presidency of the Committee on Public Instruction. In that capacity he authored a series of ambitious *mémoires* that envisioned a complete renovation of the French schools.[39]

---

[38] The ideas of the physiocrats are discussed in Keith Michael Baker, *Condorcet: From Natural Philosophy to Social Science* (Chicago, 1975), 290–2.

[39] Condorcet, *Cinq mémoires sur l'instruction publique*, vol. 1 of *Écrits sur l'instruction publique*, ed. Charles Coutel and Catherine Kintzler (Paris, 1989).

Condorcet's ideas on education faced two contrasting imperatives. First, a democratic system that offered everyone full political and civil rights was only possible if the public was educated to the rational understanding of those rights. But civil and political equality would by necessity coexist with a naturally unequal distribution of talent and mental capacities. Like Rousseau before him, Condorcet had no doubt that the distribution of talents and intellectual skills, and therefore the potential to acquire the full measure of enlightenment, was unequal.[40] In light of this situation, a new educational system would face the need to reconcile providing all the basics necessary to guarantee their rights with offering to talented individuals the opportunities they needed to develop their abilities and contribute to the progress of humanity. Condorcet's answer to this quandary was to do away completely with the existing schools that offered different kinds of teaching to students from different levels of the social hierarchy. Claiming that educational inequality was one of the principal pillars of tyranny, Condorcet insisted that the schools that had supported both the hereditary elites of the past and the "priests, jurists, men who possessed the secrets of commerce, and physicians" had to be abolished.[41]

Although Condorcet's writings on education are richly detailed and interesting, here we can only concentrate on a few main points.[42] Condorcet envisioned a universal system of primary education that would provide all students with the tools they needed to have a basic level of enlightenment. He expressed a desire that the progressive division of labor in manufactures, as described by Adam Smith, not result in workers becoming stupid by making their labor simple and repetitive, and his plan for primary education was designed to avoid this danger. Therefore all students should receive comprehensive instruction in reading, writing, mathematics, and the natural and social sciences. The aim was not to make them advanced practitioners of these subjects, but to endow them with a basic comprehension sufficient to awaken their critical faculties. Thus armed, graduates of the schools would be in a position to avoid being duped by charlatans and others because they would have the knowledge to see through their schemes.[43]

Primary education performed two other essential functions. First, as Condorcet wrote, it was the task of primary schools to select and increase that class of talented individuals "whose impartiality, disinterestedness, and enlightenment must find their [natural] end by leading opinion." Although all children must be presented the same basic course of instruction, the matter could not rest there because the unequal distribution of talents meant that ultimately those students were destined for different roles in society. Second, it was just as important that those students who were not destined for leadership or scholarly positions avoid succumbing to envy or cultivating improper ambitions. Instead, by virtue of their rational understanding of the social

---

[40] On this point, see John Carson, *The Measure of Merit: Talents, Intelligence, and Inequality in the French and American Republics, 1750–1940* (Princeton, N.J., 2007), 19–26.

[41] Condorcet, *Cinq mémoires* (cit. n. 39), 37.

[42] For a fuller exposition, see Baker, *Condorcet* (cit. n. 38), 293–303, and Francisque Vial, *Condorcet et l'éducation démocratique*, repr. ed. (Geneva, 1970). Also useful is R. R. Palmer, *The Improvement of Humanity: Education and the French Revolution* (Princeton, N.J., 1985), esp. chap. 4 on Condorcet's plan.

[43] See Condorcet, *Cinq mémoires* (cit. n. 39), 52, on the division of labor. Condorcet discussed primary education in the second *mémoire*, the longest and most detailed of the five. See 107 on how a basic knowledge of mathematics and mechanics can help citizens understand whether proposals for something like the construction of a bridge are founded on sound principles or not.

structure they must learn to accept "their modest and tranquil (*paisible*) careers."[44] This was arguably the linchpin of Condorcet's entire conception of education, for it inserted a rationally based, enlightened form of intellectual dependency into the very heart of his utopian social vision. He acknowledged this would be no easy task, requiring a long period of transition to achieve:

> That confidence in a profound reason whose steps one cannot follow, that voluntary submission to talent, that homage to renown, these cost self-love so dearly that it will be a long time before they become habitual sentiments, and not a kind of obedience forced by imperious circumstances and reserved to times of danger and trouble.[45]

In light of their crucial role in determining the truths of social science, scientists (*savants*) were accorded considerable attention by Condorcet. He refuted any suggestion that a reorganized Academy of Sciences be made directly responsible for teaching, arguing instead that it should be placed in overall control of the curriculum, but with responsibility for selecting and paying teachers left to individual departments in France. Just as important, it was up to scientists to pursue their investigations, to perfect their understanding of the natural and social sciences, and to insure that this knowledge was represented in the school curricula. More important than the details of those sciences was the precise and analytical spirit of their mode of inquiry, for, as Condorcet pointed out, all knowledge, especially knowledge of the social sciences, was provisional and imperfect. Better that students learn to reason in such sciences clearly and precisely than learn to memorize a set of facts.[46]

Thus, confronted as he was by the most radical possibilities for instituting a society based on principles of democracy and social equality, Condorcet envisioned the completion of what for earlier generations had only been a theoretical problem: how to spread enlightenment and advance the rationalization of society. His solution to this problem was a pedagogical system that would grant everyone the basic tools of reasoning and scientific knowledge, but would reserve to the scientific elite the promulgation of the basic truths of nature and society on which the progress of society depended. Far from instituting a new social hierarchy, however, Condorcet emphatically denied that his new elite resembled the old order in any way. "Truth is the sole sovereign of free people," he enthused at the conclusion of his mémoires on education, "men of genius are its ministers."[47] Selflessly dedicated to the public, the intellectual elite in Condorcet's system served as witnesses to the truth, a new social role that could only exist with the creation of an enlightened society.

Over the short term, Condorcet's proposals for a hybrid educational system perched between elitism and popular enlightenment went nowhere. They were shouted down in a National Assembly that was becoming ever more radical in its ambitions for revolutionary change, and shortly thereafter Condorcet's arrest was ordered.[48] But the ideals of enlightenment that guided his proposals had a longer history, both in France,

[44] Ibid., 53–4; translation mine.
[45] Ibid., 65–6; translation from Baker, *Condorcet* (cit. n. 38), 301.
[46] Baker, *Condorcet* (cit. n. 38), 296–301.
[47] Condorcet, *Cinq mémoires* (cit. n. 39), 239.
[48] For a discussion of educated elites and democratic principles in France after Condorcet's time, see Carson, *Measure of Merit* (cit. n. 40), 64–8 and chap. 4, and Daniela S. Barberis, "Moral Education for the Elite of Democracy: The *Classe de Philosophie* between Sociology and Philosophy," *J. Hist. Behav. Sci.* 38 (2002): 355–69.

where they informed the education of the technical elite in the nineteenth century, and also in the United States between 1945 and 1965.[49] At first glance, the United States would seem an unlikely place to nurture Condorcet's elitist vision of an enlightened society. Thoroughly anti-elitist and anti-intellectualist to boot, the United States is the very last setting where one might expect scientists to be ennobled as ideological experts and guardians of truth. Yet the peculiar circumstances of the immediate post–World War II world, with the fascist form of totalitarianism recently vanquished and the Marxist form growing ever more threatening, made Americans receptive to language that was more similar to Condorcet's than even the actors themselves might have acknowledged. As articulated in the 1940s, the connection between science and democracy was at base quite simple: as two systems of thought and practice that depended on open inquiry, rigorous analysis, and free critique, science and democracy mutually supported one another and led to the cultivation of the same social values of equality, open-mindedness, and public-spirited contribution to the commonwealth. Robert Merton had already made this very same point in the dark days of 1942, and in the immediate aftermath of the war, talk like this could be heard everywhere.[50]

The distinctive ideological setting of the Cold War offered Americans a chance to speak explicitly about science and democracy, but, as had also been true for Condorcet, no consideration of science in the service of democracy after World War II could avoid addressing the issue of pedagogy as a vehicle for enlightening the public about science. And like Condorcet, they did so in the context of the relations between experts and the public. Two well-known speakers on this theme were Vannevar Bush, the MIT electrical engineer and head of the wartime Office of Scientific Research and Development (OSRD), and James B. Conant, the president of Harvard and, as things turned out, an influential advocate for the history of science in the general university curriculum.

In his popular review of wartime scientific research, *Modern Arms and Free Men*, Bush reflected on his experience as head of the OSRD, pointing out that the rapid pace of scientific progress raised important questions about how to control large armies and new weapons and whether democratic institutions were up to the task of governance. Not surprisingly, his answer in this popular exposition was *yes*. Bush discounted the assertion that control over the applications of science in warfare should be left to specialists. "We are not a dictatorship," he wrote, "we are a free people, and as we think, so will our public servants act." Given such a requirement for the public's role in governing, however, what must it know about science? "Must every citizen, then, grasp the full nature of atomic energy, evaluate the modern submarine, predict the consequences of supersonic flight?" Such a notion, Bush replied, was "absurd." Yet his prescription for how the public was to be involved in decision making in the face of such technical difficulties was surprisingly optimistic, even naive: in some almost mysterious way, public opinion should come to an understanding of the issues at hand and thereby exert its influence over the choices made on whether and how to apply scientific knowledge.[51] Given the importance assigned by Bush to

[49] Essential guides for the discussion that follows are Rudolph, *Scientists in the Classroom* (cit. n. 13), and David A. Hollinger, "Science as a Weapon in the *Kulturkämpfe* in the United States during and after World War II," *Isis* 86 (1995): 440–54.

[50] Hollinger, "Science as a Weapon" (cit. n. 49), 441–2; Rudolph, *Scientists in the Classroom* (cit. n. 13), 51–5.

[51] Bush, *Modern Arms and Free Men* (New York, 1949), 2–3.

the workings of public opinion, it is surprising that later in *Modern Arms and Free Men*, when the discussion turned to education, he raced lightly over the democratic functions of education and focused instead on the need to recruit more well-qualified students to the scientific professions.[52] In effect, Bush repeated Condorcet's concern that talented students be allowed to rise to the top, but he avoided paying specific attention to the other half of Condorcet's plan, the spreading of enlightenment.

A more specific plan for education was presented by Conant in a set of lectures he presented at Yale University in 1945 and 1946 on the theme of "understanding science." Like Bush, Conant opened his discussion by invoking the great distance that had opened up between the scientist's knowledge of the world and that of the average American citizen. Whether it was a question of making investments and choosing business strategies or of government decisions in matters of public health and weapons systems, the need to understand science better was manifest and urgent. Yet the solution to this problem was by no means an easy one: "No magic pill can be administered to make a person capable of matching wits with an expert," he wrote.[53] Conant did not support the idea that what the public required was more information about science. Instead, what the public needed was to learn more about the kind of thinking that scientists employ in their pure and applied research. His statement on this point is quite illuminating:

> Being well informed about science is not the same thing as understanding science, though the two propositions are not antithetical. What is needed is methods for imparting some knowledge of the tactics and strategy of science to those who are not scientists. Not that one can hope by any short-cut methods to produce in the mind of a non-scientist the same instinctive reaction toward scientific problems that is the hallmark of an investigator; but enough can be accomplished, I believe, to bridge the gap to some degree between those who understand science because science is their profession and intelligent citizens who have only studied the results of scientific inquiry—in short, laymen.[54]

From this point onward in his lectures, Conant offered episodes from the history of science as a way of guiding his audience into how scientists think.

The language employed by Bush and Conant differed from Condorcet's, but the aim of their thinking was very much the same. One has only to substitute the phrase "analytical rigor" for "tactics and strategy" in the quotation from Conant above, and his statement becomes one that Condorcet could easily have understood, and even articulated for himself. Yet it was clear to Conant, as it was to Bush and also Condorcet, that the public could never equal the scientist in terms of the latter's knowledge and expertise in the natural and social worlds. At best the public could learn to understand the rational basis for how scientists do their work. There was a clear division of labor between the scientist and the layman, but it was obviously not one related to the productive exchange of goods and services. Instead, the division was one that assigned to particular actors in society the task of maintaining and expanding the public's store

---

[52] Ibid., 98–104.

[53] The lectures were published in 1947 as Conant, *On Understanding Science: An Historical Approach*, and then under the title *Science and Common Sense* (New Haven, Conn., 1961), quoted on 2. The only revision made between the first and second editions was the removal of a late chapter in the original, which Conant claimed in a new foreword was no longer relevant to the main issue of the lectures.

[54] Ibid., 4.

of knowledge. Those who dedicate themselves to science, as all three never tired of pointing out, serve the public good, even when left to their own quirky researches. They are a society's ministers of truth.

CONCLUSION: THE SWEET DESPOTISM OF REASON

It has been argued here that our modern idea of the scientist as expert conflates two separate components: the first a fairly ordinary sense of expertise as represented socially by the division of labor, and the second an ideology that places the scientist in the position of articulating the laws of nature and society for the public. In the first case of expertise, the scientist is responsible for the delivery of a product, be it a cure for a disease, a bridge, or a new silicon chip. By contrast, in the case of ideological expertise there is no contract, no exchange of goods or services, and no imputation of trust—at least not the same kind of trust that we find in occupational expertise. The scientist's role is to deliver truth, and this function succeeds best when the scientist is understood to be speaking for the truth more or less transparently, and not as a member of a particular trade, guild, or profession. To conclude this article, I want to draw out some of the consequences of this argument with respect to the STS literature on expertise and democracy.

As mentioned in the introduction, the issue of whether scientific experts pose a threat to the practice of democracy has been formulated most extensively by the political philosopher Stephen Turner, but it has also figured in the writings of Yaron Ezrahi and obviously informs Harry Collins and Robert Evans's advocacy for their "third wave" in science studies. The issue is whether experts, by virtue of their command of technical details on questions like global warming or genetically modified organisms, preempt any possibility of informed debate by the public on how to deal with such questions. Turner's formulation of the issue is an excellent one. "If experts are the source of the public's knowledge," he writes, "and this knowledge is not essentially superior to unaided public opinion, not genuinely expert, the 'public' itself is presently not merely less competent than the experts but is more or less under the cultural and intellectual control of the experts."[55] The public may shout and wheeze all it wants, but if the experts are the only ones who are in a position to say which alternatives are available and feasible, then what share does the public have in determining outcomes? Note that this situation prevails even in cases where the experts disagree. Even where urgent action is called for, we all have to wait for the experts to come to a consensus, or else move boldly, and perhaps blindly, forward.[56]

Having set the table in this way, Turner then proceeds to divide experts into several different categories. His is an interesting and thoughtful analysis, but for my purposes I need dwell only on his first group, the "Type I" experts, whom Turner exemplifies with physicists. The expertise of physicists, he writes, poses no problem for democracy—not for the reason that we might first imagine, because quarks and muons are meaningless to our social and political lives, but instead because their expertise can be readily attested to by the public:

[55] Turner, "What Is the Problem with Experts?" (cit. n. 14), 125.

[56] Collins and Evans profess that their intention is to intervene only in those not-infrequent cases where action is called for but there is no scientific consensus pertaining to the matter at hand; see Collins and Evans, *Rethinking Expertise* (cit. n. 2), intro.

We all know (or have testimony that comes from users or recipients) about the efficacy of the products of physics, such as nuclear weapons, that we do accept, and we are told that these results derive from the principles of physics, that is to say the "knowledge" that physicists certify one another as possessing. Consequently we do have grounds for accepting the claim of physicists to possess knowledge of these matters, and in this sense our "faith" in physics is not dependent on faith alone.[57]

It is noteworthy in the quoted passage how readily Turner mixes the physicists' occupational expertise with their ideological role as ministers of truth, but for the present I want to direct your attention to his claim that "we all know" certain things to be true. Ezrahi made a similar move in *The Descent of Icarus*, a study of the use of science as a rhetorical resource for governments in the United States and other liberal democracies. Ezrahi pointed to the rise of experimental natural philosophy in the seventeenth century as a source for the powerful idea that the mechanisms of nature could be made both visible and intelligible, describing this as the origin of an "attestive" culture of public ratification that would become a mainspring of liberal democratic politics. Policies that can be attested to by the public as rational, disinterested, and leading to the common good are the goal of liberal democratic governments, and Ezrahi finds the source of attestation in the political arena in the practices of the natural sciences.[58]

But what really do we all know? Turner and Ezrahi assume that the public is in a position to validate scientists' claims in an active sense of "voting for the truth," but this assumption seems dubious to me. What we all know about nuclear physics and atomic bombs is what physicists tell us. It may be reasonable to believe that the physics of the atom has something to do with the power of nuclear weapons, but we can only take the physicists' word for it that one has anything to do with the other. Theory and practice claims invariably work like that. There is a ubiquitous conviction on the part of the public that theory can and does inform practice, but our understanding of the particulars of that relationship is dependent on how it is presented by the experts. What we really know is that physicists represent certain things as being true about the world. That is, they stand for us in knowing things about nature that we could know in principle, but do not know in fact. And we also know that physicists appear to have a kind of occupational expertise that permits them to participate in designing nuclear weapons of impressive power. That is the limit of what we know.

In what respect, then, might it be accurate for scientists to claim that "we know" one thing or another such that the statement does not refer only to themselves? This is essentially the problem that Condorcet confronted in the French Revolution, quite possibly as the first person to recognize it explicitly in its full force, because he was presented with the radical potential of a society completely released from its traditional institutional and social moorings. In such a society, would knowledge become universally available? Condorcet's answer assumed first that the question had to be postponed to a more enlightened future, but even at such a utopian moment he could not accept that enlightenment would be equally distributed to all members of society. While it is true that Condorcet did not explicitly grant a distinction between occupational and ideological expertise, I believe nevertheless that his solution to the

---

[57] Turner, "What Is the Problem with Experts?" (cit. n. 14), 130.
[58] Ezrahi, *Descent of Icarus* (cit. n. 14), chap. 3.

quandary of what an enlightened society could know depended on assuming something very much like it.

Finally, what do we gain analytically and practically by dividing our understanding of expertise, as I have done here? Is the critique of what "we all know" a counsel of despair and passivity? I do not think so. In the first place, while it may be possible to differentiate occupational expertise from its ideological counterpart, it remains true that today when we think about expertise the two kinds are thoroughly conflated, with few clear demarcations. Statements of fact readily and repeatedly leak over into proposals for action, in which experts are called on to "produce" a new policy or a solution to a problem. In such discussions, the public takes an active role,[59] and for the most part I agree with Turner's fairly pragmatic description of how expertise can be reconciled with democratic practice. The fact that our core convictions about science and politics are based on ideologies does not mean that we cannot use them as the basis for action.

The vociferous and ideologically charged debates over global warming are an excellent example of this. One line of thinking about these controversies contends that the denial of global warming is the product of deliberate obfuscation and the manipulation of an unenlightened public, a claim that is hard to deny.[60] But surely there is more going on here than the distortion of the truth, for this explanation does not account for the energy and vociferousness of the debates. I would argue that debates over global warming exemplify how the two forms of expertise can come into conflict on occasion, when the representation of truth comes into direct conflict with the recruitment of scientists to propose solutions that are politically and economically contentious. In other words, when scientific matters begin to impinge significantly on policy choices for a broad swath of the population, the situation becomes contentious. I have little doubt that there is a core of basic scientific facts that all parties in the debate would agree to, such as the distinctive and powerful ability of carbon dioxide to absorb infrared radiation. I have never seen anyone dispute that fact. The argument begins over what these facts add up to with respect to the history of the Earth as a whole (history always being an object of contention, it seems), and what steps are reasonable in order to alter that history.

In understanding this debate, it might be helpful to hold the components of expertise separate with respect to how we evaluate scientists' roles in it. In our society, we require scientists to tell us things about the world that we hold to be true, but we also require them to function as occupational experts who, like butchers, bakers, and candlestick makers, have special skills that can be used to produce new technologies or implement solutions to problems. We need to recognize that those two roles can come into conflict, although they do not invariably do so, and where we suspect that such a conflict might be happening, we have the right to ask of the experts that they make their case by explaining their thinking to us. In the end, we may not all know what scientists know, but in a democratic polity we have the right to have explained to us why what they know is appropriate to the choices to be made.

---

[59] A similar point is made by Agar in his discussion of what happened to expert authority in the 1960s. The greater access to expert debates provided by the press and by the publication of important critiques such as Rachel Carson's *Silent Spring* (1962) blew open what had previously been disputes carried on in more restricted professional circles. See Agar, "What Happened in the Sixties?" (cit. n. 1).

[60] Naomi Oreskes and Erik M. Conway, *Merchants of Doubt: How a Handful of Scientists Obscured the Truth on Issues from Tobacco Smoke to Global Warming* (New York, 2010).

Thin Description:

Surface and Depth in Science and Science Studies

by Theodore M. Porter[*]

ABSTRACT

Since the time of the French Revolution, a sequence of modern thinkers has theorized the thinning of the world in relation to the growth of science. In a large, diverse, and politicized world, subtlety seems to recede into nooks and corners, while information is revered for its ready accessibility and seeming solidity. The sciences, adapting their public voice and some of their inward practices to such expectations, have over the twentieth century flourished more and more in the public sphere as preeminent sites of facts, data, and statistics. Yet the aspiration to superficiality yields up all kinds of unexpected consequences, which provide fascinating opportunities for historical and social studies of science if we can ourselves resist the siren song of thinness.

A classic historical interpretation of science tells of how it supplanted and discredited an ancestral philosophy that conceived nature as dense with moral significance. Sometimes triumphantly, sometimes regretfully, historical writers have chronicled the irresistible demise of these signs and meanings. Auguste Comte, in a classic formulation, characterized the progress of science as the abandonment of theological and metaphysical conceptions in one field after another, from ancient mathematics to Comte's brand-new field of sociology, whose founding moment was precisely the articulation of this law. In the positivist world, nature stands apart from all our values, feelings, and purposes. No longer are the signs of God's handiwork strewn across heavens and earth, and no more do humans connect sympathetically with them. Historians of science know well how much trimming and pressing was required to cast Descartes, Kepler, and Newton as heroes in the triumphalist drama of positive science. Centuries later, even among the apostles of positivism, this ideology of neutral description was often accepted with painful reluctance, and then remoralized by these same individuals in campaigns to transform society by training up citizens to become, through education in science, social and selfless or rational and independent. Disinterestedness itself requires moral cultivation, and considerable public reputations have been achieved in science by preaching and displaying the virtues of detachment.[1]

[*] Department of History, 6265 Bunche Hall, UCLA Box 951473, Los Angeles, CA 90095–1473; tporter@history.ucla.edu.

[1] For a range of views on inclusion and exclusion in positive knowledge, see Charles Gillispie, *The Edge of Objectivity* (Princeton, N.J., 1960); John Heilbron, "Fin-de-Siècle Physics," in *Science, Technology, and Society in the Time of Alfred Nobel*, ed. Carl-Gustav Bernhard, Elisabeth Crawford, and Per Sèrböm (Oxford, 1982), 51–71; Theodore M. Porter, "The Death of the Object: Fin-de-Siècle

Europeans of the early modern period, in many cases right through the nineteenth century, understood science not as value free but as resting on a proper Christian morality. Or the values of science might be more cosmopolitan if, as Leibniz (drawing inspiration from Jesuit missionaries in China) believed, every religion contains a core of inherited truth. By this reasoning, proper science was at once the product and a powerful agent of the cosmopolitanism of knowledge. Nineteenth-century liberal elites looked to rigorous moral and intellectual education in the culture and languages of the ancient world to form a value-ecumenical, though never value-free, "objectivity," understood as an attribute of forceful, if ascetic, individuals rather than weak, submissive ones. Similar aspirations underlay the campaigns for scientific culture in the era of T. H. Huxley, Claude Bernard, and Hermann Helmholtz, who insisted that science was better suited than classical languages to inculcate these rigorous and universal values. Over the eighteenth and nineteenth centuries, natural science came more and more to be seen as the very model of universal knowledge. Only with the antihistoricist movements of the fin de siècle, however, did leading intellectuals in Europe and America place it systematically at odds with history and literary culture. The significance of science for the modernist rupture with tradition is usually typified by physics, mathematics, and laboratory biology, along with philosophical commentaries, forms of science that are most readily compared to elite movements in the arts. These high-culture sciences are part of the story, but not, I think, the main part.[2]

More decisive was the rise of the new human sciences, especially of practical, social, and administrative fields such as scientific management, military and asylum medicine, insurance mathematics, institutional economics, educational psychology, and statistics in its many forms. These scientific movements promised a whole new basis for organizing social life, one to be grounded in rational thought and organization rather than in history and custom. For them, the growth of democracy was a social reality to be acknowledged but often not an ideal to be promoted. Public enlightenment, after all, seemed at best a project for the long term, one that depended not just on instruction but also on the mass inculcation of sound moral character. The new social and human sciences aimed in the here and now to ward off the dangers of mass society by means of efficient administration. This effort called for a new kind of objectivity in the guise of value freedom, the least common denominator of cultures rather than a shared aspiration to the expansion of knowledge and wisdom. The retreat from the public sphere was not simply a scientizing move, but the choice of a form of science that should keep aloof from the strident turn-of-the-century campaigns being waged by socialists, nationalists, and racialists under a different banner of science.[3]

Modernism in science meant refashioning it as a set of technical fields, which

Philosophy of Physics," in *Modernist Impulses in the Human Sciences*, ed. Dorothy Ross (Baltimore, 1994), 128–51; Lorraine Daston and Peter Galison, *Objectivity* (Brooklyn, 2007), chap. 4.

[2] Carl Schorske, *Fin-de-Siècle Vienna: Politics and Culture* (New York, 1980); Dorothy Ross, *The Origins of American Social Science* (Cambridge, 1992).

[3] On value freedom I draw inspiration from Robert Proctor, *Value-Free Science: Purity and Power in Modern Knowledge* (Cambridge, Mass., 1991), which gives in chapters 7, 8, and 10 a much richer account of the German origins of this notion. On new strategies and uses of the human sciences around 1900 see Ross, *Modernist Impulses* (cit. n. 1); Anson Rabinbach, *The Human Motor: Energy, Fatigue, and the Origins of Modernity* (New York, 1992); David Lindenfeld, *The Practical Imagination: The German Sciences of State in the Nineteenth Century* (Chicago, 1997); John Carson, *The Measure of Merit: Talents, Intelligence, and Inequality in the French and American Republics, 1750–1940* (Princeton, N.J., 2007).

should stand now outside the wider culture. The new objectivity stood conspicuously apart from the customs and institutions of ordinary life. Ironically, though not by accident, these separatist ideals contributed to a reshaping of the culture. Science, allied in practice with industry and bureaucracy, stood more than ever for a standardized world, for uniform rules that presume no need for the nurturing of traditions or the cultivation of judgment. These campaigns were most caustic and required least to be disguised in the early twentieth century. Science, drawing from a range of cultural resources, launched the American century, powered by an engine of technology that would neutralize ideology through systematic reliance on planning or markets or market-based planning. In its most naked form, however, the movement of standardization encountered definite limits. Instead of subduing the most powerful fields of professional expertise, it often was co-opted by them as a glittering language of detached reason. Assuming more subtle forms in the later twentieth century, the ethic of discipline by rules and evidence has extended more widely and penetrated more deeply. This is no victory of cunning and secretive conspirators but a heterogeneous interplay of actions and resistances. Differences of politics, culture, and values have not only persisted, but proliferated. The forces of administrative rationality, drawing back to advance more effectively, took as their watchword *objectivity*, that neutral world of cool reason, and have labored tirelessly to forge consensus with the tools of thin description.

THICK AND THIN

Clifford Geertz published his famed essay on "thick description" in 1973 as the introduction to a volume of ethnographic writings, *The Interpretation of Cultures*. Crediting the phrase to Gilbert Ryle, Geertz deployed it to confer specificity and scientific credibility on that defining anthropological object, "culture." Only through the interpretive work of thick description, he argued, can we make sense of human actions and exchanges. Thick description went beyond manifest actions and utterances to get at meanings; Geertz construed it as the proper and distinctive method of the science of anthropology. Ryle had pointed to the difference between the involuntary twitch of an eye and a conspiratorial wink, which, though perhaps identical as physical movements, are altogether different as gestures. While thick description does tend to get complicated, since meanings can be elusive and cultures are far from homogeneous, it does not refer merely to an abundance of detail. The point was not to include everything and omit nothing, still less to prohibit analysis. Geertz was articulating an alternative to behaviorism and behavioralism, unifying principles for social science in postwar America. The focus on behavior was a strategy to make social science into real science, something more like physics and less subject to values and prejudices because restricted to observable phenomena of the sort that could be registered by instruments. The sacrifice of human meaning seemed not just a price worth paying for solid results, but the liberating essence of a proper objective methodology that now would rise above stubborn tradition and invisible culture. Geertz, withdrawing from this program, insisted on interpretation, a different approach to the empirical for a different object of science.[4]

---

[4] Geertz, *The Interpretation of Cultures* (New York, 1973); Ryle, *The Concept of Mind* (Chicago, 1949).

"Thick description," then, was a battle cry against an ideal of social sciences harmonized, if not unified, by a shared commitment to the external and observable. Its more economical opposite attended rather to surfaces, either ignoring the question of meaning or taking it as self-evident. It is noteworthy that Geertz did not launch his challenge to the primacy of thin description in its stronghold, the industrialized, bureaucratized, modernized West, but among Berbers, Jewish traders, and French military officers in Morocco, at or just beyond the margins of European culture. Similarly, in historical writing this language of thickness was taken up first of all by scholars working to rediscover the profound otherness of popular culture in early modern Europe, and subsequently in other provinces of those alien lands, Long Ago and Far Away. A modernized world, by contrast, might not require the teasing out of unfamiliar meanings, but should demonstrate those paired forms of rationality emphasized by a slimmed-down Max Weber, the bureaucratic and the scientific.

Of late, cultural and intellectual historians, including most historians of science, have been skeptical of easy dichotomies like this one between romanticized tradition and stark modernity. Away from history and ethnography, on the farthest edge of social science, economists suppose that all the ostensible mystagoguery of premodern life is a kind of smokescreen, beneath which priests, peasants, and mystics, like the rest of us, carry on in pursuit of a rational maximum. An estimable tradition in social theory, backed up in certain respects by practical reformers of medicine, law courts, and all kinds of policy processes, either accepts that kind of rationality as the natural outcome of history or strives mightily to enact it. Enthusiasts of ethnography might ask, Why, in a thick world, do economists stride with heads high through the corridors of power, while cultural historians pass along their possibly profound insights to one another? Why, in the world of business and administration, are lengthy reports with all their uncertainties circulated among underlings, while the "executive summary," purged of ambiguity and detail, goes to the people at the top? Thinness is, if not the natural state of things, an appealing modern project. It beguiles us with its terse, muscular economy.

Yet we are dealing here not merely with a choice between alternative intellectual strategies of interpretation and analysis, but with practical ways of knowing the world and of shaping action. The language of thin description is supported, in the main, by the institutions of natural and social science. Its worship of ordered visibility, exemplified in the hard-edged symmetries of modernist architecture and urban planning, underlies the power of bureaucratic states.[5] As scholars engaged in social and humanistic research on science and technology, we must be aware of this duality. Thin description is not merely a methodological option for humanists, but a favored mode and resource of the people and the institutions we study. It is, in short, one of our most consequential objects of research, and it is not to be comprehended adequately in its own terms.

But what exactly is thin description? In a thin world, surfaces should be valid and deep meanings superfluous. When Mother Hubbard goes to her modern cupboard to get a thicker account, she finds that it is bare, not merely for lack of access to a deeper level of understanding, but because the world has changed. These days, she recognizes, no such profundity is wanted. "For physicalism there is no 'depth,' all is 'surface,'" wrote Otto Neurath in 1931. Leaving such concepts as "motive," "I," and

---

[5] James C. Scott, *Seeing Like a State* (New Haven, Conn., 1998).

"personality" to the magic of long ago (*Frühzeit*), science deals with behavior in a spatial-temporal order as a frame for predicting.[6] But can the thickness and depth of tradition—or of innovation—really be dead? How were we able to drink up the sea? Did science do that? And why should it act this way? For in no way is scientific knowledge intrinsically thin or superficial. Even if it declines to seek out human meaning in nature, it has supplied abundant material for more complex understandings. Science is a powerful engine of exploration and explanation, having opened up worlds of which our forefathers never dreamed. I mean quarks and black holes, radioactive elements and fields of fossils, the elemental composition and the changing proportions of the atmosphere, retroviruses and tectonic plates and the strange biology of hydrothermal vents as well as the unfathomable complexity of the human organism. Social science deals more often with a world accessible to ordinary experience, but it too possesses its observatories, its tools of calculation and theoretical traditions that diverge from common knowledge and complicate or undercut most of our easy answers.

Why, then, do these tools of analysis and discovery not enrich and deepen our understanding, and with it the world itself? The short answer is that, of course, they do, that both by making and by knowing, science and technology heighten the complexity of the world. To be sure, the world was always incomprehensibly complex. Not even Aristotle ever possessed more than a tiny fraction of the knowledge and skill of those sages, statesmen, poets, artisans, hunters, warriors, merchants, mothers, healers, and tillers of the soil, his ancient contemporaries. Still, the intellectual and technological division of labor in our world is incomparably more minute, more ramified and diversified, than Aristotle or Leibniz or Humboldt ever imagined. It supplies an inexhaustible reserve of startling observations and explanations for the curious, leaving aside that abundance of minimally meaningful information, doubling in quantity every few years, that Aristotle would have deemed unworthy of attention by the lover of wisdom. On this matter, we moderns are well equipped linguistically to share his assessment. Indeed, we effortlessly surpass him, for Aristotle would have been dazzled by the depth and richness of our science. Our culture typically writes off most of what cannot be appealingly recorded on film as merely technical.

This term *technical*, a keyword of the modern age, implies first of all that a topic has little interest except to those engaged in it by occupation or other necessity. Applied to its etymological cousin, *technology*, it licenses us to ignore the details of a thing on the supposition that its importance is merely instrumental. Technologies, being practical and designed for use, are not required to satisfy the intellect, and liberal institutions like universities have rarely acted as if knowledge of basic engineering, any more than of artisanal trades, was necessary to furnish an educated person. The shunting aside of science is another matter, since science has always been supposed to enlighten us. Its identification with technicality provides a cover of justification for those dull souls, the bane of science educators, who just don't *like* science. For if science is intrinsically specialized and technical, there may not be much reason for nonscientists even to try to comprehend it, or much hope that their success can be more than apparent. The landscape of science offers, at best, the temptation of the Garden of Eden (Revised Plan): to eat, not of the tree of knowledge, but the

---

[6] Neurath, *Empirische Soziologie: Die wissenschaftliche Gestalt der Geschichte und Nationalökonomie* (Vienna, 1931), 13, 11; see Proctor, *Value-Free Science* (cit. n. 3), 155.

fruit of fun. Richard Feynman and James Watson are prominent among the serpents who have told us so. Yet their rebuttal reaffirms the dogma it seeks to combat—that science, like technology, is for nerds.

Or, to be more polite and more sociological, we say that science is a profession, thereby abandoning the campaign for science as public reason. The classic professions are quintessentially technical, and the *Oxford English Dictionary* links early uses of this word *technical* specifically with them. Who ever argues that legal study should be included in everyone's formation as part of what makes us educated? The profession of law maintains close control over this kind of knowledge, which supports skyboxes in athletic stadiums, tasteful private art collections, luxury yachts, and lavish summer estates. The movement, largely ineffectual, to teach everyone about law is animated by a philanthropic desire to provide the lay public with some protection against lawyers. Science once perceived its mission first of all in terms of the diffusion of light, but more and more it has become, if not a profession, then a commercial enterprise. Let technicality reign, under protection now by patents, while the rest of us may attend to science for the sake of protection against scientists. For technical matters are precisely those whose inwardness concerns only experts, and the claim of technicality functions not least to fend off meddlesome outsiders. Technical knowledge in science is often not thin and yet, for just that reason, may be held closely as the preserve of specialists.

The field of history of science was itself conceived under the sign of technicality.[7] Back when radioactive fallout was still settling on exploded Pacific atolls, science seemed more and more like a problem for democracy, given that it was more and more of consequence in the everyday affairs of production and circulation, of technological, administrative, military, medical, and financial power. Often the subtle or difficult points that might be written off as technical have great moral and political significance, so that the conscientious citizen or even the self-interested worker or consumer needs to know when to draw back the curtain in order to inspect the machinery and the wizards who operate it.

More nobly, we might be moved to learn about science through intellectual appreciation rather than suspicion. The generation of historians, including some scientists, who came to maturity in the 1950s worked to demonstrate the roots of scientific understanding in the great intellectual achievements of Western civilization. They aspired to present it as less exotic and less isolated than was implied by prevailing stereotypes. As an achievement of Western Culture or, in a less parochial idiom, of Global Humanity, science may indeed deserve to be known and appreciated by all educated people, in the same way that they would take an interest in art, literature, and philosophy, or history, geography, and political economy. And science clearly does matter for history and culture. As pianos and paints participate in the evolution of music and art, so have archeology and astronomy, even anomalies in planetary curve fitting by deferents and epicycles, played their role in realignments of faith and politics.[8]

There is nothing thin or shallow about any of this, nor in the enlightened hope that a basic familiarity with scientific ways of working and reasoning might weaken the hold of prejudice and encourage open, rational, and effective public action. But the alignment of science with technological and bureaucratic expertise has situated

---

[7] Theodore M. Porter, "How Science Became Technical," *Isis* 100 (2009): 292–309.
[8] A key point in Thomas Kuhn's classic study *The Copernican Revolution* (Cambridge, Mass., 1957).

the public more often as object or obstacle than as participant in the great traditions of science.[9] Paradoxically, the public interface of science has provided a decisive stimulus to thinning. In schools, the problem solving and rote learning typical of instruction in science give way, in moments of enlarged vision, to a discourse of scientific method. The ideal here is to link science with ordinary life, into which it can infuse its spirit of rationality. Most notably among educationalist admirers of John Dewey, faith in method offered hope that any field of inquiry and almost any activity could be made more scientific. Well-established sciences—that is, technical sciences—are too forbidding for this purpose. Much expertise and discernment are required to ask a fruitful scientific question in an established discipline. But to a child equipped with the precepts of scientific method, the whole world is a catalog of questions that can be put to experimental test. Do girls like different colors from boys? Is Pepsi preferred to Coke? Do chemists vote differently from physicists? Typically, this kind of hypothesis-testing experiment follows a rudimentary statistical design. There is no reason that such procedures cannot be part of a more searching inquiry, but the insistence on a generalizable method releases us from the particular content of any discipline and hence draws emphasis away from all its recondite aspects. In times of challenge or suspicion, scientific leaders sometimes draw attention away from depth and complexity by invoking the nostrum of scientific method.

You will look long and hard among historians of science to find anyone who gives much credence to this faith in scientific method. Yet the search for a simple common denominator of science goes on, excising every thought of interpretation or meaning and displaying the same curious devotion to surface. In their 1986 postscript to *Laboratory Life*, Bruno Latour and Steve Woolgar called for a moratorium on cognitive explanations of technoscience.[10] They invited readers to revisit the problem ten years hence if anything about science or technology was left to be explained. Latour's announced methodology was simply to follow scientists around, with no regard for anything they might say about purposes or theoretical content. Following scientists around, as if this could supply an all-encompassing basis of social research, and refusing to ask about meanings, seems the very model of thin description and just as unsatisfactory for comprehending social interactions as it is for dealing with individual beliefs. Such delight in surfaces is almost indistinguishable from the asceticism of behaviorist psychology, which resolved at all costs to avoid talking about mind because it could not be experienced impersonally or subjected to experimental control. You might say that we have always been postmodern.

On these matters, though, things are not as they seem. Cognitive explanations, in the sense practiced by psychology, are often motiveless and detached from material and social circumstances, another form of thinness. By contrast, a reconstruction of the institutional settings, the financial arrangements, the suppliers and instruments, the students, technicians, and illustrators, the conversations, and the collaborators and competitors that come into view when we follow a scientist around presents us with a dense array of human relationships and of purposes, meanings, and misunderstandings. Material and social practices are notably thick, especially when these

---

[9] Deborah R. Coen describes (thickly) a strikingly different vision and organization of science in "The Tongues of Seismology in Nineteenth-Century Switzerland," *Sci. Context* 25 (2012): 73–102.

[10] Latour and Woolgar, *Laboratory Life: The Construction of Scientific Facts*, 2nd ed. (Princeton, N.J., 1986), postscript, 280.

reveal the links of science to other aspects of the world, to craft and mass production, commerce, religion, museums, music, databases, administration, art, budgeting, prognostication, and politics. Science here measures up quite well by way of thickness against the Morocco standard, and the world of science in practice is an outstanding site of meanings and interpretations. However it may be with cognitive explanations, intellectual history is fully compatible with an interest in objects and practices and should be reckoned not as a rival or competitor to cultural history, but as its best ally.

THEORIZING THE THIN WORLD

And yet it is not so easy to shake the idea that the expansion of science has made the world thinner. *The World Is Thin* is a book of superficial journalism waiting to be written. The philosophical defense of culture and tradition has developed in reaction to a sense of thinness triumphant, in the form of rationalistic science. The pioneer of such writing was Edmund Burke, who issued forth with his antimodern diatribe within months of the outbreak of revolution in France, identifying instantly the master disciplines of the new era. "But the age of chivalry is gone; that of sophisters, economists, and calculators has succeeded, and the glory of Europe is extinguished forever." The heart of the problem for Burke was the suppression of inherited experience in favor of dogmatic theory and rootless learning: "After I had read over the list of the persons and descriptions elected into the *Tiers Etat*, nothing which they afterwards did could appear astonishing. Among them, indeed, I saw some of known rank; some of shining talents; but of any practical experience in the state, not one man was to be found. The best were only men of theory." Here he referred not to savants but to lawyers; and not, he went on, renowned magistrates or professors but "obscure provincial advocates, . . . stewards of petty local jurisdictions, country attorneys, notaries, and the whole train of the ministers of municipal litigation, the fomentors and conductors of the petty war of village vexation." Also there were traders, men who "had never known any thing beyond their own counting-house," and physicians, as if "the sides of sick beds are . . . academies for forming statesmen and legislators."[11]

Burke avowed no animus toward science. Men of learning, he declared, had long enlarged the ideas and furnished the minds of the nobility and clergy, who gave in return their protection: "Happy if learning, not debauched by ambition, had been satisfied to continue the instructor, and not aspired to be the master." For it is ridiculous to treat the "constitution of a kingdom" as "a problem of arithmetic."[12] Cool neutrality, the union of abstract reasoning and unpitying analysis, is by nature corrosive, he explained. A proper science of politics should embrace feeling and devotion. Detached, deracinated science meant a loss of real knowledge in its indispensable form of wisdom and practical experience, cultivated and perpetuated through an organic linking of the generations.

Burke's association of revolution and democracy with a triumph of abstract knowledge was carried on by conservatives and cautious liberals. In *Democracy in America* (1835–40) Alexis de Tocqueville, while not altogether pessimistic, worried about

---

[11] Burke, *Reflections on the Revolution in France*, 2nd ed. (London, 1790), 59, 62–4.
[12] Ibid., 117, 76.

the democratic preference for general principles, simpler to hold in the mind and memory, over a more complex experience, built up over a whole life and passed on to children and grandchildren. His words were admired by Frédéric Le Play, whose commitment to monographic social studies emerged out of his travels as an engineer with the French Corps des mines. Le Play blamed revolution more than scientific abstraction for the dispersion of experience, but he argued also that the abyss of lost memory meant a hollowing out of science, stretching it like a skin over depths it never could plumb.

Le Play's monographs possess an ironical pertinence because he filled them with tables. Not only since Burke, but going back to Diderot and Buffon, critics of academic science had condemned numbers and mathematics as barren and superficial for omitting the forces of life. To Le Play, numbers were not merely a convenient medium in which to combine and summarize, but could be turned to exploring the very density of human interactions that, in modern times, had retreated to the shadows. He had gotten his start in political economy preparing double-entry accounts of production and trade, and these formed the template for the family budget studies that made up the essential core of his mature social science. Le Play packed his monographs with information about the moral as well as economic connections of the laborer's family with lords and patrons, revealing in many cases the persistence of old customs and reciprocal promises subsisting outside the domain of formal contracts and of what Thomas Carlyle had condemned as the cash nexus. To assign money values to these customary obligations and display them in tables was to challenge the social order of market exchange using its own tools. Beneath the thin, drab clothing of the liberal order there were yet remnants of a traditional economy based on reciprocal loyalty and obligation.

Le Play's budgetary tables formed the heart of this project, made urgent, he thought, by the cutting off of ancestral experience that was the most devastating legacy of 1789. In early 1848, just as he was beginning the systematic pursuit of social science, he proposed hopefully that instruction in statistics might provide a second-best education for leaders who could no longer rely on inherited wisdom regarding state and economy. He moved steadily rightward during the Second Empire, and after the Paris Commune of 1871 he committed his life to pursuing the grail of true social science in those societies, as far from France as possible, that, being insulated from the European revolutions, had preserved the old ways. As beneficiaries of a blessed continuity, they had no need to record their wisdom, and still less to seek a substitute in dry bookkeeping tables. It was left to Le Play and his disciples to record such wisdom for the sake of their less favored countrymen. The monographs he continued to champion became, for him, tokens of the past and hints for recovering it in the wake of a catastrophic loss of tradition.[13]

These arguments about the thinning of society and of knowledge come from nineteenth-century conservatives who objected to the modern faith in progress, reason, and revolution, whether violent or gradual. We might pursue this vein into the twentieth century, but it is more interesting to glance leftward. Critical philosophers

[13] Theodore M. Porter, "Reforming Vision: The Engineer Le Play Learns to Observe Society Sagely," in *Histories of Scientific Observation*, ed. Lorraine Daston and Elizabeth Lunbeck (Chicago, 2011), 281–302.

of the Frankfurt school such as Max Horkheimer, Theodor Adorno, and Herbert Marcuse embraced revolution, not historical continuity, yet they despised the steamroller of incessant capitalism. This they saw as allied to positivist science, its constricted rationality supporting a narrowed vision of historical possibility. Critical theory did not present social science as the center of power, which of course was in the first instance economic, but its scientific attitude bore specific responsibility for a cultural loss of thickness, a veneer of positive thinking without the much-needed deep criticism of dialectical philosophy. Like many theoretical Marxists, they found more of value in the precapitalist past than had Marx himself, who, even while loathing the destructiveness of capitalism, insisted on the historical necessity of its unadorned brutality. The Frankfurt critique identified capitalist epistemology not only with a positivist commitment to surfaces, but also with a vast domain of inauthenticity, the "culture industry." Walter Benjamin, for example, wrote resignedly of the triumph of information over storytelling.[14] Technologies of mechanical reproduction were here allied to instruments of mass diffusion such as broadcast advertising, as well as social surveys and other quantitative tools to measure and manipulate opinion. Science, too shallow in its positivist form to engage with the great historical transformations, was also involved instrumentally in manufacturing a thin gruel of cultural homogeneity.

It was not only critics of science who complained of thin, mechanical, or overspecialized knowledge production and its implications for culture. Although scientists do not often emphasize the subtlety of scientific practice and reasoning in their research publications or public pronouncements, they instinctually resist attempts to bind up their work in inflexible rules. The point is sometimes articulated where we might least expect it. Statistics is for many a model of explicit, rule-bound reasoning, but for more than a century statisticians have complained from time to time of mindlessly formulaic data analysis. Statistics, as a specialty, cuts across many specialties, providing a set of strategies and a language for coordinating collective and even multidisciplinary inquiries. The founder of statistics as a mathematical and methodological field, Karl Pearson, was particularly catholic in the range of topics he took up and particularly insistent in his denunciation of narrowness and of working by rote. For him, science was a set of skills acquired through a close relationship with a master, on the model of the medieval university. Like Huxley, who reached the peak of his reputation during Pearson's youth, he looked to science for the perpetuation and enhancement of culture. Yet he could not make the world, or even his science, as he wished, not least because his sense of craft was at odds with the claims of scientific method. Bolstered by the language of method, ambitious reformers in a wide variety of natural and social disciplines took up his mathematical tools. They provided templates for the arrangement of data and formulas for drawing conclusions, which were applied in many different subject areas. This routinized aspect, more than the exalted vision of wise and supple reasoning, was cherished by the psychologists, anthropologists, economists, engineers, and medical researchers who gave new meaning to statistical reason.[15]

---

[14] Horkheimer and Adorno, *The Dialectic of Enlightenment: Philosophical Fragments* [1944] (New York, 1972); Benjamin, "The Storyteller," in *Illuminations*, ed. Hannah Arendt, trans. Harry Zohn (New York, 1969), sec. 6, 83–109.

[15] Theodore M. Porter, *Karl Pearson: The Scientific Life in a Statistical Age* (Princeton, N.J., 2004); Gerd Gigerenzer et al., *The Empire of Chance: How Probability Changed Science and Everyday Life* (Cambridge, 1989).

LOCAL WORKSHOPS OF UNIVERSAL VALIDITY

For centuries, commentators on science have spoken of the universality of natural laws, emphasizing that scientific knowledge, and science itself, are the same everywhere. Some among them have presented this not as the suppression or bleaching out of difference, but as rising above provincialism and particularity. Nevertheless, the outcome should be a more uniform world, and in this respect the work of research is only the tip of the iceberg. Still more consequential in practice is the contribution of science to the design and propagation of standards. Uniform systems of measurement were created not merely for the sake of science, but to encourage the diffusion of technologies and, perhaps still more, to facilitate the expansion of markets and the administration of large territories. They have been allied to an ideal of rational bureaucracy, of the sort emphasized by Weber; as Burke or Le Play might have expected, the metric system was not only created by the French Revolutionary moment, but has spread in close alliance with movements of political revolution.[16] Standardized measurement has been a key ally of industrialization and mass production and, with statistics, of that great cultural shift from wisdom to information. One preeminent virtue of a uniform world is that it can more readily be grasped thinly.[17]

And yet we may doubt that science really does proceed by annihilating locality and annexing what is distinctive into an undifferentiated Nowhere. There was a time when historians regarded universality as the natural condition of science and local specificity as a temporary problem of adjustment. The scholarship of science studies has reversed field in the last generation. Steven Shapin surveys the vast sweep of science, whose validity seems to stretch uniformly over the whole globe, and observes countless workshops relying on densely moral human interactions on the most local scale—skills and expertise that can scarcely be replicated, and perhaps never fully and independently replicated. Science, whether in the seventeenth-century Royal Society or among entrepreneurial biomedical researchers and investment professionals in Silicon Valley, depends on trust built up in face-to-face interactions, drawing—as financiers and entrepreneurial scientists bond over rutted mountain-bike tracks—on gentlemanly codes of sociability.[18] If we look at science culturally, science in the making, we find that tools and concepts alike, mathematical skills as much as laboratory results, are anchored in local institutions and practices and do not travel readily or seamlessly.[19] Even when scientists at a distance are convinced, as they often are, by a published report, the work may have subtle but important differences

[16] Witold Kula, *Measures and Men*, trans. Richard Szreter (Princeton, N.J., 1986); Ken Alder, *The Measure of All Things* (New York, 2002); Charles Gillispie, *Science and Polity in France: The Revolutionary and Napoleonic Years* (Princeton, N.J., 2004).

[17] Ken Alder, *Engineering the Revolution: Arms and Enlightenment in France* (Princeton, N.J., 1999); Theodore M. Porter, *Trust in Numbers: The Pursuit of Objectivity in Science and Public Life* (Princeton, N.J., 1995); Porter, "Locating the Domain of Calculation," *Journal of Cultural Economy* 1 (2008): 39–50; Porter, "Statistics and the Career of Public Reason: Engagement and Detachment in a Quantified World," in *Statistics and the Public Sphere: Numbers and the People in Modern Britain, c. 1800–2000*, ed. Thomas H. Crook and Glenn O'Hara (London, 2011), 32–50; Lawrence Busch, *Standards: Recipes for Reality* (Cambridge, Mass., 2011).

[18] Shapin, *The Social History of Truth: Civility and Science in Seventeenth-Century England* (Chicago, 1994); Shapin, *The Scientific Life: A Moral History of a Late Modern Vocation* (Chicago, 2008). Shapin's historical arguments draw from the essays in Harry Collins, *Changing Order* (London, 1985).

[19] E.g., Andrew Warwick, *Masters of Theory: Cambridge and the Rise of Mathematical Physics* (Cambridge, 2003).

of meaning for them. Though these practices and results do not travel easily, the changes they undergo in being transplanted do not reduce to thinning. We have, instead, culture in a strong sense.

There is another implication, however. Ineffable local thickness leaves intact a view of the big modern world as thin by nature, since communities based on dense interactions are not large. If intensely local skills are fundamental to new research, and if they do not travel well, then most movement of knowledge will necessarily be thin.[20] Yet the interaction of the local and the cosmopolitan in science is far more intricate and interesting than a simple dichotomy allows. Shapin has not vanquished universality, which remains in force in the ideologies of science, and forceful also in its practices. We say now that science produces timeless truth as much as it discovers it; that science has papered the world with its universality, like the Peruvian map described by Borges, often through labors of an infrastructural kind that were not of much interest to the history of science until about a quarter century ago. The effort to create standards and produce uniformity is one of the defining features of modern science. Many who proclaim the achievement of science in demonstrating the universal laws that may be said to govern nature (and society too) would not be troubled to acknowledge that inspiration, subtle reasoning, bodily skill, and dense human interactions are required in original research. Yet they insist that the findings of good science inevitably transcend these limitations. They may speak of logics of discovery and of confirmation or emphasize how the incomparable achievements of one generation are performed by the next in student laboratories. With a bit of paradox, we can reconcile the seeming opposition. The very production of homogeneity requires the local cultivation of skill in densely interacting communities. And, conversely, the standardized world is able to mobilize localized skills in its pursuit of thinness.

We have not yet exhausted the ironical quality of this dynamic, the twisted embrace of thick and thin. Positivist social science—what the Frankfurt critics condemned—was born of an obsession with the far-flung interconnections of modernity. From about 1890, its champions invoked this complexity in support of the institutional aspirations of the rising social disciplines. The theory of modernization perpetuated this cosmopolitan, modernist impulse into the postwar world. The world of everyday experience, according to the wisdom of university-based social science, had become deceptive. Every local community was now part of a vast web of trade, production, and communication of which it knew little. To the new social scientists, social and economic knowledge based merely on experience in politics, business, law, manufacturing, banking, trade, agriculture, mining, police work, missionary activity, or poor relief was too narrow and too shallow, having failed to come to terms with the increased scale and complexity of economic and social life. A theoretically sophisticated social science, drawing from nets of information on a global scale and practiced by university-trained researchers, was now indispensable for understanding the successes and the problems of modern society. The "izations," notably modernization, globalization, and secularization, seemed to demand, as intellectual corollary, professionalization and the expertise it made possible. Only professional social

---

[20] Compare Peter Howlet and Mary S. Morgan, eds., *How Well Do Facts Travel? The Dissemination of Reliable Knowledge* (Cambridge, 2011).

science, with its data networks and its theory, could get a grip on the far-flung system of causes that typified the modern world.[21]

It seems plausible enough, and the sciences have indeed flourished in the context of large-scale production, centralized government, and international trade. But does science perform as a profession, its members supplying individualized counsel to clients with highly specific needs? Scientists more typically have struggled to overcome the specificities of place and time, the distinctiveness of institutions and cultures. Modernist science sold itself under the label of detachment and independence. Social science, in particular, strove to hold itself aloof from the government agencies, regulatory bodies, and charitable organizations that actually make policies and act in the world. In this respect, science was decidedly unlike a profession, since it aimed to keep its distance from the people and institutions that should rely on it. Also opposed to the classical ideal of a profession was the insistence on impersonal objectivity over those forms of expert judgment claimed by doctors, lawyers, and even accountants. Academic social science labored to detach itself, at least rhetorically, from the mundane practices of social life, claiming instead to ascertain facts and derive theories on the basis of research, then to apply them in the form of deductions or models to practical life. Its relations to a public it envisioned as uncomprehending encouraged thinness, as did the aspiration to placeless universality.

There remains, of course, one sense in which science did become the very paradigm of a profession. On questions scientific, the scientific establishment brooked no rivals. But what are these purely scientific topics? The vast expansion of science—and here the term *technoscience* is highly pertinent—has brought it into contact with every kind of social, economic, industrial, medical, and even moral (or moralizable) issue. Science claims every topic, but can hold none securely, for every intervention invites criticism based on its presumed practical implications. Science has increasingly sold itself as disinterested and objective as a shield against second-guessing. This sort of objectivity brings together the ideal of technical mastery with that of scientific method, now understood as something quite specific and determinate. First comes knowledge making, detached and apolitical, followed by the messier business of application to social and technological problems.[22]

This defense of science confines it to what is readily recognized as valid and shuns every hint of mere subjectivity. Sometimes, objectivity means not claiming too much. There is little room here for what Geertz called thick, that which is complex and interpretive or dense with meaning. Science should travel readily, both through space and across divides of class, ethnicity, language, educational level, and profession, and especially from the academy to government. It should be resilient in the face of challenges and should not depend on personal loyalty or trust. This capacity to travel, to cross boundaries of all sorts, is part of what makes science powerful. The advantage of traveling light—can we say "traveling thin"?—helps to explain the privileged position of numbers and calculation in modern societies. Science and statistics have become more and more tightly allied with economic forces of trade, finance, manufacturing, and marketing as well as political forces of democracy in a vast project of

[21] Thomas Haskell, *The Emergence of Professional Social Science* (Urbana, Ill., 1977).
[22] On these matters see, e.g., Sheila Jasanoff, *Designs on Nature: Science and Democracy in Europe and the United States* (Princeton, N.J., 2005).

commensuration.[23] Thin description was in this way made powerful and as such is of great interest for science and technology studies. It is important not least because of its implications for public reason. Thin description offers outsiders the opportunity to act and to choose, relying on knowledge without deep understanding. For insiders it signifies self-denying objectivity: forfeiting, if only in principle, the right to interpret.

It might be said that the thinness of science is more characteristic of its public voice than its internal workings. Indeed, my argument stresses particularly the compromises made by science in dealing with a public whose intelligence it typically does not respect. Yet a wide range of research practices are shaped at the intersection of science with medical, technological, and educational issues that include an irrepressible public dimension. While the discourse of "scientific method," for example, was elaborated primarily for schoolchildren, research scientists have found it convenient for many purposes and even have learned to believe. In the face of a mistrustful public, it may be futile to emphasize the thicker dimension of scientific practices, superb skill honestly deployed and matured by experience. A large, diverse, and politicized world has little tolerance for subtlety, which recedes into nooks and corners. It has by no means passed from this earth, but neither does it flourish in the fishbowl of audit and transparency. It must rather be fitted to a world that reveres facts for their ready accessibility and seeming solidity. Science, adapting its public voice and some of its inward practices to such expectations, flourishes now in the public sphere as the preeminent methodology for the generation of data, calculations, and statistics. It has been annexed by the information society it helped to create. Information should not be deep or thick, but suited to do-it-yourself use and adapted to a worldly politics that perceives subtlety as a cover for self-interested maneuvers tending to chicanery.[24]

This argument does not construe science as something inherently thin and universal, which has managed to reshape society in its image. Science in practice mixes locality and universality. While scientists hesitate to ascribe human meaning to experimental setups or natural systems, they certainly do practice interpretation in regard to their own work and that of their colleagues. We have, then, not intrinsic thinness, but thinning and thickening practices suited to diverse circumstances. Thick and thin have each their claims as scientific ideals, and each has uses as well as disadvantages as a mode of practical reason. Each also is inevitably turned to the pursuit of parochial interests. The idealization of thickness can protect experts by suggesting that what really matters will never be truly understood by outsiders. A faith in thinness, by contrast, relieves scientists of responsibility by implying that they are not engaged in subtle interpretation, but acting on evidence and in accordance with rules whose meaning is plain. Or, finally, a seeming transparency is often mobilized to discourage skeptics from peering into boxes that in truth are mystifyingly black, because based on arcane practices and hidden assumptions.

[23] Wendy Espeland and Mitchell Stevens, "Commensuration as a Social Process," *Annu. Rev. Sociol.* 24 (1998): 313–43.
[24] Yaron Ezrahi, "Science and the Political Imagination in Contemporary Democracies," in *States of Knowledge: The Co-production of Science and Social Order*, ed. Sheila Jasanoff (New York, 2004): 254–73.

PUBLIC REASON

It is often said that scientists, under conditions of modernity, assume many of the functions of priests. George Sarton joined Comte in hoping or wishing they would, while Friedrich Hayek condemned these technocratic pretensions as the counter-revolution of science, leading us along the road to serfdom.[25] Recent historians have tended to endorse Hayek's critical analysis of overbearing science if not his neo-liberal devotion to unfettered markets. But with the eclipse of the postwar glory of scientific planning, and with research achievement now often identified with soaring stock values, these priestly pretensions seem more and more the wrong target. The movement to privatize science, which took off in the 1980s, signifies that it should no longer be treated as a public good. Tobacco, energy, and pharmaceutical companies have become adept at the deployment of scientific claims, backed up sometimes by credentialed scientists and by papers in respected medical or scientific journals, to manipulate public opinion or bypass regulatory hurdles. The scientist-entrepreneur has emerged as a new form of cultural hero, mainly on the evidence of the accumulation of great riches or the defiance of collective assumptions. The burgeoning of interested science, which conceals many results and sometimes actively foments ignorance, has made it more difficult than ever to think of any form of science as neutral and objective, standing above private interests.[26] Thinness in this context signifies the ambition to make science (appear) transparent, or to limit what can be asserted in its name. It means a preference for neutral data over subtle expertise. The thin world hollows out the priestly function, leaving only a shell.

The category of thinness, as I have been arguing, is pertinent especially to science in the public domain. *Nullius in verba*, On No Man's Word, was the motto chosen in the 1660s by the new Royal Society of London, and in the 1830s a pioneering institution of social science, the London Statistical Society, used *Aliis exterendum*, To Be Threshed Out by Others.[27] Western political orders, perhaps most enthusiastically in America, took slogans like these very seriously. Any opening up of ordinary scientific practices, especially when the science is tied up with policy choices, now threatens to discredit the science by exposing ambiguities, choices, and judgments. Research science can scarcely ever be objective in the sense of adhering to recipes. For example, it often does not include all the data in an analysis, but makes a fresh start when the experimental system is not working properly. Scientists may articulate distinctions retrospectively, quite possibly for good reasons (but how can an outsider know that?). Only when strong political and economic interests are at play, subjecting knowledge to a battle of interests before a court of uninformed outsiders, is such adaptability and flexibility subject to prohibition under the code of correct method. Science in modern times has been compelled by commercial pressures and political

[25] Hayek, *The Counter-revolution of Science: Studies on the Abuse of Reason* (London, 1955); Hayek, *The Road to Serfdom* (Chicago, 1944); Hayek, "The Use of Knowledge in Society," *American Economic Review* 35 (1945): 519–30.

[26] Philip Mirowski, *Science-Mart: Privatizing American Science* (Cambridge, Mass., 2011); Robert N. Proctor and Londa Schiebinger, eds., *Agnotology: The Making and Unmaking of Ignorance* (Stanford, Calif., 2008).

[27] Peter Dear, "*Totius in Verba*: The Rhetorical Construction of Authority in the Early Royal Society," *Isis* 76 (1985): 145–61; Victor L. Hilts, "*Aliis Exterendum*, or, the Origins of the Statistical Society of London," *Isis* 69 (1978): 21–43.

suspicions to take more literally than it otherwise would its promises of impersonal action based on neutral facts. That kind of science, in turn, provides a model of public knowledge. It is an ideal of maximal thinness, and it does have some capacity to drink up the sea. What depths remain provoke deep anxieties and dark suspicions.

The discomfort of public science with meaning and interpretation helps to explain the ubiquity of information and calculation. Often this assumes a statistical form, referring to numbers about collectives formed of individuals who may never be contacted or observed directly by those who perform the analysis and write up the results. If those individuals are persons, there will be abundant motives and meanings that the analysis cannot incorporate. Statistics in the human domain retains an element of its primal meaning, state-istics, the descriptive science of the state, which, since about 1890, has become increasingly analytical and mathematical. States too seek access to this domain of detached objectivity. The continued prominence of statistics in the sciences of administration and regulation makes clear and explicit an alliance with the state that is now typical for much of science. The state's-eye view, looking out over a vast and heterogeneous territory that cannot be known intimately, is highly favorable to thin description. Yet we cannot forget that the statistical view, like pictures from a satellite, also enables the recognition of what is invisible to ants on the ground, and may then offer guidance on what is deserving of closer inspection. On many topics, statistics is an essential component of any complex understanding.

The idealization of thinness is fundamental to public conceptions of "evidence" in that ubiquitous slogan of our time, "evidence based." Evidence here takes on a very particular meaning as the gathering of data and algorithmic analysis. It has been applied most forcefully to medicine, revealing once again what antagonism is possible between scientific and professional ideals. This sense of evidence imposes tight constraints on the physician's judgment. Medical skill is granted an important role in the execution of a procedure with hand and eye, but not, according to the canons of evidence, in the prognosis or choice of treatment by the mind. In the early 1950s, when randomized clinical trials were still rather novel, medical statistician Austin Bradford Hill reassured practitioners that statistics was a supplement and not an alternative to clinical judgment. Since then, that rhetoric has typically hardened, offering a new conception of medical epistemology. Clinical judgment and medical experience have for centuries been predominantly case based, but evidence-based medicine is represented now by pyramidal towers of evidence, with case studies and even experiments of a physiological, biochemical, or bacteriological kind clustered at the bottom. Randomized controlled trials, which undertake specifically to measure though not to give mechanisms for the effectiveness of the treatment, are typically midway up the hierarchy of pertinence, and the top layers are dominated by metastudies, statistical syntheses of work by many authors. This perspective accents what is routine and standardized within the individual studies, which must be treated as somehow homogeneous. The meta-meta conclusion, heard often from advocates of evidence in this mold, is that every kind of practice and every kind of prediction governed by such evidence is superior to undisciplined clinical judgment, and superior even to clinical judgment that takes into account the evidence-based prescriptions.

It would be no mean feat to expel judgment from medicine. But we have come a long way. An infrastructure of blood, urine, and tissue tests, of technologies of data production and visualization, supports the onslaught of evidence. This kind of laboratory involves some of the same equipment yet is worlds apart from creative research. The procedures of pharmaceutical testing, with their randomized clinical trials, illustrate its possibilities and limits. For example, the data points stand for human patients, and when these experimental subjects vote with their feet—perhaps because they think the treatments are making them worse, perhaps because they feel fine now—this shows up as missing data, which can be very hard to interpret. Such experiments are regulated by administrative agencies and courts, no longer by scientists and doctors alone. Is it possible to reduce all the intellectual and scientific work of doctors (or technicians) to taking tissue and DNA samples and instrument readings and sending them off to a central laboratory to be analyzed and, along with the patient's answers to a completely scripted interview, fed into a uniform diagnostic and intervention module? There are excellent reasons to try to systematize diagnosis and treatment decisions. These efforts, while immodest, are far from purposeless, for thickness is allied to uncertainty and unreliability. Yet an all-purpose hammer may not be ideal for every nut, and the quest for medical effectiveness is only one element in the worship of transparency. The fatal shortcoming of expert opinion in a densely administered, legalistic world may arise not from a demonstrated inferiority to other alternatives, but from its resistance to proper regulation.

That point is still more plainly visible in the mobilization of evidence to evaluate schools and teaching programs. Here again, measurement of effectiveness is linked to an impressive infrastructure of standardized, called "objective," testing. The scientific drive for objective numbers demands a reshaping of schools into places where these numbers are meaningful. Donald Campbell, the noted social psychologist and statistician who devoted most of his career to the cultivation of such techniques, wrote trenchantly of the dangers when authority is invested in numbers to the exclusion of humanistic and qualitative insights. Without speaking of "thinness," he identified clearly its dangers in a pronouncement sometimes referred to as Campbell's law: "The more any quantitative social indicator is used for social decision-making, the more subject it will be to corruption pressures and the more apt it will be to distort and corrupt the social processes it is intended to monitor."[28] The corollary is that thin numerical indicators are never mere representations, but perform always as forms of intervention, even before the rewards and penalties set in. Another sense of "objectivity" identifies it with unencumbered independence, the importation of methodologies from outside the institutions under study. But this independence breaks down already at the moment of measurement, which brings forth irresistible efforts to bypass its purposes by exploiting its thinness. Let municipal police be judged by the percentage of cases they solve, for example, and we might anticipate a vast expansion of plea bargaining, including sentences for burglars made lighter in proportion to the number of reported cases to which they are willing to confess.[29] Assessing schools by the numbers has yielded not only an epidemic of sly statistical manipulations, but a reorientation of teaching around what can conveniently be tested and scored, efficiently and without "bias."

Adaptability and mental suppleness in pupils do not fare well as goals of education under thin testing regimes. (The astuteness of administrators, by contrast, may

---

[28] Campbell, "Assessing the Impact of Planned Social Change" (Occasional Papers Series, no. 8, Public Affairs Center, Dartmouth College, Hanover, N.H., December 1976), 49.

[29] Campbell cites J. H. Skolnick, *Justice without Trial: Law Enforcement in Democratic Society* (New York, 1966), chap. 8, 164–81.

flourish.) Yet it is immensely difficult in a politicized, legalistic culture to make the results of a thicker investigation acceptable as the basis for decisions about the fate of a school. The thinness of the testing regime, as such, is not its most troubling feature. What matters above all is its capacity to thin out programs of instruction and learning, to drink up the sea. Experts on schools are increasingly outspoken on the problems of thin indicators and can even demonstrate quantitatively some of their shortcomings. Scarcely anyone argues that numbers lack any important role for understanding the problems of schools. Designing a satisfactory measurement regime, however, is a labor of Sisyphus, especially when officials in charge may find advantage in superficiality. In New York City, a more subtle examination regime, more resistant to evasion, was recently blocked by political appointees. They preferred a thinner system of testing whose results could more readily be manipulated while circumventing the slower and more demanding process of improving education.[30]

The paradoxical trajectory of thin description with all its contradictions and unanticipated consequences provides wonderful material for historical and social studies of science, technology, and medicine. What else is so contradictory and multifaceted as the drive to simplify and routinize? What else draws so deeply on human resourcefulness, or sets into motion such complex trajectories? The appeal of thin description expands with scope, making it a key topic for the history of science on a grand scale. Yet it is worked out and confirmed—or undermined—by a profusion of microprocesses (Foucault would speak, justly, of transgressions) that also are part of this big history and that lend it its compelling if ironical thickness.

The processes of thinning depend heavily on infrastructure and are sometimes self-vindicating. Ian Hacking, who called attention to this aspect of science, has spoken of looping effects in the human sciences.[31] The trajectories of social institutions may create not self-reinforcing loops but Möbius strips, which often are overturned or inverted as a consequence of the human response to thin indicators and benchmarks. Yet we should not underestimate the triumphs of thinification and of the drive to supplant tradition with objective numbers. Thin description makes its claim as the least common denominator for a large and heterogeneous world that must be made commensurable to be regulated uniformly, that relies on far-flung bureaucracies, and that cannot put trust in professions of experts acting on local expertise. This world ruled by numbers demands of those engaged in the scholarly study of science a subtle interpretive sense in a realm of complex human interactions. The drive for thinness, while often highly technical, is dense with human meaning and leads into unfathomable depths. It poses a challenge for the history of science whose import extends well beyond the academy into all aspects of political, professional, and administrative life.

---

[30] See Jennifer Medina, "On New York School Tests, Warning Signs Ignored," *New York Times*, October 11, 2010, front page; Douglas N. Harris, "Value-Added Measures and the Future of Educational Accountability," *Science* 333 (2011): 826–7.

[31] Hacking, "The Self-Vindication of the Laboratory Sciences," in *Science as Practice and Culture*, ed. Andrew Pickering (Chicago, 1992), 29–64; Hacking, "The Looping Effect of Human Kinds," in *Causal Cognition: A Multidisciplinary Approach*, ed. Dan Sperber, David Premack, and Ann James Premack (Oxford, 1995), 351–83. Compare the related concept of performativity as used in Donald MacKenzie, *An Engine Not a Camera: How Financial Models Shape Markets* (Cambridge, Mass., 2006), 16–8.

Science, State, and Citizens:
Notes from Another Shore

by Fa-ti Fan*

ABSTRACT

The existing literature on the problem of science and the state tends to focus mainly on the relationship between the state and the scientific community, which it often sees merely in terms of conflict or collaboration. To obtain a more nuanced view, this article suggests that it is necessary and useful to include other relevant historical parameters in the picture, notably those related to expertise and citizenship broadly defined. By examining the example of mass science in Mao's China, the article demonstrates how a broad perspective on the problem of science and the state may help us better understand a seemingly unusual historical case and indicates that the approach can also be applied to other, more familiar historical cases.

SCIENCE AND THE STATE

The problem of science and the state is an old one. There has been a sizable literature on the subject, covering a wide range of scientific topics and time periods. For example, scholars have examined the close relationships between the rise of certain branches of science and early modern states. Geography, demography, statistics, botany, anthropology, and public health, to name only a few, played a notable role in the formation of modern nation-states and imperial or colonial states. Scientific and educational institutions—such as science academies, museums, geological surveys, polytechnics, and research universities—all occupied a place in the larger story of modern nation-states. Although not necessarily couched in the fashionable terms of science and state building, much of this literature tries to understand the contribution of science to the state and the state's desire to enlist the help of science and the scientific establishment.[1]

* Department of History, State University of New York, Binghamton, NY 13902; ffan@binghamton .edu.

I am grateful to the anonymous referees and the editors of this volume for their careful reading of an earlier version of this article. I would also like to thank the participants in the "Generalist Vision" conference at the Huntington Library and in the "Critical China Studies Working Group" in Toronto for their questions and comments.

[1] The literature is extensive. I can only cite a few examples here (more will be found in notes below): Eric H. Ash, ed., *Expertise and the Early Modern State*, vol. 25 of *Osiris* (2010); Theodore Porter, *Trust in Numbers: The Pursuit of Objectivity in Science and Public Life* (Princeton, N.J., 1996); Maurice Crosland, *Science under Control: The French Academy of Sciences, 1795–1914* (Cambridge, 2002); Roy Porter, ed., *The Cambridge History of Science,* vol. 4, *The Eighteenth Century* (Cambridge, 2003); Timothy Mitchell, *Rule of Experts: Egypt, Techno-politics, Modernity* (Berkeley and Los Angeles, 2002); Stuart McCook, *States of Nature: Science, Agriculture, and Environment in the*

Another major focus of studies on science and the state has been on science and "totalitarian states." Here the prime examples have been Nazi Germany and Stalin's Soviet Union. Armloads of books have been written about these topics, and we know in great detail many aspects of the relationship between science and the state in the two regimes. In Nazi Germany, racial science and eugenics thrived under state ideology. In the Soviet Union, Lysenkoism, a proletarian Lamarckian science, dominated biology and agricultural science for decades and drove modern genetics and its supporters into the cold, white wilderness. In physics, Einstein's relativity suffered political criticism in both countries. All of these stories are familiar to historians of science and have been incorporated into the standard curriculum in the history of science. In fact, they are probably the best known, and most notorious, examples of the problem of science and the state in the history of science.[2]

A third well-known topic concerning science and the state is "big science," especially in World War II and Cold War America. Studies of the Manhattan Project, the military-industrial-university complex, NASA, the supercolliders, and other large technoscientific projects have all, to some extent, raised the issue of science and the state (or, at least, the government).[3] Some have considered in depth the problems of science and war, Cold War politics, the national security state, and liberal democracy. There is also an increasing body of literature on science and international relations, usually focused on the twentieth century, that necessarily deals with the competition, rivalry, and cooperation between states. Whether these studies are about the International Geophysical Year, the scientific congresses, the Nobel prizes, or the European Union, they usually take the nation-state as the basic category in the international arena of scientific cooperation and competition.[4]

All of these are important topics, and the literature mentioned above has contributed

---

Spanish Caribbean, 1760–1940 (Austin, Tex., 2002); Gabrielle Hecht, *The Radiance of France: Nuclear Power and National Identity after World War II* (Cambridge, 1998).

[2] I do not mean that works on science in Nazi Germany and in the Soviet Union have adopted the term or concept of totalitarianism to explain the problem of science and the state. Actually, most of them have rejected it. But still the two regimes are often paired together in teaching and research on science and the state, and for good reason. The publications on these examples are legion: e.g., Monika Renneberg and Mark Walker, eds., *Science, Technology, and National Socialism* (New York, 2003); Walker, *Nazi Science: Myth, Truth, and the German Atomic Bomb* (New York, 2001); Paul Josephson, *Totalitarian Science and Technology*, 2nd ed. (Amherst, Mass., 2005); Loren Graham, *Science in Russia and the Soviet Union: A Short History* (New York, 1993); Robert Proctor, *Racial Hygiene* (Cambridge, 1988); Nikolai L. Krementsov, *Stalinist Science* (Princeton, N.J., 1996).

[3] E.g., Peter Galison and Bruce Hevly, eds., *Big Science: The Growth of Large Scale Research* (Stanford, Calif., 1992); Caroll Pursell, *The Military-Industrial Complex* (New York, 1972); Walter A. McDougall, . . . *The Heavens and the Earth: A Political History of the Space Age* (New York, 1985); Stuart W. Leslie, *The Cold War and American Science: The Military-Industrial-Academic Complex at MIT and Stanford* (New York, 1993). In a different way, the new literature on Cold War science and on social science in twentieth-century America has also deepened our understanding of relationships between science, the state, and citizenship. See, e.g., the "Focus" section "New Perspectives on Science and the Cold War," *Isis* 101 (2010): 362–411; Sarah Igo, *The Averaged American: Surveys, Citizens, and the Making of a Mass Public* (Cambridge, 2007); S. M. Amadae, *Rationalizing Capitalist Democracy: The Cold War Origins of Rational Choice Liberalism* (Chicago, 2003).

[4] E.g., James Fleming, Roger Launius, and David Devorkin, eds., *Globalizing Polar Science: Reconsidering the International Polar and Geophysical Years* (New York, 2010); Elizabeth Crawford, *Nationalism and Internationalism in Science, 1880–1939: Four Studies of the Nobel Population* (New York, 2002); Nikolai Krementsov, *International Science between the World Wars: The Case of Genetics* (London, 2005). See also the *Osiris* volumes *National Identity: The Role of Science and Technology*, vol. 24 (2009); *Intelligentsia Science*, vol. 23 (2008); and *Global Power Knowledge: Science and Technology in International Affairs*, vol. 21 (2006).

enormously to our understanding of the relationship between science and the modern state. However, there are three problematics that deserve more consideration.[5] First, the existing literature has concentrated on Europe and the United States. Little has been said with regard to the problem of science and the state in other parts of the world. This is changing, however. There is currently a growing interest in science and the nation-states in South Asia, Latin America, and East Asia. For example, in recent years, a few notable monographs have appeared on science, the state, and/or national identity concerning India, China, and Japan.[6] With such a range of studies, we can begin to ask certain important, but neglected, questions—for instance, what was the state (in a particular historical case study)? Not all modern states are the same, though as political units for administrating and regulating people, resources, defense, and traffic in the global system of modern nation-states, they must share certain similarities.[7] Therefore, we should ask to what extent and in what way what we learn from, say, European and American examples may help us understand the interactions of science and the state in other parts of the world, and vice versa. This is especially relevant because most studies of science and the state focus more on science than on the state. Apparently, we have been more willing and successful in unpacking science by invoking something called "the state" than in unpacking the state by bringing in science. It is certainly appropriate that historians of science care more about science than the state. Still, I think that it is valuable for historians of science to participate directly in the theoretical discussion of the state. How did the production of particular kinds of science alongside that of particular kinds of states actually take place? How were they mutually produced and constituted? How can the history of science contribute to theories of state formation and speak to scholars of the state?

My second problematic has to do with science itself. When we talk about science and the state, we tend to look at the scientific establishment and elite science—that which is well institutionalized and does not employ ideas and vocabulary drastically different from accepted norms in science. To be sure, there have been conflicts within

---

[5] I have generalized and simplified the existing literature on science and the state, though I hope I have not been unfair. I am certainly not alone in making this assessment of the literature. For a similar opinion, see, e.g., the introduction to a recent *Osiris* volume: Carol E. Harrison and Ann Johnson, eds., "Science and National Identity," *Osiris* 24 (2009): 1–14, on 8–9.

[6] E.g., James Bartholomew, *The Formation of Science in Japan: Building a Research Tradition* (New Haven, Conn., 1989); Hiromi Mizuno, *Science for the Empire: Scientific Nationalism in Modern Japan* (Stanford, Calif., 2008); Morris Low, *Science and the Building of a New Japan* (New York, 2005); Stuart Leslie and Dong-Won Kim, "Winning Markets or Winning Nobel Prizes: KAIST and the Challenges of Late Industrialization," *Osiris* 13 (1998): 154–85; Jahnavi Phalkey, "Science, State-Formation and Development: The Organization of Nuclear Research in India, 1938–1959" (PhD diss., Georgia Institute of Technology, 2006); Gyan Prakash, *Another Reason: Science and the Imagination of Modern India* (Princeton, N.J., 1999). There are many studies of nuclear research and South Asian states, and the number is growing fast. A very recent example is Robert S. Anderson's *Nucleus and Nation: Scientists, International Networks, and Power in India* (Chicago, 2010), which is a conventional biographic and institutional history. See also Julia Rodriguez, *Civilizing Argentina: Science, Medicine, and the Modern State* (Chapel Hill, N.C., 2006); Zuoyue Wang, "Science and the State in Modern China," *Isis* 98 (2007): 558–70; Cong Cao, *China's Scientific Elite* (London, 2004); H. Lyman Miller, *Science and Dissent in Post-Mao China: The Politics of Knowledge* (Seattle, 1996); Laurence Schneider, *Biology and Revolution in Twentieth-Century China* (Lanham, Md., 2005); Susan Greenhalgh, *Just One Child: Science and Policy in Deng's China* (Berkeley and Los Angeles, 2008).

[7] On the typology and evolution of modern states, we still often rely on works by historical sociologists, such as Perry Anderson, Michael Mann, Charles Tilly, and Theda Skocpol. For an introduction, see John A. Hall and G. John Ikenberry, *The State* (Minneapolis, 1989). The problem with this body of work is that it often leaves out certain vital components of the state, such as micropolitics, everyday life, social interactions, and cultural practice.

the scientific community, and there can be power struggles between the scientific establishment and those who want to fight their way in or take over the reins. And many of these contestations have to be understood within the context of state politics. To make this point, we need go no further than the familiar examples of Nazi Germany and the Soviet Union. However, even Lysenkoism did not break from the tradition of biology and agricultural science, despite its idiosyncrasies and hostility toward modern genetics.[8] (I should note that in the rest of the article, I will not compare communist China to the Soviet Union and Nazi Germany, because adopting that approach can easily slide back to the lazy habit of lumping them together under totalitarianism or other such notions. The differences between science in Mao's China and in Stalin's Soviet Union were profound and must not be elided.)[9]

When we consider such issues as science, expertise, and the state, again we tend to look at professional or vocational expertise—for example, in engineering, statistics, cartography, and biology. But the question of science, expertise, and the state is much broader than this conventional focus. Modern states have been interested in and had to engage with various forms of knowledge, not only the ones readily recognized as expert scientific knowledge. Imperial and colonial officials tried to contain or utilize indigenous knowledge, including knowledge about plants and animals.[10] In implementing public health measures, European states in the nineteenth century had to battle popular notions and practices of hygiene and sanitation with new, "modern" ones. The point here is not to treat the state as a self-contained entity and pit it against the people and their knowledge traditions. Rather, it is simply to point out that the state, however one defines it, often must handle bodies of knowledge that are related to scientific matters, but are not necessarily produced or approved by scientists. It is therefore important to broaden our view and include other forms of knowledge in considering the topic of science and the state.

To achieve the goal of nation-state building and modernization, most modern states pursued projects of disseminating or popularizing science. The projects might take different forms—for example, school education; public institutions such as museums; popular media, such as newspapers, magazines, and television; and nationwide events, such as science weeks. Not surprisingly, many discussions about science, the state, society, and public policy are about science communication, science outreach, science literacy, science for the public, and the like.[11] The state utilizes its appara-

---

[8] Nils Roll-Hansen, *The Lysenko Effect: The Politics of Science* (Amherst, Mass., 2004).

[9] Nevertheless, it will become clear that my research into Mao's China has benefited from certain recent literature on Soviet history; e.g., Stephen Kotkin, *Magnetic Mountain: Stalinism as a Civilization* (Berkeley and Los Angeles, 1995); Michael Gordon, "Was There Ever a 'Stalinist Science'?" *Kritika* 9 (2008): 625–39; John Connelly, "Totalitarianism: Defunct Theory, Useful Word," *Kritika* 11 (2010): 819–35; Susan Gross Solomon, "Circulation of Knowledge and the Russian Locale," *Kritika* 5 (2008): 9–26; Francine Hirsch, *Empire of Nations: Ethnographic Knowledge and the Making of the Soviet Union* (Ithaca, N.Y., 2005). I have also, in a different way, benefited from Loren Graham's *What Have We Learned about Science and Technology from the Russian Experience?* (Stanford, Calif., 1998).

[10] E.g., David Arnold, *Colonizing the Body: State Medicine and Epidemic Disease in Nineteenth-Century India* (Berkeley and Los Angeles, 1993); Warwick Anderson, *Colonial Pathologies: American Tropical Medicine, Race, and Hygiene in the Philippines* (Durham, N.C., 2006); Londa Schiebinger and Claudia Swain, *Colonial Botany: Science, Commerce, and Politics in the Early Modern World* (Philadelphia, 2004).

[11] One of the most noted examples is the "public understanding of science" movement in Britain starting in the 1980s. It did not originate from the government, but the movement quickly involved a wide range of actors and venues, including state-owned or -sponsored institutions; see

tuses and technologies of rule to mold the kind of modern citizenry it wants. For certain nations, the mission to modernize the population by science or to erect science as one of the nation's central pillars was paramount; notable examples include the Soviet Union, Japan, and communist China.[12] Hence, my third problematic is about science and citizenship. The main task of the rest of the article is to demonstrate that more attention to these three problematics will enable us to gain a better understanding of the relationship between science and the modern state.

SCIENCE AND STATE IN MODERN CHINA

In modern East Asia, the relationship between science/technology and the state has been important and complex. It simply will not do to examine one and ignore the other, for the process of building modern nation-states there has been deeply intertwined with the ideology and the development of science and technology. This is not to say that the state has controlled science or science has simply served the state, nor does it suggest a state-centered perspective or narrative of history of science. In fact, we should see the interaction of the state and science as a dynamic composite of multiple actors—scientists, the public, political institutions, and so on—forming alliances or jostling for position.[13] In other words, we must unpack the black boxes of both science and the state and examine how they were mutually constituted.[14] Let me illustrate this approach with an American example, if only because Chinese scientists of the Republican period (1912–49) thought to learn from it. The Army Corps of Engineers and the US Geological Survey mapped much of the American West during the time when the United States was expanding westward.[15] Here scientific

---

the Royal Society's report "The Public Understanding of Science," at http://royalsociety.org/Public-Understanding-of-Science/ (accessed 28 September 2011); Wellcome Trust, "Science and the Public: A Review of Science Communication and Public Attitudes to Science in Britain," at http://www.wellcome.ac.uk/stellent/groups/corporatesite/@msh_peda/documents/web_document/wtd003419.pdf (accessed 5 October 2011). See also the journals *Science and Public Policy* and *Public Understanding of Science*. Cf. David Dickson, "Science and Its Public: The Need for a 'Third Way,'" *Soc. Stud. Sci.* 30 (2000): 917–23; Jane Gregory and Steve Miller, *Science in Public: Communication, Culture, and Credibility* (New York, 2000).

[12] James Andrews, *Science for the Masses: The Bolshevik State, Public Science, and the Popular Imagination in Soviet Russia, 1917–1934* (College Station, Tex., 2003); Mizuno, *Science for the Empire* (cit. n. 6). It should be noted that the state was not always at the center of such enterprises. There were organizations in civil society that took it upon themselves to promote science, modernity, and citizenship as they defined them. Their ideas could resonate, complement, or compete with those of the state. In Republican China, scientists and intellectuals who were eager to promote science and modernize China and its citizens often complained about too much state and too little state at the same time; Zuoyue Wang, "Saving China through Science: The Science Society of China, Scientific Nationalism, and Civil Society in Republican China," *Osiris* 17 (2002): 291–322.

[13] To be sure, the state does not exist other than through people and institutions, but the state cannot be reduced to its components, either. It would be like taking a car apart and calling the pile of parts on the ground a car.

[14] Sheila Jasanoff, ed., *States of Knowledge: The Co-production of Science and Social Order* (London, 2004), uses the language of coproduction to capture this relationship.

I should clarify that I am not saying that science and the state always work together and reinforce each other. In fact, there are always tensions, contestations, and conflicts. It is, however, useful to turn away from the old perspective of seeing the state as an outside influence—be it positive or negative—on science.

[15] Robert Kohler, *All Creatures: Naturalists, Collectors, and Biodiversity, 1850–1950* (Princeton, N.J., 2006); Philip Pauly, *Biologists and the Promise of American Life* (Princeton, N.J., 2000); William Goetzmann, *Exploration and Empire: The Explorer and the Scientist in the Winning of the American West* (New York, 1966); Thomas Manning, *Government in Science: The U.S. Geological*

activities were actually part and parcel of the nation-state-building process. Here neither science nor the state can be seen as a finished product. They were components of the same process.

Few will dispute that nation-state building is a fundamental component of modern Chinese history, and science was a significant part of the story. Geology, anthropology, archaeology, and biology all contributed to the enterprise of nation-state building in twentieth-century China. The historical actors participating in the diverse but related enterprises included both government and nongovernment organizations. Founded in 1928, Academia Sinica, the highest-level national research institute, eagerly pursued archaeological excavations in North China in part with the aim to reconstruct the origins of the Chinese nation and civilization, which would become a cornerstone of Chinese national identity. Chinese historians and archaeologists at the institute felt a sense of responsibility to define and guard national history. Their particular national narrative, in turn, informed their research and interpretations in archaeology. Similarly, the Geological Survey of China and the Fan Memorial Institute of Biology (which was modeled after the Wistart Institute and the US Biological Survey) had the ambition of mapping the nation's nature—not only its natural resources, such as coal, minerals, and botanical riches, but also the less definable geobody of the nation or the "national land" (*guotu*). Not surprisingly, the ambition also extended to the mapping of the people and citizens. For example, Academia Sinica and other institutions conducted ethnographic fieldwork in Manchuria and South China, in an attempt to study the ethnic minorities and incorporate them into "the Chinese people." The enterprise took another form in the 1950s after the communist revolution. The new regime launched a large project to (re)classify all ethnic groups. The result—fifty-six ethnic groups, or fifty-five ethnic minorities plus the Han group (by far the largest)—became a national system with real political, social, cultural, and economic consequences. All of these scientific projects attempted to define China's history, nature, and citizens. Thus, they were efforts to build a nation-state, but at the same time they were also shaped by particular notions and institutions of the nation-state.[16]

Of course, the very thing called "the state"—its definition and its relationship with science, nation, society, and citizens—varies significantly from case to case. From this perspective, twentieth-century China experienced a variety of states, many of

---

Survey, 1867–1894 (Lexington, Ky., 1967). Historians of scientific imperialism have shown similar instances of the coproduction of science and empire; Roy McLeod, ed., *Nature and Empire: Science and the Colonial Enterprise, Osiris* 15 (2000); Fa-ti Fan, *British Naturalists in Qing China: Science, Empire, and Cultural Encounter* (Cambridge, 2004). See also Suman Seth's historiographic essay "Putting Knowledge in Its Place: Science, Colonialism and the Postcolonial," *Postcolon. Stud.* 12 (2009): 373–88.

[16] Tong Lam, *A Passion for Facts: Social Surveys and the Construction of the Chinese Nation-State, 1900–1949* (Berkeley and Los Angeles, 2011); Grace Shen, *Unearthing the Nation: Modern Geology and Nationalism in Republican China, 1911–1949* (Chicago, forthcoming); Thomas Mullaney, *Coming to Terms with the Nation: Classification in Modern China* (Berkeley and Los Angeles, 2010); Fa-ti Fan, "How Did the Chinese Become Native? Science and the Search for National Origins in the May Fourth Era," in *Beyond the May Fourth Paradigm: In Search of Chinese Modernity*, ed. Kai-wing Chow et al. (Lanham, Md., 2008), 183–208; Fan, "Nature and Nation in Chinese Political Thought: The National Essence Circle in Early Twentieth-Century China," in *The Moral Authority of Nature*, ed. Lorraine Daston and Fernando Vidal (Chicago, 2004), 409–37; Yuehtseng Juliette Chung, *Struggle for National Survival in Sino-Japanese Context* (London, 2002); Sean Hsiang-lin Lei, "When Chinese Medicine Encountered the State, 1910–1949" (PhD diss., Univ. of Chicago, 1999).

which were weak and short-lived, and which sometimes coexisted with each other.[17] It is therefore necessary to analyze the composition of a state in considering the relationship between science and the state in twentieth-century China. To this end, I shall break down the vision and discourse of a particular Chinese nation-state, Maoist China. This will allow me to investigate the three problematics described above in relation to the enterprise of state making through science making and vice versa. By examining the case of science in communist China—specifically, a large-scale project of mass science during the Cultural Revolution—I will demonstrate the importance of maintaining sight of a broad scope of historical actors and their knowledge in the study of science and the state. To many readers, mass science in communist China might seem unfamiliar or downright idiosyncratic. But I believe that precisely for this reason, this case can shed light on certain general characteristics of the relationship between science, the state, and citizenship in the modern world.

MASS SCIENCE AND MAOISM

One common way of studying science in Mao's China is to place it within the framework of the Cold War, where the main focus is on the role of science and technology in national defense and state rivalry, as in the case of the development of nuclear weapons.[18] Another way is to examine the tensions between science (usually as embodied by the scientific establishment) and the state (often couched in such terms as *totalitarianism* and *control*). This latter approach is concerned primarily with the relationships between the scientific establishment and the high politics of the state.[19] Both approaches have created opportunities for discussion of science in political regimes both similar to and different from communist China, and scholars have pondered the implications of democracy, professional autonomy, and ideological control for the development of science. Such directions of research are generally amenable to the dominant narrative of science and the Cultural Revolution, which emphasizes the violent intrusion of the state into the scientific community and the disruption of scientific research during that period. Although this narrative captures much tragic truth, it oversimplifies. Not only did the state need the scientific elite for certain political or practical reasons—including the development of military technology, efforts to synthesize insulin, and rice breeding—but the practice of science involved a large population far beyond the traditional scientific community.[20] The national projects of earthquake monitoring during the Cultural Revolution (1966–76) provide a case in point.

---

[17] Following the conventional political typology, they have been described, with varying degrees of accuracy, as the imperial dynastic state, the parliamentary republic, the constitutional monarchy, the warlord governments, the federalist state, the Leninist state, the Confucian-fascist state, the Stalinist state, the Maoist state, and so on. See, e.g., John Fitzgerald, *Awakening China: Politics, Culture, and Class in the Nationalist Revolution* (Stanford, Calif., 1998); Prasenjit Duara, *Rescuing History from the Nation: Questioning Narratives of Modern China* (Chicago, 1997); Frederick Wakeman, "A Revisionist View of the Nanjing Decade: Confucian Fascism," *China Quart.* 150 (1997): 395–432. For an interesting forum on the Chinese state, see the special issue "The Nature of the Chinese State," *Modern China* 34 (2008).

[18] E.g., John Lewis and Litai Xue, *China Builds the Bomb* (Stanford, Calif., 1991).

[19] E.g., Laurence Schneider, *Biology and Revolution in Twentieth-Century China* (Lanham, Md., 2005); Cao Cong, *China's Scientific Elite* (London, 2004).

[20] The "Focus" section "Science and Modern China" in *Isis* 98 (2007): 517–86 may serve as an introduction to the historiography of science in twentieth-century China.

It just so happens that more than ten major earthquakes (roughly magnitude 7 or above) took place during the ten years of the Cultural Revolution, starting with the Xingtai earthquake in March 1966 and culminating with the Tangshan earthquake in July 1976, which killed more than a quarter of a million people. Because much of China is located between two major earthquake zones—one running roughly along the Pacific Rim and the other stretching from the Himalayas to Southeast Asia— earthquakes are not uncommon in many parts of the country. Nevertheless, it is unusual for a series of major earthquakes to hit China proper with such frequency and intensity as in the 1960s and '70s.[21] As this period of natural disasters coincided with the Cultural Revolution, fear and anxiety about earthquakes were fomented in the environment of social and political upheavals. On the one hand, the Chinese experienced violent social and political storms; on the other, they worried that earthquakes might strike at any time.

Up to the mid-1960s, seismology was not well developed in China, but the Xingtai earthquake jolted the political and scientific leadership into action, and consequently the field expanded rapidly. The State Seismological Bureau was established in 1971. The number of seismological stations increased from a handful in the early 1960s to about 250 by the mid-1970s, not including a variety of observatories associated with mass science projects. According to one estimate, there were also "several tens of thousands" of lay participants in the program.[22] With ever more sophisticated technology, modern seismology depended principally on microscopic measurements, using highly sensitive instruments to monitor seismic activities. Although Chinese seismologists did not have the most advanced equipment, they similarly pursued this area of research. However, they also devoted much attention to research on macroscopic phenomena (viz. premonitory phenomena that could be observed without using instruments) as well as other phenomena that could be observed with very simple tools. Their approach sprang from three sources. First, as noted, China did not have a lot of advanced equipment, nor did it, at first, have many seismological stations. Since it was impossible to cover a large country like China with so few advanced observatories, Chinese scientists welcomed research directions that would make up for this shortage. Second, they believed that macroscopic phenomena could lead to fairly accurate predictions of earthquakes, whereas in technologically more developed countries such as the United States, the scientific establishment tended to presuppose that advanced technology rendered observations of macroscopic phenomena superfluous. Third, the political content of this approach was attractive, as it called for mass participation, mass mobilization. Since Chinese scientists believed that certain macroscopic phenomena could be symptomatic of earthquakes, they naturally wanted to have many observers out there looking for them. And, politically, if science was indeed of the masses and for the masses (as demanded by Maoism), then people should learn about science and do science. Earthquake monitoring, therefore, provided a perfect opportunity for mass science.[23]

It is necessary to spell out the ideological connections between mass science and Maoism here. Maoism served as the guiding ideology of the Cultural Revolution.

---

[21] Li Shanbang, *Zhongguo dizhen* [Earthquakes in China] (Beijing, 1981).

[22] Barry Raleigh et al. (the Haicheng Earthquake Study Delegation), "Prediction of the Haicheng Earthquake," *Eos* 58 (1977): 236–72, on 237.

[23] My discussion of earthquake monitoring and defense in China is based on my ongoing research on the topic. For a more detailed treatment, please see my article "'Collective Monitoring, Collective

Political propaganda was a fact of life in communist China, but during the Cultural Revolution, the apotheosis of Mao and his thought reached a new, feverish pitch. Under these circumstances, "ideologically correct" science had to be grounded in official Marxism and Mao's thought. Science had to be true to dialectical materialism because this philosophy correctly described the fundamental principles of how nature works. Friedrich Engels was frequently cited as a philosophical authority on nature. The goal of science was to conquer nature, to free oneself from nature. This notion echoed Engels's theory that labor creates humanity. Mao declared, "Natural science is one of man's weapons in his fight for freedom. For the purpose of attaining freedom in society, man must use social science to understand and change society and carry out social revolution. For the purpose of attaining freedom in the world of nature, man must use natural science to understand, conquer and change nature and thus attain freedom from nature." Indeed, science and technology were the results of and the tools for human struggles against nature. Ultimately, however, Maoism was less about historical materialism than about moral imperatives and the power of human will. When Mao proclaimed, "Humans will triumph over nature" (*ren ding sheng tian*), he was talking about fortitude, perseverance, and a fearless determination to win.[24]

Maoist thought also insisted that science must serve the people. To achieve that end, science had to be practical, empirical, and utilitarian. Abstract theoretical science came too close to idealism, which was not only philosophically wrong, but also scientifically unsound. For these reasons, Einstein's theory of relativity ran into trouble during the Cultural Revolution.[25] Sciences that did not have any obvious utilitarian function were at best intellectual toys of the elite. Maoism had a strong strain of antielitism. Mao himself had said that "the lowliest are the smartest; the highest the most stupid." The people learned from their experience and labor, from their long struggle with nature. Such hard-earned knowledge—concrete, reliable, and often ingenious despite its humble origins—was truly useful and valuable. Thus, science was inherently political. It was objective, but it was neither neutral nor value free. It was, de facto, class based, and good science required mass participation. Maoist thought insisted that science "walks on two legs."[26] One leg was the masses, and the other the experts or specialists. Experts and the people ought to work together. Moreover, elite scientists must learn from the people. During the Cultural Revolution, many intellectuals and scientists were sent down to the countryside for this purpose. In

---

Defense': Science, Earthquakes, and Politics in Communist China," *Sci. Context* 25 (2012): 127–54. Here I simply use the case study to illustrate my general argument about the problem of science and the state.

[24] The quotation about natural science is from Mao Zedong, *Mao zhuxi yulu* [Quotations from Chairman Mao Tse-tung] (Tianjing, 1966), 175. Stuart Schram, *The Thought of Mao Tse-Tung* (Cambridge, 1989); Maurice Meisner, *Marxism, Maoism, and Utopianism* (Madison, Wisc., 1982). Meisner's *Mao's China and After: A History of the People's Republic* (New York, 1999) provides a good overview of the Mao period. See also Sigrid Schmalzer, "Labor Created Humanity: Cultural Revolution Science on Its Own Terms," in *Cultural Revolution as History*, ed. Joseph Esherick, Paul Pickowicz, and Andrew Walder (Stanford, Calif., 1977), 185–210.

[25] Danian Hu, *China and Albert Einstein: The Reception of the Physicist and His Theory in China, 1917–1979* (Cambridge, 2005); Hu, "The Reception of Relativity in China," *Isis* 98 (2007): 539–57.

[26] This was a powerful idea and profoundly influenced the Science for the People group in Boston in the early 1970s. See Dan Connell and Dan Gover, *China: Science Walks on Two Legs* (New York, 1974); Sigrid Schmalzer, "On the Appropriate Use of Rose-Colored Glasses: Reflections on Science in Socialist China," *Isis* 98 (2007): 571–83.

the meantime, farmers and factory workers were called on to participate in scientific work at research institutes.

Furthermore, Maoist science also emphasized the need to combine both indigenous and Western science. Mao himself believed that there was much unique knowledge in the Chinese lore on medicine. This idea was reflected clearly in the "barefoot doctors" program.[27] The doctors—often villagers or sent-down youths who had received brief training—used traditional medicine and healing practices together with Western-style drugs. The program provided health care for rural areas, where medical resources were limited. To some extent, the program aimed to serve practical needs. Nevertheless, it was also emphatically ideological and embodied such political ideas as communist society, national tradition, and mass science. Along a similar line, in earthquake studies in the 1960s and '70s, the state championed the integration of folk knowledge and more technical seismology.

SCIENCE MAKING AND STATE MAKING

The party-state mobilized the masses in the national enterprise of earthquake monitoring and defense. The shortage of seismological stations and a belief in the possibility of short-term earthquake prediction explained only part of the ambition. The enterprise constituted at the same time scientific research, disaster control, national defense, a political campaign, and nation-state building. The basic tenets of the program, called "Collective Monitoring, Collective Defense," are well captured in a poster dating from the mid-1970s.

The picture displays an array of people who participated in the program (fig. 1). The slogan on the poster reads: "Under the unitary leadership of the party, and concentrating mainly on prevention, we combine the experts and the masses as well as indigenous [science] and foreign [science], rely on the masses, and do well the task of premonitoring and prevention." The man towering above the crowd, representing a party secretary, holds up a copy of the *Selected Writings of Chairman Mao*. (So, Mao Zedong's thought and the party serve as the teacher, the guiding light, to the people.) The three people lining up in front of him are a worker, a peasant, and a soldier (who together make up the backbone of the communist society). Note that the peasant woman carries an ampere meter and a clipboard of paper. The meaning is clear: even a peasant woman, humble and lowly in a traditional feudal society, can participate in and contribute to a national program of paramount importance in this new, communist society. Behind and next to her is a woman scientist, wearing glasses and a white coat. (Representations of women scientists were common in propaganda posters from the Cultural Revolution era, due to the Maoist ideology of gender equality at the time—in Mao's words, "Women hold up half the sky.")[28] Under the pointing

---

[27] For non-Chinese accounts, see John Fogarty, *A Barefoot Doctor's Manual: The American Translation of the Official Chinese Paramedical Manual* (Philadelphia, 1977); Fang Xiaoping, "From Union Clinics to Barefoot Doctors: Village Healers, Medical Pluralism and State Medicine in a Chinese Village," *Journal of Modern Chinese History* 2 (2008): 221–37.

[28] Wang Zheng, "'State Feminism?' Gender and Socialist State Formation in Maoist China," *Feminist Stud.* 31 (2005): 519–51. On the issue of gender in modern China, with a helpful discussion on women and gender in the Mao era, see Gail Hershatter, *Women in China's Long Twentieth Century* (Berkeley and Los Angeles, 2007). On representations of women in propaganda posters during the Mao era, see Tina Mai Chen, "Female Icons, Feminist Iconography? Socialist Rhetoric and Women's Agency in 1950s China," *Gend. & Hist.* 15 (2003): 268–95. I do not have space here to discuss the

Figure 1. *Picture taken from Guangdong sheng gemin weiyuanhui dizhen bangongshi and Guojia dizhenju Guangzhou dizhen dadui [Earthquake Office, Revolutionary Committee, Guangdong Province and Guangzhou Earthquake Brigade, State Seismological Bureau],* Dizhen zhishi huace [An illustrated pamphlet about earthquakes] *(Beijing, 1977). In addition to the characters in the front (discussed in the main text), there is also a row of five small images of people engaged in earthquake monitoring in the background, slightly above and more distant than the front crowd. The images are, from left to right, of a man reining in a horse standing on its hind legs; two people, possibly a worker and a peasant, discussing some scientific equipment; a male scientist, again with glasses and in a lab coat, looking at a seismograph; three young women checking the water of a well; and a young woman launching a weather balloon while a man records data. All of the images represent indicators commonly used in China at the time for monitoring and predicting earthquakes (e.g., anomalous animal behavior, groundwater variation, and weather changes).*

hand of the worker are two students—a Red Guard armed with a chart of seismological analysis who is a middle-school student and a Young Pioneer girl holding a bullhorn and an earthquake education manual who is an elementary school student. Evidently, even children play a notable role in the program.

Further right is a woman in an ethnic outfit, who is holding a piece of electrical equipment. The presence of a representative of ethnic minorities conveys multiple meanings. The obvious significance is that one of the active earthquake zones during that period was in Sichuan and Yunnan Provinces, which were home to many ethnic minorities. Therefore, the ethnic minorities had to be part of the earthquake monitoring and defense program. However, there is another, even more significant

---

genre(s) of Cultural Revolution posters. See, e.g., Lincoln Cushing and Ann Tompkins, *Chinese Posters: Art from the Great Proletarian Cultural Revolution* (San Francisco, 2007); Stefan Landsberger and Marien van der Heijiden, *Chinese Posters* (London, 2009); Melissa Chiu, *Art and China's Revolution* (New Haven, Conn., 2008).

meaning: integrating ethnic minorities into the nation-state was a major enterprise of modern Chinese nationalism throughout the twentieth century. This included portraying the hearty support of ethnic minorities for the national program and, equally important, the success of the nation-state in advancing them from backwardness to scientific and political modernity, represented here by their ability to use scientific equipment. Thus, the picture as a whole projects a vision of the communist Chinese nation-state (comprising certain conceptions of class, gender, ethnicity, polity, science, and modernity). It reveals that the Collective Monitoring, Collective Defense program was as much a political as a scientific enterprise: it was at the same time a project of nation-state building, mass scientific participation, and organized defense against natural disasters.

China devised an earthquake monitoring system that aimed to integrate both the experts, who staffed the seismological bureaus and stations, and the masses, who also played a crucial role. The observation activities of the masses were supervised by the local party committees, schools, communes, and work units rather than the seismological bureaus. Since there were thousands of these observation units in the Collective Monitoring, Collective Defense program, it is not surprising that their scientific and technological sophistication varied. The most advanced among them were well equipped and did not look very different from regular seismological observatories. But far more common were simple observation units attached to middle schools, factories, post offices, communes, and telephone stations. One unit or team usually consisted of several individuals. Some of the more common objects of observation were ground tilt, geostress, telluric currents, geomagnetism, and well-water variations.[29] In the late 1960s and early '70s, many indigenous instruments were invented for making these observations. A young farmer invented the method of measuring telluric currents by using a simple device that consisted of an ampere meter and two electrodes planted underground at a distance apart. The intensity and variation of underground electric currents were thought to indicate earthquake activity. Through official channels, the inventor requested advice from the Seismological Bureau, which then sent a scientist to evaluate the instrument. After some examination, the scientist confirmed the effectiveness of the device. This method was reported in an influential journal on earthquake studies, and soon it became very popular across the nation.[30] Similarly, ground tilt could be measured by simple instruments, such as a pendulum hanging from the ceiling and a piece of chalk.[31]

Most of the masses were not organized into active groups or teams manning observation points. But they were still participants in the earthquake-prediction program. They received information about earthquakes through political and educational channels (party offices, neighborhood organizations, production brigades, commune units, etc.) and through propaganda vehicles (pocket-size pamphlets, mimeographed handouts, colorful posters, exhibitions, film screenings, school competitions, and

---

[29] These and other, similar methods are discussed in middle-school textbooks on earthquake monitoring; e.g., Shaonian dizhen huodong bianxie zu [Editorial team of *Activities of Monitoring Earthquakes for Young People*, ed.], *Shaonian dizhen cebao huodong* [Activities of monitoring earthquakes for young people] (Shanghai, 1978).

[30] Beijing shi Yanqin xian gewei hui, Yanqing xian Zhangshanying diqu dizhen lianhe diaocha zu [United Investigative Group, Yanqing County, Zhanghanying, the Revolutionary Committee of Yanqing County, Beijing], "Tu didian fangfa jieshao" [Introducing the telluric method], *Dizhen zhanxian* 4 (1969): 29–31.

[31] *Dizhen changshi* [Common sense about earthquakes] (Beijing, 1975), unpaginated.

propaganda tours to the countryside). They learned about the basic science of earthquakes, methods of earthquake prediction, emergency defense, and so on. Although most people usually did not use instruments or keep a routine record of what they observed, they were urged to pay attention to macroscopic seismic phenomena. When they noticed any unusual phenomenon that might indicate the coming of an earthquake, they had to immediately report it to the local offices. It was their duty to do so. Together the various levels of mass science would form a blanket warning system for earthquakes.

Because the Chinese believed that some of the precursory signs of earthquakes were macroscopic phenomena that could be observed by common people, the program of earthquake prediction incorporated folk wisdom and everyday observations that described anomalous natural phenomena that might indicate the coming of an earthquake—aberrant animal behavior, unseasonable weather, well-water variations, unusual underground temperature, and so on.[32] These anomalous signs indicated a forthcoming earthquake just as symptoms suggest a hidden but menacing disease. Some of the anomalies were obvious, such as foreshocks; others were obscure, such as cloud formations. Seen as a science of reading anomalous signs or symptoms, the Chinese method of earthquake prediction may be likened to a semiotics of earthquakes. However, observing and reading signs is often difficult. What an experienced or perceptive doctor can find out about the bodily disorder of a patient by checking the pulse or looking into the ears may escape a novice completely. Similarly, the precursory signs of earthquakes could be obscure or elusive, and most of the lay participants in the prediction program had only the most basic knowledge of geology and seismology or, in remote areas, hardly any at all. To illustrate how this mass participation in scientific observation worked, I shall look at one particular area of observation: animal behavior.

Chinese seismologists accepted that aberrant animal behavior could be premonitory of an earthquake. There was a rationale behind this belief: animals are often more sensitive than humans to changes in natural environments. The Chinese found in historical records and folk wisdom much evidence to support this notion, although they did not do controlled research into the connections between earthquakes and animal behavior and physiology until the late 1970s.[33] Animals thought to be useful for earthquake prediction included a wide range of pets, livestock, domestic pests, and wild animals; dogs, cats, cows, horses, pigs, chickens, pigeons, parrots, ducks, mice, rabbits, fish, and snakes all seemed able to detect and respond to minute changes before an earthquake. One frequently cited example was the bizarre behavior of snakes in the weeks before the Haicheng earthquake in 1975. There were eyewitness accounts of snakes crawling out of their hibernation holes in the February cold of North China and freezing to death. Some described vivid images of snakes struggling to come out despite the fact that their front halves were already numb from

---

[32] E.g., Dizhen wenda bianxiezu [Editorial team of *Q&A's about Earthquakes*], *Dizhen wenda* [Q&A's about earthquakes] (Beijing, 1977); Guo Qinhua, *Zhenqian qiguan* [Unusual phenomena before earthquakes] (Beijing, 1982); Shaanxi sheng geweihui dizhenju bian [Seismological Bureau of the Revolutionary Committee of Shaanxi Province], *Dizhen yuce yufang zhishi* [Knowledge about predicting and defending against earthquakes] (Xian, 1977).

[33] Jiang Jinchang, *Qiyi de benling: Qiantan dongwu yuzhi dizhen* [Amazing abilities: Introduction to (the topic of) animals predicting earthquakes] (Beijing, 1980); Jiang and Chen Deyu, *Dizhen shengwuxue gailun* [General treatise on seismological biology] (Beijing, 1993). See also Helmut Tributsch, *When the Snakes Awake: Animals and Earthquake Prediction* (Cambridge, 1982).

the cold.[34] Burrowing animals were believed to be particularly sensitive to changes underground. However, livestock and other animals that people more usually came into contact with could also serve as convenient instruments.

This method was crucial to country folks. They lived in places not covered by formal seismological networks, and they had to predict earthquakes in order to protect themselves. They could contribute to the Collective Monitoring, Collective Defense program by utilizing their special expertise. Farmers were particularly equipped to use animals as seismic detectors, because they handled livestock and came across many wild animals every day. So, more than most people, they were able to notice aberrant behavior by animals when they saw it. They were good observers because they were familiar with the routine or normal behavior of these animals. (Similarly, in urban areas, zookeepers were good observers of anomalous animal behavior.) This kind of expertise came from people's daily labor, from their daily struggle with nature. It was based on an epistemology that derived from practice, work, and tradition.

EXPERTISE, FOLK KNOWLEDGE, AND THE STATE

Thus, monitoring macroscopic phenomena relied very much on everyday experience. Everyday experience was crucial, because it provided the background of normality against which the aberrant and the anomalous emerged. Here, everyday experience was no less a social than an individual acquirement. Most people did not experience significant earthquakes often enough to be able to compile premonitory symptoms; they had to start with the existing lore. Folk knowledge—traditional knowledge that had been passed on to and circulated among a local population—was often community-based knowledge and skill. Together with other social and individual knowledge, folk knowledge, experience, and skill constituted a kind of "lay expertise."[35] Although it did not come from formal, institutional education and training, it could be just as specialized as any other expertise. Being able to make a good observation required one to have accumulated sufficient experience, knowledge, and practice. Only when one was familiar with the routine and the normal—either a particular weather pattern or the typical behavior of particular animals at a certain time and place and in a certain situation—could one immediately notice the more subtle kind of anomalies. From the standpoint of Maoism, this expertise was not only social, but also thoroughly political—because it was embedded in class and derived from labor. There was another dimension to this expertise: it was also often tied to a particular place, for this kind of everyday knowledge was intensely local. Indeed, the value of the knowledge owed much to its localness. Since the aim of the Chinese approach to earthquake prediction was to pinpoint an earthquake rather than to pro-

---

[34] Zhongguo Kexueyuan Shengwu Wuli Yanjiusuo Dizhenzu [Seismology Group, Institute of Physical Biology, Chinese Academy of Sciences], *Dongwu yu dizhen* [Animals and earthquakes] (Beijing, 1977), 66–7.

[35] On lay expertise, see, e.g., Steven Epstein, "The Construction of Lay Expertise: AIDS Activism and the Forging of Credibility in the Reform of Clinical Trials," *Sci. Tech. Hum. Val.* 20 (1995): 408–37. The phrase is now widely used in studies on science, medicine, law, and the public. Harry Collins and Robert Evans's controversial book *Rethinking Expertise* (Chicago, 2007) sounds the alarm on the possibility of too much public participation in science and defends science against unqualified interventions by charting a "periodic table of expertises."

vide long-term forecasting or hazard assessment, precision was key. Knowledge that corresponded precisely to a place was useful for accurately predicting an earthquake.

But here the state confronted a problem. In reality, everyday knowledge, folk wisdom, and what was considered by the state to be superstition were not clearly separated bodies of knowledge. They often blurred into each other. This posed difficulties for the state's attempt to control and utilize folk knowledge. Traditionally, earthquakes and other natural calamities possessed political and cultural significance. A big natural disaster, especially something dramatic like a major earthquake, was considered an ominous sign of major political change, such as the end of a reign. Anxiety caused by such events could easily be amplified by the social and political commotion during the Cultural Revolution. Not surprisingly, the party-state was eager to educate the masses that earthquakes were natural phenomena, that they could be predicted, and that people should not have superstitious thoughts about them. The party-state did not want its political legitimacy to be questioned or challenged.[36] Campaigns against superstitions had been a long-time policy of the Chinese Communist Party. Styling itself as a modernizing force, the party—just as its predecessor and rival, the Guomindang or Nationalist Party, had done—made efforts to stamp out traditional ideas and customs it deemed to be superstitions.[37] In its view, China could advance into the modern world only if its people shed their backward culture, customs, and beliefs and replaced them with a modern, scientific—that is, Marxist, materialist—worldview. Indeed, the "Smash the Four Olds" (old customs, old culture, old habits, and old ideas) campaign during the Cultural Revolution was only the most vehement incarnation of this belief and policy.

As mentioned above, however, Maoist ideology also emphasized the value of folk knowledge—empirical and practical knowledge gained by people of lower social status from their direct contact and struggle with nature. Traditionally, the literati, who would not condescend to get their hands dirty, looked down upon this kind of knowledge. Now, under the new regime, it was believed that folk knowledge should receive due attention because of its practical value and class character. Thus, folk knowledge possessed a mixed character in the eyes of the Chinese communist

[36] The dynamic between the state and what the state regarded as folk knowledge was complex. The attempt of a Chinese state to control, contain, and appropriate folk knowledge was not new. In his study of the "popular hysteria" over the so-called soulstealers—evildoers who clipped people's hair and stole their souls—of eighteenth-century China, Philip Kuhn discusses how the emperor and the provincial bureaucrats approached the social disturbance. Kuhn's case study is relevant here because he contrasts the different attitudes of the emperor and the officials toward the popular fear of soulstealers. The emperor took it seriously and urged the officials to crush the soulstealers. In contrast, the officials preferred a less aggressive approach and were able to resist the emperor's demands by dragging their feet. Kuhn notes that the proud tradition of civil bureaucracy left enough room for the officials to balance, deflect, or even resist the emperor's demands. And he asserts that this tradition and its ability to check and balance the power of the court or its equivalent was lost in twentieth-century China. Kuhn's conclusion reminds us that the state was not a coherent whole. It was a complex of different—sometimes competing, contending, or contradictory—ideas, traditions, forces, institutions, and actors. See Kuhn, *Soulstealers: The Chinese Sorcery Scare of 1768* (Cambridge, 1990). See also Kenneth Pomeranz, "Ritual Imitation and Political Identity in North China: The Late Imperial Legacy and the Chinese National State Revisited," *Twentieth Century China* 23 (1997): 1–30.

[37] Chang-tai Hung, "The Anti-Unity Sect Campaign and Mass Mobilization in the Early People's Republic of China," *China Quart.* 202 (2010): 400–20; Steve A. Smith, "Local Cadres Confront the Supernatural: The Politics of Holy Water (Shenshui) in the PRC, 1949–1966," *China Quart.* 188 (2006): 999–1022; Smith, "Talking Toads and Chinless Ghosts: The Politics of 'Superstitious' Rumors in the People's Republic of China, 1961–1965," *Amer. Hist. Rev.* 111 (2006): 404–27.

party-state. The state faced the question of how to manage a body of knowledge that was at the same time problematic (e.g., when it took the form of superstition) and valuable (both scientifically and politically). In collecting, scientizing, and disseminating folk knowledge, the state attempted to control the knowledge for practical as well as political purposes and to transform traditional folk practice into something radically modern in its political meaning and scientific application.[38] In doing this, the state de facto claimed the ownership of a body of knowledge that had previously been in possession of the common folk. It tried to discipline this knowledge by exercising its political authority.[39]

But the state's desire to select, control, and define this body of knowledge and its efforts in disseminating scientific knowledge about earthquakes could produce unexpected results. In a village in Hubei Province, in the early 1970s, a rumor circulated among the young teenagers that under a mountain distantly visible from their village was an underground sea, in which there was so much kelp growing that people in the area could not exhaust the supply in their lifetime. They had been taught about earthquakes and were aware of basic methods of earthquake prediction, though their knowledge about the topic was fragmentary. They had vague notions about geological strata and changes. In any case, this strange mixture of geological knowledge and fantasy about exotic food (Hubei was landlocked and far away from the sea) showed how the young villagers produced their own interpretations and meanings from the knowledge transmitted to them from above.[40] Impressionistic notions about geological formations were grafted onto real everyday wishes, desires, and imagination. Ultimately, the state could not fully maintain its hegemony over meaning making and monopolize a regime of knowledge.

ALL EYES, ALL EARS

Indeed, earthquakes were not simply natural events that had to be dealt with. They were also full of political meanings. The state wanted to eliminate the tradition of regarding natural disasters, including earthquakes, as ominous. At the same time, it also tried to instill political meanings into how people responded to earthquakes, at least as told in the "true stories," or instructive lessons, circulated in various forms of publications and propaganda. We can analyze these "true stories" as narratives that functioned as morality tales. The following anecdote is a rather typical one:

[38] Jacob Eyferth, in *Eating Rice from Bamboo Roots: The Social History of a Community of Handicraft Papermakers in Rural Sichuan, 1920–2000* (Cambridge, 2009), discusses how the Chinese communist state similarly tried to claim and utilize the traditional skill of papermaking in a rural community.

[39] In his acclaimed book *Seeing Like a State: How Certain Schemes to Improve the Human Condition Have Failed* (New Haven, Conn., 1998), James C. Scott argues that the combination of an administrative ordering of nature and society, high modernism, an authoritarian state, and a weak civil society led to many disasters in the twentieth century. He is probably right, though his categories—state, civil society, etc.—are conventional and his argument is not strikingly new. (His examples are fascinating, though.) However, an auxiliary argument of his—about the power of practical knowledge or what he calls *metis*—raises interesting issues when we think of mass science in Mao's China. Maoist ideology pursued an unusual vision of modernity, and mass science drew heavily on the lore of practical knowledge. Unfortunately, Scott does not say much about how the state might have tried to appropriate this kind of knowledge. He simply calls attention to metis as an alternative and crucial body of knowledge.

[40] On the authority of a friend of mine from the Hubei village. The interview took place in Beijing in 2007.

Livestock keeper Zhang of the First Production Brigade, Erjiegou Commune, Panjin District, was responsible for raising twenty-one piglets, all of which were born in October 1974. On February 2 [two days before the Haicheng earthquake], he noticed that the pigs in the sty were biting each other. The tails of nineteen pigs were already bitten and chewed off; the stubs were dripping blood. Some pigs pushed their backsides against the wall to hide their tails. When the livestock keeper took them out of the sty, they stopped attacking each other. . . . [After receiving the report], the earthquake observer of the commune carried out a special investigation [into the matter]. . . . Accordingly, the commune undertook measures in anticipation of an earthquake and therefore greatly reduced the damage.[41]

In the narrative, the protagonist was doing his routine work when he noticed something unusual. Rather than purposefully looking for anomalies, he was simply being alert and vigilant. Those higher up on the chain of command all responded in a timely and responsible manner. As a result, more serious damage was averted. To be a good comrade, therefore, was to be a vigilant soldier in the people's war against earthquakes. Such language and morals were familiar to people in communist China. For years they had been constantly admonished to spot and report on rightists, counter-revolutionaries, class enemies, and anyone who allegedly posed a threat to communist society and the party-state. Just as in political campaigns, ignoring the directions of the party-state about earthquake monitoring was not only foolish, but also politically damnable. The Haicheng earthquake of 1975 supplied the best lesson for all. The day before the powerful earthquake occurred, the local officials had already mobilized to evacuate people from their homes. Yet, there were those who did not obey the order to evacuate or, after staying in the cold for a while, returned to their houses. Consequently, they were killed in the earthquake.[42] The lesson of the incident was clear: the ones who were killed or injured in the earthquake were the ones who did not believe that the party-state could predict earthquakes. They endangered themselves because they failed to put full faith in the communist government and because they did not absolutely follow the leadership of the party-state. The lesson was not simply about predicting and escaping a natural disaster; it was also about political loyalty and the authority of the party-state.

Arguably, the warning system was an extension of the surveillance society under the communist rule, for, as noted, people had been constantly admonished to be vigilant, to keep an eye out for enemies of the party-state, and to report any suspicious activities.[43] Only now, the enemy was a natural force. Of course, one should not assume

[41] Zhongguo Kexueyuan Shengwu Wuli Yanjiusuo Dizhenzu, *Dongwu yu dizhen* (cit. n. 34), 72; translation mine.

[42] Jiang Fan, *Haicheng dizhen* [Haicheng earthquake] (Beijing, 1978), 89.

[43] Mass observation is not always led or imposed by the state. A project of mass observation can be in a complex relationship to the state. The mass observation movement in Britain that began in the 1930s provides an interesting example. The movement was conceived and organized by a group of left-leaning intellectuals. They recruited hundreds of untrained observers to record what they saw and heard in everyday life, which naturally means that they observed and recorded what their neighbors, colleagues, people they encountered on the street, etc., did and expressed. The assumption was that with enough information, an accurate picture of the everyday life and mood of the people would emerge. The results often raised questions about the picture the British government wanted to present to the people about itself and about the state of affairs. In this sense, by recruiting public participation, the mass observation movement challenged the image making of the state. See Nick Hubble, *Mass Observation and Everyday Life: Culture, History, Theory* (Basingstoke, 2010). A somewhat comparable example from China is the "One Day in China" project. Led by the renowned writer Mao Dun, a group of leftist intellectuals called for submissions from ordinary people of their activities, thoughts,

that this system worked seamlessly. Political instabilities during the Cultural Revolution could be disruptive.[44] Factional fighting was rampant. Besides, people always had their own ideas, desires, aims, and lapses. Boredom, negligence, and indifference seeped in after the initial excitement. How many people could keep on checking the same unchanging meter and recording the readings day after day, three times a day, without occasionally slacking? Symbolic incentives—stationery or a little money— were provided. At the year's end, a celebratory meeting gathered people together; prizes, certificates, and the like were handed out. Even so, it could not be guaranteed that people would remain vigilant. Nevertheless, we should not underestimate the seriousness and dedication many people felt toward earthquake monitoring. What they were expected to do was, after all, what they would want to do for themselves—that is, protect themselves and their families. In an area where major earthquakes were possible, this was a matter of life and death. At the risk of overgeneralization, it may be said that the operation of the program was not primarily a matter of coercion, nor was it simply a matter of the state trying to produce good citizens (or, in this particular political formulation, good comrades) by employing techniques of governance, because its scientific-political program and indeed its survival depended on them. The enterprise was also premised on the fact that many people—good comrades or not—shared some of its goals.

CITIZEN SCIENCE AND THE STATE

Maoist mass politics and mass science challenged the Weberian idea of technocracy and modernity, which the scientific establishment in Western societies held up as the true characterization of itself: detached, objective, rational, impersonal, apolitical, governed by rules, and run by experts. Maoist thought and science attacked this notion of technocracy and insisted on the primacy of politics—that is, constant class struggle and revolutionary creativity from the masses. Thus, Maoism erected a different vision of political and scientific modernity. The masses were a huge reservoir of knowledge, creativity, and energy. Once they were mobilized and their resources were unleashed, they would create a new world (consider the Great Leap Forward). Yet ultimately, the Collective Monitoring, Collective Defense program depended, in principle, on a highly regimented society. It was thus not entirely different from the city, commune, and factory militia programs that aimed to maintain communal security.

With its close ties to the state and its didactic political function, Maoist mass science cannot be readily compared to "popular science" in democratic societies. It was also different from "public science," as it is commonly understood, for the latter assumes a separation between a public realm (say, the public sphere) and a private

---

and observations during a single specific day, May 21, 1936. Their goal was to compile a book that would "reveal the entire face of China during one day." The editors hailed this project as a successful "general mobilization of minds." The final product reflected a leftist view of the masses, everyday life, and social realism. It was a view very different from that of the conservative Guomindang government. Parts of the book have been translated and reorganized in Sherman Cochran's *One Day in China: May 21, 1936* (New Haven, Conn., 1985).

[44] Roderick MacFarquhar and Michael Schoenhals's *Mao's Last Revolution* (Cambridge, 2006) provides a detailed chronicle of the Cultural Revolution.

realm.[45] Maoist ideology left little room for such a separation. Nor can Maoist science be reduced to "amateur science," "lay science," "vernacular science," or "science dissemination and popularization," although it contained elements of all of them. One could perhaps simply leave it as a singular entity, calling it "mass science" or "people's science," and not try to see how it relates to science in other countries. Yet I think that approach would be unfortunate. Not only would it risk relegating science in communist China to a corner in the historiography of modern science, turning it into a mere oddity, an object of curiosity, but it would also miss an excellent opportunity to ask important questions about science and modernity.

Fundamentally, Mao's mass science was defined by the relationship between science and legitimate membership in a particular political community. In this sense, it may be fruitfully compared to what might be called "citizen science." I suggest that insofar as it tries to characterize the relationship between science and citizens, this term has three different, but related, meanings. None of them captures the full meaning of Maoist mass science, but all of them provide grounds for comparison. "Citizen science" has usually been used in a positive sense. It refers to voluntary, collaborative, complementary, and broad-based participation in scientific activities. In this version of citizen science, there have been different, but not mutually exclusive, emphases on the concept. The participation may be initiated by the scientific establishment. The reasons for this can be manifold. The scientific community may promote citizen science because it is good for public relations, because it can lead to better use of resources, and/or because the scientific community really needs assistance. In this scenario, the experts can devote their time, energies, and resources to specialized research while the hundreds or thousands of volunteers can help solve other, less technically demanding problems. Many science projects that enlist backyard astronomers fall into this category.[46] Another example is the Christmas Bird Count.[47] A wide network of observers is necessary for a reliable survey of bird populations. There is no way that professional ornithologists can accomplish the feat, but a system of numerous volunteer bird watchers is perfect for this important task. In these cases, the volunteers are amateurs and their role is to assist, rather than to challenge or critique, the scientific establishment. Of course, there can be tensions or conflicts between the scientists and the amateurs, but such incidents are accidental rather than intrinsic to this particular notion of citizen science.

---

[45] For a recent discussion on "popular science" and related concepts, see the "Focus" section "Historicizing 'Popular Science,'" *Isis* 100 (2009): 310–68. Here it might be helpful to mention James Andrews's *Science for the Masses* (cit. n. 12), which relies on the notions of "public science" and "the popularization of science." It focuses on the role of print culture and public discourse in science popularization. This focus has to do with the author's approach and perspective, but it also has to do with the historical fact that even the Soviet Union did not go as far as Mao's China in pursuing the kind of mass science discussed in this article. On communist China, see Sigrid Schmalzer, *The People's Peking Man: Popular Science and Human Identity in Twentieth-Century China* (Chicago, 2008).

[46] Alex Wright, "Managing Scientific Inquiry in a Laboratory the Size of the Web," *New York Times*, December 28, 2010, D3; Timothy Ferris, *Seeing in the Dark: How Amateur Astronomers Are Discovering the Wonders of the Universe* (New York, 2003). Operation Moonwatch, however, has to be understood within the Cold War context; see W. Patrick McCray, *Keep Watching the Skies! The Story of Operation Moonwatch and the Dawn of the Space Age* (Princeton, N.J., 2003).

[47] E.g., Mark V. Barrow, *A Passion for Birds: American Ornithology after Audubon* (Princeton, N.J., 1998), chap. 7.

There is another usage of the term "citizen science." This usage strongly encourages democratic participation in science or matters concerning science. The central argument is that science is too important to be left to experts. The public has a stake in science and what scientists do.[48] Therefore, it has the right and responsibility to take part in certain decisions about science. This usage is common in the field of science and technology studies. Employing this notion of citizen science, scholars have, on the one hand, effectively critiqued establishment science and technology in modern societies and, on the other hand, suggested a better mode of incorporating science and technology into society—one that is more responsive to public needs and concerns and better grounded in society. The essence of this formulation is that science and technology should be an integral part of the public domain of a democracy. This is an admirable goal. However, much of this literature takes for granted or leaves unexamined the notion of citizenship. Often the starting point or default assumption is a commonsense definition of participatory liberal democracy. This is fine inasmuch as the scholars are arguing for what they think is a good sociopolitical model for responsible science and technology. To historians, however, this perspective lacks sufficient historical sensitivity. The main problem is that citizenship has not been the same thing across times, places, and political communities. Indeed, who gets to be a citizen has often been determined by science (e.g., racial science). Hence, one must historicize citizenship and the citizenry before investigating the matrix of science, the state, and citizenship.[49]

Finally, there is a third implication of "citizen science" that also deserves attention. In fact, it is the most relevant to the topic of this article. A citizen is a member of an organized political entity, most likely a state. (Much of the literature on citizen science and related topics leaves the state in the background or underestimates the historical centrality of the state in any definition of citizenship.) Citizen science is, therefore, not simply any form of broad participation in science, but one that has to do with the ideology, institutions, and functions of a state. Thus defined, citizen science is inextricable from the politics of the state, which can mean that citizen science comprises measures of state control, nation-state building, surveillance, discipline, and political participation through the operations of the state apparatus or through the constitution and practice of citizenship. To be sure, the peculiar political system of Mao's China produced a distinct kind of citizenship and citizen science.[50] One can perhaps even go so far as to say that in the case of earthquake monitoring, the communist state attempted to mobilize as well as regulate its people by telling them what and how to see, hear, smell, and taste (in order to detect the anomalous phe-

[48] The literature on this topic is growing fast; e.g., Alan Irwin, *Citizen Science: A Study of People, Expertise, and Sustainable Development* (London, 1995); Melissa Leach, Ian Scoones, and Brian Wynne, eds., *Science and Citizens: Globalization and the Challenge of Engagement* (London, 2005); Joseph Wachelder, "Democratizing Science: Various Routes and Visions of Dutch Science Shops," *Sci. Tech. Hum. Val.* 28 (2003): 244–73; Wiebe Bijker, Roland Bal, and Ruud Hendriks, *The Paradox of Scientific Authority: The Role of Scientific Advice in Democracies* (Cambridge, 2009). Generally speaking, historians of science have lagged seriously behind science studies scholars in thinking about issues concerning science, the state, and citizens.

[49] F. Charvolin, A. Micoud, and L. Nyhart, eds., *Des sciences citoyennes? La question de l'amateur dans les sciences naturalistes* (La Tour d'Aigues, 2007), includes several excellent essays by historians of science.

[50] It has been called "revolutionary citizenship," which befits the Maoist idea of perpetual revolution; see, e.g., Elizabeth J. Perry, *Patrolling the Revolution: Worker Militias, Citizenship, and the Modern Chinese State* (Lanham, Md., 2006).

nomena in their natural environment) and by keeping them constantly alert and vigilant. Thus, the state defined a kind of "biocitizenship"—in addition to other political characterizations of citizenship, such as those based on class, ethnicity, and political pedigree—by disciplining the bodies of its people: a good citizen was someone who could detect threats, be they natural or political, to the communist nation-state.[51] Arguably, in their own ways, all modern states strive to produce ideal citizens through practices and techniques of governmentality, for this is what modern states are and this is what they do.[52] To this end, states have also deployed other technologies of rule, such as the use of maps, museums, urban planning, and other instruments in pursuing what has been called, especially in the radical high modernist cases, "seeing like a state."[53]

But, of course, China's response to earthquakes at the time also has to be understood in the context of possible wars with the Soviet Union and the United States; that is, a situation in which the state perceived itself as being under serious threat from other, hostile states. The border conflict with the Soviet Union in 1969 and the escalating American involvement in Vietnam sent the Chinese leaders into frantic preparation for war.[54] One of the most ubiquitous propaganda slogans at the time was "Prepare for war, prepare for famine, for the people" (*beizhan, beihuang, wei renmin*). Taking this into account, we can see that the war against earthquakes became part of a larger effort at national defense. People were kept on high alert for imminent attacks from within and without; class enemies, foreign aggressors, and natural disasters might strike anytime. In this sense, one can compare the Collective Monitoring, Collective Defense program to the civil defense projects in Cold War America. In the early 1950s, Operation Skywatch of the Ground Observation Corps enrolled tens of thousands of volunteers in an effort to sight enemy aircraft. Civilians manned watchtowers on night shifts, went through drills, and were taught the science needed for civil defense projects.[55] In both the Chinese and American cases, science, citizens, and the state were tightly connected (even though they defined these terms very differently or even in opposition to each other). Given this, Maoist science was perhaps not so unique. It reflected the intimate relationships among science, citizenship, and the macropolitics of the modern state and society.

---

[51] See, e.g., Adriana Petryna, *Life Exposed: Biological Citizens after Chernobyl* (Princeton, N.J., 2002); Aihwa Ong and Stephen Collier, eds., *Global Assemblages: Technology, Politics, and Ethics as Anthropological Problems* (Hoboken, N.J., 2004), esp. Nikolas Rose and Carlos Novas, "Biological Citizenship," 439–63. But of course these ideas are all indebted to Michel Foucault's conception of biopolitics.

[52] Graham Burchell, Colin Gordon, and Peter Miller, eds., *The Foucault Effect: Studies in Governmentality* (Chicago, 1991). For a popular introduction to the concept of governmentality, see Mitchell Dean, *Governmentality: Power and Rule in Modern Society* (Thousand Oaks, Calif., 1999). Prakash (*Another Reason*, cit. n. 6) uses the concept of governmentality extensively.

[53] Scott, *Seeing Like a State* (cit. n. 39). See also Benedict Anderson, *Imagined Communities: Reflections on the Origin and Spread of Nationalism* (London, 1991), chap. 10; Thongchai Winichakul, *Siam Mapped: A History of the Geo-body of a Nation* (Honolulu, 1997).

[54] Chen Jian, *Mao's China and the Cold War* (Chapel Hill, N.C., 2001); Judith Shapiro, *Mao's War against Nature: Politics and the Environment in Revolutionary China* (New York, 2001), chap. 4.

[55] Sharon Ghamari-Tabrizi, *The Worlds of Herman Kahn: The Intuitive Science of Thermal Nuclear War* (Cambridge, 2005), 98–103; Tracy Davis, *Stages of Emergency: Cold War Nuclear Civil Defense* (Durham, N.C., 2007); Laura McEnaney, *Civil Defense Begins at Home* (Princeton, N.J., 2000); Guy Oakes, *The Imaginary War: Civil Defense and American Cold War Culture* (New York, 1995). See also Matthew Grant, *After the Bomb: Civil Defense and Nuclear War in Britain, 1945–68* (Basingstoke, 2010).

CONCLUSION

I have argued for a broad perspective on the problem of science and the state. It is my view that the existing literature on science and the state tends to focus on the relationship between the government and the scientific establishment. As a result, both the state and the scientific community are rather narrowly defined and are largely isolated from other historical actors and from the broad sociocultural context. Moreover, viewed in this way, the historical picture looks very much like a variety of conflicts or collaborations between state authorities and the scientific establishment, possible frictions within either or both groups notwithstanding. I used the historical case of mass science in Mao's China to illustrate the need for a more nuanced and comprehensive approach to the problem of science and the state. I demonstrated that it would be impossible to understand the ways in which mass science and the communist Chinese nation-state coproduced each other without looking beyond the scientific elite and their scientific expertise. Furthermore, it would be difficult to understand the vision and the making of the Chinese nation-state during that period without taking into account such mass scientific-political projects as the Collective Monitoring, Collective Defense program.

With regard to the particular historical case, I introduced two critical components to the picture—a broadly defined category of expertise and scientific knowledge and an analysis of the interrelations among science, state, and citizens/citizenship. I tried to show that the state and the scientists were interested in a diverse body of knowledge about earthquakes, ranging from modern seismology to folk knowledge, because they believed that all of it would be valuable to earthquake monitoring and prediction. In order to understand how their project worked, therefore, we cannot limit our attention to professional scientists and their scientific expertise. Other forms of knowledge about the environment constituted part of the body of useful knowledge about earthquakes. Whether one calls them folk or vernacular knowledge, they have to be a vital element of our analysis of the relationship between science and the state. In this case, the attempt to appropriate, modernize, and distribute the knowledge was part and parcel of a broader agenda to produce proper members of a certain kind of modern polity. But it was also an attempt to urge them to share the stakes in nation-state building and communal defense against natural disasters.

To call attention to the complex relationships among multiple historical actors in the configuration of science and the state is not to endorse a state-centered narrative.[56] This perspective does not privilege the state as the prime subject of history. Rather, it is simply to recognize the fact that the state was an important force in the history of modern science and to call for better approaches to analyzing and historicizing the

---

[56] There are possible pitfalls in "bringing the state back in"—e.g., squeezing the history of science into the tunnel-visioned perspective of national history, recycling the notions of national styles (without first problematizing the nation-state), returning to the history of political institutions with science mixed in, and reducing science to the terms of the state. In addition, there is also a particular risk in calling attention to the state in discussing China, as historians have already put too much emphasis on the state in explaining Chinese history (whether talking about the imperial bureaucratic state or the communist authoritarian state). For a classic example, see Karl Wittfogel, *Oriental Despotism: A Comparative Study of Total Power* (New Haven, Conn., 1957). None of these pitfalls should prevent us from taking seriously the problem of science and the state, as long as we do not see history through the lens of the state or see it as a historical actor with coherent and autonomous subjectivity; Fa-ti Fan, "Redrawing the Map: Science in Twentieth-Century China," *Isis* 98 (2007): 524–38; Fan, "National Narrative and the Historiography of Chinese Science," *Gujin lunheng* 18 (2008): 199–210.

relationship between science and the state. It might be objected that this perspective excludes transnational history, which is so necessary and popular in many subfields of history today. But I do not think it does so. In fact, not only does it not exclude the transnational perspective, but it forms an integral part of it. For instance, only when we take into consideration the relationships among science, the state, and citizenship can we properly understand the role of Chinese American scientists in American science, Chinese science, and the transnational flow and blockage of science.[57] To ignore these relationships is to ignore the global system of nation-states, which plays a major role in setting boundaries, making connections, distributing powers, and maintaining social and political order and hierarchies.[58]

When talking about science, I think we still tend not to relate it to macropolitical topics and to skate over the state, citizenship, the nation, the public, and other related topics.[59] We often simply take these terms for granted and assume that our readers will understand what they mean. This assumption is in part due to the fact that our intended audience is usually Western (particularly Anglo-American). The problem with this assumption is twofold. First, even in the Western tradition or traditions, these concepts have been heatedly debated. They are anything but straightforward. And, second, it is a fact that not all states and societies were organized according to the categories taken for granted in the dominant historiography. States and their political principles and organizations have always been diverse and changing, and so have the accompanying notions of citizenship, political authority, and legitimation. Mao's China may have been an unusual case. But still it helps to remind us that when we use such terms as *the state* and *citizenship*, we must be careful about the historical context and the cases at hand. In other words, we must not treat these historical factors simply as the backdrop to the history of science.

[57] See Zuoyue Wang's illuminating essay "Transnational Science during the Cold War: The Case of Chinese/American Scientists," *Isis* 101 (2010): 367–77.

[58] I do not mean to say that we should embrace the system as it is. James C. Scott, in *The Art of Not Being Governed: An Anarchist History of Upland Southeast Asia* (New Haven, Conn., 2009), offers a case study and some reflections on this issue. But see also Sanjay Subrahmanyam's critical review of the book in *London Review of Books*, December 2, 2010, 25–6.

[59] It is very encouraging that many recent *Osiris* volumes have tackled these big issues: vol. 25 (cit. n. 1); vols. 24, 23, and 21 (cit. n. 4); vol. 20 (2005), *Politics and Science in Wartime: Comparative International Perspectives on the Kaiser Wilhelm Institute*; and vol. 17 (2002), *Science and Civil Society*. On East Asia, see, e.g., the special issue "Public Participation in Science and Technology," *East Asian STS* 1, no. 1 (2007).

Wissenschaft and *Kunde*:
The General and the Special in Modern Science

*by Lynn K. Nyhart**

ABSTRACT

In this article I argue that the general and the particular (or "special") have been fundamental categories in the conceptual organization of modern science. Using nineteenth-century German zoology as a case study, I argue for the importance of two basic relationships between them. One was incorporated into the dominant concept of *Wissenschaft* and became an international paradigm for academic, disciplinary science. The other is best captured in the term *Kunde* (knowledge), often used as a suffix (*-kunde*) to mean "studies"; it flourished especially outside the universities. Attending to these two approaches offers new insight into understanding conceptual change in science. I further suggest that the Kunde approach represents a way of knowing much more widespread than we have hitherto recognized; moreover, it bears intriguing parallels to certain kinds of interdisciplinary studies today.

In 1840, William Whewell wrung his hands over the nature of scientific knowledge, wondering "how we [can] have any *general* knowledge if our thoughts are fixed on particular objects; and, on the other hand,—how we can attain to any *true* knowledge of nature by contemplating ideas which are not identical with objects in nature."[1] Whewell famously solved his problem through a reformed method of induction. In his rendering, knowledge rises from the particular to the general through a stepped "colligation" of facts and ideas. Facts are joined together by means of unifying ideas at a low level, then groups of these higher-level, complex facts in turn may be joined together by a higher-level generalization, and so forth. The result, Whewell wrote, was "*real general knowledge* respecting the external world. . . . It is *real*, because it arises from the combination of real facts, but it is *general*, because it implies the possession of general ideas. Without the former, it would not be knowledge of the external world; without the latter, it would not be knowledge at all."[2]

To the extent that historians of science think about the philosophy of science, or

* Department of the History of Science, University of Wisconsin–Madison, 1225 Linden Drive, Madison, WI 53726; lknyhart@wisc.edu.

For their helpful comments on various versions of this article I would like to thank Rodolfo John Alaniz, Elihu Gerson, Andrew Mendelsohn, Kathryn Olesko, Tania Munz, David Sepkoski, and two anonymous reviewers for *Osiris*. I owe particular debts to Robert Kohler for his steadfast encouragement in the face of many messy versions of this article, and to Thomas Broman, Judith Kaplan, Scott Lidgard, and Daniel Liu for their stimulating conversation and concrete suggestions on the general and the special while I have been wrestling with this topic.

[1] Whewell, *Philosophy of the Inductive Sciences, Founded upon Their History*, 2 vols. (London, 1840), 2:311–2; emphasis in the original.
[2] Ibid., 212.

about things as abstract as "the general" and "the particular," we tend to point to this inductive method as a foundational one (more often using the version of Whewell's distant predecessor Francis Bacon, and contrasting it to Descartes's deductive method). There is good reason for this: the inductive approach to generalization has historically held a prominent place in the Anglo-American philosophy of science, and it remains familiar to scientists today. The quarrels among Whewell, John Stuart Mill, and John Herschel over induction in the 1830s and 1840s, when this modern style of the philosophy of science emerged, have received corresponding attention in the history of philosophy.[3] However, in recent decades, mainstream history of modern science has been somewhat disinclined to treat seriously the history of scientific ideas and their philosophical implications, presumably as too internalist, too old-fashioned, and too much (merely) serving the interests of scientists. As a result, we have largely ceded this territory to the philosophers of science. It is time to take it back. If the history of science is to exist as a separate field and not be subsumed by general history, then we need to deal with "science" not only as a social and cultural phenomenon but also as an intellectual one.[4] If we accept this, then such questions as how scientists conceived of the general, the particular, and their relations become newly worthy of exploration, even foundational. A closer examination of the relationship of the general to the particular in modern science shows that induction and deduction are by no means the whole story; these categories also bore other relationships to one another that have been left in history's shadows. The aim of this article is to bring two of these relationships, prominent in nineteenth-century Germany, out into the light for examination.

In this paper I argue that the relationship between the general and the particular has been fundamental to the conceptual organization of modern science—but not only in the way that Whewell and his British rivals understood it. Focusing on Germany, which offered an important model of science to the Western world beginning in the nineteenth century, I argue that there have been at least two other basic ways of understanding these relations.[5] One became incorporated into *Wissenschaft*, of the sort that was institutionalized in the German university system and that became an international paradigm for disciplinarity. The other, less familiar to historians, is captured in the term *Kunde* (knowledge), often used as a suffix (*-kunde*) to mean "studies"; it flourished largely outside the universities. Attending to these two approaches, Wissenschaft and Kunde, offers a new way of reconstructing the basic conceptual structures of scientific knowledge in the nineteenth and early twentieth centuries.

To get there, I first characterize the distinction between the general and the particular (or the "special," as it was termed in both German and English) as categories of knowledge. I then use the particular case of zoology to show the different ways in

---

[3] See Laura J. Snyder, *Reforming Philosophy: A Victorian Debate on Science and Society* (Chicago, 2006), and sources cited therein.

[4] I do not wish to say that the history of ideas or philosophical questions precedes the social and cultural or proceeds autonomously, as anyone familiar with my work will understand. On "science" (vs. science), see Peter Dear's article in this volume. On the history of science's disciplinary or anti-disciplinary identity, see Paul Forman's article in this volume and the conclusion of the present article.

[5] These are assuredly not the only other ideas about relations between the general and the particular. For a brilliant dissection of the meaning and construction of a "general observation" in eighteenth-century medicine and meteorology, see J. Andrew Mendelsohn, "The World on a Page: Making a General Observation in the Eighteenth Century," in *Histories of Scientific Observation*, ed. Lorraine Daston and Elizabeth Lunbeck (Chicago, 2011), 396–420.

which scientists could relate these categories, which included both the Wissenschaft approach and the Kunde approach. I then suggest some benefits that come from attending to these as two distinctive ways of making scientific knowledge. First, as the example of zoology shows, it can help us understand conceptual change at the disciplinary and subdisciplinary levels. Second, as we will see, Kunde represents a form of knowledge that was much more widespread than we have hitherto recognized; moreover, it bears intriguing parallels to certain kinds of interdisciplinary studies today. Attention to these two categories thus opens up a number of new questions for the history of science. At the same time, it brings a *longue durée* historical perspective to today's turmoil over disciplinarity and interdisciplinarity in academia and may help us clarify our own place within it.

THE GENERAL AND THE PARTICULAR, IN GENERAL

The inductive approach and disputes around it have focused on the *process* of generalization as a grouping of phenomena into successively larger categories. There the particular and the general are relative categories: one level of fact-idea combinations may be more or less general than another. By contrast, other approaches to the general and the particular have placed a categorical line between them, treating them as different kinds. This is an ancient distinction. One version of it may be traced back to Aristotle, who in the *Nicomachean Ethics* divided the work of the rational soul into two parts, "one by which we contemplate the kind of things whose originative causes are invariable, and one by which we contemplate variable things."[6] The former Aristotle called the "scientific," the latter the "calculative." An early modern version familiar to historians of science is Francis Bacon's distinction between "natural philosophy" (the study of constant causes in nature) and "natural history" (the study of particulars)—a distinction that would linger, especially in Britain, well into the nineteenth century. In these two cases, the distinction is between constant causes (a kind of generality) and things that express nature's manifoldness.

A late nineteenth-century German formulation of this divide was famously undertaken by the philosopher Wilhelm Windelband, in an 1894 address titled "History and Natural Science." Here he spelled out a distinction between what he called the "nomothetic" sciences, which sought the general laws—the constants—underlying phenomena, and the "idiographic" sciences, which focused on unique historical events and sought "a complete and exhaustive description of a single, more or less extensive process." These, he declared, were two fundamentally different kinds of knowledge, and he argued that each legitimately contributed to complete knowledge. In Windelband's scheme, then, the general is the category of constant laws and principles—the nomothetic, or as Windelband put it, "what is invariably the case." This is much the same as the older natural philosophy, or Aristotle's scientific perspective. The special or particular refers to our apprehension of the specific events in the world, produced uniquely by history and nature—what Windelband called the idiographic.[7] Although this seems akin to the old Baconian natural history, Windelband did not see classifi-

---

[6] Aristotle, *Nicomachean Ethics*, trans. W. D. Ross, book 6, sec. 1, as posted on the Internet Classics Archive at http://classics.mit.edu/Aristotle/nicomachaen.6.vi.html (accessed 3 January 2011).

[7] Windelband, "Rectorial Address, Strasbourg, 1894" [translation by Guy Oakes of "Geschichte und Naturwissenschaft"], *Hist. & Theory* 19 (1980): 165–85, on 174–5. For a similar perspective expressed just a little earlier, see Franz Boas, "The Study of Geography" (1887), reprinted in *Volksgeist*

cation as the main task of the scientist here. Rather, he hoped to develop a more sophisticated approach to understanding unique, unrepeatable events.

Part of Windelband's aim was to provide a logically grounded foundation for a philosophy of the *Geisteswissenschaften* (the human sciences), which would place them on the same footing as the much better analyzed *Naturwissenschaften* (the natural sciences). It would also provide a new field of study for philosophy itself and help shore up its then-teetering standing as the queen of the sciences.[8] (However, his sharp distinction between natural-scientific knowledge as causal and deterministic versus historical-scientific knowledge as descriptive did not well serve scientists working in the natural history disciplines, who—if they paid attention to him—were placed, like social scientists, in an uncomfortable middle location.)

Windelband's speech drew on observations of the academic world around him, and his distinction may be seen as a philosopher's gloss on a division between the general and the special that was commonplace in nineteenth-century German academia, expressed in university courses, textbooks, and bibliographical reviews, especially in medicine and the natural sciences. The distinction can be found as early as the seventeenth century, when it was used to divide the general from the special in theology, metaphysics, and ethics.[9] It came into academic medicine by the 1710s: in pathology and therapeutics, the general referred to overall attributes of disease (and their remedies) and the special referred to how to recognize and treat particular diseases. By the 1820s, most medical school subjects and some other natural sciences were divided into general and special courses, as evidenced in lectures and textbooks of anatomy, physiology, pathological anatomy, surgery, psychiatry, chemistry, and botany.[10] By the later nineteenth century, the same distinction can be found in textbooks of geography, finances, and history.[11]

---

as *Method and Ethic: Essays on Boasian Ethnography and the German Anthropological Tradition*, ed. George W. Stocking Jr. (Madison, Wisc., 1996), 9–16.

[8] Windelband, "Rectorial Address" (cit. n. 7), intro. by Oakes, 165–8.

[9] Stephen Penton, *Apparatus ad Theologiam in Usum Academiarum: I. Generalis; II. Specialis* (London, 1688); in Abraham Calov's *Scripta Philosophica* (Lübeck, 1651), part 3 was titled "Metaphysicae Divinae Pars Generalis" and part 4 "Metaphysicae Divinae Pars Specialis"; Peter Schwaan divided his book *Ethica Stylo Scholastico Concinnata, et Notis Variis Illustrata* (Frankfurt, 1771) into general and special ethics.

[10] On the prominence of the general in German academia from the early nineteenth century to the mid-twentieth, see Michael Hagner and Manfred Laubichler, eds., "Vorläufige Überlegungen zum Allgemeinen," in *Der Hochsitz des Wissens: Das Allgemeine als wissenschaftlicher Wert* (Zurich, 2006), 7–21. On pathology and therapy, see Thomas H. Broman, *The Transformation of German Academic Medicine, 1750–1820* (Cambridge, 1996), 28–9. For the division in an even earlier therapeutics text, see Johann Samuel Carl, *Praxeos Medicæ Therapia Generalis et Specialis Pro Hodego: Tum Dogmatico tum Clinico, in Usum Privatum Auditorum Ichnographice* ([Halle?]: Litteris et impensis Orphanotrophei, 1718). See the titles of lecture notes in Hans Carl Leopold Barkow, "Medical Lecture Notes Taken at the Universities of Berlin and Greifswald, 1815–1820" (MS B 222), and Moritz Leo-Wolf, "Notes from Medical Lectures Given in Berlin, 1822–25, and in Heidelberg, 1825–26" (MS B 203), both held at the National Library of Medicine, Bethesda, Md. On anatomy, see K. F. T. Krause, *Handbuch der menschlichen Anatomie* (Hanover, 1833–8). Volumes 1 and 2 of Richard von Krafft-Ebing's three-volume *Lehrbuch der Psychiatrie auf klinischer Grundlage* (Stuttgart, 1879–80) addressed the general and special "pathology and therapy of insanity [*des Irreseins*]." Lists of lectures at several German universities in 1842 indicate the continued ubiquity of the *allgemeine-specielle* distinction. See, e.g., "Literarische Nachrichten," in *Intelligenzblatt der "Allgemeinen Literatur-Zeitung"* (1842), nos. 10 (Giessen) and 11 (Tübingen). Outside Germany, the most prominent example of the division was certainly Buffon's *Histoire naturelle, générale et particulière* (Paris, 1749–1804).

[11] The geographical handbook *Unser Wissen von der Erde*, ed. Alfred Kirchhoff (Prague, 1886–93) was divided into *allgemeine* and *specielle Erdkunde* (geography), the latter synonymized with

In the later nineteenth and early twentieth centuries, authors used it to organize textbooks in a range of fields in (at least) the French, Italian, and English languages.[12] As this listing makes clear, the division was ubiquitous. It was portable among disciplines, becoming conventional across the academic spectrum, and it was not confined to German universities.

In these works, the general referred to that which transcended and therefore united all special categories; it typically also included whatever were considered to be the principles of the field. As to the special: whereas Windelband's idiographic sciences referred mainly to the understanding of particular historical *events*, in most fields the special referred to *objects* that reflected the diversity of the physical and intellectual worlds. Special anatomy, for instance, was organized internally according to tissue types: bones, ligaments, nerves, blood vessels, viscera, and so forth. Special geography and history typically divided the world by broad geographic regions, then countries, and then regions within countries. A late nineteenth-century American textbook divided biblical hermeneutics into general and special, where general hermeneutics presented the overall principles of interpretation, and special hermeneutics applied these principles to particular classes of writing, such as "historical, poetical, philosophical, and prophetical," such that each required its own particular methods of interpretation. An early twentieth-century American education textbook categorized higher education into a general form (i.e., liberal arts) and "special education" forms, which encompassed professional education, technical and agricultural education, teacher education, arts and manual education, commercial education, the education of women, Negroes, and Indians, and that of "defectives." (Only in 1884 did Alexander Graham Bell co-opt the term *special education* for students with disabilities, and the broader range of applications persisted for some time.)[13]

Overall, then, the further divisions of the special were classificatory, ordering that diversity, whether it was reflected in anatomical tissue types, physiological functions, diseases, zoological or botanical taxa, geographic regions, or techniques. The categories of the general and the special were ways to manage two primary values pertaining to knowledge and its acquisition: the search for unity and transcendence and the acknowledgment of diversity in the real world.

Even in the conventions of the general and special marked out by university courses and textbooks, however, the general often entailed more than just "that which holds

---

Länderkunde and further subdivided by regions and countries. Adolph Wagner's *Finanzwissenschaft* (Leipzig, 1883–1901) contained sections on general and special theories of taxation; the latter section was further subdivided by country. On the general among German historians of the nineteenth century, see Martin A. Ruehl, "Kentaurenkämpfe: Jacob Burkhardt und das Allgemeine in der Geschichte," in Hagner and Laubichler, *Der Hochsitz des Wissens* (cit. n. 10), 23–72.

[12] Based on WorldCat searches of cognates of the phrase "general and special" in different languages. A Google Ngram sampling of "general and special" (in English, with and without capitals; see books.google.com/ngrams) shows a major peak in the 1880s, and another one after about 1905; browsing the titles in the sample suggests that the latter was due in part to references to Einstein's theories of general and special relativity.

[13] Milton Spenser Terry, *Biblical Hermeneutics: A Treatise on the Interpretation of the Old and New Testaments* (New York, 1883), 17; Edwin Grant Dexter, *A History of Education in the United States* (New York, 1906), table of contents. Although Bell has been credited with introducing the term *special education* to apply to the education of the disabled, he was not the first to use the term, and the broader list of special educations persisted well into the twentieth century, referring in the higher education literature to the idea of having special tracks (or majors) as well as general education at the college level.

true for all the special categories." It also often involved claims about theoretical generality or lawfulness in Windelband's sense of the nomothetic. Thus by the mid-nineteenth century, in the medical field of anatomy, general anatomy had come to refer primarily to cellular anatomy, both because cells were common to all different tissues and organs and also because they were argued to carry the essential qualities of life. This convention would also be instantiated in laboratory courses as a division into microscopic (general) and gross (special) anatomy. The greater prestige value of microscopic anatomy in the later nineteenth century derived from a threefold association: its productivity in terms of new research, its claim to universal coverage, and its basis in cell theory as a unified theoretical foundation for anatomy.[14]

As the case of anatomy suggests, the division between the general and the special signified something more than a common convention for organizing knowledge. Its prominence in nineteenth- and early twentieth-century German academia warrants our attention, for this setting was a key locus of modern science as it emerged in the early nineteenth century along with the research-based university system and its organization into disciplines.[15] The ideology of the nineteenth-century German research university valued the expansion of knowledge through empirical research along "special" lines. The new emphasis on research (as opposed to mere teaching) required a certain degree of specialization, gradually shutting the door on an older scholarly teaching tradition in which topics could be traded around among professors, and raising the expectation that a scholar would specialize in one area (if a broad one) over the course of his career.[16] At the same time, the concept of Wissenschaft as the goal of scholarly understanding placed a premium on general knowledge, thereby creating a productive tension between the two. This tension was intrinsic to the ideology of the nineteenth-century German university, with its commitments to the union of teaching and research and (at least partially) to "basic" knowledge as a foundation for practical improvement. This same dynamic is visible in other countries as well, reinforced in the later nineteenth century as these fundamental elements of the German university ideology were taken up elsewhere.

As developed in the German-speaking states, university disciplines themselves marked a compromise between the general and the special, representing an array of scholarly fields, each the province of a single full professor at any given university. These fields were in a continual state of dynamic tension, especially at their boundaries. Research opened up new areas of knowledge whose producers vied for their inclusion and dominance within a limited institutional domain (although, to be sure, it

---

[14] See Lynn K. Nyhart, *Biology Takes Form: Animal Morphology and the German Universities, 1800–1900* (Chicago, 1995), esp. 218–28.

[15] On the German universities as a primary context for articulations of the general, see Hagner and Laubichler, "Vorläufige Überlegungen" (cit. n. 10), 13. On the beginning of this period, see R. Steven Turner, "The Great Transition and the Social Patterns of German Science," *Minerva* 25 (1987): 56–76, and sources cited therein; Andrew Cunningham and Perry Williams, "De-centring the 'Big Picture': *The Origins of Modern Science* and the Modern Origins of Science," *Brit. J. Hist. Sci.* 26 (1993): 407–32; and David Cahan, ed., *From Natural Philosophy to the Sciences: Writing the History of Nineteenth-Century Science* (Chicago, 2003), especially the chapter by Cahan, "Institutions and Communities," 291–328.

[16] For a discussion of this process in German medicine, see Broman, *Transformation of German Academic Medicine* (cit. n. 10), esp. 177–9. For an analysis of the rise of medical specialties (a different but related phenomenon) in early nineteenth-century France, see George Weisz, "The Emergence of Medical Specialization in the Nineteenth Century," *Bull. Hist. Med.* 77 (2003): 536–75.

was growing for much of the nineteenth and early twentieth centuries).[17] Scholars in numerous disciplines struggled over what would constitute general knowledge within their domains, over how to attain it, and over how general "the general" would be.[18] This had significant institutional consequences, for special topics could be experimented with through courses, often taught by private lecturers, or *Privatdozenten*, but for a topic to gain institutional permanence (marked by a dedicated professorship or chair) it had to gain recognition as important—perhaps even general. Relatedly, in some disciplines, "many-sidedness" was considered a necessary virtue in a professor, and the characterization of being "one-sided" was a slur that could quash a job candidate's hopes.[19]

To summarize: nineteenth-century scientists regularly worked with two different renderings of the relationship between the general and the special. The Windelbandian version, let's call it, served as an organizational framework structuring the basic categories of existing knowledge: here the general and the particular were different kinds of knowledge. But scientists also sought to *generalize*, moving from the particular to some larger, more scaled-up claim that might also be called the general, via the process of induction. To see how the relations between the general and the special worked in practice, and how scientists used these categories in the process of generalizing or scaling up their claims, let us consider the particular case of zoology. Through this example, we will see how the two approaches of Wissenschaft and Kunde represent different relations between the categories of the general and the special, and thus different ways of making scientific knowledge.

THE GENERAL AND THE SPECIAL, IN PARTICULAR: ZOOLOGY

Like other academics, late nineteenth-century working zoologists in Germany divided their subject into the general and the special, and as in most other academic fields, the distinction they drew was looser than Windelband's. In their usage, the special referred to their empirical objects of study—the different kinds of animals, which differed from one another and so required special examination. The general comprised the questions and topics that could be applied generally, that is, to all objects in the special, empirical domain of study. This distinction is rendered visible in the structure of textbooks of the 1870s–90s. Two of the leading zoology texts of the period, which divided themselves into general and special parts, were Carl Claus's *Fundamentals of Zoology* (four editions, 1868–80) and Richard Hertwig's *Textbook of Zoology* (1892).

The special part, which in each case takes up the vast bulk of the book, is devoted to taxonomic description. This is divided by phyla and further subdivided into classes and orders (genera and species were too detailed to go into in a basic textbook), and typically describes the groups by what distinguishes them taxonomically—the traditional descriptive stuff of classificatory zoology since Linnaeus. This was the main

---

[17] The locus classicus of this argument is Joseph Ben-David, "Scientific Productivity and Academic Organization," *Amer. Sociol. Rev.* 25 (1960): 828–43; see also Awraham Zloczower, *Career Opportunities and the Growth of Scientific Discovery in Nineteenth-Century Germany, with Special Reference to Physiology* (New York, 1981).

[18] For examples from a wide variety of disciplines, see the essays in Hagner and Laubichler, *Der Hochsitz des Wissens* (cit. n. 10).

[19] This was truer in zoology than in anatomy; see Nyhart, *Biology Takes Form* (cit. n. 14), 173–4.

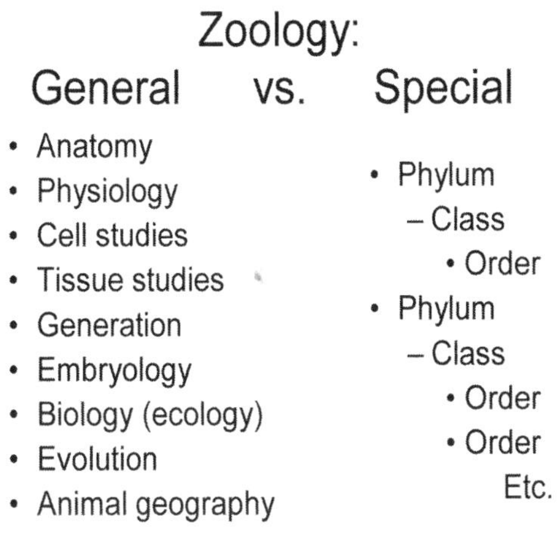

Zoology:
General   vs.   Special

General	Special
• Anatomy	• Phylum
• Physiology	– Class
• Cell studies	• Order
• Tissue studies	• Phylum
• Generation	– Class
• Embryology	• Order
• Biology (ecology)	• Order
• Evolution	Etc.
• Animal geography	

Figure 1. *Categories of the general and the special in zoology*

substance of zoology (just as actual diseases and their treatment formed the main empirical substance of pathology and therapeutics). On the general side of the ledger lies everything else: comparative anatomy and morphology, cell and tissue study, physiology, generation, embryology, evolution, animal geography, and biology (the usual late nineteenth-century term for the analysis of organisms in relation to their environments, or ecology). All were basic research problems of zoology, and the basis of subdisciplinary specialization at the time (fig. 1).

Given that systematics was as much a subdiscipline of zoology as any of these other topics—indeed, long a dominant one, with problems of its own to solve—this sharp division might seem jarring. But if we understand the special as the category covering the diversity found empirically in nature (and, by the way, its classification), then it makes sense. The general side, though taking up much less space in the textbooks, involved all the different angles from which a zoologist might illuminate the objects described on the special side—all the different parts that together would constitute their complete understanding. Here the general did not necessarily refer to causes or constants, as in Windelband's nomothetic sciences, but to perspectives by which one could characterize or analyze the special objects of attention.

General introductions were a long-standing feature of zoology treatises and textbooks (even when the term *general* was not used). Georges Cuvier's standard-setting *Animal Kingdom, Arranged in Conformity with its Organization* (1817) opened with a compelling thirty-two-page introduction addressing the nature of life and the substances that composed it, the importance of organization, the consequent distinctions between natural history and other sciences, and the methods of comparative anatomy that underlay his system. Then he moved right into describing the animals, a task that took four volumes (and eventually, in the posthumous edition of 1836–49, eleven volumes).[20]

---

[20] Cuvier, *Le règne animal distribué d'après son organisation*, 4 vols. (Paris, 1817), vol. 1; Cuvier et al., *Le règne animal distribué d'après son organisation*, edition with engraved plates, 11 vols. in 20 (Paris, 1836–49). More expansively, Friedrich Siegmund Voigt's six-volume contribution to the encyclopedic project *Naturgeschichte der drei Reiche* devoted its first 150 pages to "general zoology," comprising mainly discussions of organ systems, membranes, fluids, forces, and senses; Voigt, *Lehrbuch der Zoologie*, 6 vols. plus atlas (Stuttgart, 1835–42).

The first edition of Claus's textbook, in 1868, followed a similar approach: a twenty-eight-page introduction covered the distinctions between life and nonlife, animals and plants, comparative anatomy and development, and the history of ideas about classification systems, including the transformation of their meaning by Darwin's theory. This was followed by some eight hundred pages of material divided taxonomically and largely devoted to the description of differentia—the anatomical characteristics that divided one group from another.[21] In the third edition of 1876, however, Claus expanded what had been an introduction and retitled it the "General Part" (*Allgemeiner Teil*). This took up the first 136 pages of a tome that now extended to over 1,250. It rose from the most basic division between organic and inorganic bodies, through the distinction between animals and plants, to the nature of organic individuality. It then launched into cells and tissues, organs and organ systems, reproduction, development, the history of systems, the definition of species and the explication and defense of evolutionary theory, with critical discussion of various different versions that were challenging Darwin's by that time. Hertwig's 1892 textbook covered much the same terrain (in a slightly different order), with the addition of brief sections on ecology and geographic distribution.[22]

Clearly, the general here did not comprise a single unifying theory or integrated theoretical framework for zoology. Rather, all of the topics in the general constituted focused problem areas. Cytology, for example, shed light on one aspect of understanding an organism. So did biogeography and evolution. They were partial angles on the material. What made them general was not that they posited constant laws understood to hold true for all animals, but rather that they constituted those categories of questions that could be asked of all organisms.

If special zoology covered the taxonomic diversity of nature and its classification, and general zoology covered all the ways of approaching animals, how did zoologists actually relate the two in their research? Four kinds of relationships are logically possible, with different implications for scaling up, or generalizing in the Whewellian sense (which of course was still important for claims to significance).

One approach would be to stay on the special side of the ledger and work purely on classificatory questions from a strictly anatomical analysis of external features— the least complicated kind of criteria for sorting organisms, with little recourse to broader questions. This had been an especially prominent approach in entomology, a subfield that, not incidentally, remained wide open to amateurs. In this case, scaling up would be a matter of moving up taxonomic levels, of discovering those characteristics that were common to different forms. Most classificatory work in the later nineteenth century, however, was not that "pure," but required engagement with general problems of comparative anatomy, functional morphology, geographic distribution, and/or biology (understood in its common late nineteenth-century sense as ecology), because these helped make taxonomic determinations. Though most late nineteenth-

---

[21] Claus, *Grundzüge der Zoologie* (Marburg, 1868).

[22] Claus, *Grundzüge der Zoologie*, 3rd ed. (Marburg, 1876); Hertwig, *Lehrbuch der Zoologie* (Jena, 1892). Carl Theodor Ernst von Siebold's 1869/70 course in zoology, as recorded by L. v. Ammon (Deutsches Museum Munich, Handschriftenabteilung/Sondersammlung, N 12/186 [8677]), was also divided into general and special parts, though in the latter section Siebold descended from the most complex down to the least, rather than using the upward-moving scheme followed by Claus and Hertwig.

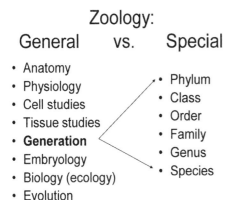

Zoology:
General   vs.   Special

- Anatomy
- Physiology
- Cell studies
- Tissue studies
- **Generation**
- Embryology
- Biology (ecology)
- Evolution
- Animal geography

- Phylum
- Class
- Order
- Family
- Genus
- Species

Figure 2. Combined approach no. 1: start from the general. Scaling up involves greater taxonomic coverage.

century German zoologists did devote some time during their careers to classificatory work, it almost inevitably was combined with one or more of these problem areas.[23]

A second approach would be to work only the general side of the ledger, seeking to integrate the different perspectives. This would entail explicitly making theory, and in the late nineteenth century, much energy went into this side, especially following Ernst Haeckel's 1866 *Generelle Morphologie* (note that word "general" again), which sought to create a general theory of form that connected individual and phylogenetic development, anatomy, and physiology in a synthesis that claimed to be causally deterministic.[24] As Haeckel's powerful example suggests, scaling up on this side focused on connecting more of the problem areas to one another, often by subsuming the one under the other. The more problem areas covered, the greater the explanatory scope of the theory. (Haeckel's was a morphological synthesis, not a complete biological one. It did not claim to capture all the topics on the general side of the zoological ledger—for example, although he invented the word *ecology* in this work, he did not treat it—but it did cover most of them.)

Most actual empirical research *combined* the general and the special at a less sweeping level, tilted toward either the general or the special (figs. 2 and 3). One combining approach was to focus on a question or angle, making the general the object of attention, and taking instances of the special to exemplify the general properties (or test claims about them). Thus one might ask, "What is the nature of fertilization?" and investigate this across numerous different organisms. The result would be, perhaps, general knowledge about this process. Here, taxonomic coverage was important: the more different groups a particular characterization extended to, the more general it would be. This approach is the one more commonly associated with Wissenschaft in the second half of the nineteenth century.

The other combining approach started from the special, the object side. In this

---

[23] Nyhart, *Biology Takes Form* (cit. n. 14); see also Peter J. Bowler's *Life's Splendid Drama* (Chicago, 1996), which, however, pursues these questions mainly in the Anglo-American context. See also Nyhart, *Modern Nature: The Rise of the Biological Perspective in Germany* (Chicago, 2009), esp. chap. 9.

[24] Haeckel, *Generelle Morphologie der Organismen: Allgemeiner Grundzüge der organischen Formen-Wissenschaft, mechanisch begründet durch die von Charles Darwin reformirte Descendenz-Theorie*, 2 vols. (Berlin, 1866).

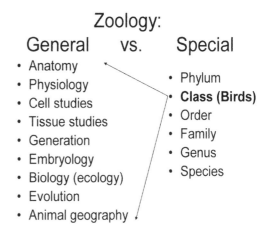

Zoology:

General   vs.   Special

- Anatomy
- Physiology
- Cell studies
- Tissue studies
- Generation
- Embryology
- Biology (ecology)
- Evolution
- Animal geography

- Phylum
- **Class (Birds)**
- Order
- Family
- Genus
- Species

Figure 3. Combined approach no. 2: start from the special. Scaling up involves greater explanatory richness.

case, the general is what illuminates the empirical thing. That is, we understand the object of our attention (and here it is an object—a kind of organism) by dint of knowing all the general features of it. Suppose one wants to know about spiders. Here one would want to understand their anatomy, physiology, behavior, classification, where they live (both ecologically and geographically), what they eat, how they build webs and capture their prey, and so forth. (Note that this is far more than classification, though it still pertains to the special object of study.) The more different aspects one knows about spiders, the greater one's knowledge of spiders might be said to be. This approach exemplifies what I am calling Kunde, for reasons that will be made clear in a later section.

Both combining approaches connect the general and the special, and both imply a comparative perspective, but they make different kinds of knowledge. In the first, scaling up privileges the epistemic value of empirical coverage or breadth; in the second, the value of explanatory richness.[25] That is, when a zoologist claims to say something big in the Wissenschaft mode, he is claiming that his generalization covers many diverse taxa. When he claims to say something big in the Kunde mode, he is claiming that he knows many different things about the organism (or higher-level taxon).

Under what conditions was each of these approaches favored? We get some clues from Hertwig's 1886 *éloge* of Carl Theodor Ernst von Siebold. To Hertwig, a leading laboratory zoologist who concentrated on the processes of early development, Siebold's wide-ranging research seemed undirected, *absichtslos*. Modern-day research in the mid-1880s, by contrast, started from a particular question and used clearly articulated methods to answer it: "Once it has been established, through numerous comparative investigations, how one is to research, which characteristics of structure and development are significant, it is often only necessary to apply the worked-out [*ausgebildete*] norm to a particular case. Thus will the particular [*Einzelne*] carry the spirit of Wissenschaft." Once the structure is set, Hertwig wrote,

---

[25] Alan Love develops a similar contrast between what he calls "theoretical generality" (i.e., applicability to many different species—my "taxonomic coverage") and "explanatory completeness" (coverage of "all relevant aspects of phenomena under scrutiny"—my "explanatory richness"). I find my terms a little more transparent. Love, "Marine Invertebrates, Model Organisms, and the Modern Synthesis: Epistemic Values, Evo-Devo, and Exclusion," *Theory Biosci.* 128 (2009): 19–42, on 33.

even those with only moderate talent for scientific research "can produce competent and rounded works, which not that much earlier would have required an outstanding and original gift."[26] By contrast, in the first half of the century, scientific zoology was still a "fallow field"; it was easier then to find new facts but harder to know their significance.

If historians wanted to, it would be easy to find in Hertwig's description of "modern" zoology T. S. Kuhn's "normal science," the day-to-day working out of small problems within an existing framework, and in his characterization of the older style something like a "pre-paradigmatic" situation. But I want to direct our attention to a different point, which comes in the next part of Hertwig's speech. Hertwig insisted that in modern, disciplined research "the object of investigation has value for the researcher only insofar as it contributes to solving his problem." Indeed, one might not even need to study the whole organism but just focus on those few parts that address the question at hand. By contrast, Siebold and "the older zoologists" were interested in all aspects of the animals they studied. Hertwig expressed amazement at Siebold's "sympathy with the object of his research"—his publications ostensibly on parthenogenesis in wasps also threw in "an abundance of details about the life-habits of these animals, how the wasps build their nests, how they protect themselves against moisture and enemies, how they behave when faced with unexpected attack, and more."[27] "What an undisciplined, amateurish approach to studying nature!" Hertwig seems to be saying.

Hertwig presents here exactly the contrast between the two combined relations of the general to the special just alluded to—the first in which the empirical investigation is treated as an instance contributing to the general problem, and the second in which the empirical case is the object illuminated by the general angles. He pegs the first as modern and directed, and the second as almost unintentional, driven by undisciplined curiosity. Yet in his day, Siebold was a leader of scientific zoology. What are we to take away from this?

First, Hertwig's statement clearly reflects the situation of a younger up-and-comer of the mid-1880s, looking back on a historical period of zoology he had not experienced. Starting in the mid-1840s, Siebold and a number of like-minded colleagues sought to render zoology "scientific" by searching for laws and causes, largely via questions of functional morphology.[28] In this respect, Hertwig's characterization of Siebold as "undirected" did the latter an injustice. To be sure, Siebold's later work on parthenogenesis shows him to have been deeply engaged in his wasps' lives. This represented more than just a personal interest in their habits, however. His full descriptions represented a commitment to a rounded understanding of the living creatures he worked on, even while focusing in on a particular aspect. Indeed, in the mid-1870s Siebold fretted about the younger generation's overly narrow emphasis on anatomy and development, with its attendant loss of appreciation for *living* creatures. Hertwig's generation might be seeking the general, but in Siebold's perspective, theirs was a narrow construction of the general, lacking the explanatory richness necessary to understand living things.[29]

[26] Hertwig, *Gedächtnisrede auf Carl Theodor von Siebold* (Munich, 1886), 13; translation mine.
[27] Ibid., 14–5.
[28] Nyhart, *Biology Takes Form* (cit. n. 14), esp. 94–6.
[29] Ibid., 202–5.

This narrowing of focus reflected two shifts in the German zoological community since Siebold's younger days. First, the discipline had become more disciplined, as the intellectual territory of animal morphology became filled in and routinized. Research became more intensive than extensive (not unlike settlement or agriculture after a certain stage of development)—hence Hertwig's premium on a really strong question and clear methods to get at it. Second, the pressure toward narrowness was increased by the rapid expansion of the population of zoologists. Not only was knowledge growing in the 1870s and 1880s, but the number of people seeking to make significant claims in zoology (and therefore gaining a chance at a job) was climbing at least as fast.[30] Specialization was a consequence not just of the growth of knowledge but also of the need to get results. This setting favored making theoretical claims on the basis of limited research, and then seeing further research as testing those claims. It favored the nomothetic stance (viewing the empirical as merely a case of the general), and especially a variant of that stance that tilted hard toward unifying science around a small number of integrative principles or laws.

A model for that tightly governed unity had existed since the late 1830s in the cell theory, which posited that all the functions of life were carried out at the cellular level. From that time onward, increasing numbers of life scientists, whether working as anatomists, physiologists, botanists, or zoologists, studied cell function and the formation of tissues in development. In the 1870s and 1880s, attention to these aspects intensified, with a special focus on the processes of fertilization and early development. By the turn of the twentieth century, some scientists would claim, under the banner of "general physiology" or "general biology," that an understanding of all of life would derive from greater knowledge of the cellular substance.[31] Academic zoology in Germany was not wholly taken over by this new wave. It remained one option—an increasingly prestigious one, to be sure—among numerous approaches to understanding organisms in the German academic setting.[32] However, in the early twentieth century, it was probably the most aggressive claimant to the meaning of "the general" in the life sciences.

What about the Kunde approach, focused on multifaceted knowledge of one kind of living organism (like Siebold's wasps)? Although Hertwig judged this approach harshly as amateurish, and Siebold worried it was slipping away, it would soon revive as "biology," or early ecology. It would flourish not in lab-oriented university institutes, but in settings where comparison and taxonomy were highly valued: civic and university zoology museums. Here, too, a strong framework was in place, namely, taxonomy—"special zoology" and its subdivisions. As civic museums expanded in the 1890s and early 1900s, and the number of curatorships increased, these were

[30] Ibid., 193–5.

[31] The literature on the history of cell theory in the nineteenth century is sizable. See François Duchesneau, *Genèse de la théorie cellulaire* (Montreal, 1987); Henry Harris, *The Birth of the Cell* (New Haven, Conn., 1999); and, for an excellent brief overview citing the major earlier sources, Jane Maienschein, "Cell Theory and Development," in *Companion to the History of Modern Science*, ed. R. C. Olby et al. (New York, 1990), 357–73. On "general biology" as cell biology, see Manfred Laubichler, "Allgemeine Biologie als selbständige Grundwissenschaft und die allgemeinen Grundlagen des Lebens," in Hagner and Laubichler, *Der Hochsitz des Wissens* (cit. n. 10), 185–205. Major primary sources include Max Verworn, *Allgemeine Physiologie: Ein Grundriss der Lehre vom Leben* (Jena, 1895), and Oscar Hertwig, *Allgemeine Biologie: Zweite Auflage des Lehrbuchs "Die Zelle und die Gewebe"* (Jena, 1906).

[32] See Nyhart, *Biology Takes Form* (cit. n. 14).

tied to increasingly narrow (though still pretty big) taxonomic groups. The research carried out by these somewhat more specialized curators also deepened. It focused in more closely on their preferred organisms, but addressed multifaceted aspects of those organisms, considering them as living things shaped by their many relations to their environment. Such was especially the case at the museums with the largest staffs and the greatest division of labor—Hamburg, Frankfurt, and the Berlin Museum für Naturkunde, where from the late 1880s on, systematics was joined with evolution, biogeography, and biology. These zoologists, too, were working both sides of the ledger; they too were deepening their focus, only with the organisms as their starting point.[33]

This was where things stood at about the turn of the century. Seeing the characteristics of the general and the special, and the different ways in which they could be related, is worthy in itself, because it gives us a handle on the different epistemic values cultivated within zoology—Wissenschaft's high valuation of empirical coverage (theory that covers many taxa) and Kunde's high valuation of explanatory richness (multifaceted, synthetic knowledge of its object). It further shows how those values mapped, at least roughly, onto institutes and museums as sites where they were put into practice. Let us press a little further on these categories, both inside German zoology and beyond, to see how they were involved as disciplinary knowledge changed.

FINDING THE GENERAL IN THE SPECIAL

By the early twentieth century, something interesting was taking place in the zoology literature reviews and bibliographical literature. Like textbooks, these were forms in which scholars had to provide an explicit scheme for organizing knowledge in their field. It is especially clear in the *Archiv für Naturgeschichte*, which since the 1830s had listed and abstracted the literature in the different taxonomic areas of zoology. Traditionally, each annual report was organized taxonomically into the subcategories of special zoology but began with a general introductory essay. Such essays typically reviewed the major synthetic works in the field as well as problems or questions that crossed taxonomic areas. (Reviewers covering different taxonomic areas, for example, addressed Darwin's *Origin* after 1859 and Haeckel's *Generelle Morphologie* after 1866.) By the turn of the century, however, the literature had expanded to the extent that abstracting individual works was next to impossible. A new strategy was required. In 1908, when Georg Seidlitz published the "general" review of literature on insects, covering works that treated more than one taxonomic order of insects, he first listed the 553 titles alphabetically by author. He then included a series of cross-reference lists that organized the information by content: literary and technical aids; systematics; evolution; morphology, histology, physiology, and embryology; biology; economic entomology; geographical distribution; and paleontology—all the problem areas that people in the field were working on (except, notably, heredity, which was subsumed under the category of biology). Seidlitz also wrote that year's article on Coleoptera (beetles), which followed this same organization, first with reference to Coleoptera in general, and then reproducing it at the level of each family within the order. (Beetles were popular: the article covered 1,044 works and ran to

[33] Nyhart, *Modern Nature* (cit. n. 23).

over 350 pages.) What had been the categories of the general now reappeared inside the special, at each level. Systematics still provided a basic structure, and the largest set of literature within the overall review, but it was visibly infused, at every level, with all of these perspectives.[34]

How are we to interpret this? Assuming that such literature reviews were organized with the aim of helping working zoologists find literature on topics they were interested in, it is clear that they were expected to be interested in all these different perspectives on their taxonomic category—and that they in fact were, since the existing literature could be divided up that way. The separation between the general and the special was dissolving. Or rather, the entities formerly categorized as the general and the special had become the warp and weft of natural historical research. By the early twentieth century, systematists in Germany specialized in a particular group (warp) while also seeking to contribute to general biological problems (weft).

Was this really new? I think it was, for zoology,[35] and I suggest that it was part of a longer trajectory, in which epistemic value in zoology gradually shifted from the general illuminating the special to the other way around. In the 1870s, systematics still provided the main framework for zoology as a whole: zoologists worked on "problems," but systematics organized how they did so. Increasingly, however, the categories of the general were becoming valued on their own terms, leading to increasing intellectual and institutional autonomy, both from one another and from their visible dependence on the special objects that constituted the empirical substrate of their work. By the early twentieth century, the balance had shifted.

Although both zoologists and many "general biologists" still united the two sides of the ledger, the value placed on the direction of generalization was asymmetrical. Physiologists, cell biologists, and, after 1900, geneticists—all of whom laid claim to being general biologists—did not typically claim to be experts on a particular taxon. That would be too specific, and it ran against the grain of the kind of generalization they were promoting.[36] By contrast, most zoologists continued to specialize in particular taxonomic areas, while offering their research as contributions to larger problems. Museum-based zoologists, aware that their systematic work could be viewed as merely particular, were especially conscious of the need to show relevance to broader problem areas—which they typically drew in relation to ecology and biogeography.[37] Indeed, in the 1920s, Willy Kükenthal instructed the authors of his edited *Handbuch der Zoologie* to focus on special zoology, and so to omit general zoology, excluding theorizing and reducing morphological description. Instead, authors were to empha-

---

[34] Seidlitz, "Insecta: Allgemeines für 1908" and "Coleoptera für 1908," *Archiv für Naturgeschichte*, pt. 2 (1909): 1–60 and 61–416.

[35] Windelband wrote, as if it were a matter of course, that "all *wissenschaftliche* work has the purpose of putting its special problems into a wider framework and resolving specific questions from the standpoint of more general perspectives" (translation adapted from Windelband, "Rectorial Address" [cit. n. 7], 169). Yet there is ambiguity here: does this mean that understanding all aspects of a spider's life is not wissenschaftlich, or does Kunde-style explanatory diversity qualify as a "more general perspective"?

[36] On the aspirations and claims to general biology in the late nineteenth and early twentieth centuries, see Laubichler, "Allgemeine Biologie" (cit. n. 31).

[37] See, e.g., Karl Kraepelin, "Die Bedeutung der naturhistorischen, insonderheit der zoologischen Museen," *Naturwissenschaftliche Wochenschrift* 3 (1888–9): 74–6, 85–6, 90–3; August Brauer, "Biogeographie," in *Die Kultur der Gegenwart*, pt. 3, sec. 4, vol. 4, ed. Richard Hertwig et al. (Leipzig, 1914), 176–85, esp. 177.

size "physiology, ecology, geographical distribution and systematics."[38] Thus, in this standard-setting handbook, these formerly general categories appear to have been fully incorporated into special zoology itself, exploding the long-conventional equation of special zoology with the systematic categories alone. The larger intellectual structure of zoology, and of the life sciences more broadly, was becoming reorganized, even if the institutional structure of the university-based disciplines was not.

This trend is clearly visible beyond Germany as well. Already in 1907, the pioneering British geneticist William Bateson could comment at the International Congress of Zoologists, with a kind of false retrospection, that zoologists used to be "students of Coelenterata, insects, Vertebrata, or whatever it might be." Now, however, they organized their research "in new formations," around broad problems that reached across taxa. In investigating the new field of heredity, especially, Bateson asserted, "We take facts wherever we can find them. We are botanists to-day, zoologists tomorrow."[39]

Bateson's position was hardly that of a neutral observer: he was pushing hard for this transformation, and in 1907 he was premature in relegating the taxonomic organization of the life sciences to the past tense. But the "new formations" of zoology were real, and were reflected in the organization of the congress. The 1907 *Proceedings* was divided into sections on animal behavior, comparative anatomy, comparative physiology, cytology and heredity, embryology and experimental zoology, entomology and applied zoology, general zoology, paleozoology, systematic zoology, and zoogeography—a list remarkably similar to those in the "general" categories of textbooks and literature reviews.[40] This contrasts sharply with the organization of the sixth international congress, which took place three years earlier in Berne, Switzerland. There the sections were general zoology, vertebrates (systematics), vertebrates (anatomy and embryology), invertebrates (exclusive of arthropods), arthropods ("most of the papers in this section dealt with ants"), applied zoology, and zoogeography.[41]

While it is true that the organization of the 1907 meeting represented an aggressive move by the organizers and their American hosts to present a new, problem-based order for zoology, we know they had the material to work with, because they were in fact able to organize the sections this way.[42] I would suggest that this was because even more traditional zoologists already were practicing a twofold organization of their material—contributing simultaneously to knowledge of their chosen taxon and

[38] Printed instructions to authors of the *Handbuch der Zoologie*, ed. Kükenthal and Thilo Krumbach (Berlin, 1927–34), quoted in Jürgen Haffer, Erich Rutschke, and Klaus Wunderlich, "Erwin Stresemann (1889–1972)—Leben und Werk eines Pioniers der wissenschaftlichen Ornithologie," vol. 34 of *Acta Historica Leopoldina* (2000), on 271.

[39] Bateson, "Facts Limiting the Theory of Heredity (Address)," in *Proceedings of the Seventh International Zoological Congress, Boston, 19–24 August, 1907* (Cambridge, Mass., 1912), 306–19. I thank Kristin Johnson for alerting me to this source.

[40] *Proceedings of the Seventh International Zoological Congress* (cit. n. 39).

[41] "The Sixth International Congress of Zoology," *Nature* 70, no. 1819 (1904): 473–5, on 474.

[42] The international congresses also point us toward another important issue: the role of international contact and cooperation in connecting the different national traditions in zoology and biology. Especially relevant here is the first part of Nikolai Krementsov's *International Science between the World Wars: The Case of Genetics* (New York, 2005), on international congresses. Other models of international contact and exchange were also significant. See Josef Karl Partsch, *Die Zoologische Station in Neapel: Modell internationaler Wissenschaftszusammenarbeit* (Göttingen, 1980).

to a prominent question area. It was a strategy to maintain significance in the face of specialization around new topics and the declining prestige of the taxonomist's commitments to Kunde-style understanding. This was a lasting change, marking a shift in what counted as significant knowledge in the life sciences.

As historians of biology will recognize, I have just re-described what is often referred to as the "transformation of the life sciences" of the late nineteenth and early twentieth centuries. Historians have emphasized a variety of different features of this complex transformation: the specialization and fragmentation of problems, a new prominence accorded to philosophical reductionism, the rise of new experimental approaches and techniques, and an accompanying prioritization of lab over field.[43] While the details of the story are different in different countries (in ways that still need to be articulated more clearly and comparatively), it is fair to say that together these factors both caused and expressed a change in the conceptual structure of the life sciences (and especially zoology, the field on which most historians of biology have focused). Research on particular taxonomic groups did not go away, but the stuff of taxonomy—the organisms and their classification—gradually lost its primacy as the raison d'être for the big questions and became increasingly merely the vehicle through which new questions were asked and answered. Kristin Johnson put her finger on the heart of this conceptual transformation when she characterized it as a "transition from nineteenth century taxa-defined disciplines such as zoology, botany, entomology, and ornithology, to the problem-based disciplines of twentieth-century biology" (implied here are such areas as physiology, anatomy, genetics, embryology, and evolution).[44]

Historians of biology have often told this story in the service of a longer narrative that points toward the putative reunification of biology in the evolutionary synthesis from the late 1930s to the early 1950s—a powerful if incomplete and deeply teleological story that has been much criticized but is hard to escape fully.[45] I wish instead to point us toward more structural questions of conceptual and disciplinary change. We might, for instance, compare biology to other disciplines, focusing on relations among their problem areas and subordinate subject matters to see how they changed over time. As seen through its literature reviews, for instance, early twentieth-century

[43] The literature on this topic is enormous, if mainly focused on the United States. Accounts emphasizing quite different perspectives agree on these basic points: e.g., Garland E. Allen, *Life Science in the Twentieth Century* (New York, 1975); H. Penzlin, "Die theoretische und institutionelle Situation in der Biologie an der Wende vom 19. zum 20. Jahrhundert," in *Geschichte der Biologie*, 3rd ed., ed. Ilse Jahn (Heidelberg, 2000), 431–41; Jan Sapp, *Genesis: The Evolution of Biology* (Oxford, 2003); Lynn K. Nyhart, "Natural History and the 'New' Biology," in *Cultures of Natural History*, ed. Nicholas Jardine, James Secord, and Emma C. Spary (Cambridge, 1996), 426–43. For an especially apt characterization of the specialization and fragmentation of the life sciences, see Jonathan Harwood, "Universities," in *The Cambridge History of Science*, vol. 6, *The Modern Biological and Earth Sciences*, ed. Peter J. Bowler and John V. Pickstone (New York, 2009), 90–107.

[44] Johnson, "The Return of the Phoenix: The 1963 International Congress of Zoology and American Zoologists in the Twentieth Century," *J. Hist. Biol.* 42 (2009): 417–56, on 417. The continued fragmentation of zoology is a central theme of Johnson's article.

[45] On the evolutionary synthesis triumphant, see esp. Ernst Mayr and William Provine, *The Evolutionary Synthesis: Perspectives on the Unification of Biology* (Cambridge, Mass., 1980), as well as much of Mayr's other historical work. Critical perspectives include Stephen Jay Gould, *The Structure of Evolutionary Theory* (Cambridge, Mass., 2002), esp. chap. 7; Michael R. Dietrich, "Richard Goldschmidt's 'Heresies' and the Evolutionary Synthesis," *J. Hist. Biol.* 28 (1995): 431–61; Vassiliki Betty Smocovitis, *Unifying Biology: The Evolutionary Synthesis and Evolutionary Biology* (Princeton, N.J., 1996); and Joe Cain and Michael Ruse, eds., *Descended from Darwin: Insights into the History of Evolutionary Studies, 1900–1970* (Philadelphia, 2009).

German geography had a very similar structure to late nineteenth-century zoology, with general geography (*allgemeine Erdkunde*) referring to many different subdisciplines or problem areas (geophysics, terrestrial magnetism, geognosy, oceanography, meteorology, plant geography, animal geography, ethnology, map projection, land measurement, and the physics and mechanics of the Earth)—all distinct from regional (special) geography or *Länderkunde*, which was divided into a hierarchy of geographical units. Just as zoology did in the early twentieth century, in the 1920s and 1930s geography too merged the general and special categories, applying the former to each level of the latter.[46] Did this happen in other disciplines as well? And how was it interpreted at the time?

Returning to zoology, we might also ask, how did all this turning inside out of category relationships shape disciplinary structure? It would appear to depend on what one means by "discipline." Mario di Gregorio wrote in 2009 that "zoology, the study of the animal kingdom, is no longer seen as a coherent branch of science. The specialization of the twentieth century has seen zoology's territory divided among a host of separate disciplines." At the same time, new approaches "made 'zoology' a less relevant category."[47] Indeed, the renaming of the American Society of Zoologists as the Society for Integrative and Comparative Biology in 1997 would seem to confirm Di Gregorio's statement (though the Deutsche Zoologische Gesellschaft and other national zoological societies still exist). Certainly a number of important biology departments reorganized themselves in the 1980s and 1990s to reflect a new clustering of problem areas, resulting in divisions of departments into new departments (typically two) with names like "cell and molecular biology" and "evolution, ecology, and behavior." (Note that these often have not one name but multiple ones, suggesting that they are something more like coalitions of problems than integrated disciplines.)[48] Yet even these changes took around a century to occur, and separate departments in zoology and botany with PhDs, undergraduate majors, and courses persist in many American universities. So what might be meant by "disciplines" in Di Gregorio's comment about "separate disciplines"? In large part, he is talking about the problem areas that once inhabited the general side of the ledger (along with new ones that have developed since then), whether or not they are actually called disciplines in formal university department names.

The lag time between conceptual and institutional change in universities suggests that in analyzing disciplinary change, we must consider the interactions of conceptual, demographic, and institutional structures over the *longue durée*. A necessary part of that complex project is to attend to fundamental conceptual structures such as the general, the special, and their relationships to help us locate the changing epistemic values that (often invisibly) accompany the big sea changes of science. A dynamic historical sociology of science needs intellectual history.[49]

---

[46] See the annual literature reviews in the *Geographisches Jahrbuch*; the clearest instance of this blending of the general with the special comes in the 1930 review for North America: B. Dietrich, "Nordamerika (1916–30)," *Geographisches Jahrbuch* 45 (1930).

[47] Di Gregorio, "Zoology," in Bowler and Pickstone, *Modern Biological and Earth Sciences* (cit. n. 43), 205–24, on 205, 221.

[48] On this aspect of the biomedical disciplines, see the cogent points made by Stephen Turner in "What Are Disciplines? And How Is Interdisciplinarity Different?" in *Practising Interdisciplinarity*, ed. Peter Weingart and Nico Stehr (Toronto, 2000), 46–65, esp. 58.

[49] Contrary to a sociological perspective that black-boxes the content of disciplines when analyzing them. See, e.g., ibid., esp. 51–2.

KUNDE AND INTERDISCIPLINARITY

In addition to helping us think about change within disciplines, attention to the relationships between the general and the special invites us to ask what we can learn
from examining the Kunde approach to knowledge. As I have argued elsewhere,
museum-based zoologists were not isolated in developing the Kunde approach; it
was widespread outside the universities in the late nineteenth century, often (although
not always) associated with museums of many kinds.[50] Simply listing the pursuits
ending with "-kunde" that emerged in the nineteenth century, we get an impressive
list: *Völkerkunde* (ethnology), *Volkskunde* (folklore/folklife studies), *Orientkunde*
(Oriental studies, later *Orientalistik*), *Heimatkunde* (regional or homeland studies),
Meereskunde (marine studies), and *Handelskunde* (commerce studies). To these we
may add the older enterprises of classical and German *Altertumskunde* (archaeology),
Erd- and *Länderkunde* (geography and its subfield regional studies), *Bücherkunde*
(bibliology or book connoisseurship), *Naturkunde* (natural history), *Heilkunde*
(medicine), and *Arzneikunde* (pharmacy).

What did all these terms have in common? Dictionaries from the eighteenth century on equate Kunde with Wissenschaft, and the existence on this list of venerable
fields like medicine and pharmacy helps us see that close connection. However, in
the nineteenth century, as I have suggested, a real distinction emerged. Whereas Wissenschaft, especially as it developed in the universities, emphasized the search for
general unifying laws, and thus stressed the primacy of theory (of which particular
objects were merely instances), the Kunde projects were characterized by different
attributes. First, they were encyclopedic, in the way that I have described the study of
spiders: they sought to know everything about their object of attention. What unified
their investigations was not the search for an overarching theory, but rather the fullness and multifacetedness of understanding about them to be reached. (Note that this
is different from Windelband's idiographic science, which stressed the unique and
causally complex character of events over the multifaceted approach to understanding them.)

Second, the Kunde projects tended to be organized around material objects and
to emphasize hands-on knowledge of them. Here we might see connections to both
linguistic roots of the term: *kennen*, "to know," and *können*, "to do." Thus Heilkunde
and Arzneikunde were hands-on arts of healing, and not only theory-based sciences.
Natur-, Völker-, Volks-, Heimat-, and Handelskunde all had material *objects* as their
objects of study: natural history specimens, weapons, household and agricultural
tools, religious masks, textiles, building structures, and value-added commercial
products. Classics and Oriental studies, while warring for primacy as the sources for
cultural origin stories of the history of civilization, shared the characteristic that in
situ studies of places and the objects associated with them—including their natural
history, their sculptural and archaeological leavings, and their coins and inscriptions
pounded into stone—came to challenge the study of textual documents. They further
shared an emphasis on place with other geographical studies: Heimatkunde, Länder-

---

[50] Nyhart, *Modern Nature* (cit. n. 23), chap. 7, esp. 253–6; Nyhart, "Kundekunde; oder, das Allgemeine im Museum" [Knowledge of knowledge; or, the general in the museum], in Hagner and Laubichler, *Der Hochsitz des Wissens* (cit. n. 10), 207–37.

kunde, and Meereskunde all sought to bring together the myriad physical, natural, cultural, and historical aspects of a particular region or place.[51]

Third, these projects tended not to be exclusively academic, university-based sciences. They fostered knowledge that might be useful beyond academia, or they welcomed the interest and expertise of nonacademics, or both. Experience counted in these projects, even un- or undercredentialed experience. Here collectors of objects could speak with some authority, in ways that might not be heard otherwise. In conjunction with the first two attributes, we often see in museums and societies organized around the Kunde projects a more open social space of collaboration, involving people with different skill sets. Heimatkunde was the province of teachers, hikers and beautifiers, natural history collectors, racial heritage researchers, and local historians.[52] As Suzanne Marchand has said of those involved in Oriental studies, "Some would have described themselves as theologians, classicists, historians, geographers, archaeologists, or art historians." She might have added diplomats, travelers, and naturalists to her list as well.[53]

Whereas some of the Kunde projects merely brought together different kinds of people with common interests in a shared object, others were more explicitly collaborative and interdisciplinary. Notable in this regard were the ambitions raised for Meereskunde at the turn of the twentieth century. Expanding upon the sorts of collaborations developed in the long tradition of ocean exploration, its founders imagined Meereskunde as a new science that would unite physical oceanographers, geologists, chemists, biologists, geographers, and even (in one prominent version) historians or anthropologists, to generate a collective general understanding of the ocean and the relation of humans to it. While it failed to unite as a field in the way its most prominent members hoped it would, the museum that this vision produced, which did in fact include all these different facets, survived for over thirty years, until its destruction in World War II.[54]

[51] On the common orientation of the geographical sciences, and their connection to a common object rather than a search for natural law, see Boas, "Study of Geography" (cit. n. 7).

[52] Here see esp. Alon Confino, *The Nation as a Local Metaphor: Württemberg, Imperial Germany, and National Memory, 1871–1918* (Chapel Hill, N.C., 1997); Nyhart, *Modern Nature* (cit. n. 23), esp. 163–4, 192–5, 268–78.

[53] Marchand, *German Orientalism in the Age of Empire: Religion, Race, and Scholarship* (Cambridge, 2009), xxiii; Richard Pischel et al., *Die Deutsche Morgenländische Gesellschaft, 1845–1895* (Leipzig, 1895). On the role of collectors and other liminal figures in Oriental studies, see Judith R. H. Kaplan, "Language Science and Orientalism in Imperial Germany" (PhD diss., Univ. of Wisconsin–Madison, 2012), esp. chap. 2. My thanks to Judy for allowing me to cite this unpublished material. On naturalists exploring the Orient, see Gabriel Finkelstein, "'Conquerors of the Künlün'? The Schlagintweit Mission to High Asia, 1854–57," *Hist. Sci.* 38 (2000): 179–218, and Lynn K. Nyhart, "Emigrants and Pioneers: Moritz Wagner's 'Law of Migration' in Context," in *Knowing Global Environments: New Historical Perspectives in the Field Sciences*, ed. Jeremy Vetter (New Brunswick, N.J., 2010), 39–58.

[54] One version is outlined in Johannes Walther, *Allgemeine Meereskunde* (Leipzig, 1892); another is presented in Ferdinand Freiherr von Richthofen, "Die Begründung eines oceanographischen Instituts nebst Meeresmuseum in Berlin," manuscript memo in Geheimes Staatsarchiv Preussischer Kulturbesitz, Berlin, Rep. 76 V[a] Sekt. 2, Tit. X, Nr. 158, Bd. 1, and was realized in the Berlin Museum für Meereskunde. See Nyhart, *Modern Nature* (cit. n. 23), 278–88. It is ironic that Stephen Turner ("What Are Disciplines?" cit. n. 46, on 57) chooses marine science to exemplify the division of labor within a (presumably unified) field that houses some "discipline-like subfields"—without any recognition of the work and history that was required to turn them from a collaborative interdisciplinary project into a (mega?)discipline.

The Kunde projects of nineteenth-century Germany thus provided a complement to Wissenschaft in multiple ways. In addition to positing a different relation of the general to the special, they also provided an alternative to the logocentrism and credential exclusivity of the universities, appropriately finding their characteristic locus in the museum, an institution that was more directly open to public participation than the university. Although never a serious direct competitor to Wissenschaft, they did present an approach to knowledge that is sufficiently distinctive to be ranged somewhere alongside John Pickstone's natural historical, analytical, experimental, and technoscientific "ways of knowing" and that might extend this array further into the social and human sciences.[55]

How, more specifically, might thinking about Kunde and its attributes help us understand scientific knowledge in new ways, in the life sciences and beyond? How might it capture modes or patterns of scientific investigation to which we have not attended sufficiently? We might attend especially to its central attribute of bringing different perspectives to bear on a common physical or material object of investigation. Daniel Liu has recently investigated protoplasm as an object of common attention among biochemists, cell biologists, general physiologists, and plant physiologists in the late nineteenth and early twentieth centuries, in changing configurations over time. In Liu's analysis, protoplasm supplied a "conceptual disciplinary nexus." While it would be possible to talk about protoplasm as a "boundary object," the emphasis is different. Whereas Susan Leigh Star and James Griesemer's boundary objects provide a common unit of exchange across different meaning systems and social worlds, here the argument is that looking at the same object from many different perspectives pulled practitioners of these different fields "towards common commitments, ideas, and vocabulary."[56]

A more self-conscious twentieth-century version of this multifaceted, multidisciplinary Kunde approach toward a life-science object appears in an episode in the late 1970s, when the biologist Peter Beaconsfield convened a conference on the placenta as a "neglected experimental animal." He brought together scientists interested in "biochemistry, cell replication, cancer, immunology and ageing" with the explicit goal of getting them to collaborate on making the placenta into a common object of study.[57] (As it turns out, the placenta did not become the common object that Beaconsfield hoped for—hence its obscurity as an experimental "animal.") Beaconsfield's use of the term *experimental animal* for the placenta is certainly idiosyncratic, but it suggests his desire to make it something like a model organism. Yet here, too, the existing analytical term *model organism* seems not quite right, at least as historians of science have used it, because in most cases these organisms have his-

[55] Pickstone, *Ways of Knowing: A New History of Science, Technology and Medicine* (Chicago, 2000).

[56] Liu, "The Shoggoth of Science: Protoplasm and the Biological Disciplines, 1880–1940" (master's paper, Univ. of Wisconsin–Madison, April 2011), 4–5; Star and Griesemer, "Institutional Ecology, 'Translations' and Boundary Objects: Amateurs and Professionals in Berkeley's Museum of Vertebrate Zoology, 1907–39," *Soc. Stud. Sci.* 19 (1989): 387–420. I thank Dan for permission to quote from and discuss his unpublished paper.

[57] Beaconsfield, "Editor's Note," in *Placenta: A Neglected Experimental Animal*, ed. Beaconsfield and Claude Villee (Oxford, 1979), 3. This episode has been examined by Jennifer Kaiser in "'A Neglected Experimental Animal': The Placenta as Model Organism" (seminar paper, History of Science 909, Univ. of Wisconsin–Madison, spring 2011); I thank Jen for allowing me to use this example.

torically been picked as models for studying one particular system or attribute, rather than as a device for consciously uniting disparate perspectives.[58]

Both protoplasm and the placenta, like nineteenth-century Meereskunde, suggest how the Kunde aim of gaining multifaceted knowledge of one object might morph into the goal of interdisciplinary collaboration (or be hoped to). And it is this aspect of the Kunde projects that has perhaps the greatest resonance with more recent intellectual trends. Despite the many differences between nineteenth-century German universities and American universities since World War II (the greatest of which is the rampant academicization and credentialization of seemingly all knowledge), there is a family resemblance between the Kunde projects and certain modern modes of interdisciplinarity captured in the term *studies*. Area studies, American studies, African American studies, women's studies (now more often "gender and women's studies"), lesbian, gay, bi-, transsexual, and queer (LGBTQ) studies, environmental studies, and science and technology studies (STS) each have their own trajectories in the American academic context. Of course there are major differences among these studies projects—area studies, for example, benefited from considerable federal funding during the Cold War era, which then bled away in the 1990s;[59] American studies was more obviously grounded in the desire to cross the lines of just two disciplines, English and history, rather than many;[60] women's studies, LGBTQ studies, African American studies, and other ethnic studies programs have identity politics at their centers in ways that others in the broader list do not; and STS and environmental studies tend to cross disciplinary lines among the humanities, social sciences, and natural sciences more and in different ways than do the other fields.

Nevertheless, certain parallels between the nineteenth-century Kunde projects and twentieth-century "studies" are striking. The first and most important is the value they place on multifaceted knowledge of one thing: what studies do is examine their object from many directions. In our present parlance, they are multi- or interdisciplinary, training their different perspectives on one (if perhaps vaguely bounded) object of attention, whether it is women, or the Middle East, or the environment. What they have in common is a commitment to understanding that object in all its many facets.

[58] The literature on the history of model organisms has mushroomed in recent years. For some key works, see Adele E. Clarke and Joan H. Fujimura, eds., *The Right Tools for the Job: At Work in Twentieth-Century Life Sciences* (Princeton, N.J., 1992); Bonnie Tocher Clause, "The Wistar Rat as a Right Choice: Establishing Mammalian Standards and the Ideal of a Standardized Mammal," *J. Hist. Biol.* 26 (1993): 329–49; Doris T. Zallen, "The 'Light' Organism for the Job: Green Algae and Photosynthesis Research," *J. Hist. Biol.* 26 (1993): 269–79; Robert E. Kohler, *Lords of the Fly: Drosophila Genetics and the Experimental Life* (Chicago, 1994); Angela N. H. Creager, *The Life of a Virus: Tobacco Mosaic Virus as an Experimental Model, 1930–1965* (Chicago, 2002). A charmingly accessible historical survey of model organisms is Jim Endersby, *A Guinea Pig's History of Biology* (Cambridge, Mass., 2007).

[59] See Mitchell L. Stevens and Cynthia Miller-Idriss, "Academic Internationalism: U.S. Universities in Transition; A Report on Consultations Convened at the Social Science Research Council, New York City, 14 & 17 October 2008" (unpublished manuscript, March 2009), posted on the Social Science Research Council website at http://www.ssrc.org/workspace/images/crm/new_publication _3/%7Bc22d385c-d25a-de11-bd80-001cc477ec70%7D.pdf, and the related "Thematic Guide to SSRC Bibliography on Area Studies, October 2008" at http://www.ssrc.org/publications/view /E21D2895-D45A-DE11-BD80-001CC477EC70/ (accessed 11 January 2012); see esp. the section "Academic Disciplines and Area Studies," 3–5.

[60] See Jerry A. Jacobs, *Interdisciplinarity, Specialization and a Defense of Liberal Arts Disciplines* (Chicago, forthcoming), chap. 6, "American Studies: Interdisciplinarity over Half a Century." My thanks to Professor Jacobs for allowing me to see and cite his book manuscript.

At the risk of overrepetition, this is not nomothetic knowledge—not oriented toward general laws instantiated by particular objects. Hence the anxiety level expressed in controversies over area studies in political science, which seems to be a somewhat extreme case of the status anxiety that is fairly high in the social sciences. Area studies are widely associated with a perception that a multidimensional focus on a particular place and its various cultural, geographical, and other contexts is "soft," in contrast to the "hard" (quantitative or law-bound, or wissenschaftlich) social sciences.[61] While it might be easy for humanists (and especially historians of science) to look down on a scientism that seems so obviously misguided, for my purposes here, the more interesting point is that area studies represent a version of Kunde that has been remarkably durable, uniting a form and object of knowledge that has been found worthy since the early nineteenth-century classicists and Orientalists.

A further parallel lies in the openness of both nineteenth-century Kunde and contemporary studies to people who are not academically credentialed experts. This takes two forms. One is the strong connection with the world outside academia. Area studies have always had a significant connection to national and international politics—as both classics and Oriental studies did in the nineteenth century—and both the cluster of identity-politics studies and environmental studies have typically had an activist dimension. The second form of openness to uncredentialed participants is a function of modern interdisciplinarity itself: we might call it the value of ignorance.

Or rather, the value of other people's knowledge. Studies programs embody the faith that it is worth knowing the views of people educated in different disciplines and consequently holding different perspectives on the object of study. Attached to that relationship is also an acknowledgment of one's relative ignorance of the other approaches. That is, a full understanding of women (to take one example) requires the collaboration of people educated in literary, philosophical, historical, biological, sociological, political, and anthropological approaches, none of whom will be expert in all of these fields. Like the physical oceanographer contributing to Meereskunde or the Heimatkundler expert on local mushrooms, the participant in modern interdisciplinary studies acts in many ways as an interested amateur—credentialed in her own field, to be sure, but not in the other ones needed for a full understanding of her object. This amateur quality is also what keeps such interdisciplinary studies both "interdisciplinary" and "studies." Once everyone shares a common canon and methods, a field is no longer interdisciplinary: it has become disciplinary (whether or not it is enshrined as a department or a discipline in universities, and whether or not it would be recognized as Wissenschaft in some nineteenth-century sense). This, then, is another mode of academic discipline development—out of previously interdisciplinary "studies" fields, as embodied by such fields as women's studies—and the history of science.

WISSENSCHAFT, KUNDE, AND THE HISTORY OF SCIENCE

What, then, are the implications of these findings for our field? How might this attention to the general and the special, and to Kunde as a foil to Wissenschaft, be made

---

[61] See Loren Graham and Jean-Michel Kantor, "'Soft' Area Studies versus 'Hard' Social Science: A False Opposition," *Slav. Rev.* 66 (2007): 1–19; Stevens and Miller-Idriss, "Academic Internationalism" (cit. n. 59), esp. 7.

more broadly useful to the history of science? By way of discussion and conclusion, let me suggest two different ways to think about this.

One is empirical and analytical: we can look for new cases. How have claims about the general, the special, and their relationship been involved in the development of other sciences, and of fields beyond the natural sciences? When and where have these categories gained significance; when and where not? Do the clustered attributes of Kunde-type knowledge—multifaceted, often interdisciplinary perspectives on a single object of study, openness to nonacademics, connections to material objects and real-world problems—appear in other times and places, in other relations to academic culture? What sorts of work do these attributes do, and what niche do Kunde projects hold in the broader ecology of knowledge production in modern societies? How do their stories differ across the natural sciences, social sciences, and humanities? There is a wide field here for investigation, and potential for reinterpreting older findings.

The other way is more reflexive. How does the history of science itself look in light of the general and the special, of Wissenschaft and Kunde? First: like zoologists in the early twentieth century, we are always weaving together the special and the general, gaining knowledge of our particular places and subjects while seeking to integrate them into broader questions that might have salience to others' particulars. If this seems completely normal to us—not even worth remarking upon—it is because we are at a point of discipline development and specialization where we have completely assimilated this way of making our special topics gain general meaning.

Second: although in this particular article I call for greater attention to the problem of knowledge and changing conceptual structures in science, this itself is, of course, just part of understanding the development of science. Historians of science rightly wish for a multidimensional, Kunde-like understanding of science. This gives us license—and obliges us—to consider all manner of issues: the funding patterns and sources, relations to technology, instrumental and natural philosophical goals, institutional forms and structures, geographical differences and commonalities, cultural assumptions and confrontations, politics, and epistemic and social values that together form modern science. Illuminating the dynamics of scientific change from these many perspectives is our raison d'être, and we need to stay open to as many as we are able to. Our task, then, must be a collective one, of pooling our knowledge of the different aspects of our own large object of study, science, to try to gain a full understanding of it.

Finally, given the parallels I have drawn between Kunde and post–World War II interdisciplinary studies, it is worth asking how our field fits into the current turbulent state of academic knowledge. As anyone looking around a university today knows, interdisciplinarity is now ubiquitous. Analysts of the topic point out that many sciences since World War II have been organized around "problem-oriented approaches" that draw on multiple disciplines; many of the problems are applied ones, whether the applications are in engineering, commerce, or policy. Government funders have long encouraged this approach and continue to do so. More recently, interdisciplinarity has honeycombed the humanities, yielding up new centers, programs, and workshops left and right.[62]

---

[62] For general overviews of interdisciplinarity that include problem-oriented research as a major strand, see Julie Thompson Klein, "A Conceptual Vocabulary of Interdisciplinary Science," in

Have we completely lost our discipline(s)? Are universities becoming nothing but a collection of pods that work on problems, shifting with the demands of commerce and the whims of intellectual fashion? Despite both apocalyptic and utopian declarations of the "end of disciplinarity,"[63] history makes me skeptical. The parallel to the transformation of the life sciences around 1900 is too striking. In both cases, there is a shift from "traditional" categories of knowledge (taxon-based ca. 1900, "discipline-based" ca. 1975) to a problem-based order of knowledge that draws from many traditional categories at once. In both cases, while the problem-based order is innovative and exciting, it is also discomfortingly disunified.

One might then reasonably view the current tumult not as the overthrowing of the modern discipline-based order of knowledge, but rather as a particularly intense episode of a pattern often repeated since the early nineteenth-century establishment of the modern disciplines. In this pattern, older institutionalized categories gradually become intellectually hollowed out as their original problem complexes are no longer seen as significant ones, while new categories emerge that do not fit neatly into the old institutional units but rather push and link in new directions. If some of these become big enough to stabilize around common tools, methods, and canons, produce students for whom there is a market demand, and successfully create a historical trajectory that places them in a lineage of continuity and change leading to themselves, then they might be called disciplines. That is what happened to molecular biology, women's studies, and the history of science. Meanwhile, other departmentalized disciplines, even venerable ones like zoology, anatomy, and physiology, are gradually fading away.[64] From this point of view, we are witnessing a process that has long taken place. It is only the lack of a longer-term historical perspective that has prevented us from seeing the continuity.

I am not counseling complacency here. Problem-oriented research usually prioritizes social and commercial problems, and its penetration into the universities seriously threatens both the humanities and basic research in all fields. We need to defend

---

Weingart and Stehr, *Practising Interdisciplinarity* (cit. n. 48), 3–24, and Jacobs, *Interdisciplinarity* (cit. n. 60). A broad and much-cited picture of problem-oriented research that makes claims for its "transdisciplinary" and commercially influenced character, which its authors call "Mode 2" knowledge production, is in Michael Gibbons et al., *The New Production of Knowledge: The Dynamics of Science and Research in Contemporary Societies* (Stockholm, 1994). For one of many examples drawing from a particular discipline (political economy), see Ngai-Ling Sum and Bob Jessop, "On Pre- and Post-disciplinarity in (Cultural) Political Economy," 2003, available on the Lancaster University website at http://eprints.lancs.ac.uk/229/1/E-2003c_sum-jessop2003.pdf (accessed 8 January 2011), 6, 23. On recent US government pushes in the direction of problem-oriented research as a way to synthesize fragmented research communities while addressing major social problems, see National Research Council, Committee on a New Biology for the 21st Century, *A New Biology for the 21st Century: Ensuring the United States Leads the Coming Biology Revolution* (Washington, D.C., 2009), available on the National Academies Press website at http://www.nap.edu/catalog/12764.html (accessed 11 January 2012).

[63] For an early expression of the utopian view, see the editorial by F. A. Long, "Interdisciplinary Problem-Oriented Research in the University," *Science* 171, no. 3975 (1971): 961; see also the sources cited in n. 62 above. For a well-worked-out apocalyptic argument that nevertheless underestimates the amount of interdisciplinarity that existed, especially in the sciences, before the 1980s, see Paul Forman's article in this volume, together with Forman, "The Primacy of Science in Modernity, of Technology in Postmodernity, and of Ideology in the History of Technology," *Hist. & Tech.* 23 (2007): 1–152.

[64] Susan C. Lawrence, "Anatomy, Histology, and Cytology," in Bowler and Pickstone, *Modern Biological and Earth Sciences* (cit. n. 43), 265–84, esp. 284; Richard L. Kremer, "Physiology," ibid., 342–66; Di Gregorio, "Zoology" (cit. n. 47). At the University of Wisconsin–Madison, the medical school's departments of physiology, anatomy, and pharmacology were officially reorganized in February 2011 into new departments of "neuroscience" and "cell and regenerative biology."

the value of both of these. But we do not need to tie that defense to disciplinarity. Another way, as the scientists long ago figured out, is to show how our research—our basic, humanistic research (not to mention our teaching!)—actually contributes to solving problems beyond our own academic interests. What should be the balance between democracy and expertise in solving social problems that involve science? How should our society respond to looming environmental disasters that require scientific expertise to fully understand? Historical analysis can help us understand the very parameters of these questions and how the choices apparent to us now have been constrained by past events and choices. How does the flow of money and patronage shape the directions and content of scientific research? What should be the role of undirected exploration of new knowledge areas in modern society, and how has basic research actually contributed to solving social problems? Historical perspectives on these questions can illuminate how these factors have helped to form the very kinds of knowledge we call "science," and on multiple timescales: the answers may look different when viewed over the last quarter century, the last century, the modern era, or the period since the medieval origin of the universities. These are hardly new questions, but they remain compelling and relevant, and historians of science are better equipped to answer them than anyone else.

Indeed, the history of science is well positioned to participate in the world of problem-oriented research. We are an interdisciplinary discipline. Connected in three directions—history, science, and science studies—we are deft with multiple ways of knowing. What we most need to do is to make sure that we have a seat at the table, so that our historical ways of knowing, and the empirical things we know about, become part of the general conception of what the big problems of our time are and how they might be tackled.

Booms, Busts, and the World of Ideas:

Enrollment Pressures and the Challenge of Specialization

*by David Kaiser**

ABSTRACT

Historians of recent science face a daunting challenge of scale. Local case studies—our principal means of interrogating past scientific practices—fail to capture
the sweep and texture of some of the most dramatic changes in scientific life since
World War II. During that period, most facets of research grew exponentially, from
numbers of practitioners to numbers of research articles published in any given
specialty. The explosive growth placed unprecedented pressure on the intellectual
coherence of various disciplines. Drawing on examples from the postwar physics
profession in the United States, I suggest that simple quantitative methods can aid
in elucidating patterns that cut across isolated case studies, suggesting themes and
questions that can guide close, archival research.

CYCLES, PATTERNS, AND THE CHALLENGE OF SCALE

For some time, historians of science have recognized a mismatch between many of
our most prized methodological approaches and whole classes of phenomena that
demand scrutiny. The challenge seems especially acute for sciences of the past sixty
years. There is a problem of scale. Close-focus case studies, deep archival excavations, microhistories, and comparable investigations inspired by the sociological and
ethnographic experiments of the 1970s and 1980s have enlivened our understanding
of the cultures and practices of science enormously. Never again should historians
assume that knowledge claims or laboratory techniques traveled effortlessly from
off-scale mind to off-scale mind through the ages; our eyes are rightly focused on the
local and contingent. Yet these tools of inquiry seem to be no match for the brute fact
of exponential growth—the extraordinary expansion of people, places, and papers
that has marked the scientific life at least since World War II. To date, no deeply satis-

* Program in Science, Technology, and Society and Department of Physics, Massachusetts Institute
of Technology, 77 Massachusetts Avenue, Cambridge, MA 02139; dikaiser@mit.edu.
My thanks to Robert Kohler, Kathryn Olesko, Bruno Strasser, and two anonymous referees for helpful suggestions on an earlier draft, and to Alex Wellerstein for preparing the figures. The following
abbreviations are used for archival sources: AIP-EH—Elmer Hutchisson papers in NBL; AIP-EMD—
Education and Manpower Division records in NBL; AIP-HAB—Henry A. Barton papers in NBL;
CIT—California Institute of Technology archives, Pasadena; LIS—Leonard I. Schiff papers, Stan-
ford University Archives, Palo Alto, Calif.; NBL—Niels Bohr Library, American Institute of Physics,
College Park, Md.; PR-AR—*Physical Review* annual reports, in the editorial office of the American
Physical Society, Ridge, N.Y.; UCB—Bancroft Library, University of California, Berkeley.

fying synthesis has emerged from averaging over dozens of small-scale studies; no clear synthesis has coagulated from the clutter of individual case units.[1]

The challenge of scale is hardly new. Decades ago, Lucien Febvre, Fernand Braudel, and their colleagues in the *Annales* school sought some means to capture the *longue durée*. Within the history of science, scholars like Derek Price pursued their own methods to fathom large-scale phenomena and long-term patterns. Unlike in their day, vast digital databases are now available, huge storehouses of information on dissertations filed, articles published, and grants received. The databases are no panacea—number crunching will never replace the careful sifting of meanings from subtle and dense sources. But let us not be afraid to count. Quantitative tools can complement historical analysis and help to direct it, opening up new questions, suggesting new patterns, and helping to spot those topics, people, or places on which close-focus scrutiny might yield the richest rewards. With their power to telescope between scales of activity, from the lab bench to the nation-state, these tools offer one way to try to piece together a new amalgamated account of scientific change.

Back in the 1960s, Price found his favorite graph: the logistic (or S-shaped) curve. Whether measured in terms of numbers of authors or numbers of publications, many scientific fields began with a burst of exponential growth followed by saturation and an eventual steady state. The same pattern held for many other features of modern science, ranging from the number of known chemical elements at a given time to the energies achieved by particle accelerators. Price got a lot of mileage from his simple curve; he came to see it everywhere. The ubiquity of these logistic curves, and their repeated appearance from the age of Galileo and Newton to the present day, suggested to Price that there might exist some universal, underlying structures of science. Though the methods were distinct, Price's quest for universality was shared by other eager modelers of science from that time, including Thomas Kuhn.[2]

Price's model was focused mainly on stasis and equilibrium—long-term periods of relative stability—not unlike Kuhn's picture of normal science punctuated by the occasional revolution. Perhaps inevitably, given more recent world events, I have become interested in a rather different defining shape. When we start to count things from the postwar era, a stark pattern quickly emerges: the speculative bubble. The shape itself has become familiar to many of us. Choose your (least) favorite example: stock prices or real estate values in recent times, the fabled tulip craze that gripped Amsterdam in the 1630s, or the South Sea Bubble of 1720 that nearly bankrupted Isaac Newton (among many others).[3] Such bubbles share a common form. They are bid up by earnestness combined with runaway hype and hope and amplified by various feedback mechanisms until the whole house of cards comes tumbling down. Like Price's logistic curve, bubbles begin with an exponential climb. Unlike in Price's pet

---

[1] See, e.g., articles by Robert Kohler and others in the "Focus" section "The Generalist Vision in the History of Science," *Isis* 96 (2005): 224–51, and references therein. See also Lorraine Daston and Peter Galison, *Objectivity* (New York, 2007), 35–6, 47–50.

[2] Price, *Little Science, Big Science* (New York, 1963), chap. 1. See also Price, *Science since Babylon* (New Haven, Conn., 1961); Kuhn, *The Structure of Scientific Revolutions* (Chicago, 1962).

[3] On the early modern bubbles, see esp. Simon Schama, *The Embarrassment of Riches: An Interpretation of Dutch Culture in the Golden Age* (Berkeley and Los Angeles, 1988), chap. 5; Anne Goldgar, *Tulipmania: Money, Honor, and Knowledge in the Dutch Golden Age* (Chicago, 2007); Richard S. Westfall, *Never at Rest: A Biography of Isaac Newton* (New York, 1980), 861–2; Thomas Levenson, *Newton and the Counterfeiter: The Unknown Detective Career of the World's Greatest Scientist* (Boston, 2009), 244–5.

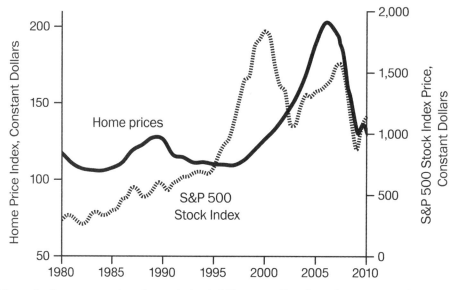

Figure 1. *Recent examples of speculative bubbles, revealing their characteristic shape.*
The black line shows the inflation-adjusted US home price index, and the dashed gray line
shows the inflation-adjusted Standard & Poor's 500 stock index. Based on data compiled by
the Yale economist Robert Shiller, available at http://www.econ.yale.edu/~shiller/data.htm
(accessed 2 January 2011).

graph, however, no sustainable stasis is achieved; the fall is as sharp as the rise (see fig. 1).

We all are painfully aware of speculative bubbles in the financial world these days. Though slower moving, similar processes have characterized academic life as well, especially (though not exclusively) in the United States since World War II. The pattern is particularly clear in the case of student enrollments. The classrooms of American colleges and universities bulged like never before following World War II. Several major changes, including the GI Bill, brought over two million veterans into the nation's institutions of higher education. Enrollments in nearly every field rose exponentially. Just as quickly, enrollments across nearly all disciplines in the United States faltered in the early 1970s, amid the earliest stirrings of stagflation, détente, and massive cuts in education and defense spending. As we will see in the following section, one case illustrated the general pattern in starkest form: graduate-level enrollments in physics. Physicists encountered the vast shifts first and most acutely— they experienced the extremes of what quickly became the norm. Their enrollments served as a bellwether in good times and bad. Rising fastest and falling sharpest, physicists' enrollment trends heralded systemic transitions throughout American intellectual life (fig. 2).

Simple time-series graphs like this one elicit several follow-up questions. Who wanted all these physics students, and why? As I discuss in the third section, the suggestive similarity in shape between figures 1 and 2 is no mere coincidence. The dynamics behind the physicists' enrollment curve bore all the classic features of a speculative bubble. The dramatic oscillations in student numbers, meanwhile, point beyond questions of policy and recruitment. What effects, if any, did these sharp

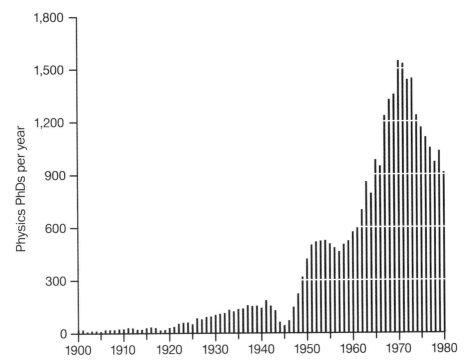

Figure 2. *Number of physics PhDs granted by US institutions per year, 1900–80. Based on data available in National Research Council,* A Century of Doctorates: Data Analysis of Growth and Change *(Washington, D.C., 1978), 12; National Science Foundation, Division of Science Resources Statistics,* Science and Engineering Degrees, 1966–2001, *report no. NSF 04–311 (Arlington, Va., 2004).*

swings have on physics itself—on the intellectual landscape of the field? The fourth section of this article illustrates how the sudden pressure of numbers after the war threatened physicists' own vision of their field. Abstract concerns over specialization collided with the practical requirements of processing and publishing the latest research findings; unprecedented pedagogical pressures reshaped physicists' long-standing habits of communicating results and organizing knowledge. Finally, in the closing section, I note that the Cold War bubble of figure 2 was no isolated event. Rather, boom-and-bust cycles became a repeating phenomenon. The fact that most academic fields had enrollment curves similar to figure 2 suggests looking for comparable epistemic effects in other disciplines as well.

Unlike Price, with his high hopes for his logistic curve, I do not believe that speculative bubbles have been lurking always and everywhere in the conduct of science. They are no silver bullet, unlocking hidden patterns in all places and eras. Their partial character is part of their appeal. I am fascinated precisely by the modest range of scales they seem to capture: patterns and cycles that unfolded over a few decades, rather than centuries; phenomena that seemed to affect many fields of inquiry but not all of them, in similar but not identical ways. On this point I draw particular inspiration from literary scholar Franco Moretti. As Moretti has emphasized, "cycles constitute *temporary structures within the historical flow.*" He continues: "The short span is all flow and no structure, the *longue durée* all structure and no flow, and cycles are the—unstable—border country between them. Structures, because they intro-duce repetition in history, and hence regularity, order, pattern; and temporary, be-cause they're short (ten, twenty, fifty years . . .)."[4] Much as Moretti has argued, one-dimensional plots like figure 2 will never be the last word in our efforts to understand the past. But they can be a productive starting point, prompting new questions as we aim to make sense of broad patterns in the recent history of intellectual life.

THE POSTWAR POPULATION EXPLOSION

A well-known historiographical arc has traced the transition from the gentleman-amateurs of natural philosophy to professional scientists, with a convenient-enough pivot marked by William Whewell's invention of the term *scientist* in 1840. To this genealogy we may add a third phase: the mass-produced scientist. The new creature was at least as different from the nineteenth-century "man of science" as that ideal-ized figure had been from the early modern philosopher-courtier. They differed in self-perception as well as means of production. Robert Kohler once likened graduate training in the United States during the late nineteenth century to a "PhD machine." Yet as Frederick Taylor and Henry Ford knew so well, not all machines operate on the same scale or with comparable efficiencies. During the 1950s, the mechanisms of graduate training evolved from cottage industry to factory-scaled production.[5]

The United States underwent a massive experiment in social engineering during the decades after World War II, in what might be called the credentialing of America. Higher education was booming. The proportion of twenty- to twenty-four-year-olds who received a bachelor's degree doubled between the early 1950s and the early 1970s, while the proportion of twenty-five- to twenty-nine-year-olds who received a PhD quadrupled. The increases were hardly distributed evenly across fields. Between 1950 and 1963, for example, the nation's population increased by 25 percent (from around 152 million to 190 million); its total labor force grew by just shy of 17 percent (from 65 million to 76 million workers); while its pool of PhD-trained scientists and engineers grew by a whopping 136 percent (from 45,000 to 106,000). That is, the PhD-level scientific and technical workforce grew eight times more quickly than the total labor force.[6]

[4] Moretti, *Graphs, Maps, Trees: Abstract Models for a Literary History* (New York, 2005), 14; em-phasis in the original. My thanks to Michael Gordin for bringing Moretti's book to my attention.

[5] See esp. Kohler, "The Ph.D. Machine: Building on the Collegiate Base," *Isis* 81 (1990): 638–62; Steven Shapin, "The Image of the Man of Science," in *The Cambridge History of Science,* vol. 4, *Eighteenth-Century Science,* ed. Roy Porter (New York, 2003), 159–83; Shapin, *The Scientific Life: A Moral History of a Late Modern Vocation* (Chicago, 2008).

[6] On credentialing, see also Frank Newman, "The Era of Expertise: The Growth, the Spread, and Ultimately the Decline of the National Commitment to the Concept of the Highly Trained Expert, 1945 to 1970" (PhD diss., Stanford Univ., 1981). The statistics on degrees per age cohort were cal-culated from data in the following sources: on US bachelor's degrees, Douglas L. Adkins, *The Great American Degree Machine: An Economic Analysis of the Human Resource Output of Higher Edu-cation* (Berkeley, Calif., 1975), 190–4, with additional data supplied by Roman Czujko, director of the American Institute of Physics (AIP) Statistical Research Center, personal communication to the author, 19 January 2005; on US PhDs, National Research Council, *Century of Doctorates* (cit. fig. 2 caption), 7; on US population cohorts, Population Division of the Department of Economic and Social Affairs of the United Nations Secretariat, *World Population Prospects: The 2004 Revision*, avail-able at http://www.un.org/esa/population/publications/WPP2004/WPP2004_Volume3.htm (accessed 30 January 2012), and *World Urbanization Prospects: The 2003 Revision*, available at http://www .un.org/esa/population/publications/wup2003/WUP2003Report.pdf (accessed 30 January 2012). On changes in the US labor force between 1950 and 1963, see *Impact of Federal Research and Develop-*

Within this constellation, physics grew fastest of all. According to data collected by the National Science Foundation's National Register of Scientific and Technical Personnel—a register created during the early 1950s to facilitate the federal government's mobilization of scientists in times of war—between 1954 and 1970 the number of professional physicists employed in the United States grew substantially faster than the numbers of all other scientific professionals: 210 percent faster than earth scientists, 34 percent faster than chemists, 22 percent faster than mathematicians, and so on.[7]

Physics had not always led the pack. Averaged over the period 1890–1941, the annual number of physics PhDs granted in the United States doubled every thirteen years—slower than chemistry and mathematics; slower, too, than history, English, and foreign languages. During the Depression years, the growth rate for physics slowed even further. On the eve of World War II, it would have taken eighteen years to double the annual output of physics PhDs in the United States, based on the pattern set during 1930–9. The situation changed immediately after the war. Between 1945 and 1951, the annual output of physics PhDs from US institutions doubled every 1.7 years—a tenfold increase in rate. No other field came close: physics grew nearly twice as quickly as chemistry, for example, and fully 12 percent faster than its nearest competitors, engineering and psychology. By the mid-1950s, American institutions required just two years to graduate as many physics PhD recipients as the entire country had produced between 1861 and 1929.[8]

The rapid rise in numbers of young physicists sprang from much more than simple demographics. From the bumps and wiggles in the running tally of new physicists one may read off the changing political economy of the postwar years: the flood of returning veterans from World War II, the impact of the Korean War, the hardening of the Cold War and the Sputnik surprise, and the dramatic reversal of national priorities years into the slog of Vietnam. At best, the baby boom—which first hit American colleges and universities in 1964, driving first-year undergraduate enrollments up by 37 percent from the previous year—played a supporting role in setting the pace of change (figs. 3 and 4).[9]

Developments in other countries help to put the American situation in context. Consider the United Kingdom and Canada, wartime allies of the United States and partners in the Manhattan Project to design and build nuclear weapons. By the postwar years, Britain and Canada also shared a system of higher education and advanced

---

ment Policies on Scientific and Technical Manpower: Hearings Before the Subcomm. on Employment and Manpower of the S. Comm. on Labor and Public Welfare, 89th Cong. 142 (1965) (testimony of Bowen C. Dees, associate dean of planning, National Science Foundation).

[7] Based on data in the series of National Science Foundation reports titled *American Science Manpower* (Washington, D.C., 1959–71). On the National Register, see the form letter from Henry A. Barton (director, AIP), 16 November 1950, a copy of which is in LIS, box 1, folder "Amer. Inst. of Physics (AIP), General"; "Scientific Manpower Studies," in *The Third Annual Report of the National Science Foundation* (1953), 26–33, available at http://www.nsf.gov/pubs/1953/annualreports/start .htm (accessed 18 January 2012).

[8] David Kaiser, "Cold War Requisitions, Scientific Manpower, and the Production of American Physicists after World War II," *Historical Studies in the Physical and Biological Sciences* 33 (2002): 131–59, on 134–6, 157–9. See also Spencer Weart, "The Physics Business in America, 1919–1940: A Statistical Reconnaissance," in *The Sciences in American Context: New Perspectives*, ed. Nathan Reingold (Washington, D.C., 1979), 295–358.

[9] On the baby boom, see Todd Gitlin, *The Sixties: Years of Hope, Days of Rage*, rev. ed. (New York, 1993), 11–21, 164.

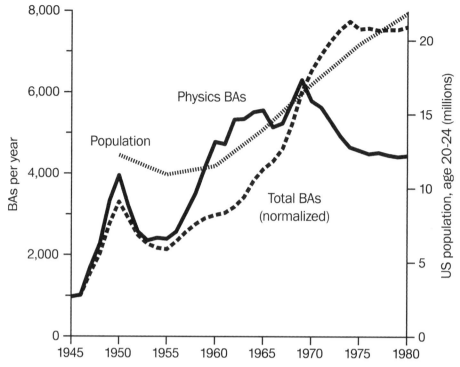

Figure 3. *Number of physics bachelor's degrees granted by US institutions per year, 1945–80. Also shown are the total number of US bachelor's degrees granted in all fields per year, normalized to the number of physics degrees in 1945 (to show relative rates of change), and the portion of the US population between the ages of twenty and twenty-four. To convert the normalized BAs to total BAs, multiply by 122. The data come from Adkins,* Great American Degree Machine, *190–4; Czujko, personal communication; and United Nations Secretariat,* World Population Prospects *(All cit. n. 6).*

degrees broadly similar to that of the United States, with the PhD degree increasingly seen as a necessary prerequisite for professional scientific employment.[10] As in the United States, those countries likewise saw very rapid growth in the numbers of new physics PhDs granted immediately after the war; there, too, the pace set by physics exceeded that of most other fields. Yet unlike in the United States, physics quickly settled into the pattern set by higher education more generally. In Britain, for example, the number of physics PhDs granted each year doubled at precisely the same rate as for all fields combined during the decade after Sputnik; the Soviet satellite elicited no new burst in physicist training. In Canada, meanwhile, the doubling rate for physics PhDs slipped to nearly 20 percent slower than the rate for all fields combined during the same time interval. In other nations with similar Anglo-American education systems, such as Australia, the growth in advanced physics training re-

[10] Burton R. Clark, *Places of Inquiry: Research and Advanced Education in Modern Universities* (Berkeley and Los Angeles, 1995); Yves Gingras, *Physics and the Rise of Scientific Research in Canada*, trans. Peter Keating (Montreal, 1991). See also Clark, ed., *The Research Foundations of Graduate Education: Germany, Britain, France, United States, Japan* (Berkeley and Los Angeles, 1993), pt. 2; Jean Babcock et al., "American Influence on British Higher Education: Science, Technology, and the Problem of University Expansion, 1945–1963," *Minerva* 41 (2003): 327–46.

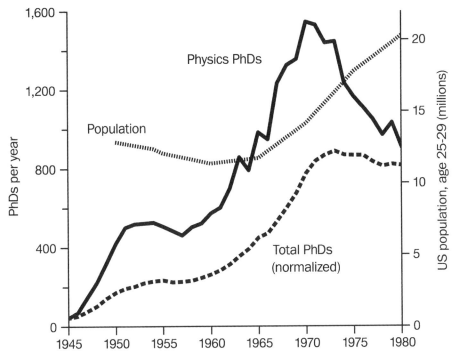

Figure 4. *Number of physics PhDs granted by US institutions per year, 1945–80. Also shown are the total number of US PhDs granted per year in all fields, normalized to the number of physics degrees in 1945 (to show relative rates of change), and the portion of the US population between the ages of twenty-five and twenty-nine. To convert the normalized PhDs to total PhDs, multiply by 38. The data come from National Research Council,* Century of Doctorates, *7, 12; National Science Foundation,* Science and Engineering Degrees *(Both cit. fig. 2 caption); United Nations Secretariat,* World Population Prospects *(cit. n. 6). See also the AIP graduate student surveys, 1961–75, available in AIP-EMD; additional data supplied by Roman Czujko, personal communication to the author, 11 April 2002.*

mained a largely demographic effect, mirroring overall population growth—even after Australia inaugurated its own nuclear power program in the mid-1950s.[11]

The country with the pattern closest to that of the United States proved to be the Soviet Union. In both countries, physics grew faster than any other field. In the Soviet Union, new physicists entered the labor force at an average annual growth rate of 10.7 percent between 1951 and 1974: more than one-and-a-half times faster than the pace for all scientists combined, and fully 15 percent faster than the growth rate for engineers. In fact, the stocks of professional physicists and mathematicians in both Cold War superpowers grew at nearly the same steep pace for a quarter century, far exceeding rates of overall population growth (fig. 5).

All told, between 1945 and 1975, 124,000 individuals completed undergraduate degrees in physics in the United States, while 24,000 completed PhDs in the subject.

---

[11] Based on data in Roger Bilboul, ed., *Retrospective Index to Theses of Great Britain and Ireland, 1716–1950* (Santa Barbara, Calif., 1975); Geoffrey M. Paterson and Joan E. Hardy, *Index to Theses Accepted for Higher Degrees by the Universities of Great Britain and Ireland and the Council for National Academic Awards* (London, 1951–); Gingras, *Physics and the Rise* (cit. n. 10), appendixes; *The Union List of Higher Degree Theses in Australian University Libraries* (Hobart, 1961–91).

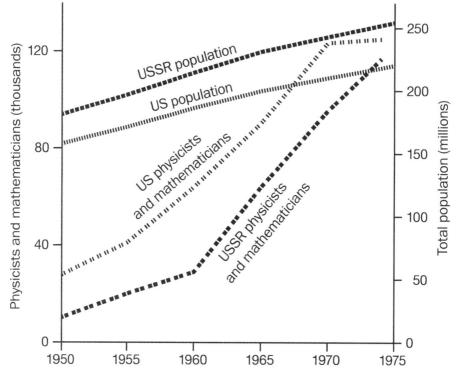

Figure 5. *Total number of physicists and mathematicians in the United States and Soviet Union, 1950–74. (The Soviets did not distinguish between physicists and mathematicians for such accounting; all were graduates of the physico-mathematical faculties in the Soviet universities. Hence, both groups have been included in the American tally as well.) Also shown are the total populations of both countries. The data on physicists and mathematicians are from Catherine P. Ailes and Francis W. Rushing,* The Science Race: Training and Utilization of Scientists and Engineers, US and USSR *(New York, 1982), 107; the population data are from United Nations Secretariat,* World Population Prospects *and* World Urbanization Prospects *(cit. n. 6). Ailes and Rushing used US Bureau of Labor Statistics data for the US case.*

More and more of the nation's universities retooled, adding graduate education in physics to their roster of offerings. In 1950, 52 institutions in the United States conferred PhDs in physics. In 1960 the number had risen to 78; by 1970, it was 148.[12] Ultimately, such runaway growth could not be sustained; or, as physicists might say, for every action there must be an equal and opposite reaction. And so there was. Physics PhDs in the United States peaked in 1971, then fell precipitously.

A painful conjunction triggered the fall. Internal audits at the Department of Defense began to question whether the postwar policy of funding basic research on university campuses—which had underwritten the education of nearly all physics graduate students since the war—had produced an adequate return on investment. As the Vietnam War raged, campus protesters grew equally dissatisfied with the Penta-

[12] The numbers of US institutions granting physics PhDs were computed from data in the ProQuest Dissertations and Theses Database, accessed via http://www.proquest.com.libproxy.mit.edu/en-US/products/dissertations/ and available from http://www.proquest.com.

gon's presence on campus and often targeted physicists' facilities. To supply troops for the escalation of fighting, meanwhile, military planners began to revoke draft deferments—first for undergraduates in 1967, then for graduate students two years later—reversing a twenty-year policy that had kept science students in their classrooms. Détente with the Soviets and the onset of stagflation in the early 1970s exacerbated the situation, as each induced substantial cuts in federal spending for defense and education.[13]

The fast-moving changes affected nearly every field across the universities, but none more severely than physics. While annual conferrals of PhDs across all fields slid by a modest 8 percent between their peak in the early 1970s and 1980, physics PhDs plummeted by fully 47 percent. Several fields experienced sharp downturns—mathematics went down by 42 percent, history 39 percent, chemistry 31 percent, engineering 30 percent, and political science 20 percent from their early 1970s highs—but physics led the way.[14] Demand for young physicists vanished even more quickly. Whereas more employers than student applicants had registered with the Placement Service of the American Institute of Physics (AIP) throughout the 1950s and into the mid-1960s, by 1968 young physicists looking for jobs outnumbered advertised positions by nearly four to one. Three years later, the gap had widened much further: 1,053 physicist job seekers registered, competing for just 53 jobs.[15]

SPECULATIVE BUBBLES

Economists have developed sophisticated models to try to understand such vacillations in the scientific labor market. The most prominent has been nicknamed the "cobweb" model, based on the pattern sketched out on a graph of wages versus workforce. Increase demand for specialists in a particular field, and wages for those workers should rise as well. The upturn in wages will encourage more students to flock to the discipline, increasing the labor pool available, which will ultimately overshoot demand. The glut in supply will in turn lead to lower wages, discouraging some students in the next cohort from pursuing that line of work, and so on, keeping the pattern oscillating, cobweb-like, around an idealized market equilibrium. Fancier models have replaced present-day wages (at the time a student needs to decide on a course of study) by projected (future) wages, to take into account the long delays incurred during the training process itself, and other bells and whistles have been added to the basic cobweb model to try to improve its accuracy.[16]

[13] See esp. Daniel Kevles, *The Physicists: The History of a Scientific Community in Modern America*, 3rd ed. (Cambridge, 1995), chap. 25; Stuart W. Leslie, *The Cold War and American Science: The Military-Industrial-Academic Complex at M.I.T. and Stanford* (New York, 1993), chap. 9; Roger Geiger, *Research and Relevant Knowledge: American Research Universities since World War II* (New York, 1993), chaps. 8–9; Kelly Moore, *Disrupting Science: Social Movements, American Scientists, and the Politics of the Military, 1945–1975* (Princeton, N.J., 2008), chaps. 5–6.

[14] Calculated from data in National Science Foundation, *Science and Engineering Degrees* (cit. fig. 2 caption).

[15] "Placement Register, Statistical Comparison, APS-AAPT Annual Joint Meeting, 1963 to 1969," n.d., ca. 1970, and "Summary: Placement Service Register, the American Physical Society Meeting, Washington, D.C.," 3 June 1971, both in AIP-EMD, box 13, folder "Placement Literature"; "Supply and Demand," October 1970, in AIP-EMD, box 13, folder "Placement Service Advisory Committee." See also Kaiser, "Cold War Requisitions" (cit. n. 8), 151–2.

[16] See esp. Kenneth J. Arrow and William M. Capron, "Dynamic Shortages and Price Rises: The Engineer-Scientist Case," *Quarterly Journal of Economics* 73 (1959): 292–308; Richard B. Freeman, *The Market for College-Trained Manpower: A Study in the Economics of Career Choice* (Cambridge,

A fine model, it has been applied successfully to many sectors of the labor market. Despite decades of concerted effort, however, neither the model nor its many variants have *ever* produced accurate predictions of the bulk flows of supply and demand for scientists and engineers in the United States. An expert panel of economists, statisticians, and public-policy specialists convened by the National Research Council recently concluded:

> Interest in predicting demand and supply for doctoral scientists and engineers began in the 1950s, and since that time there have been repeated efforts to forecast impending shortages or surpluses. As the importance of science and engineering has increased in relation to the American economy, so has the need for indicators of the adequacy of future demand and supply for scientific and engineering personnel. This need, however, has not been met by data-based forecasting models, and accurate forecasts have not been produced.[17]

Even dressed in the restrained language of a blue-ribbon technical report, this is a striking admission of failure.

The tremendous surges in American physics enrollments suggest a different economic metaphor: a speculative bubble. Economist Robert Shiller defines a speculative bubble as "a situation in which temporarily high prices are sustained largely by investors' enthusiasm rather than by consistent estimation of real value."[18] Shiller emphasizes the roles of hype, amplification, and feedback loops in driving the dynamics of such bubbles. Consumers' enthusiasm for a particular item—be it a hot new tech stock or a hip loft near Central Park—can attract further attention to that item. Increased media attention, in turn, can elicit additional consumer investment, and the rise in demand will drive up prices. The price increase will become a self-fulfilling prophecy, drawing still more fawning from commentators and investment from consumers. "As prices continue to rise, the level of exuberance is enhanced by the price rise itself," as Shiller explains. Shiller likens the process to naturally occurring Ponzi schemes, which can sharply boost prices—if only for a while—even in the absence of outright fraud or deliberate deception. Donald MacKenzie likewise emphasizes the feedback dynamics of performativity: the fact that financial models act back on the very markets they are meant to simulate can increase financial markets' susceptibility to boom-and-bust cycles.[19]

As with stock prices or the housing market, so with graduate training. The Cold War bubble in physics enrollments was fed by earnest decisions based on incomplete or imperfect information, intermixed with hope and hype that had little discernible grounding in fact. Feedback loops between scientists, policy makers, and journalists kept the market for American physicists (and specialists in related fields) artificially inflated. Faulty assumptions that could easily have been checked assumed a seeming

---

1971); Glen G. Cain, Freeman, and W. Lee Hansen, *Labor Market Analysis of Engineers and Technical Workers* (Baltimore, 1973); Larry R. Leslie and Ronald L. Oaxaca, "Scientist and Engineer Supply and Demand," in *Higher Education: Handbook of Theory and Research*, vol. 9, ed. John Smart (New York, 1993), 154–211; National Research Council, *Forecasting Demand and Supply of Doctoral Scientists and Engineers: Report of a Workshop on Methodology* (Washington, D.C., 2000), chap. 2.

[17] National Research Council, *Forecasting Demand and Supply* (cit. n. 16), 1.

[18] Shiller, *Irrational Exuberance*, 2nd ed. (Princeton, N.J., 2005), xvii.

[19] Ibid., chap. 4, quotation on 81; MacKenzie, *An Engine, Not a Camera: How Financial Models Shape Markets* (Cambridge, 2006), chap. 7.

naturalness, hardened by prevailing geopolitical conditions. When those conditions changed abruptly, physics had nowhere to go but down.

Though hardly unique for the time period, my favorite example of the hype-amplification-feedback process concerns a series of reports that were commissioned during the 1950s on Soviet advances in training scientists and engineers. Three major reports were released between 1955 and 1961 to assess the Soviet threat: Nicholas DeWitt's *Soviet Professional Manpower*, Alexander Korol's *Soviet Education for Science and Technology*, and DeWitt's *Education and Professional Employment in the USSR*. Both DeWitt and Korol were Russian expatriates who had relocated to Cambridge, Massachusetts: DeWitt to the new Russian Research Center at Harvard, and Korol to the equally new Center for International Studies at MIT. As we now know, both centers were secretly funded by the CIA.[20]

DeWitt and Korol each urged caution in the interpretation of statistics like annual Soviet degree conferrals—in part because of basic definitional mismatches between types of academic degrees in the Soviet Union and the United States, in part because of serious questions about academic standards at some of the Soviet training centers, and most of all because the Soviet rolls were bloated by correspondence-school students. The latter earned their degrees in science and engineering by sending homework assignments through the mail to overworked instructors, with no benefit of laboratory work or face-to-face instruction. Indeed, the potential for mistaken impressions seemed so serious that Korol refused to tabulate enrollment data from the Soviet and American tallies side by side, in order to avoid "unwarranted implications."[21] DeWitt printed such comparative tables only after emphasizing all the caveats at length, and even affixing a lengthy appendix on what he called the "perplexities and pitfalls" of interpreting Soviet education statistics. Nonetheless, when he counted up annual degrees in the two countries, it appeared that the Soviets were graduating two to three times more students per year in science and engineering than were American institutions.[22]

That ratio—"two to three times"—quickly took on a life of its own. DeWitt's and Korol's reports had been careful, lengthy, serious affairs. The journalistic coverage, on the other hand, leaned toward the sensationalistic. "Russia Is Overtaking U.S. in Training of Technicians," announced a typical front-page headline in the *New York Times*; "Red Technical Graduates Are Double Those in U.S.," echoed the *Washington Post*. Leading spokespeople from the CIA, the Department of Defense, the Joint Congressional Committee on Atomic Energy, and the Atomic Energy Commission routinely trotted out the same stripped-down number ("two to three times") in public speeches and congressional testimony, with no trace of DeWitt's caveats or cautions. Each proclamation elicited further hand-wringing in the newspapers—all before the

---

[20] DeWitt, *Soviet Professional Manpower: Its Education, Training, and Supply* (Washington, D.C., 1955); Korol, *Soviet Education for Science and Technology* (Cambridge, 1957); DeWitt, *Education and Professional Employment in the USSR* (Washington, D.C., 1961). On the composition of the reports, see also David Kaiser, "The Physics of Spin: Sputnik Politics and American Physicists in the 1950s," *Soc. Res.* 73 (2006): 1227–9. For other examples of hype over "scientific manpower" and its effects on enrollments during the Cold War, see Kaiser, "Cold War Requisitions" (cit. n. 8), 142–51.

[21] Korol, *Soviet Education* (cit. n. 20), 407–8.

[22] DeWitt, *Soviet Professional Manpower*, viii, xxvi–xxxviii, 259–61; DeWitt, *Education and Professional Employment*, xxxix, 549–53; Kaiser, "Physics of Spin," 1229–30 (All cit. n. 20).

surprise launch of the Sputnik satellite in October 1957.[23] Here, in raw form, was the first step in Shiller's model: hype.

Next came amplification. Sputnik helped here; as luck would have it, Korol's long report appeared just two weeks after the Soviet launch. Enterprising physicists leaped on the unforeseen opportunity, flogging DeWitt's number everywhere from emergency meetings with the president to syndicated radio programs and beyond. I. I. Rabi, who had known President Dwight Eisenhower at Columbia University, urged Eisenhower to use Sputnik as a pretext to spur further scientific training in the United States. Elmer Hutchisson, director of the AIP, encouraged his peers to use the "almost unprecedented opportunity" presented by Sputnik to "influence public opinion greatly." Hans Bethe, past president of the American Physical Society (APS), found himself repeating DeWitt's ratio of "two to three times" to journalists and in radio addresses without knowing (as his handwritten notes on typewritten speeches indicate) whence the number had come or how it had been computed. Eager journalists soaked it all up.[24]

Most significant of all, lawmakers and their physicist consultants used the launch of Sputnik and the purported "manpower gap" in science and engineering training to push through the massive National Defense Education Act, signed into law in September 1958. The act unleashed about $1 billion in federal spending on education (nearly $8 billion in 2011 dollars), restricted to critical "defense" fields such as science, mathematics, engineering, and area studies. The act represented the first significant federal aid to education in a century: not since the Morrill Land-Grant Colleges Act of 1862 had the federal government intervened so directly in educational matters, which had traditionally been considered the prerogative of state and local governments. One close observer of the legislative wrangling behind the National Defense Education Act concluded that opportunistic policy makers had used the Sputnik scare as a "Trojan horse": the act's proponents had been "willing to strain the evidence to establish a new policy."[25]

Passing legislation is usually a messy affair. The effects, in this case, were crystal clear. During its first four years, the National Defense Education Act supported 7,000 new graduate fellowships, or about 1,750 per year. On the eve of the bill's passage, American institutions had been producing only 2,500 PhDs per year across all of engineering, mathematics, and the physical sciences. The huge federal outlay, in other words, amounted to an overnight increase of 70 percent in the nation's funding capacity to train graduate students in the physical sciences. During that same period,

[23] Benjamin Fine, "Russia Is Overtaking U.S. in Training of Technicians," *New York Times*, November 7, 1954; "Red Technical Graduates Are Double Those in U.S.," *Washington Post,* November 14, 1955. The *New York Times* article responded to a preview of DeWitt's first book-length report: DeWitt, "Professional and Scientific Personnel in the U.S.S.R.," *Science* 120, no. 3105 (1954): 1–4. On further repetition of the "two to three times" claim by public officials and journalists, see Kaiser, "Physics of Spin" (cit. n. 20), 1231–3.

[24] Hutchisson, memorandum to American Institute of Physics Advisory Committee on Education, 4 December 1957, in AIP-EH 3:3; Bethe, "Notes for a Talk on Science Education," n.d., ca. April 1958, in Hans A. Bethe Papers, box 5, folder 4, Division of Rare and Manuscript Collections, Cornell University, Ithaca, N.Y. On Rabi's meeting with Eisenhower on October 15, 1957, see Barbara Barksdale Clowse, *Brainpower for the Cold War: The Sputnik Crisis and the National Defense Education Act of 1958* (Westport, Conn., 1981), 11; Robert Divine, *The Sputnik Challenge: Eisenhower's Response to the Soviet Satellite* (New York, 1993), 12–3; Zuoyue Wang, *In Sputnik's Shadow: The President's Science Advisory Committee and Cold War America* (New Brunswick, N.J., 2008), 75–7.

[25] Clowse, *Brainpower for the Cold War* (cit. n. 24), 91, 87.

the act funded half a million undergraduate fellowships as well as block grants to institutions and added incentives to states to increase science enrollments.[26] Hence the final element in Shiller's model: feedback.

As Shiller is quick to note, speculative bubbles can take hold even without outright chicanery. Such was the case here. The influential physicists who used Sputnik to argue for increased graduate training were not acting inappropriately: it was their job to lobby on behalf of the profession. Increased funding for higher education, moreover, is hardly an evil thing. Yet the cycle of hype, amplification, and feedback quickly came unmoored from any reasonable assessment of the underlying situation. Careful readers of DeWitt's massive reports would have noticed that his data only supported the rallying cry of "two to three times" if one lumped together degrees in engineering, agriculture, and health—leaving out science and mathematics altogether, and ignoring all the important stipulations about different types of degrees, uneven quality, and the predominance of correspondence students. If one dropped agriculture and health and included science and mathematics, the Soviets' numerical advantage fell by a factor of ten. Moreover, if one looked squarely at degrees in physics—the field usually hailed as most important, rightly or wrongly, amid the hue and cry over "scientific manpower"—then DeWitt's tables indicated that the United States held a two-to-one *lead* over its rival, rather than a deficit. (Later assessments confirmed that ratio.) None of those points were buried in classified reports; all were as plain on the page as the "two to three times" data.[27] Yet physicists, policy makers, and journalists traded sober analysis for the giddy flights of a speculative bubble— and all trundled along just fine until the bottom fell out.

THE CENTER WILL NOT HOLD

Speculative bubbles like the physics enrollment curve interest me not only because of the changes they induce in who enters the field, what they seek in a physics career, and what jobs they receive after graduation. Other questions follow as well. Did the brute-force demography on display in figure 2 shape the intellectual history of the discipline—that is, did the sudden changes in educational infrastructure affect physics itself?[28]

Consider specialization. Before World War II, physics departments across the United States had commonly held doctoral students responsible for "a good general knowledge of the entire field of physics," as a Berkeley memorandum explained in 1928.[29] A few years later, after completing his qualifying examination at Caltech, a graduate student mused about the "certain satisfaction" he had attained from "knowing

[26] On grants and fellowships funded by the act, see ibid., 151–5, 162–7; Divine, *Sputnik Challenge* (cit. n. 24), 164–6; Geiger, *Research and Relevant Knowledge* (cit. n. 13), chap. 6. The data on PhDs in physical sciences and engineering come from National Research Council, *Century of Doctorates* (cit. fig. 2 caption), 12.

[27] Kaiser, "Physics of Spin" (cit. n. 20), 1237–9.

[28] On employment patterns and changing personae, see David Kaiser, "The Postwar Suburbanization of American Physics," *Amer. Quart.* 56 (2004): 851–88. On epistemic effects of the bubble, see Kaiser, "Whose Mass Is It Anyway? Particle Cosmology and the Objects of Theory," *Soc. Stud. Sci.* 36 (2006): 533–64; Kaiser, "Turning Physicists into Quantum Mechanics," *Phys. World* 20 (May 2007): 28–33; Kaiser, *How the Hippies Saved Physics: Science, Counterculture, and the Quantum Revival* (New York, 2011).

[29] Unsigned memorandum, "Requirements for the Degree of Doctor of Philosophy, Department of Physics, August 1928," in Department of Physics records, 4:22, in UCB.

all of physics at one time."[30] Soon after the war, however, few departments retained the language of "the entire field of physics." Indeed, faculty across the country debated just what physics students should be expected to know. The old quest for comprehensive coverage (even on "comprehensive" exams) struck many physicists as unworkable. As early as 1951, Karl K. Darrow, long-time executive secretary of the APS, lamented that no one could fulfill the task he had been assigned as keynote speaker for the upcoming meeting: to address "the whole of physics."[31]

Trends like these seem obvious enough to understand in the abstract. Scholars have long noted the general process by which specialization unfolds. As the number of practitioners goes up, so too does the volume of research output, making it necessary for individuals to narrow their focus—to "know everything about nothing." Where once there was natural philosophy, by the nineteenth century a patchwork had emerged of physics, chemistry, biology, geology, and so on. With continued growth, these fields, too, underwent their own internal divisions, a kind of mitosis of the scholarly mind. Looking back over the span of centuries, the process can seem unavoidable—nothing more than "natural evolution," as Samuel Goudsmit, the long-time editor of the *Physical Review*, noted in the early 1970s.[32]

If the process was obvious in outline, however, its pace caught many physicists off guard after the war and sent them scrambling for some means of redress. *Physics Abstracts* illustrates the trend. The London Physical Society and the British Institution of Electrical Engineers established *Physics Abstracts* in 1898; by 1903, the APS and most European groups aided in the endeavor. From the start, a full-time staff of physicists made regular surveys of the world's physics journals—over one hundred journals in 1900, two hundred by 1940, and nearly five hundred by 1965—collecting and publishing abstracts of the articles in monthly installments of *Physics Abstracts*. Although the number of abstracts had grown steadily during the 1920s and 1930s, the floodgates opened soon after the end of World War II. The number of abstracts published in 1949 (7,500), for example, was nearly twice that published in 1948 (4,090). The numbers continued to climb more than twice as quickly as during the interwar period. *Physics Abstracts* published more than 10,000 abstracts per year for the first time in 1954; in 1971, the journal printed more than 84,000 abstracts. Just as quickly, after the Cold War bubble burst, the rate of growth slouched by nearly a factor of four (fig. 6).

The worldwide acceleration of physics publications was especially marked in the United States. In fact, half of all entries published in *Physics Abstracts* during the

[30] Martin Summerfield, entry of 10 March 1939, in Caltech "Bone Books," box 1, 3:60, in CIT.

[31] Darrow, "Physics as a Science and an Art," *Phys. Today* 4 (November 1951): 6–11, on 6. On changing comprehensive exams, see also P. C. Martin et al., "Proposed Changes in Oral Examination Procedures," 10 November 1959, in Department of Physics records, "Correspondence: 1958–60," box A-P, folder "1959–1960 agenda, dept. luncheon," Harvard University Archives, Cambridge, Mass.; Charles Zemach to Burton Moyer, 29 July 1963, in Department of Physics records, 4:22, in UCB; correspondence and exams from 1948–9 in LIS box 9, folder "Misc. problems"; comprehensive exams from 1962 in Felix Bloch papers, 10:19, Stanford University Archives, Palo Alto, Calif., and in W. C. Kelly, "Survey of Education in Physics in Universities of the United States," unpublished report, 1 December 1962, app. 19, available in AIP-EMD, box 9.

[32] Goudsmit, 1972 annual report, in PR-AR. See also Price, *Science since Babylon* (cit. n. 2), chap. 5; Diana Crane, *Invisible Colleges: Diffusion of Knowledge in Scientific Communities* (Chicago, 1972); John Ziman, *Knowing Everything about Nothing: Specialization and Change in Research Careers* (New York, 1987); Tony Becher, *Academic Tribes and Territories: Intellectual Enquiry and the Cultures of Disciplines* (Bristol, 1989).

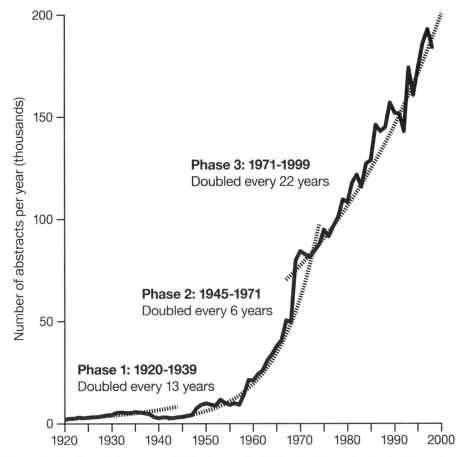

Figure 6. *Number of abstracts published annually in* Physics Abstracts. *Dashed lines show best-fit exponential growth curves for each period.*

1950s and 1960s stemmed from journals published in the United States and the Soviet Union, the countries with the fastest-growing ranks of professional physicists. Even though staffers at *Physics Abstracts* scrutinized comparable numbers of journals from each of the Cold War superpowers, American journals accounted for nearly twice as many entries as their Soviet counterparts (consistent with the two-to-one American lead in new physics degrees granted per year). During this period, the single largest source of entries in *Physics Abstracts* was the American workhorse of a journal, the *Physical Review*, published by the APS.[33]

The *Physical Review* swelled like no other journal after the war. The cause seemed

[33] Stella Keenan and Pauline Atherton, *The Journal Literature of Physics: A Comprehensive Study Based on "Physics Abstracts,"* report no. AIP-DRP PA1 (New York, 1964); Keenan and F. G. Brickwedde, *Journal Literature Covered by "Physics Abstracts" in 1965*, report no. ID 68–1 (New York, 1968); L. J. Anthony, H. East, and M. J. Slater, "The Growth of the Literature of Physics," *Reports on Progress in Physics* 32 (1969): 709–67. In 1965, *Physics Abstracts* included complete coverage of 82 journals, of which 17 were published in the United States and 13 in the Soviet Union; the remaining 413 journals scanned by the *Physics Abstracts* staff contributed less than 23 percent of all the abstracts published that year. See Keenan and Brickwedde, *Journal Literature*, 6, and app. 4, on 22.

clear, even at the time. "Graduate school enrollment should be watched as basis for prediction of size of future issues" of the *Physical Review*, concluded the APS Advisory Committee on Publication Policy in the mid-1950s.[34] Goudsmit agreed. In his annual reports he took to graphing the number of articles published in the journal alongside the growing membership of the APS, driven up each year by the new crop of PhDs. During his first year as editor, in 1951, the *Review* published fewer than five thousand pages. During his final year as editor, in 1974, the journal published more than thirty thousand pages.[35]

The lightning-fast expansion affected every aspect of the journal. Shortages of paper and labor—especially of skilled operators who could handle the journal's sophisticated mathematical typesetting requirements—continued to hamper its production well into the late 1950s.[36] Staffing the editorial office also became a challenge. When Goudsmit became editor of the *Review* in 1951, his position was part-time, so he could continue research at Brookhaven National Laboratory. Other than Goudsmit, the editorial office included two secretaries and one full-time assistant. By the time Goudsmit retired in the early 1970s, the editorial staff had swelled to thirty people: ten PhD physicists working as full-time editors, two more assisting as part-time editors, plus eighteen people working in full-time clerical and administrative capacities. All those hands were kept busy. Virtually every year between 1951 and 1969, Goudsmit reported an increase over the previous year in the number of submissions received, articles published, and pages printed. Whereas the journal processed 1,379 article submissions in 1955—averaging 115 new submissions each month— the number had doubled by 1965 and nearly tripled by 1968. Given the number of transactions, Goudsmit explained to a colleague in 1966, the journal "is no longer similar to the neighborhood grocery store where old customers get personal attention." Instead it had become "more like a supermarket where the manager is hidden in an office on the top floor. As a result, lots of things are just done by routine rather than by human judgment." He meant it literally: by that time, the office was experimenting with a new punch-card computer system to mechanize the process of matching referees with submissions and to track the progress of referee reports received, responses sent to authors, and so on.[37]

Physicists could not fail to notice the effects of all those efficiencies. Goudsmit himself noted in the mid-1950s that each issue of the journal had become "almost too bulky to carry."[38] One physicist working at the National Bureau of Standards complained to Goudsmit on behalf of "the poor over-burdened members" of the

[34] Minutes of Advisory Committee on Publication Policy of the American Physical Society, 6 January 1956, in AIP-HAB 57:7.

[35] Goudsmit, 1951–74 annual reports, in PR-AR.

[36] R. P. Rohrer (president of Lancaster Press) to Henry Barton, 27 May 1949, in AIP-HAB 79:13; Elmer Hutchisson, memorandum of 20 September 1957, and unsigned memorandum, "Study of the Physics Publishing Problem," 18 February 1958, both in AIP-EH 14:3; R. E. Maizell, "Minutes of AIP Committee to Study Physics Publishing Problems," 18 April 1959, in AIP-EH 14:2.

[37] The quotations are from Goudsmit to Leonard Schiff, 2 September 1966, in LIS box 4, folder "Physical Review." The data on numbers of submissions and pages published come from PR-AR. On staffing, see E. L. Hill (editor of *Physical Review*) to Karl K. Darrow (secretary of the APS), 14 October 1950, in AIP-HAB 32:6; Goudsmit, 1973 annual report, in PR-AR; APS Council meeting minutes of 27 November 1953 and 26 November 1954, in AIP-HAB 79:1. See also Simon Pasternack (assistant editor, *Physical Review*) to Leonard Schiff, 22 January 1958 and 27 June 1963, in LIS box 4, folder "Physical Review."

[38] Goudsmit, 1956 annual report, in PR-AR.

APS "whose book-shelves and closets are beginning to burst with the vast amount of paper that you are helping to distribute every year"; American physicists now needed to "cope with scientific literature by the ton."[39] A senior physicist at Berkeley lodged a similar complaint:

> Having been a subscriber of the *Physical Review* from 1913 on, I had to sell my back numbers up to 1947, simply because I had no space to store them. Even today, with the limited space in a comfortable, but small 8 room house, I am already finding that the growing files of journals take up so much space that I am just at my wit's end as to what to do. My own office, which is sufficiently commodious in the new building, can no longer house the journal either.[40]

Goudsmit found such complaints a bit silly. Given the rapid pace of research in physics, he reasoned, the journal's contents became obsolete relatively quickly. "There is really little reason to keep more than about 'six feet' of *The Physical Review* at home," he explained in his 1955 annual report. To the physicist from the National Bureau of Standards, Goudsmit recommended an even more direct method for keeping the journal's bulk in check: physicists should stop being "overly sentimental" and simply rip out those articles they wanted upon each issue's arrival, throwing away the rest. His correspondent was not impressed by the suggestion. "One revolts against such destruction of the printed word, even if bound in nothing more than paper, and moreover if one throws away something one later needs it cannot possibly be replaced." Yet ten years later, a Caltech physicist reported to Goudsmit on his own efforts to do just as Goudsmit had recommended. He reduced the two feet of shelf space taken up by his copies of the 1963 *Physical Review* to a few inches by ripping each issue apart, throwing away those articles of no interest to him, and stapling the rest back together. The arts-and-crafts project had worked, but it had not been easy. Perhaps, the physicist suggested, the journal could be bound with a different kind of glue, to better facilitate such scan-and-tear operations.[41]

Goudsmit, too, wondered what was keeping the journal together, literally and figuratively. In 1962 he reported with exasperation that the journal was bumping up against the printer's limit: the press could only bind individual issues that contained fewer than five hundred pages. "We are rapidly reaching this technical limit," Goudsmit noted, "but have already long ago passed the psychological limit above which the subscriber is overwhelmed by the bulk and looks only at the few articles in his own narrow field." Not long after that he inserted an editorial into the journal, entitled simply "Obscurantism." Most articles struck him as having been written for "a few specialists only," or, even worse, as a kind of "memorandum to [the author] himself or merely for the benefit of a close collaborator."[42]

[39] W. B. Mann to Goudsmit, 11 January 1955, in AIP-HAB 79:14.

[40] Leonard B. Loeb (Berkeley) to Goudsmit, 19 April 1955, in Raymond Thayer Birge papers, box 19, folder "Loeb, Leonard Benedict," in UCB.

[41] Goudsmit, 1955 annual report, in PR-AR; Mann to Goudsmit, 11 January 1955 (cit. n. 39); Thomas Lauritsen to Goudsmit, 27 December 1968, in Thomas Lauritsen papers, 12:14, in CIT. Goudsmit's recommendations were remarkably similar to those of sixteenth-century scholars Conrad Gesner and Girolamo Cardano; see Ann Blair, "Reading Strategies for Coping with Information Overload ca. 1550–1700," *J. Hist. Ideas* 64 (2003): 11–28, on 25–7; Blair, *Too Much to Know: Managing Scholarly Information before the Modern Age* (New Haven, Conn., 2010).

[42] Goudsmit, 1962 annual report, in PR-AR; Goudsmit, "Obscurantism," *Physical Review Letters* 13, no. 17 (1964): 519–20.

As research journals like the *Physical Review* grew fatter and fatter, and editors like Goudsmit recommended that readers conquer the heft with scissors and glue, others began to offer even more brazen suggestions about the future of scientific publishing. Some proclaimed that the entire system of scientific journals would need to be scrapped and replaced by some radical alternative. As early as June 1949, physicists debated the publication problem at an annual meeting. One idea that emerged was to stop printing journals like the *Physical Review* altogether and to replace them with a weekly newspaper that would contain the titles and abstracts of all physics articles received. Subscribers could then write directly to the APS editorial office to request only those articles in which they were interested; photo-offset copies of specific articles could then be printed and mailed on demand. Although Henry Barton, the director of the AIP, remained skeptical of the idea—he suggested that the finances and logistics of such a plan be thoroughly analyzed "before it gains too much headway in the minds of our members"—the London Physical Society in fact converted to this system for several years during the early 1950s before concluding that it was too expensive, and the idea kept resurfacing among American physicists well into the 1960s.[43]

One leading physicist offered a different suggestion a few years later. The *Physical Review* should become a kind of "greatest hits" journal, surrounded by several smaller, specialized journals catering only to narrow subfields. The *Review* would then consist entirely of reprints of specially selected articles deemed most important or most interesting in the specialist literature. Photo-offset printing (rather than retypesetting each article) could keep the costs low, and the *Review*'s size could be strictly controlled to allow the average physicist to be able to read or skim the best work from the whole of physics. Still others suggested converting the *Review* into a *Reader's Digest* of physics, publishing specially commissioned, short and accessible versions of specialized articles that appeared elsewhere.[44]

Yet the dream of sampling the best work from the whole of physics faded fast. The subject index to *Physics Abstracts* illustrated the problem. Throughout the 1950s, physicists representing the APS and the AIP negotiated with counterparts in Britain over the length and arrangement of the index. Both groups agreed that the index required constant revision, with new levels of detail added to keep up with the fast pace of specialization.[45] In 1930, the index featured eight main categories—general physics; meteorology, geophysics, and astrophysics; light; radioactivity; heat; sound; electricity and magnetism; and chemical physics and electrochemistry—only one of

---

[43] Barton to John T. Tate (editor, *Physical Review*), 20 June 1949, in AIP-HAB 79:13; Mann to Goudsmit, 11 January 1955 (cit. n. 39); Goudsmit, 1955 annual report, in PR-AR. In fact, the idea had been advanced in the early 1930s and resurfaced after the war. See Ralph H. Phelps and John P. Herlin, "Alternatives to the Scientific Periodical: A Report and Bibliography," *UNESCO Bulletin for Libraries* 14 (March–April 1960): 61–75; my thanks to Michael Gordin for bringing this reference to my attention. The 1949 suggestion eventually became the operating principle behind the first electronic physics preprint system during the early 1990s, now available at http://www.arXiv.org.

[44] H. A. Barton, "Thoughts on the Physics Publication Problem," unpublished memorandum, 7 March 1956, in AIP-EH 14:1; Goudsmit, "Editorial: The Future of Physics Publications; A Proposal," *Physical Review Letters* 10, no. 2 (1963): 41–2.

[45] E. Hutchisson, "Report of the Joint APS-AIP Committee on Science Abstracts," n.d., ca. March 1951; W. K. Brasher (secretary, British Institution of Electrical Engineers) to Hutchisson, 29 March 1951; S. Whitehead (deputy chair, Committee on Management, British Institution of Electrical Engineers) to Hutchisson, 29 March 1951, all in AIP-EH 15:1. See also the other correspondence in AIP-EH 15:1 and 15:3.

which, electricity and magnetism, required a further division into four subcategories. By 1955, major fields like nuclear physics, separated into six subcategories, had been added to the list. Ten years later, nuclear physics had been carved up into thirty-five distinct subcategories, and solid-state physics into thirty-eight.

The subdivisions and rearrangements continued but did not converge. The 1960 subject index featured eighty-four subject headings in all (major and minor); simply reading through the list of topics had become a chore. By 1967, the AIP experimented by printing separate lists of subfields by specialty in its own indexes; it was no longer able to fit all the categories and subject headings into a single unified list.[46] Scanning abstracts had long since become infeasible, so physicists tried an even sharper condensation. For a brief time in the late 1960s, the publishers of *Physics Abstracts*, in conjunction with the AIP, printed *Current Papers in Physics*, a biweekly newsletter arranged in tabloid-newspaper format that simply listed author names and article titles, by subject category, for items that were due to appear in forthcoming issues of *Physics Abstracts*.[47]

Several physicists feared that the baroque complexity of the subject index would carry pedagogical ramifications. The joint committee of the APS and AIP on *Physics Abstracts* suggested that all competing index schemes should be judged by how effectively graduate students could use them. The AIP actually conducted tests of "index efficiency" in January 1960 by running thirty-six graduate-student volunteers through time trials. "The attempt was made to simulate a real life situation in information retrieval," the report began. Groups of students were assigned one of five indexes: some used the experimental indexes based on permutations of keywords from titles, while others used the existing subject indexes in the *Physical Review*, *Physics Abstracts*, *Nuclear Science Abstracts*, or *Chemical Abstracts*. Each student received copies of the first page of fifteen different articles with title and author lines blacked out. Using only their assigned index, they had to identify each article and find at least one subject heading under which it was classified. The report then listed the average number of articles located (of the original fifteen) by students in each group, the average time taken per article (ranging from 1.4 to 6.2 minutes), and the average number of false leads (from 10 to 18.2).[48]

The physicists' journals did not just swell from the enrollment pressures; they cracked. As early as 1949, Berkeley physics professor Emilio Segrè suggested that the *Physical Review* should be split into two journals, one aimed at experimentalists

[46] Dwight E. Gray, minutes on meeting of joint APS-AIP Abstracting Committee, 24 November 1958, in AIP-EH 15:3; A. A. Strassenburg (director, AIP's Office of Education and Manpower) to Robert B. Leighton, 12 July 1967, in Robert B. Leighton papers, 1:12, in CIT. See also Simon Pasternack to Leonard Schiff, 27 June 1963, in LIS box 4, folder "Physical Review."

[47] Maizell, "Physics Abstracting Services," unpublished report, 15 February 1960, in AIP-EH 15:3; Van Zandt Williams, Elmer Hutchisson, and Hugh C. Wolfe, "Consideration of a Physics Information System," *Phys. Today* 19 (January 1966): 45–52; Franz L. Alt and Arthur Herschman, *Plans for a National Physics Information System*, report no. ID 68–6 (March 1968), in LIS box 1, folder "AIP Information Program"; see also the questionnaire dated 15 November 1968 from the AIP Information Division in the same folder and the correspondence in AIP-EH 15:3, 16:10.

[48] R. E. Maizell, minutes from meeting of APS-AIP Committee on Science Abstracts, 27 April 1960, in AIP-EH 15:3; unsigned report, "Investigation of Index Efficiency as Based on Tests with Graduate Physics Students," 19 January 1960, in AIP-EH 16:10. See also Philip Morse, *Library Effectiveness: A Systems Approach* (Cambridge, 1968); Erik P. Rau, "Managing the Machine in the Stacks: Operations Research, Bibliographic Control, and Library Computerization, 1950–2000," *Library History* 23 (June 2007): 151–68.

and the other at theorists. Others thought that the *Review* should be split along subject lines, one *Review* for each topical division of the APS. Goudsmit dismissed the idea as uneconomical: such divisions would needlessly duplicate editorial office effort, requiring the maintenance of separate subscription lists and mailing labels, and so on. To officers of the AIP, the problem ran deeper than economics. They feared such splitting "would be dangerous," furthering an intellectual balkanization that the institute had been founded to fight.[49]

As a compromise, Goudsmit agreed to begin arranging the articles within each issue of the *Review* by topic early in 1953, "without any general announcement" that he was doing so. Yet with submissions to the *Review* growing by as much as a third in a single year—as they did between 1952 and 1953—Goudsmit felt compelled to return to the idea of splitting the journal. He included a ballot on the back cover of the APS *Bulletin*, mailed out to members in advance of the society's January 1955 meeting in New York, asking them to vote on whether the *Review* should be split into two journals, one catering to solid-state physics and the other to nuclear and high-energy physics. Within weeks hundreds of ballots poured in, most of them favoring the split—although, Goudsmit was quick to note, most of those in favor had come from solid-state physicists, who thought that their specialty was unduly crowded by nuclear topics in the *Review*.[50]

Beyond the ballots, impassioned letters began to circulate; the question of whether or not to split the *Review* had clearly touched a nerve. Norman Ramsey, for example, Harvard physicist and member of the APS Executive Council, wrote to his fellow council members to reiterate his "personal preference" for "no splitting of the Physical Review whatsoever": such a split presented too many "evils," and "compartmentalization in physics should be discouraged." Views like Ramsey's on the council carried the day. As the society's Advisory Committee on Publication Policy noted, the decision not to split the *Review* was taken "for ideological rather than other reasons. Influential Council members deplored any tendency to compartmentalize physics."[51]

But the matter would not go away. In 1962 the editorial office suggested a compromise. The *Review* could be split into four separate sections, covering nuclear physics, high-energy physics, solid-state physics, and atomic physics. Each section would come out biweekly and still be sent to all subscribers; individual physicists could then keep their preferred sections and give away (or toss away) the others. Half of the plan soon went into action. Beginning in 1963, the *Physical Review* was printed in two sections, *A* on solid-state and atomic physics and *B* on nuclear and high-energy physics—but, to appease those who had argued against splitting, the two sections were paginated continuously (so libraries could bind them together as a single vol-

---

[49] Segrè to Henry A. Barton, 14 February 1949, in AIP-HAB 28:5; Goudsmit, 1952 annual report, in PR-AR; Elmer Hutchisson to Barton, 14 February 1951, in AIP-EH 15:1. On the founding of the AIP and the issue of fragmentation, see Spencer Weart, "The Solid Community," in *Out of the Crystal Maze: Chapters from the History of Solid-State Physics*, ed. Lillian Hoddeson et al. (New York, 1992), 617–69.

[50] Goudsmit, 1952–4 annual reports, in PR-AR; John C. Slater to Clyde A. Hutchison Jr., 4 February 1953, and Slater, meeting minutes of APS Committee on Publication Policy, 15 April 1953, in AIP-HAB 57:7; Karl K. Darrow, APS Council meeting minutes, 26 January 1955, in AIP-HAB 79:1.

[51] Ramsey to APS Council members, 6 June 1956, in AIP-HAB 79:14; minutes of Advisory Committee on Publication Policy, APS (cit. n. 34). See also Leonard Loeb to Raymond Birge, 27 October 1954, and Loeb to Goudsmit, 19 April 1955, in Raymond Thayer Birge papers, box 19, folder "Loeb, Leonard Benedict," in UCB; Birge to Goudsmit, 5 April 1955 in Birge papers, box 40, folder "Letters written by Birge, January–April 1955," in UCB; Mann to Goudsmit, 11 January 1955 (cit. n. 39).

ume) and covered by a single index. Two years later, members were given the option of only subscribing to one half or the other, and only 45 percent chose to continue receiving both sections. The uproar against the evils of specialization that had erupted among council members a decade earlier had faded, no longer much of a match against the journal's exponential growth.[52]

Even these changes rapidly proved insufficient. With the editors again citing the need for "drastic changes" if the journal were to remain of any use to researchers, the inexorable divisions continued. In 1966, the *A* and *B* sections were themselves each divided in half—atomic and molecular physics separated from the rest of solid-state physics; nuclear physics separated from high-energy particle physics—and the following year the editors introduced a five-part division. Yet even with the latest splits, the size of individual issues remained unmanageable. Finally the realities of operating a journal that received more than ten new submissions every day of the year overwhelmed those advocates who had hoped to stem the tide of specialization. In 1970 the *Physical Review* was divided into four separate biweekly journals, each paginated and indexed independently. Both *Physical Review B* (on solid-state physics) and *Physical Review D* (on particle physics) were themselves further subdivided, those issues appearing on the first of each month catering to a different set of topics than the issues appearing two weeks later. In their annual report, the editors noted that a sizable proportion of subscribers to *Physical Review B* were so specialized that they chose to subscribe only to one or the other of these subsections, and the same held true among subscribers to *Physical Review D*—a trend that continued throughout the 1970s.[53] Similar pressures affected other leading journals. Elsevier divided both its *Nuclear Physics* and *Physics Letters* into separate *A* and *B* sections in 1967, and the *Zeitschrift für Physik* split into three separate journals in 1970.

Ironically, after wrestling with the issue of how to balance ballooning size with intellectual coherence for twenty years, the *Physical Review* split into separate journals just as its massive growth ground to a halt. When plans were laid to divide the journal back in 1968 and 1969, the increases still seemed unstoppable—after growing by an average of 280 submissions per year for each of the previous seven years, receipts leapt by an additional 420 submissions in 1968 alone. Yet by 1970, when the journal finally split, the growth had definitely stopped: the total number of submissions actually fell by 115 that year, and it remained flat for the remainder of the decade. The centrifugal pressures had mounted during the feverish rise of the Cold War bubble, but the *Physical Review* split just when it might have finally achieved some stability.[54]

To Goudsmit and others, the great challenges facing scientific publishing were at root pedagogical, both in cause and effect. All those graduate students needed to publish their work somewhere, and the journals had ballooned in response. In turn, the bulge affected what got published, and in what manner. No student (nor any practitioner) could engage the full range of research on, say, nuclear physics if several hundred dissertations and articles on the topic had been filed the previous year, with several hundred also filed the year before that, and so on. Of course the *Physical Review* had outstripped all previous growth rates and the *Physics Abstracts* subject index had

[52] Goudsmit, 1962–5 annual reports, in PR-AR.

[53] 1966–79 annual reports (by Goudsmit and others), in PR-AR. See also S. Pasternack and A. Herschman, "Editorial: A Proposal for Changing the *Physical Review*," *Physical Review* 137, no. 7AB (1965): AB1.

[54] 1961–79 annual reports (by Goudsmit and others), in PR-AR.

become so unwieldy—how else to make room for all those new dissertations? Along the way, the physics landscape itself had changed, its internal divisions and units of currency (research articles, dissertations) arrayed in a dramatically different fashion than what had seemed natural just a few decades earlier.

OTHER BUBBLES

The physicists' bubble, so sharply pronounced between 1945 and 1975, was not a one-shot deal. In fact, graduate-level physics enrollments rebounded during the 1980s in the United States, bid higher and higher by many of the same mechanisms that had inflated the first bubble. A resurgence of defense-related spending under the Reagan administration—including the sprawling Strategic Defense Initiative, or "Star Wars"—combined with new fears of economic competition from Japan drove enrollments in physics and neighboring fields up exponentially once more, nearly matching the late-1960s peak. They fell sharply a decade later with the end of the Cold War. Just as during the early 1970s, shared conditions across fields led to an overall decline in graduate-level enrollments.[55] By the time PhD conferrals bottomed out in 2002, annual PhDs across all fields had fallen by more than 6 percent from their 1990s peak; annual PhDs in science and engineering had fallen by 10 percent; while annual PhDs in physics had plummeted by 26 percent. Once again, dire predictions of shortfalls in the scientific labor supply had been stupendously mistaken; once again, physics marked the extremes of a general pattern throughout American universities (fig. 7).[56]

The dynamics behind the second bubble were remarkably similar to the earlier example. Beginning in 1986, the director of the National Science Foundation and colleagues sounded the alarm again that the United States would soon face a devastating shortage of scientists and engineers. Foundation projections indicated that there would be 675,000 too few scientists and engineers in the United States by the year 2010. Just as in response to the DeWitt and Korol studies from the 1950s—especially the stripped-down ratio of two to three times more science and engineering graduates per year in the Soviet Union than in the United States—the dramatic projections of shortages helped to unleash generous federal spending.[57]

Unlike the DeWitt and Korol studies, the 1980s study by the National Science Foundation did not impress many close observers. In keeping with broader economic modeling during the Reagan administration, the study had neglected to consider demand at all, sticking with only supply-side variables. Yet few skeptics came forward until the early 1990s, after the Soviet Union dissolved and the Cold War ground to an

[55] See, e.g., Juan Lucena, *Defending the Nation: U.S. Policymaking to Create Scientists and Engineers from Sputnik to the "War against Terrorism"* (New York, 2005), chap. 4.

[56] The declines in annual PhD conferrals across each category were calculated from data tabulated in the annual National Science Foundation reports titled "Science and Engineering Doctorate Awards," 1994–2006, available at http://www.nsf.gov/statistics/doctorates (accessed 10 January 2011).

[57] David Berliner and Bruce Biddle, *The Manufactured Crisis: Myths, Fraud, and the Attack on America's Public Schools* (New York, 1995), 95–102; Daniel Greenberg, *Science, Money, and Politics: Political Triumph and Ethical Erosion* (Chicago, 2001), chaps. 8–9; Eric Weinstein, "How and Why Government, Universities, and Industry Create Domestic Labor Shortages of Scientists and High-tech Workers" (unpublished working paper), available at http://www.nber.org/~peat/Papers Folder/Papers/SG/NSF.html (accessed 10 January 2011); Lucena, *Defending the Nation* (cit. n. 55), 104–12, 133.

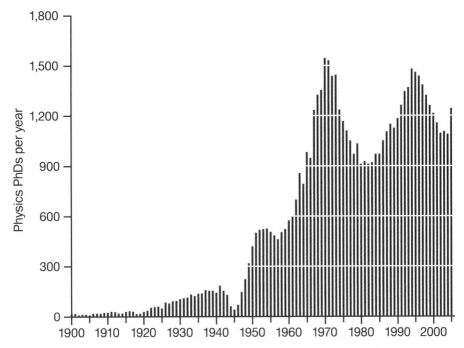

Figure 7. *Number of physics PhDs granted by US institutions, 1900–2005. Based on data from the AIP Statistical Research Center, available at http://www.aip.org/statistics (accessed 10 January 2011).*

unexpected halt. Just as in the earlier era, reality checks that could easily have been tried were not, while the scarcity talk looped from hype to amplification to feedback all over again. And just as in the early 1970s, the second bubble burst, triggering double-digit unemployment among PhD-level scientists and mathematicians. The glut of freshly minted scholars—many of whom had been lured to graduate school with federally funded fellowships and promises of plentiful academic jobs to come— occasioned testy hearings in Congress. The push-back led ultimately to the dismantling of the Policy Research and Analysis Division within the National Science Foundation, which had developed the faulty supply projections.[58]

So much for repeating bubbles over time. What about comparable effects on other disciplines, which like physics were caught up in boom-and-bust cycles after the war? Consider the field of history. Unlike for physics, few calls had rung out to boost annual production of history PhDs to help prosecute the Cold War. Yet the field (like most others) had been buoyed by the general expansion of the infrastructure of American higher education, a side effect of the speculative bubble and the harried calls for increased "scientific manpower" before and after Sputnik. As a result, PhDs in history grew rapidly during the 1960s, only to peak in the early 1970s and fall sharply—the pattern should by now be familiar (fig. 8).

[58] See esp. Berliner and Biddle, *Manufactured Crisis*; Greenberg, *Science, Money, and Politics*; Weinstein, "How and Why Government" (All cit. n. 57); Lucena, *Defending the Nation* (cit. n. 55), and references therein.

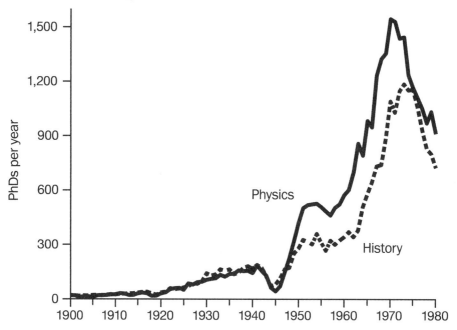

Figure 8. *Number of PhDs granted by US institutions per year in physics and history, 1900–1980. Based on data from National Research Council,* Century of Doctorates, *13, and National Science Foundation,* Science and Engineering Degrees *(Both cit. fig. 2 caption).*

Just as for the physicists, years of overproduction were met by a crushing contraction. Physicists seeking jobs had outnumbered positions posted with the AIP by a factor of twenty to one in 1971. The ratio for young historians competing for interviews at the American Historical Association meeting around that time was only marginally better: 2,481 applicants for 188 positions, or about thirteen to one. Intellectual fragmentation, and not just job-market prospects, elicited impassioned concern from leading historians throughout the 1970s and 1980s, echoes of the physicists' losing battle against specialization.[59]

Equipped with time-series graphs like figure 8, we may return to a question raised thirty years ago by Robert Darnton. Darnton had noted a remarkable trend among history dissertations completed in the United States. Once-dominant specialties like political history and intellectual history had fallen consistently in terms of their share of all history dissertations. By Darnton's reckoning, political history had accounted for 34.3 percent of all history dissertations in 1958, 33.4 percent in 1968, and 23.7 percent in 1978. During that same interval, intellectual history had fallen from 10.5 to 9.5 to 8.8 percent. Meanwhile, social history grew by leaps and bounds, from just 6.8 percent of all dissertations in the field in 1958, to 10.4 percent in 1968, and 27.1 percent in 1978—a fourfold increase in just two decades.[60]

[59] Peter Novick, *That Noble Dream: The "Objectivity Question" and the American Historical Profession* (New York, 1988), 574 (on historians' job market), 577–92 (on specialization and fragmentation).

[60] Darnton, "Intellectual and Cultural History" (1980), repr. in Darnton, *The Kiss of Lamourette: Reflections in Cultural History* (New York, 1991), 191–218, on 202.

Darnton gestured toward the likely suspects to try to account for the dramatic trends. Clearly the social movements of the 1960s and 1970s, from civil rights to the women's movement and gay liberation, had left their mark on the discipline, encouraging the study of peoples and events that had barely rated notice during earlier time periods. But in the end Darnton threw up his hands. No matter how riveting those social and political movements had been, no clear causal arrows seemed to connect the swirling Zeitgeist with graduate curricula. For such trends, Darnton had to conclude, "their origin remains a mystery."[61]

As Darnton surmised, we need not downplay the impact that broader social and political conditions likely had on the intellectual direction of the history profession. Yet graphs like figure 8 suggest a plausible means to begin filling in the gears that Darnton feared were missing from such explanations. Just as for the physicists, broad changes in method and "acceptable" topics to study coincided with steep changes in enrollments. No discipline can sustain itself by pumping out five hundred dissertations each year reanalyzing Descartes's *Meditations* or the military strategies of World War I. Faced with record-breaking numbers of dissertations to advise—and certainly against a backdrop of broader social movements—historians may well have welcomed social history not just as a new set of methods, but as a limitless source of new topics. That would certainly help account for the timing and acceleration inherent in Darnton's numbers. Social history had made only modest inroads by the end of the turbulent 1960s. On the other hand, students who completed their dissertations in 1978 would have entered graduate school (on the average) five or six years earlier— right at the peak of the historians' curve in figure 8.

Obviously simple correlations like these do not guarantee cause. Rather, they spur further close-up inquiry. Just as figure 2 inspired questions of how physicists struggled to manage the contours and content of their discipline, we might ask comparable questions about how historians handled the massive expansion in their own ranks. Likewise, Charles Newman and Russell Jacoby may have glimpsed an important truth when they sneered that postmodern literary theory served as "an infinitely expendable currency, the ultimate inflation hedge."[62] The enrollment curve for PhDs in English literature in the United States looks remarkably similar to that of history in figure 8.[63] Might the heat and light behind the debates over multiculturalism and the broadening of the Western literary canon during the 1980s have owed something to the same kinds of enrollment pressures?

My interest is not to develop a hydraulic theory of scholarly production, some one-dimensional account of institutional pushes and pulls that might determine thought patterns or research trends. Many ingredients shape the contours of intellectual life, from budget lines to political exigencies, cultural cues, and shifting enrollments. Not all are independent of each other, nor are their interactions simple to disentangle. From their combination, however, certain patterns often emerge. The physicists' bubble has the potential to illuminate comparable shifts in other fields across the natural sciences, social sciences, and humanities. Disciplines as varied as chemistry,

[61] Ibid., 205.

[62] Newman, "The Post-modern Aura: The Age of Fiction in an Age of Inflation" (1984), as quoted in Jacoby, *The Last Intellectuals: American Culture in the Age of Academe*, 2nd ed. (New York, 2000), 173.

[63] The enrollment curve for English departments in the United States may be produced from data in the sources listed in the fig. 2 caption.

biology, computer science, psychology, literature, and history each went through boom-and-bust cycles during the postwar decades. None was a carbon copy of the physicists' example; they all occurred later in time and remained smaller in magnitude. But each suggests how attending carefully to the ebb and flow of student numbers can help us understand the rhythms of disciplinary change. They might even point the way back to a robust mesoscopic account of scholarly life, informed by the finest local case studies but not limited to them.

Collecting Nature:
Practices, Styles, and Narratives

*by Bruno J. Strasser**

ABSTRACT

The standard narrative in the history of the life sciences focuses on the rise of experimentalism since the late nineteenth century and the concomitant decline of natural history. Here, I propose to reexamine this story by concentrating on a specific set of material and cognitive practices centered on collections. I show that these have been central for the production of knowledge not only in natural history, from the Renaissance to the present, but also in the experimental sciences. Reframing the history of the life sciences in this way makes historical continuities visible and raises new possibilities to contextualize recent developments in science, such as the proliferation of databases and their growing use.

INTRODUCTION

The "data deluge" (sometimes the "data tsunami") is upon us, and the resulting "data storms" and "floods" threaten to "drown" all those who have not learned to "swim in a sea of data."[1] From the pages of the magazine *Wired* to those of the *Economist*, *Nature,* and *Science,* an abundance of aquatic (and biblical) metaphors describe a new era in which humanity seems to be threatened by an unprecedented amount of data. The July 2008 cover of *Wired* went so far as to announce that the so-called deluge signified "the end of science" and to explain that "the quest for knowledge used to begin with grand theories," but now "it begins with massive amounts of data."[2] Numerous scientists in the natural and social sciences have announced the coming of age of a

* Yale University, Program in the History of Science and Medicine. Current address: University of Geneva, IUFE, Pavillon Mail, 40, Bd du Pont-d'Arve, CH—1211 Geneva 4; bruno.strasser @unige.ch.

I would like to thank Robert E. Kohler and Kathryn Olesko for their encouragement, comments, and infinite patience, and Robin Scheffler, Rachel Rothschild, and especially Helen Curry, as well as two anonymous referees for *Osiris*, for helpful suggestions.

[1] The expression "data deluge" is widely used, at least since the early 1990s; see P. Aldhous, "Managing the Genome Data Deluge," *Science* 262 (1993): 502–3, and more recently, G. Bell, T. Hey, and A. Szalay, "Computer Science: Beyond the Data Deluge," *Science* 323 (2009): 1297–8. A special report featured on the cover of the *Economist* was titled "The Data Deluge and How to Handle It" (February 27, 2010); *Wired* titled its issue on the subject "The End of Science" (July 16, 2008); *Nature* had a cover on "Big Data" (vol. 455, no. 7209 [2008]). For "data tsunami," see Anthony J. G. Hey, Stuart Tansley, and Kristin Tolle, *The Fourth Paradigm: Data-Intensive Scientific Discovery* (Redmond, Wash., 2009), 117, 131; for "floods," Bell, Hey, and Szalay, "Computer Science," 1297; for "swim in a sea of data," David S. Roos, "Bioinformatics—Trying to Swim in a Sea of Data," *Science* 291 (2001): 1260–1, on 1260.

[2] *Wired*, "End of Science" (cit. n. 1), cover.

"data-driven science" representing a "fourth paradigm" that follows the empirical, theoretical, and computational paradigms.[3] The value of data-driven science is being vigorously debated and contested in the scientific community. These discussions are all the more relevant because science funding agencies, at least in the United States, rely explicitly on a "hypothesis-driven" model of scientific research in evaluating research grant proposals.[4] The key assumption shared by almost all participants in the current debate over the value of data-driven sciences is that experimentation is the benchmark for what constitutes science. As the director of a National Institutes of Health (NIH) biomedical center put it in *Science*, the hypothesis-driven experimental method is "the scientific method," which "has driven conceptual inquiry for centuries and still forms the basis of scientific investigation."[5]

This assumption is also embedded in a narrative that dominates "generalist" histories of the life sciences.[6] According to this standard story, the study of living nature, from ancient Greece to the late Enlightenment, focused on naming, classifying, and describing the outer and inner morphology of plants and animals (*nommer, classer, décrire*, as Georges Cuvier put it). This project, known as "natural history" since at least the ancient Roman days of Pliny the Elder, initially was not concerned with change over time, as the modern sense of "history" would imply, but with the establishment of a "systematic" (and static) account of nature.[7] This project developed most forcefully during the Renaissance—thanks to the expansion of travel and the development of print—and found its home in wonder cabinets, gardens, herbaria, and eventually in natural history museums. In the mid-nineteenth century, this project was challenged by the rise of experimentalism, especially in physiology and embryology, which proposed to reveal the functions of life in the laboratory through the use of instruments and controlled experiments. Experimentation had occasionally played a role in the production of biological knowledge prior to that time—for example, in the seventeenth century in the work of William Harvey and Santorio Santorio, in the eighteenth with Abraham Trembley and Charles Bonnet—but unlike in the physical sciences, it remained marginal until the nineteenth century.[8] Then experimentalism began to displace natural history as the center of the life sciences. After 1900, the development of genetics, biochemistry, and crystallography hailed the newly recognized power of experimentation in unlocking the secrets of life and led to the ultimate triumph of experimentalism in the life sciences with the rise of molecular biology and the demise of natural history.

This narrative captures crucial aspects in the history of the life sciences since the late nineteenth century, such as the growing authority of the experimentalist discourse and the rise of particular experimental sciences. Yet, like its sister narrative

[3] For the social sciences, see Gary King, "Ensuring the Data-Rich Future of the Social Sciences," *Science* 331 (2011): 719–21.

[4] M. A. O'Malley et al., "Philosophies of Funding," *Cell* 138 (2009): 611–5.

[5] G. J. Nabel, "The Coordinates of Truth," *Science* 326 (2009): 53–4, on 53.

[6] E.g., Lois N. Magner, *A History of the Life Sciences* (New York, 2002); Peter J. Bowler and Iwan Rhys Morus, *Making Modern Science: A Historical Survey* (Chicago, 2005); Peter Dear, *The Intelligibility of Nature: How Science Makes Sense of the World* (Chicago, 2006).

[7] On the history of natural history, see Nicholas Jardine, James A. Secord, and Emma C. Spary, eds., *Cultures of Natural History* (New York, 1996); Paul Lawrence Farber, *Finding Order in Nature: The Naturalist Tradition from Linnaeus to E. O. Wilson* (Baltimore, 2000).

[8] Mirko Grmek, *La première révolution biologique: Réflexions sur la physiologie et la médecine du XVIIe siècle* (Paris, 1990).

on the laboratory revolution in medicine, it invites renewed questioning. Indeed, like all big narratives, this one excludes or obscures particular historical events whose significance might look different today than it did almost half a century ago when this narrative was first proposed. Detailed historical studies on the development of recent biological science (e.g., the changes that have led to the recent data deluge) offer good opportunities to test the explanatory power of this narrative, which was largely crafted from historical material about the specific changes that took place at the juncture of the nineteenth and twentieth centuries. Further motivating a reassessment of this narrative is the fact that some of the historical assumptions (the progress of experimentalism) and historiographic categories (experimentalism, morphology, and natural history) embedded in the narrative seem less obvious today than they did a few decades ago.

If historians of science want to remain relevant to current conversations about the changing nature of scientific research (including the data deluge and the data-driven sciences), they need to offer intellectual frameworks that can bring these changes into historical perspective, raise fresh questions, and challenge received views.[9] This article proposes such a framework, centered on the roles of collecting and collections in the production of knowledge in the life sciences. As I argue here, the development of the life sciences since the second half of the twentieth century is at odds with the standard narrative of the rise of experimentalism. Too much recent biological research, centered on the collection, comparison, and computation of biological data, does not fit well in a story focused on the triumph of experimentation over other ways of producing knowledge. To make sense of these recent historical transformations, I propose a different set of analytical categories, focused on a specific cognitive and material practice centered on the constitution and use of collections, which can in turn help us revisit the longer history of the life sciences, from the Renaissance to the present. It can also, as I suggest in the conclusion, offer promising venues to draw new connections between the life sciences and other sciences in which collecting and collections have played an equally important role. This offers new vistas to revisit the history of the sciences more generally, and in particular to view more critically the claims that data-driven science represents a turning point in the history of science.

GENERALIST VISIONS IN THE LIFE SCIENCES

Writing a generalist or "big-picture" history of biology poses particular challenges.[10] To begin with, despite the many books on the subject, there was no such word as *biology* for most of the subject's history. The study of nature was pursued from many different avenues, sometimes grouped under the name *natural history*, but the extension and meaning of this term has changed profoundly since the Renaissance. The word *biology* began to be used consistently around 1800, but did not enjoy wide currency until half a century later.[11] Institutionally, biology became a recognizable

[9] On the necessity to offer larger intellectual frameworks in the history of science in order to reach a broader readership, see Steven Shapin, "Hyperprofessionalism and the Crisis of Readership in the History of Science," *Isis* 96 (2005): 238–43.

[10] Ludmilla Jordanova, "Gender and the Historiography of Science," *Brit. J. Hist. Sci.* 26 (1993): 469–83; James A. Secord, "Introduction," *Brit. J. Hist. Sci.* 26 (1993): 387–9; Robert E. Kohler, "A Generalist's Vision," *Isis* 96 (2005): 224–9.

[11] Peter McLaughlin, "Naming Biology," *J. Hist. Biol.* 35 (2002): 1–4.

discipline in the United States from the late nineteenth century, but botanists and zoologists (and anthropologists and bacteriologists) in Europe usually enjoyed separate
institutional and intellectual lives well into the mid-twentieth century.[12] Even after a
comfortable institutional home was finally created for biology in universities, noisy
infighting made amply clear that "biology" did not refer to a happy and united family.
Its members disagreed on most things, including intellectual agendas, methodological approaches, and even the question whether it was a pursuit suited to both sexes.

The great diversity of biological practices certainly matched the diversity found
within biology's objects of study, but this brings little comfort to the historian who
would like to write a broad story of its development. Understandably, most historians
of biology have fallen back on more limited projects and let themselves be guided by
the disciplinary divisions of biology, such as physiology, embryology, immunology,
systematics, evolution, and molecular biology. Writing within or about such categories makes sense: they were recognized by the historical actors themselves and
structured their intellectual goals, research practices, social communities, and institutional homes. Yet taking these as units of analysis has tended to isolate historical
subjects from other scientific endeavors and has limited historical explanations of
their development to factors internal to a given discipline.

In the 1970s, a few historians attempted to write more encompassing stories. For
example, William Coleman and Garland Allen each covered an entire century (the
nineteenth and the twentieth, respectively) in their histories of biology and incorporated different approaches to the study of life into their analyses.[13] They organized
their narratives around intellectual concerns such as "form," "function," and "evolution" (for Coleman), or disciplines such as "embryology" and "genetics" (for Allen).
Overall, they told similar stories. In the late nineteenth century, studies of function
(through experimentation) began to take over studies of form (i.e., "morphology"),
and in the twentieth century, the experimental method triumphed in the life sciences,
culminating in the success of molecular biology. For Coleman, "In its name—experiment—was set in motion a campaign to revolutionize the goals and methods
of biology." For Allen, "It was the twentieth century that saw the fanning out of the
experimental method in all areas of biology."[14]Although they differed in their explanation of what drove this transformation—the import of new chemical and physical
methods from the "outside" (Coleman) or a revolt against morphology from the "inside" (Allen)—both focused on the broad category of experimentalism, as a material
practice, a physical place, and an intellectual program. Coleman and Allen's story
has served as a framework for most subsequent histories of the life sciences.

To fully understand the origins of the historical categories that structured the narratives proposed by Coleman and Allen, it is helpful to bring them into the context of
the development of the life sciences at the time of their production. In the late 1960s
and 1970s, when Coleman's and Allen's works were written, biology departments
had been carved into battlegrounds between proponents of different research agendas, yet it was already clear that the future belonged to the experimental life sciences.

---

[12] William V. Consolazio, "Dilemma of Academic Biology in Europe," *Science* 133 (1961): 1892–6.
[13] Coleman, *Biology in the Nineteenth Century: Problems of Form, Function and Transformation*
(Cambridge, 1971); Allen, *Life Science in the Twentieth Century* (Cambridge, 1978); Allen, "Morphology and Twentieth-Century Biology: A Response," *J. Hist. Biol.* 14 (1981): 159–76, as well as
the other contributions to that special issue of the *Journal of the History of Biology*.
[14] Coleman, *Biology*, 2; Allen, *Life Science*, xvi (Both cit. n. 13).

Beginning in 1962, the Nobel Prize committee rewarded molecular biologists year after year for their experimental work. Universities around the world created institutes of molecular biology, testifying to the achievements and promises of this new science.[15] Molecular biologists launched journals devoted to their field, beginning with the *Journal of Molecular Biology* in 1959. Naturalists complained bitterly about the excessive (in their view) attention and support given to experimental approaches.[16] In his autobiography, revealingly titled *Naturalist,* the evolutionary biologist E. O. Wilson recalled the epic wars in Harvard's corridors and faculty meetings that pitted him against another young faculty member, the molecular biologist James D. Watson (the "most unpleasant person" Wilson had ever met).[17] Other naturalists, such as Ernst Mayr and George Gaylord Simpson, waged a war along the same battle lines (Allen was a student of Simpson at Harvard in 1966).[18] They criticized the new molecular evolutionists, who claimed to reconstruct the history of life through the comparison of single molecules, for being overly simplistic. In response, the molecular evolutionists ridiculed the traditional evolutionists, whose work focused on examination of skeletons and fossils, as outdated museum workers who toiled on characters too subjective to be of any scientific value. To mock their opponents' morphological methods and illustrate their limited scope, two molecular evolutionists asked in 1967, "How many vertebrae does a sponge have?"[19] Mayr attempted to draw up a peace settlement, which divided biology's territory into complementary fields, organismic and reductionist biology, concerned with asking historical and functional questions and answering them by appealing to ultimate and proximal causes, respectively.[20] The institutional and intellectual successes of molecular biology in the 1960s provided an end point to Coleman's and Allen's narratives, while the boundaries that were drawn by biologists themselves provided the outline for categories such as *experimentalist* and *naturalist.*

This is not to say that the methodological and disciplinary distinctions drawn by Coleman and Allen, based on their later twentieth-century experiences, were fully at odds with the distinctions made by scientists using the same terms at the turn of the century, nor that these distinctions did not capture some of the important tensions running though the life sciences around 1900. But seeing today what their categories meant to the state of the life sciences and the debates among life scientists in the 1960s invites us to consider a different reading of the broad history of the life sciences. This article is an experiment in *not* taking experimentalism as the end point

[15] On molecular biology as a postwar science, see Soraya de Chadarevian, *Designs for Life: Molecular Biology after World War II* (Cambridge, 2002); Bruno J. Strasser, "Institutionalizing Molecular Biology in Post-war Europe: A Comparative Study," *Stud. Hist. Phil. Biol. Biomed. Sci.* 33 (2002): 533–64.

[16] Michael R. Dietrich, "Paradox and Persuasion: Negotiating the Place of Molecular Evolution within Evolutionary Biology," *J. Hist. Biol.* 31 (1998): 85–111; Joel B. Hagen, "Naturalists, Molecular Biology, and the Challenge of Molecular Evolution," *J. Hist. Biol.* 32 (1999): 321–41.

[17] Wilson, *Naturalist* (Washington, D.C., 1994).

[18] Mayr, "Cause and Effect in Biology," *Science* 134 (1961): 1501–6; Mayr, "The New versus the Classical in Science," *Science* 141 (1963): 765; Simpson, "Organisms and Molecules in Evolution," *Science* 146 (1964): 1535–8; Theodosius Dobzhansky, "Biology, Molecular and Organismic," *American Zoologist* 4 (1964): 443–52; Dobzhansky, "Taxonomy, Molecular Biology, and the Peck Order," *Evolution* 15 (1961): 263–4.

[19] W. M. Fitch and E. Margoliash, "Construction of Phylogenetic Trees: 2. How Well Do They Reflect Past History?" *Brookhaven Symposia in Biology* 1 (1969): 217–42, on 238.

[20] Erika Lorraine Milam, "The Equally Wonderful Field: Ernst Mayr and Organismic Biology," *Hist. Stud. Nat. Sci.* (2010): 279–317.

of the story and *not* taking the contrasts drawn by Mayr between organisms and molecules, and by Allen and Coleman between morphology (and natural history) and experimentalism, as the defining tensions in the history of the life sciences.

GENERALIST VISIONS, MORE GENERALLY

Taking this approach to a generalist vision for the history of the life sciences creates new challenges. Unless one's goal is a history of great men, from Aristotle to, say, J. Craig Venter, it is necessary to choose some high-level analytical category to structure a narrative that will highlight the profound changes and continuities in the history of the life sciences. The goal is not to find categories that enable us, as Plato would suggest, to better carve history "at its joints" (the past is not a chicken), but to find categories that make meaningful connections between historical practices, actors, and events and that can become, at least temporarily, powerful heuristic tools to understand the past and the present. The most common categories used by historians of science have included periods (e.g., romanticism), spaces (the Atlantic world), nations (Germany), disciplines (physiology), ideas (extinction), styles (holistic), and places (laboratories). A combination of styles and places could be particularly promising to structure a generalist vision.

Styles of scientific reasoning, under various names, have been used productively by authors as different as Ludwig Fleck (*Denkstil*), Thomas Kuhn ("paradigms"), Michel Foucault ("episteme"), Gerald Holton ("themata"), Alistair Crombie ("styles"), Jon Harwood ("national styles"), John Pickstone ("ways of knowing"), and others.[21] These analytical categories help historians make sense of the development of science, capturing some of its methodological texture with more nuance than a unique and atemporal "scientific method" would allow. Crombie, for example, distinguishes six styles of scientific reasoning, including elementary postulation, comparative ordering, analogical modeling, statistical analysis, historical derivation, and experimental control.[22]

Because they focus on cognitive practices in science, narratives structured around styles usually pay little attention to material practices and thus tend to separate activities that are linked by common material practices (e.g., agricultural breeding and academic genetics) if they reflect different cognitive goals. Some authors, such as Pickstone, have considered this difficulty and have attempted to use categories that reflect to some extent both the cognitive and the material.[23] "Analysis," one of Pickstone's ways of knowing, for example, is understood as both a mental operation of analyzing abstract ideas into more specific components and a material operation of physically dividing scientific objects into their constitutive parts. Another strength of Pickstone's approach is that even though his ways of knowing each have their own historicity and have, for example, enjoyed their greatest successes at different times,

[21] For a discussion of Pickstone's categories, see Bruno J. Strasser and Soraya de Chadarevian, "The Comparative and the Exemplary: Revisiting the Early History of Molecular Biology," *Hist. Sci.* 49 (2011): 317–36.

[22] Crombie, *Styles of Scientific Thinking in the European Tradition: The History of Argument and Explanation Especially in the Mathematical and Biomedical Sciences and Arts*, 3 vols. (London, 1994).

[23] Pickstone, "Museological Science? The Place of the Analytical/Comparative in Nineteenth-Century Science, Technology and Medicine," *Hist. Sci.* 32 (1994): 111–38; Pickstone, *Ways of Knowing: A New History of Science, Technology and Medicine* (Manchester, 2000); Pickstone, "Working Knowledges before and after circa 1800: Practices and Disciplines in the History of Science, Technology and Medicine," *Isis* 98 (2007): 489–516.

they do not replace each other, as do Kuhnian paradigms, but add new layers to the makeup of science, technology, and medicine. An approach like Pickstone's, which considers styles of reasoning as individual components, makes it possible to analyze scientific practices in terms of these different elements, present in various proportions, without reducing scientific practices to any one of them, like Crombie's styles do. Chunglin Kwa's recent history of science uses Crombie's taxonomy to *classify* different sciences into his six styles; by contrast, I propose here, in line with Pickstone's ways of knowing, to *analyze* different sciences in terms of the material and cognitive practices on which they rely, focusing specifically on practices centered on collections.[24] This approach illuminates the heterogeneity of cognitive and material practices within disciplines and the similarities among disciplines, whereas disciplinary histories have tended to stress the unity of cognitive and material practices within disciplines and the differences among disciplines. One does not need to adopt Pickstone's categorization of the various ways of knowing to appreciate the value of analyzing sciences according to different kinds of practices, such as "collecting" and "comparing."

This is not necessarily a simple task: identifying the role of epistemic practices from the testimonies of historical actors can be challenging. In this respect, life-science historians have faced the same problem as medical historians exploring the "laboratory revolution" in medicine.[25] A standard account, supported by abundant documentary evidence, gave voice to the physicians who, since the late nineteenth century, emphasized how much their profession had been transformed by the laboratory sciences. In short, they claimed that medicine had finally become "scientific," thanks to the introduction of experimental methods. But as a number of studies since the 1980s have convincingly shown, much of medical practice remained unaffected by the new experimental sciences.[26] This does not mean that the laboratory was irrelevant for medicine's transformation, simply that it was more of a rhetorical tool, at least in the United States, used by physicians for the social elevation of their profession. Similarly and at the same time, the repeated claims by scientists about the superiority of the experimental over the natural historical method did not necessarily reflect actual changes in their research practices. To be sure, Coleman, Allen, and many others have amply documented that the discourse about the power of experimentation was not mere rhetoric. It did indeed capture a significant transformation and reflected the growing importance of those sciences that relied most on experimental methods, such as genetics or biochemistry. Nevertheless, a key question remains: Did the unquestionable intellectual successes of the experimental sciences in the twentieth century result from their reliance on the experimental approach alone and its superiority over other forms of inquiry, such as those based on collections? It is difficult to assess solely from the statements made by scientists that this approach had played some role in the successes of their investigative enterprises, because in a cultural context where the association with experimentation carried such authority and prestige and where other approaches, such as those based on collections, were derided as old-fashioned,

[24] Kwa, *Styles of Knowing: A New History of Science from Ancient Times to the Present* (Pittsburgh, 2011).

[25] Andrew Cunningham and Perry Williams, *The Laboratory Revolution in Medicine* (Cambridge, 1992).

[26] John Harley Warner, "Ideals of Science and Their Discontents in Late Nineteenth-Century American Medicine," *Isis* 82 (1991): 454–78.

one can easily understand how researchers might have downplayed the role of the latter in their work, or even denied that it played any significant role at all.

To explore the possibility that collecting practices have been important, not only for the field and museum sciences but also for the laboratory sciences, it is first necessary to question the common conflation by historical actors and scholars alike of places and practices. The "geographical turn" in science studies brought a welcome focus on neglected issues, such as the circulation of people, things, and knowledge,[27] but unfortunately, it has also reified places, such as laboratories and museums, as spaces where similar kinds of practices are supposed to have occurred.[28] It is too often assumed that the laboratory was necessarily a place of experimentation, in the same way that the field was a place of collection and the museum a place of comparison. Even though these have been the key practices in each of these three places, all witnessed a variety of other practices carried out in service of scientific research. The laboratory, in the nineteenth century alone, was a place for teaching morphology, for preparing specimens, and for conducting experiments;[29] the field was a place of collecting, but also of experimentation, observation, and other practices;[30] and the museum (the key target of ridicule for the proponents of the laboratory) hosted a variety of epistemic practices, including experimentation. For example, in 1928, the American Museum of Natural History established its own laboratory devoted to experimental research.[31] Surprisingly, historical studies that take "place" as their focus have not fully taken advantage of one of cultural geography's key insights: places are not passive receptacles for practices; instead, practices generate places (i.e., a café is a place where people drink coffee). This insight offers promising venues to develop analytical categories connecting places and practices, paying attention to both.[32]

A NEW GENERALIST VISION FOR THE LIFE SCIENCES?

The remainder of this article focuses on a specific set of epistemic and material practices, based on collecting things and data, for the constitution of collections, and their use, especially as tools for comparative studies. This account takes a broad view from the Renaissance to the present, but emphasizes the developments in the twentieth century. Few would challenge the fact that from the fifteenth to the nineteenth centuries,

---

[27] David N. Livingstone, *Putting Science in Its Place: Geographies of Scientific Knowledge* (Chicago, 2003); D. A. Finnegan, "The Spatial Turn: Geographical Approaches in the History of Science," *J. Hist. Biol.* 41 (2008): 369–88; Simon Naylor, "Introduction: Historical Geographies of Science— Places, Contexts, Cartographies," *Brit. J. Hist. Sci.* 38 (2005): 1–12; S. Shapin, "Placing the View from Nowhere: Historical and Sociological Problems in the Location of Science," *Trans. Inst. Brit. Geogr.* 23 (1998): 5–12.

[28] Ian F. McNeely and Lisa Wolverton, *Reinventing Knowledge: From Alexandria to the Internet* (New York, 2008).

[29] On the different uses of the laboratory, see the "Focus" section "Laboratory History," *Isis* 99 (2008): 761–802.

[30] Robert E. Kohler, *Landscapes and Labscapes: Exploring the Lab-Field Border in Biology* (Chicago, 2002).

[31] Gregg Mitman and Richard W. Burkhardt Jr., "Struggling for Identity: The Study of Animal Behavior in America, 1930–1950," in *The Expansion of American Biology*, eds. Keith R. Benson, Jane Maienschein, and Ronald Rainger (New Brunswick, N.J., 1991), 164–94.

[32] On places and practices, see Henri Lefebvre, *The Production of Space* (Oxford, 1991). For a successful attempt at this in the history of science, see Robert E. Kohler, "Place and Practice in Field Biology," *Hist. Sci.* 40 (2002): 189–210, and Richard W. Burkhardt Jr., "The Leopard in the Garden: Life in Close Quarters at the Muséum d'Histoire Naturelle," *Isis* 98 (2007): 675–94.

collections such as cabinets of curiosities and wonder cabinets, museums and zoos, and gardens and herbaria were all, in widely different ways, central to the production of knowledge about nature. The key historical question is, What happened to collecting practices after these institutions became less prominent in academic science? The standard account emphasizes that as the experimental sciences (such as genetics, biochemistry, and eventually molecular biology) gained importance, collecting practices and collections associated with museums and other institutions devoted to the pursuit of natural history lost their central role in the production of knowledge in the life sciences. This article argues that, on the contrary, collecting and collections have remained essential in the experimental sciences for the production of biological knowledge. The proliferation of databases of experimental data is the most visible sign of the key role played by collections in the life sciences (and elsewhere). If this is so, one can explore the development of the life sciences from the early modern period to the present in a new way.

One might immediately object that early collections of specimens differ too much from twenty-first-century "biobanks" (such as collections of molecules) or databanks (such as collections of data about molecules) to be productively compared to them. One should remember, however, that early modern and modern collections were not simply storehouses for whole organisms. Zoological museums often stored only bones and skins, and herbaria only parts of plants. Other collections of plant and animal parts became particularly important beginning in the nineteenth century, especially collections of seeds, blood, tissues, and cells.[33] Furthermore, natural history collections always included data, in the form of drawings or verbal descriptions of species, alongside material specimens. As Martin Rudwick has so beautifully shown, Cuvier's work relied not only on the fossils and bones present at the Muséum d'Histoire Naturelle in Paris, but also on the drawings of his own "paper museum."[34] Thus, it should not seem too big a step to argue that collections of data about biological molecules (such as DNA sequences, protein-structure coordinates, or functional MRI images), when gathered in electronic databases, can be subsumed under the same category of *collection* as collections of plants, animals, fossils, and so on. This is an expansion of the usual meaning of *collection* in the life sciences, but it is one that seems not only conceptually plausible, but also historiographically useful.

A more detailed historical, epistemological, and ontological justification for subsuming all of these—museums, herbaria, biobanks, and databases—under a unique category lies beyond the scope of this article. For the time being, let us assume that these assemblages share a sufficient number of similarities to be plausibly brought together under the single analytical category of *collection*, and that this, in conjunction with practices of collecting and comparing, can be used as the basis to reexamine the history of the life sciences. As I will show, foregrounding the history of collections, collecting, and comparing, instead of the categories of museum and laboratory, or natural history and experimentation, offers a different perspective on the history of the life sciences. It allows us to make connections and see continuities otherwise obscured, to ask new questions about the transformations of the life sciences in the

---

[33] Jack Ralph Kloppenburg, *First the Seed: The Political Economy of Plant Biotechnology, 1492–2000* (Cambridge, 1988); Susan E. Lederer, *Flesh and Blood: Organ Transplantation and Blood Transfusion in Twentieth-Century America* (Oxford, 2008); Bronwyn Parry, *Trading the Genome: Investigating the Commodification of Bio-information* (New York, 2004).

[34] Rudwick, "George Cuvier's Paper Museum of Fossil Bones," *Arch. Natur. Hist.* 27 (2000): 51–68.

twentieth century, and to use the rich historical literature on collections up to the nineteenth century to explore and inform our understanding of recent developments.

THE PRACTICE OF COLLECTING

To show the connections between the practice of collecting as it occurred in the twentieth century and in earlier centuries, I will now provide a cursory and necessarily incomplete overview of collecting in natural history from the Renaissance to around 1900 (later periods are territories largely uncharted by historians). Constrained by space and the limited existing literature, I address only three questions: How was it done? By whom? And for what purpose?

One of the first volumes to focus on collecting as a way of knowing, *Sammeln als Wissen,* brought several contributors together to reflect on particular collecting practices.[35] All of them were associated with early modern natural history, yet most of the contributions fell back on the traditional themes of museum history—the exploration of the collections themselves and their cultural meanings—rather than considering collecting as a practice. The richest volume on collecting practices in the life sciences, *Cultures of Natural History*, offers unique insights to understand natural history collecting, but only until the nineteenth century.[36] Some equally revealing examples of particular collecting practices from the mid-nineteenth to the twentieth century include studies of naturalists, especially botanists, in Britain; the Smithsonian's scramble to secure artifacts from American Indians in the Northwest; geologists' search for fossils in England; the efforts of museums and government agencies to survey the biodiversity of the United States; and the attempts of medical researchers to obtain brains from kuru patients in Papua New Guinea.[37] Each study highlights different aspects of collecting, but taken together, they can serve as a basis for outlining some of the common features of collecting.

Robert Kohler seems to be the first historian to have explicitly attempted to analyze collecting as a practice and to focus on the modern, rather than the early modern, period.[38] He argues that "collecting sciences" have included not only systematic biology, but also anthropology and ethnology, geology and mineralogy, and even, at some point, pathology and chemistry. He resolutely takes "a generous view of collecting practices" and convincingly argues that the collecting sciences should not be limited to the natural sciences, but should also include the social sciences and other sciences involved in collecting data.[39] Nonetheless, he draws a sharp distinction between the collecting sciences that focus on things and those that do not. Kohler sees the most characteristic aspect of collecting sciences as being the materiality, the

---

[35] Anke te Heesen and Emma C. Spary, eds., *Sammeln als Wissen: Das Sammeln und seine Wissenschaftsgeschichtliche Bedeutung* (Göttingen, 2001).

[36] Jardine, Secord, and Spary, *Cultures of Natural History* (cit. n. 7).

[37] David Elliston Allen, *The Naturalist in Britain: A Social History* (London, 1976); Douglas Cole, *Captured Heritage: The Scramble for Northwest Coast Artifacts* (Norman, Okla., 1995); Simon J. Knell, *The Culture of English Geology, 1815–1851: A Science Revealed through Its Collecting* (Aldershot, 2000); Robert E. Kohler, *All Creatures: Naturalists, Collectors, and Biodiversity, 1850–1950* (Princeton, N.J., 2006); Warwick Anderson, *The Collectors of Lost Souls: Turning Kuru Scientists into Whitemen* (Baltimore, 2008).

[38] Kohler, "Finders, Keepers: Collecting Sciences and Collecting Practice," *Hist. Sci.* 45 (2007): 428–54.

[39] Kohler, *All Creatures* (cit. n. 37), 433.

"thing-y" nature, of the objects it deals with. He defines collecting scientists as not just "finders," because "all scientists are finders (in one way or the other)," but as "keepers," because "only collecting scientists are also keepers."[40] For Kohler, finding material objects and keeping them defines the collecting sciences. Although I take issue with some aspects of Kohler's definition, especially with respect to the "found" nature of things, my attempt to characterize collecting practices takes a similarly broad approach, from the fifteenth to the end of the nineteenth century.

Naturalists throughout this period are generally recognized as having been active collectors. Although their collecting practices took extremely diverse forms, the key challenge of collecting, and of establishing a collection, remained the same over time: how to bring spatially dispersed objects to a central location and make them commensurable. Consequently, collecting was (and is), above all, a spatial practice. Renaissance collections were filled first with objects coming from the immediate environment.[41] Local plants, animals, and minerals, especially, were brought into closer proximity with one another. Collectors, after having exhausted the diversity of their local surroundings, embarked on the more ambitious goal of filling their collections with objects found far beyond their everyday reach.

Establishing this kind of collection, like establishing empires, required the mastery of space. Collectors produced a movement of natural things, which were often dispersed across the world, toward central locations, just as empires produced movements of goods from colonies to metropoles. Unsurprisingly, colonial powers were collecting powers, and colonies constituted rich collecting grounds.[42] The geographical reach of an empire represented an immense field for collecting. The objects in the collections of Kew Gardens or the British Museum, for example, came from the same places and followed the same routes as the other goods circulating through the British Empire. And collecting, just like the imperial enterprise, required domination over people, not just things. Indeed, most collecting was done by proxy. Collectors in the metropole relied on local naturalists, hunters, and gatherers in the colonies, although some collectors did go into the field themselves to collect specimens (and were sometimes carried on comfortable chairs by the locals).[43]

But collecting should not be reduced, as it sometimes has been, to the history of colonial exploitation. It also followed the lines of gift economies, as in Renaissance Italy or the French republic of letters.[44] For example, in eighteenth-century France,

[40] Ibid., 432.

[41] Paula Findlen, *Possessing Nature: Museums, Collecting, and Scientific Culture in Early Modern Italy* (Berkeley and Los Angeles, 1994); Brian W. Ogilvie, *The Science of Describing: Natural History in Renaissance Europe* (Chicago, 2006).

[42] Lucile H. Brockway, *Science and Colonial Expansion: The Role of the British Royal Botanic Gardens* (New York, 1979); Richard W. Burkhardt Jr., "Naturalists' Practices and Nature's Empire: Paris and the Platypus, 1815–1833," *Pacific Sci.* 55 (2001): 327–41; Londa L. Schiebinger and Claudia Swan, eds., *Colonial Botany: Science, Commerce, and Politics in the Early Modern World* (Philadelphia, 2005); Daniela Bleichmar and Peter C. Mancall, eds., *Collecting across Cultures: Material Exchanges in the Early Modern Atlantic World* (Philadelphia, 2011).

[43] Londa L. Schiebinger, *Plants and Empire: Colonial Bioprospecting in the Atlantic World* (Cambridge, Mass., 2004), chaps. 1–2. See also Harold John Cook, *Matters of Exchange: Commerce, Medicine, and Science in the Dutch Golden Age* (New Haven, Conn., 2007).

[44] Paula Findlen, "The Economy of Scientific Exchange in Early Modern Italy," in *Patronage and Institutions: Science, Technology, and Medicine at the European Court, 1500–1750*, ed. Bruce T. Moran (Rochester, N.Y., 1991), 5–24; Findlen, *Possessing Nature* (cit. n. 41).

where the possession of natural objects became a sign of distinction, there were coveted cultural objects to be offered to collectors, often rich patrons who would reciprocate by offering other natural objects or patronage in return.[45] These practices created what Emma Spary has so appropriately called, referring to the network of collectors around the French Jardin des Plantes in the eighteenth century, a system of "polite indebtedness."[46] Furthermore, although collecting centers have often been imperial capitals (e.g., London, Amsterdam, and Paris), this was not always the case (e.g., Geneva, Kew, and Montpellier). It might thus be more productive to think that collections only became "centers" once they succeeded in generating a "periphery." By convincing naturalists around the world to send them specimens, museums and botanical gardens became centers for the production of natural knowledge.

Natural history objects not only traveled as gifts along existing social networks, but also moved as commodities that could simply be purchased by collectors. This practice of collecting was particularly important for those who wished to obtain specimens beyond the frontiers of the empire or who did not have access to colonial networks of power. In the busy merchant port at Canton in the nineteenth century, for example, British collectors purchased animals and plants from the luxuriant markets, and so the city itself became the "field" for these second-order collectors.[47] At the same time in the United States, animal dealers, often hunter-entrepreneurs, offered wild animals for sale to zoos and natural history museums.[48]

When the routes of empire or commerce were unavailable, collectors mounted their own expeditions. In the late nineteenth and early twentieth centuries, the major natural history museums in Europe and in the United States commissioned expeditions to Asia, Africa, and South America. The American Museum of Natural History, for example, sent groups of naturalist-collectors to Congo between 1909 and 1915 to survey the local fauna and bring back mammals for the museum's African Hall.[49] Expeditions were not always successful in bringing back animals, especially live animals, but they always succeeded in returning with stories, and these stories shaped the image of the naturalist-collector into one of an explorer-adventurer.[50] In the same period, as Kohler has shown, a new way of collecting developed under the auspices of museums and governmental agencies: systematic survey collecting. Thanks to a specific nexus of environmental, technological, and cultural factors, this produced an "inner frontier," where biologically rich collecting spaces were never too far from civilization.[51]

Many collectors associated with these different modes of collecting were what Londa Schiebinger has aptly called "armchair collectors," in that they relied exclu-

[45] Philipp Blom, *To Have and to Hold: An Intimate History of Collectors and Collecting* (Woodstock, N.Y., 2003).

[46] Spary, *Utopia's Garden: French Natural History from Old Regime to Revolution* (Chicago, 2000), 77.

[47] Fa-ti Fan, *British Naturalists in Qing China: Science, Empire, and Cultural Encounter* (Cambridge, Mass., 2003).

[48] On the animal dealers, see Mark Barrow, "The Specimen Dealer: Entrepreneurial Natural History in America's Gilded Age," *J. Hist. Biol.* 33 (2000): 493–534; on hunters, Elizabeth Hanson, *Animal Attractions: Nature on Display in American Zoos* (Princeton, N.J., 2002), chap. 3.

[49] Lyle Rexer and Rachel Klein, *American Museum of Natural History: 125 Years of Expedition and Discovery* (New York, 1995).

[50] Douglas J. Preston, *Dinosaurs in the Attic: An Excursion into the American Museum of Natural History* (New York, 1986).

[51] Kohler, *All Creatures* (cit. n. 37).

sively on a network of individuals to gather materials for them.[52] Some had participated in collecting expeditions in the field early in their careers, but later became
curators at institutions such as natural history museums or botanical gardens, where
they directed their collecting enterprises from their desks through correspondence
networks. These armchair collectors painstakingly attempted to coordinate and discipline field collectors in making standardized observations of the specimens they
collected, a condition for successfully creating a reliable "collective observer" and
a useful collection.[53] The fact that such collectors often published in their own name
faunas and floras that were based on specimens and data gathered by numerous collaborators maintained the illusion that the knowledge produced through collecting
practices was, like that produced through experimentation, the result of an individu-
alist endeavor. This conformed to the persistent ideal that only individuals are the
creators of knowledge.[54]

Because collecting was essentially a collective practice, carried out by very different actors, the issue of epistemic and social coordination was essential.[55] When
Western physicians collected brains from patients who had recently died of kuru in
New Guinea, they had to negotiate over the status of the brain and their own status as
collectors; they were considered physicians by some, sorcerers by others.[56] Similarly,
when envoys from the Smithsonian collected canoes and other artifacts on the Northwest coast, they bargained over the authenticity and value of these artifacts with the
natives, who had sometimes created them especially for the collectors.[57] Even when
all the individuals in the collective were of a similar professional background—botanists, for example, in the case of the Kew Gardens naturalist Joseph Hooker's collecting plants from his New Zealand correspondents—intense arguments took place
between the field and the institutional collector over the status of a rare find and its
relation to the attribution of credit and authorship.[58]

Given this diversity of modes of collecting, what were the main characteristics of
collecting as a unified practice? First, collecting was a spatial practice, always negotiating problems of position, scale, and reach. Second, and following on the first,
collecting was a local practice. Even global collecting efforts, at some level, required
local collectors to gather materials in the field. Third, it was a collective practice. Few
individuals assembled their collections alone; almost all relied on extended networks
of people. As Spary has put it, natural history in the eighteenth century, a science
highly reliant on collecting, was "a science of networks."[59] Finally, given the heterogeneity of these networks, bringing together naturalists, hunters, and merchants (to

---

[52] Schiebinger, *Plants and Empire* (cit. n. 43), chap. 1.

[53] Peter Galison and Lorraine Daston, "Scientific Coordination as Ethos and Epistemology," in *Instruments in Art and Science: On the Architectonics of Cultural Boundaries in the 17th Century*, ed. Helmar Schramm, Ludger Schwarte, and Jan Lazardzig (Berlin, 2008), 296–333.

[54] Mario Biagioli and Peter Galison, eds., *Scientific Authorship: Credit and Intellectual Property in Science* (New York, 2003).

[55] For a useful perspective on the translation of various interests in a collecting network, see Susan Leigh Star and James R. Griesemer, "Institutional Ecology, 'Translations' and Boundary Objects: Amateurs and Professionals in Berkeley's Museum of Vertebrate Zoology, 1907–1939," *Soc. Stud. Sci.* 19 (1989): 387–420.

[56] Anderson, *Collectors of Lost Souls* (cit. n. 37).

[57] Cole, *Captured Heritage* (cit. n. 37).

[58] Jim Endersby, *Imperial Nature: Joseph Hooker and the Practices of Victorian Science* (Chicago, 2008).

[59] Spary, *Utopia's Garden* (cit. n. 46), 97.

name a few) required the translation of very diverse interests. Because the objects of collections meant very different things to these different people, collectors negotiated complex issues of credit, translating among often incommensurable values to keep objects flowing toward their collection.

WHO THE COLLECTORS WERE

Still unanswered in this overview of collecting as a practice are the questions, Who was doing the collecting, how did these individuals characterize themselves, how were they characterized by others, and what was their social position among scientists? Answering these questions offers illuminating material for comparisons between old and new collectors. In his *Collectors and Curiosities*, Krzysztof Pomian provides a wonderfully rich account of the world of collectors—a group that included far more than natural history collectors—in France and Italy between the sixteenth and the eighteenth centuries.[60] He shows how collectors who had amassed medals, paintings, instruments, and natural specimens and stored these together in collections such as wonder cabinets began, in the seventeenth century, to specialize by focusing on a single kind of object. Antiquaries favored artifacts reflecting the life of the ancients; savants favored the collection of natural history objects. Within each category of collected object, the collectors were a highly heterogeneous mix of people with different interests, a characteristic that has continued to define collectors to the present day.[61] As a result, individual collectors have had very diverse and unstable social identities.

This diversity of social identities certainly applies to those who collected natural objects. Beginning in the Renaissance, when the social identity of "naturalist" was solidifying, most naturalists were collectors of some sort, but not all collectors of natural objects were naturalists. Natural objects were collected by all kinds of people for all kinds of reasons. But it is possible to identify some of the categories according to which collectors came to be identified. These included, since the nineteenth century at least, the amateur and the professional, and the field and the museum collector.

Since the nineteenth century, amateur naturalists, who often had as much expertise in their field of specialty as professionals, were indispensable to collecting enterprises, from surveys of local floras to expeditions to remote places where only local inhabitants possessed knowledge of their natural environment.[62] The enthusiastic participation of amateurs in natural history collecting proved to be a mixed blessing for the professional naturalists, such as Hooker at Kew Gardens. He could count on local collectors to provide specimens from the other side of the Earth (in this case, New Zealand), but tensions arose over issues of credit, especially over the right to name new species. Amateurs were generally unpaid, so they sought remuneration in other forms, such as the right to name species for posterity. But Hooker and other naturalists at the metropole thought that naming was the privilege of the

---

[60] Pomian, *Collectors and Curiosities: Paris and Venice, 1500–1800* (Cambridge, 1990).

[61] For a rich account of cabinet collecting in the Renaissance and early modern period, see Blom, *To Have and to Hold* (cit. n. 45).

[62] David Elliston Allen, "Amateurs and Professionals," in *The Cambridge History of Science,* vol. 6, *The Modern Biological and Earth Sciences*, ed. Peter J. Bowler and John V. Pickstone (Cambridge, 2009), 15–33.

professional who entered the specimen into the formal scientific literature.[63] The association between collecting and amateur science has also been at times a curse for naturalists in their quest for professional respectability in the sciences, especially where amateur collecting became an accepted leisure activity. For example, in Victorian Britain, collecting ferns was a popular hobby for the rising bourgeoisie, just as collecting plants became part of a middle-class ideal of vacationing in the United States at the turn of the twentieth century.[64] Because the late nineteenth century was also a moment when the sciences were becoming increasingly professionalized, the association of amateur activity with collecting was detrimental to the development of sciences dependent on collecting. As Kohler has put it, "Scientific collecting was just too much like camping and sport hunting to be taken seriously by guardians of the public purse—too much like plain fun."[65]

Collecting, especially in botany, was not only associated with amateurs; it was to some extent gendered as a female pursuit. Even before the Victorian days of the great fern craze, when women eagerly collected specimens, botanists tried to dispel the idea that in Britain the pursuit was "of so low a character, as to be calculated for the amusement of women," as one commentator put it in 1831.[66] Half a century later, the author of a letter published in *Science* was still trying to counter the widespread idea that botany and the collecting of botanical specimens were "suitable enough for young ladies and effeminate youths, but not adapted for able-bodied and vigorous-brained young men who wish to make the best use of their powers."[67] In the early twentieth century, when women became an increasingly important "workforce" in science, they were predominantly relegated to subaltern and repetitive tasks, such as data and specimen collection; this division of labor reinforced the gendering of the sciences that depended on collecting practices.[68] Into the twentieth century, the gendering of natural history collecting continued to affect all of biology. The molecular biologist Sydney Brenner, before the rise of his new discipline, stated, "Biology, I am sorry to say, was a subject for girls."[69]

The association of natural history collecting with amateurs limited the professional opportunities of collectors. Among professional collectors, a few found coveted

[63] On how nomenclature rules shifted the power between museum and field collectors, and between European and New World collectors, see Christophe Bonneuil, "The Manufacture of Species: Kew Gardens, the Empire and the Standardisation of Taxonomic Practices in Late 19th Century Botany," in *Instruments, Travel and Science: Itineraries of Precision from the Seventeenth to the Twentieth Century*, ed. Marie-Noëlle Bourguet, Christian Licoppe, and Heinz Otto Sibum (New York, 2002), 189–215; Sharon E. Kingsland, *The Evolution of American Ecology, 1890–2000* (Baltimore, 2005), chap. 2; and Endersby, *Imperial Nature* (cit. n. 58), chap. 8.

[64] On the former, see Allen, *Naturalist in Britain*; on the latter, Kohler, *All Creatures*, chap. 2 (Both cit. n. 37), and Mark V. Barrow, *A Passion for Birds: American Ornithology after Audubon* (Princeton, N.J., 1998).

[65] Kohler, *All Creatures* (cit. n. 37), 93.

[66] Cited in Endersby, *Imperial Nature* (cit. n. 58), 39.

[67] J. F. A. Adams, "Is Botany a Suitable Study for Young Men," *Science* 9 (1887): 116–7, on 117. This source is cited in Philip J. Pauly, "Summer Resort and Scientific Discipline: Woods Hole and the Structure of American Biology," in *The American Development of Biology*, ed. Ronald Rainger, Keith R. Benson, and Jane Maienschein (Philadelphia, 1988), 121–50.

[68] On women as a workforce, see Margaret W. Rossiter, *Women Scientists in America: Before Affirmative Action, 1940–1972* (Baltimore, 1995), chap. 3; on women in surveys, Kohler, *All Creatures* (cit. n. 37), 215–20; and on the gendering of biology at the turn of the century, Pauly, "Summer Resort," (cit. n. 67), 129.

[69] Cited in de Chadarevian, *Designs for Life* (cit. n. 15), 89.

positions in natural history museums as directors, like Louis Agassiz and later Mayr at Harvard's Museum of Comparative Zoology, or as curators of specific collections, like Simpson, who was in charge of the American Museum of Natural History's Department of Geology and Paleontology.[70] These positions might have represented professional stability for those who held them but, due to the cultural significance of museums, they also reinforced the association between collectors and leisure activities, especially since the late nineteenth century. The place of collecting has, in large part, defined the identity of the collectors.

As Paula Findlen has so eloquently shown, early modern wonder cabinets, such as that of Ulisse Aldrovandi in Bologna, which contained several thousands of specimens, were hybrid places.[71] Aldrovandi's cabinet, as all others, served many functions. It was a display of his patron's power, a place of civil conversation, and a repository that he used as a basis for the descriptions of animals he published in his numerous natural history books. The leisurely and the scholarly lived side by side. Only in the second half of the nineteenth century did museums begin to make a clearer division between the two. This new "dual arrangement" physically separated the spaces devoted to research and those intended for public displays.[72] Yet, because natural history museums were mainly funded by public monies, philanthropies, and ticket sales, they often emphasized their role as places of education and amusement, rather than research. These museums, with their emblematic dinosaur skeletons, became increasingly associated with other institutions of bourgeois entertainment, such as movie theaters, restaurants, and zoos. This did not raise the scientific stature of the research carried out in rooms behind the museums' lavish dioramas.[73] Even the "research" expeditions of natural history museums were framed as enterprises of exploration and adventure (rather than as purely scientific pursuits) to attract the public attention required to fund these costly endeavors.[74] As a result, collectors were once again trapped by their association with amateurish pursuits. Museums solely devoted to research, such as Berkeley's Museum of Vertebrate Zoology, remained exceptional, far too rare to affect the cultural meaning of sciences dependent on collecting.

Regardless of its association with amateurish activities, within the naturalist community, personal field-collecting experience seems to have been indispensable to making a career.[75] Even naturalists who directed collecting enterprises from their

[70] On the former, see Mary P. Winsor, *Reading the Shape of Nature: Comparative Zoology at the Agassiz Museum*, Science and Its Conceptual Foundations (Chicago, 1991); on the latter, Léo F. Laporte, *George Gaylord Simpson: Paleontologist and Evolutionist* (New York, 2000).

[71] Findlen, *Possessing Nature* (cit. n. 41).

[72] On the dual arrangement, see Lynn K. Nyhart, *Modern Nature: The Rise of the Biological Perspective in Germany* (Chicago, 2009), chap. 6, and Mary P. Winsor, "Museums," in Bowler and Pickstone, *Modern Biological and Earth Sciences* (cit. n. 62), 60–75.

[73] On dioramas at the American Museum of Natural History, see Donna Haraway, "Teddy Bear Patriarchy: Taxidermy in the Garden of Eden, New York City, 1908–1936," *Social Text* 11 (1984–5): 20–64; Stephen C. Quinn, *Windows on Nature: The Great Habitat Dioramas of the American Museum of Natural History* (New York, 2006); Karen Wonders, *Habitat Dioramas: Illusions of Wilderness in Museums of Natural History* (Uppsala, 1993).

[74] On the funding of these expeditions, see Michael Kennedy, "Philanthropy and Science in New York City: The American Museum of Natural History, 1868–1968" (PhD diss., Yale Univ., 1968); on expeditions, Lyle Rexer and Rachel Klein, *American Museum of Natural History: 125 Years of Expedition and Discovery* (New York, 1995).

[75] Ernst Mayr, E. Gorton Linsley, and Robert Leslie Usinger, *Methods and Principles of Systematic Zoology*, McGraw-Hill Publications in the Zoological Sciences (New York, 1953), chap. 4.

desks had been active in collecting specimens in the field at the beginning of their careers. In the nineteenth century, Hooker accompanied a polar expedition to gather the material for his *Botany of the Antarctic Voyage* and traveled to India and the Himalayas before he got a position at Kew Gardens.[76] Similarly, in the twentieth century, Mayr collected thousands of bird skins in Papua New Guinea before he joined the American Museum of Natural History.[77] When these professionals became directors of garden or museum collections, where they relied almost exclusively on existing collections or on other naturalists to gather specimens in the field, they nonetheless claimed an identity of (former) field collectors,[78] unlike theoretical physicists, for example, who would pride themselves on never performing experiments.

THE USES OF COLLECTIONS

The perception that collectors were solely interested in the accumulation of specimens, rather than in the production of knowledge, left most of them without a job, at least in academia. The caricature of the naturalist as a "stamp collector," an expression used at least since the mid-nineteenth century, or the physicist Ernest Rutherford's comment in the first decades of the twentieth century that "all science is either physics or stamp collecting," illustrates the low scientific standing attributed to collecting, especially in the twentieth century.[79] Yet, for most naturalists, collecting was a means to an end with recognizable scientific value: the constitution of a collection that would serve as the basis for their production of biological knowledge.[80]

Even though the institutions hosting collections—museums, gardens, and zoos—have been, since the nineteenth century, places of public enlightenment, moral education, and entertainment, they were (and are) also key places for the production of scientific knowledge. Historians who have worked on natural history museums have emphasized these institutions' role in the collection and display of specimens, but paid less attention to how they were used for the production of knowledge. This circumstance necessitates exploring the role of collections on a more general than specific level, with reference especially to how collection both mirrors and differs from experimentation in practice.

At least since the beginning of the early modern period collections were used to gain insight into the natural world. In wonder cabinets, such as Ferrante Imperator's seventeenth-century cabinet in Naples, the juxtaposition of widely different specimens served to highlight their uniqueness, rarity, or wondrous character.[81] After the

[76] Endersby, *Imperial Nature* (cit. n. 58), chap. 1.

[77] On Mayr's travel, see Jürgen Haffer, *Ornithology, Evolution, and Philosophy: The Life and Science of Ernst Mayr, 1904–2005* (New York, 2007), chap. 2; on his use of collections, Kristin Johnson, "Ernst Mayr, Karl Jordan, and the History of Systematics," *Hist. Sci.* 43 (2005): 1–35.

[78] See, e.g., Mayr's self-characterization in Mayr, Linsley, and Usinger, *Methods and Principles* (cit. n. 75), chap. 4.

[79] On the history of "stamp collecting," see Kristin Johnson, "Natural History as Stamp Collecting: A Brief History," *Arch. Natur. Hist.* 34 (2007): 244–58.

[80] For a good example of the role of collections for systematic work in the twentieth century, see Johnson, "Ernst Mayr" (cit. n. 77).

[81] Findlen, *Possessing Nature* (cit. n. 41); Lorraine Daston and Katharine Park, eds., *Wonders and the Order of Nature, 1150–1750* (Cambridge, Mass., 1998); Robert John Weston Evans and Alexander Marr, *Curiosity and Wonder from the Renaissance to the Enlightenment* (Aldershot, 2006).

collapse of "emblematic natural history,"[82] collections continued to be essential tools for the production of natural knowledge, but in a different epistemological setting. Louis XV's natural history collection, for example, was cataloged by Georges Louis Leclerc Buffon in what eventually became his thirty-six-volume description of the natural objects known to the eighteenth century, the *Histoire naturelle, générale et particulière*.[83] After the Revolution, when the collection was incorporated into the Muséum d'Histoire Naturelle in Paris, it served as a basis for Cuvier's masterful natural histories of quadrupeds and fishes and for his theories of animal anatomy and extinction.[84] In the twentieth century, the elaboration of the evolutionary synthesis by Simpson and Mayr resulted from their extensive examination of the collections of fossils and bird skins of the American Museum of Natural History.

Although these various naturalists' collections were composed according to widely different rules, they all served the same purpose: making systematic comparisons possible by physically juxtaposing different objects. Since the Renaissance at least, collections seem to have worked as material representations of nature, as a "second nature" that could be described, measured, analyzed, and compared in order to generate natural knowledge. These collections can be considered to have been representations of nature, like paintings, because they reflected an intentional perspective, embodied in a narrow selection of natural objects. Furthermore, they were groupings of things as made by collectors, not as found in nature. A collector isolated a thing in nature (say, a bird), stripped it of its relations to its surroundings (the forest), left behind most of its properties (such as being alive), and turned it into a specimen embedded in a new system of relations with other specimens in a collection. Birds could be found in trees, but in collections, there were only specimens. One only needs to think of the indispensable role of taxidermists in preparing specimens for museum conservation to realize how much these are also human artifacts. In this sense, the production of knowledge from collections was no different from the production of knowledge from experiments. The objects of knowledge in the experimental sciences, the "epistemic things" that Hans-Jörg Rheinberger has so productively explored, were not found in nature either; they were made through the human creation of "experimental systems" and the production of controlled phenomena.[85]

Importantly, collections differed from catalogs or repositories of identical things, in that they embodied the idea that the objects they contained were related in some natural (or supranatural) way that the comparative perspective would reveal. After the seventeenth century, natural objects were collected separately from human artifacts, such as scientific instruments, because they were believed to be related in a unique way.[86] Though based on widely different assumptions, Richard Owen's search

[82] William B. Ashworth, "Emblematic Natural History of the Renaissance," in Jardine, Secord, and Spary, *Cultures of Natural History* (cit. n. 7), 17–37.

[83] Buffon, *Histoire naturelle, générale et particulière, avec la description du cabinet du roy* (Paris, 1749).

[84] Dorinda Outram, *Georges Cuvier: Vocation, Science, and Authority in Post-revolutionary France* (Manchester, 1984); Toby A. Appel, *The Cuvier-Geoffroy Debate: French Biology in the Decades before Darwin* (New York, 1987); M. J. S. Rudwick, *Georges Cuvier, Fossil Bones, and Geological Catastrophes: New Translations and Interpretations of the Primary Texts* (Chicago, 1997).

[85] Rheinberger, *Toward a History of Epistemic Things: Synthesizing Proteins in the Test Tube* (Stanford, Calif., 1997); Ian Hacking, "The Self-Vindication of the Laboratory Sciences," in *Science as Practice and Culture*, ed. Andrew Pickering (Chicago, 1992), 29–64.

[86] Pomian, *Collectors and Curiosities* (cit. n. 60).

for an archetype and the post-Darwinian search for common descent both proceeded through the comparison of specimens in collections that were believed to share common properties.[87] The notion of homology served as a guiding principle to organize collections and make comparisons. This helps clarify the conceptual limits of the category of *collection*. Collections assembled things that were believed to be related in nature, not just in a researcher's mind. This also justifies why they can be considered representations, because they bore this kind of epistemological relationship to the natural world.

Collections can also be conceptualized as "relational systems." Unlike experimental systems, which offer the possibility to manipulate and create differences in a single object, relational systems make the systematic comparison of many objects possible. Natural history collections were composed not only of individual things, but also of all the many relations among the things they contained. As a result, their epistemic potential was understood to grow exponentially with their size and was driven by the ideal of "completedness." As Buffon put it, "At each sight, not only does one gain a real knowledge of the object considered, but furthermore one discovers the relationships it can have with those around it."[88]

Comparisons such as those made by Buffon and other naturalists required more than the spatial concentration of things. Scientific collections, unlike many others, made the comparison of apples and oranges (or of bacteria and elephants) possible by performing two operations: an ontological reduction and a formal standardization. Elements in a collection were each reduced to a common set of properties (bones or skins) and then were formatted identically (as mounted specimens). Taken alone, these operations might seem trivial, yet together they potentiated the epistemic function of collections. They made the production of general knowledge possible through the comparison of numerous items (a logically invalid, but practically valuable, form of induction). For example, from the comparison of numerous bird specimens with one another, researchers drew general conclusions about classes of things—types of birds—and about their structure, function, and history.

Experimentalists achieved the aim of producing general knowledge in a different way: they paired the use of a few carefully selected (mainly for practical reasons) "exemplary" model systems with the assumption that these systems were representative across broader classes of things. The exemplary perspective has been as important for experimentalists as the comparative perspective has been for naturalists.[89] From Claude Bernard's use of dogs as models of human physiology to geneticists' use of fruit flies as models of genetic transmission, the growth of the experimental life sciences depended on the development of model organisms.[90] These provided the laboratory researchers' "second nature." Experimentalists firmly believed that the knowledge produced with these few model species was of universal validity. The French molecular biologists Jacques Monod and François Jacob (paraphrasing Albert Kluyver) put it best in 1961: "[What is] true of *E. coli* must also be true of

[87] Nicolaas A. Rupke, *Richard Owen: Biology without Darwin* (Chicago, 2009).

[88] Buffon, *Histoire naturelle* (cit. n. 83), 4; translation mine.

[89] For a broader discussion of this distinction, see Strasser and de Chadarevian, "Comparative and the Exemplary" (cit. n. 21).

[90] Frederic Lawrence Holmes, *Claude Bernard and Animal Chemistry: The Emergence of a Scientist* (Cambridge, 1974); Robert E. Kohler, *Lords of the Fly: Drosophila Genetics and the Experimental Life* (Chicago, 1994).

Elephants."[91] Needless to say, Monod and Jacob never brought an elephant into the laboratory to check this assumption.

Experimental systems and relational systems both produced universal knowledge, via not only an abstract intellectual operation, but also a material transformation. As Bruno Latour has suggested, knowledge produced in a laboratory is made universal by extending the conditions of the laboratory to the outside world, including to other laboratories.[92] Facts produced experimentally in a laboratory somewhere can be replicated in laboratories anywhere, not only because of underlying regularities in nature, but also because laboratories have been made almost identical to one another through the standardization of instruments, protocols, and skills. Laboratories can thus be pictured as "centrifugal places": facts travel outward from their initial site of production. Collections, by contrast, can be pictured as "centripetal places": they concentrate objects often otherwise dispersed around the world (such as plants and animals) and partially standardize them in order to make them more easily comparable. When objects become accessible in a single place, in a single format, they can be arranged to make similarities, differences, and patterns apparent to the eye of a single human investigator; collections concentrate the world, making it accessible to the limited human field of view. As Buffon put it in 1749, "The more you see, the more you know."[93]

This brief overview of collecting practices, the identity of collectors, and the epistemic uses of collections serves as a backdrop to my historical reconstruction of the surprising development of collections of biological things and data in the twentieth century. Later collectors faced some of the same challenges as their predecessors, but in a very different context. As briefly outlined in the introduction of this article, historians of the life sciences have explored in great detail the rhetorical and institutional battles between experimentalists ("laboratory men") and naturalists ("museum men") at the turn of the twentieth century and between molecular biologists and evolutionary biologists at midcentury.[94] Historians of the life sciences have also investigated how naturalists responded to the dominance of experimentalism and the transformations that took place within natural history.[95] But what historians have not yet explored is the fact that the same debates took place not just between experimentalists and naturalists or within natural history, but also between experimentalists and within the experimental life sciences themselves. The stellar rise of the experimental life sciences in the twentieth century obscured the fact that their success was not

[91] Monod and Jacob, "General Conclusions: Teleonomic Mechanisms in Cellular Metabolism, Growth, and Differentiation," *Cold Spring Harbor Symposia on Quantitative Biology* 21 (1961): 389–401.

[92] Latour, "Give Me a Laboratory and I Will Raise the World," in *The Science Studies Reader*, ed. Mario Biagioli (New York, 1999), 258–75.

[93] Buffon, *Histoire naturelle* (cit. n. 83).

[94] Allen, "Morphology and Twentieth-Century Biology" (cit. n. 13); Dietrich, "Paradox and Persuasion"; Hagen, "Naturalists" (Both cit. n. 16).

[95] Keith Vernon, "Desperately Seeking Status: Evolutionary Systematics and the Taxonomists' Search for Respectability, 1940–60," *Brit. J. Hist. Sci.* 26 (1993): 207–27; Joel B. Hagen, "Experimental Taxonomy, 1920–1950: The Impact of Cytology, Ecology, and Genetics on the Ideas of Biological Classification" (PhD diss., Oregon State Univ., 1984); Hagen, "Experimentalists and Naturalists in 20th-Century Botany—Experimental Taxonomy, 1920–1950," *J. Hist. Biol.* 17 (1984): 249–70; Bruno J. Strasser, "Laboratories, Museums, and the Comparative Perspective: Alan A. Boyden's Serological Taxonomy, 1925–1962," *Hist. Stud. Nat. Sci.* 40 (2010): 149–82.

necessarily the result of experimental practices, but emerged also, as I argue, from practices centered on collections.[96]

COLLECTING EXPERIMENTS IN THE TWENTIETH CENTURY

As an initial step in considering the place of collecting and collections in the twentieth-century experimental life sciences, let us reexamine the paradigmatic example of the experimentalists' triumph: molecular biology. The stories of the greatest successes of molecular biology—determining the three-dimensional structure of proteins, understanding the structural basis of their function, and deciphering the genetic code—have all been told as having resulted from experimental virtuosity (generally leading to Nobel Prizes). In the last case, the narrative is made all the more poignant by the success of two relatively unknown researchers in 1962 in cracking the code experimentally after a number of great minds, including Francis Crick, had tried unsuccessfully for years to find a solution theoretically. But as Soraya de Chadarevian and I have shown elsewhere, these achievements were due to a combination of experimental and comparative approaches, not to experimental breakthroughs alone.[97] Frederick Sanger, who determined for the first time the sequence of a protein (insulin isolated from an ox), was at a loss to identify which part of the molecule played a significant role for its biochemical function—that is, at a loss until he sequenced insulin molecules from several other species, compared the sequences, and identified specific regions that had remained constant throughout evolution. Similarly, to understand the structural basis of the hemoglobin molecule's function, Max Perutz relied on an extensive collection of hemoglobin molecule variants, which he compared systematically.[98] Finally, the first codon of the genetic code was determined thanks to an ingenious experiment, but in the determination of the remaining sixty-three codons, collections of protein sequences from various organisms proved to be a tremendous asset.

These examples were not isolated cases in the history of molecular biology. Scientific recognition and public visibility have usually gone to the authors of audacious experiments performed on "exemplary" cases in model organisms and model systems. But these achievements were often made possible by the comparison of experimental data from a much wider range of organisms and systems. The results obtained on these other systems were not mere repetitions of the initial finding; instead, the accumulation of results opened up new epistemic possibilities—namely, systematic comparisons. In the second half of the twentieth century, public collections of experimental data became increasingly common, providing researchers with the material they needed for making such comparative studies.

In the scholarly discussion that followed the publication of Allen's history of the twentieth-century life sciences, one point became clear. Even though experimentalism grew tremendously in this period and captured most of the attention, natural

---

[96] For an examination of this claim for the history of molecular biology, see Strasser and de Chadarevian, "Comparative and the Exemplary" (cit. n. 21).

[97] Ibid.

[98] Soraya de Chadarevian, "Following Molecules: Haemoglobin between the Clinic and the Laboratory," in *Molecularizing Biology and Medicine: New Practices and Alliances, 1910s–1970s*, ed. de Chadarevian and Harmke Kamminga (Amsterdam, 1998), 171–201.

history did not disappear. Lynn K. Nyhart argued that natural history might have become secondary to the experimental life sciences, but it kept growing with the general expansion of biology, while Keith R. Benson claimed that natural history remained "alive and well," although "primarily within museums."[99] In the twentieth century, collecting in the field (e.g., to provide specimens for natural history museums or data for ecological studies) remained an important activity, and naturalist collections, of specimens and data alike, continued to grow—and historians of science have hardly begun to explore this area.

I go one step beyond this argument that natural history (with natural history collecting) was "alive and well" in the twentieth century, to contend that collecting was also an essential practice for the experimental sciences in the twentieth century and that they, too, relied on collections for the production of knowledge. As I have shown in several prior papers, numerous collections of data about the structure and function of molecules began to be assembled in the 1960s.[100] Almost half a century later, these collections not only still exist, but they have become indispensable tools for most laboratory researchers. Before discussing the historiographic benefits of considering these collections in the same light as the naturalists' collections, a brief overview of their development is in order.

The development of data collections in the twentieth century resulted from an increasing rate in the production of data, the perception of an "information overload," the intellectual opportunities offered by systematic comparisons of data, and the power offered by computers and eventually computer networks to conduct these on a large scale. The accumulation of data not only made the creation of collections possible, it made them increasingly useful. At the same time, these collections often facilitated experiments that produced even more data.[101]

The *Atlas of Protein Sequences and Structure*, for example, a collection of data about protein sequences, was first published in 1965 by the physical chemist Margaret O. Dayhoff.[102] She justified the creation of her collection with this statement: "There is a tremendous amount of information regarding evolutionary history and biochemical function implicit in each sequence and the number of known sequences is growing explosively." She felt that it was "important to collect this significant information, correlate it into a unified whole and interpret it."[103] Indeed, starting in the 1960s, the pace of protein sequencing was becoming "fast and furious."[104] In 1968, an editorial in *Science* made the point that the determination of protein sequences

---

[99] Nyhart, "Natural History and the 'New' Biology," in Jardine, Secord, and Spary, *Cultures of Natural History* (cit. n. 7), 426–43, on 422; Benson, "From Museum Research to Laboratory Research: The Transformation of Natural History into Academic Biology," in Rainger, Benson, and Maienschein, *American Development of Biology* (cit. n. 67), 49–83, on 77.

[100] See esp. Bruno J. Strasser, "Collecting, Comparing, and Computing Sequences: The Making of Margaret O. Dayhoff's Atlas of Protein Sequences and Structure, 1954–1965," *J. Hist. Biol.* 43 (2010): 623–60, and Strasser, "The Experimenter's Museum: GenBank, Natural History, and the Moral Economies of Biomedicine," *Isis* 102 (2011): 60–96.

[101] For an example of the same dynamic in early modern collections, see Isabelle Charmantier and Staffan Müller-Wille, "Natural History and Information Overload: The Case of Linnaeus," *Stud. Hist. Phil. Biol. Biomed. Sci.* 43 (2012): 4–15.

[102] Strasser, "Collecting, Comparing, and Computing" (cit. n. 100).

[103] Dayhoff to Carl Berkley, 27 February 1967, National Biomedical Research Foundation Archives, Georgetown University, Washington, D.C. (hereafter, NBRF Archives).

[104] "Proteins: Yet More Sequences," *Nature* 224 (1969): 313.

was "one of the most important activities today."[105] The "explosion" in sequence data that Dayhoff and others observed resulted from several factors, including the development in 1967 of Pehr Edman's Sequenator, a rapid and efficient automatic protein sequencer. The availability of this machine emboldened researchers to take on the challenge of larger and more complex proteins. The rising interest in molecular evolution also led a number of researchers to sequence proteins from ever more diverse species. The *Atlas* itself further facilitated these sequencing efforts by offering researchers a number of homologous sequences with which they could compare their partial experimental results, thus contributing to the growth of sequence data that it was supposed to tame.

Similarly, the creation in 1973 of the Protein Data Bank (PDB), a collection of data about the three-dimensional structure of proteins, followed the announcement at the Cold Spring Harbor Symposia on Quantitative Biology two years earlier that several new protein structures had been solved thanks to improved crystallographic methods.[106] Expecting a rapid growth in the number of protein structures solved experimentally and counting on the possibilities for comparison offered by a collection, the crystallographers Helen M. Berman, Edgar F. Meyer, and Walter C. Hamilton established the PDB at Brookhaven National Laboratory and began to distribute the data describing the structure of proteins. Like the *Atlas*, the PDB greatly facilitated the determination of new protein structures, thus adding to the growth of crystallographic data.

A decade later, the European Molecular Biology Laboratory and the NIH sponsored the creation of DNA sequence collections.[107] Again, the creation of these collections was prompted by breakthroughs in the methods to produce data. In this case, two new methods to sequence DNA were developed in 1977 that led to an exponential increase in the amount of sequence data and plans to organize them in a collection. Yet, as two molecular researchers put it shortly afterward, "the rate limiting step in the process of nucleic acid sequencing is now shifting from data acquisition towards the organization and analysis of that data."[108] When the European Molecular Biology Laboratory's DNA sequence library and the NIH's GenBank became publicly available in 1982, they too fueled the explosion of data.

These collections and the many others that were created in the same period grew rapidly in size and popularity. Only seven years after its first edition, the *Atlas* had grown tenfold, and five years later, it was among the fifty most cited scientific items of all time.[109] Similarly, GenBank, today's largest collection of biological information, has grown to contain over one hundred billion As, Ts, Gs, and Cs, which amounts to as many letters as are found in two thousand copies of Buffon's thirty-six-volume *Histoire naturelle*. In 2011, more than twenty thousand computers connected directly to GenBank every day, indicating an even larger number of actual users.[110]

[105] Philip H. Abelson, "Amino Acid Sequence in Proteins," *Science* 160 (1968): 951.

[106] Helen M. Berman, "The Protein Data Bank: A Historical Perspective," *Acta Crystallographica A* 64 (2008): 88–95.

[107] Strasser, "Experimenter's Museum" (cit. n. 100).

[108] Thomas R. Gingeras and Richard J. Roberts, "Steps toward Computer Analysis of Nucleotide Sequences," *Science* 209 (1980): 1322–8.

[109] Margaret O. Dayhoff to Donald DeVincenzi, 10 June 1980, NBRF Archives.

[110] Dennis Benson (National Center for Biotechnology Information), personal communication to the author, 3 October 2011.

The key point is that these collections are part and parcel of experimental research carried out within laboratories. One researcher reported to Dayhoff that the *Atlas* "is the most heavily used book in our lab," while another confessed, "We use your book like a bible!" (a strange comparison indeed).[111]

If we want to take seriously the resemblances between these more recent databases and earlier collections, we need to ask some of the same questions about the collecting practices that led to their establishment and that support their continued growth as we asked earlier about older practices. How was collecting performed? Where did the items in the collections come from? Who were the collectors? Why did they collect? What were the epistemic and social rewards of their collecting enterprises? And finally, how were the collections used for the production of knowledge?

TAKING SPECIMENS FROM THE FIELD TO THE LABORATORY

The history of the experimental life sciences has been told from the vantage point of the few organisms—*Drosophila*, corn, and mice—that have served as model organisms.[112] Yet experimentalists produced data about a much broader range of species, including wild ones, such as the badger, bison, fox, green monkey, guinea pig, llama, mink, red deer, and reindeer. How did these organisms of the field become laboratory objects? The animal materials used in modern protein research came from sources both close to home and far away; they were obtained from laboratory researchers occasionally venturing into the field and from professional naturalists, animal dealers, and zookeepers. In their variation and extent, these modes of collecting are suggestive in many ways of early natural history networks.

In most cases, researchers investigating proteins obtained their material from local slaughterhouses where they could purchase large amounts of tissues for a low price, most often organs that were not sold for human consumption. Proteins were then carefully extracted and purified in the laboratory. As a result of this particular economy, many studies were conducted on cows, pigs, horses, and chickens. Biochemists purified cytochrome *c* proteins, for example, from beef, using a few "freshly minced" heart muscles.[113] The meatpacking industry also provided material for scientists. For example, the Chemical Research and Development Department of Armour and Company (best known in the postwar United States for its hot dogs with "open fire flavor") purified ribonuclease, lysozyme, and other proteins from bovine pancreases and put them up for sale to researchers.[114] The whaling industry was another source; it provided sperm whale meat for Perutz's studies of hemoglobin in Cambridge and supplied other studies of insulin carried out in Japan, where the whaling industry was well developed.[115]

Human samples came from equally diverse and extended networks. Pathological hemoglobins in humans, for example, were provided by clinics in regions where the

---

[111] Allen B. Edmundson to Robert S. Ledley, 25 November 1969, and Oliver Smithies to Winona Barker, 5 October 1970, NBRF Archives.

[112] See, e.g., Jim Endersby, *A Guinea Pig's History of Biology* (Cambridge, 2007).

[113] David Keilin, "Preparation of Pure Cytochrome *c* from Heart Muscle and Some of Its Properties," *Proceedings of the Royal Society of London B—Biological Sciences* 122 (1937): 298–308.

[114] "Science Exhibition," *Science* 106 (1947): 567–75.

[115] Georgina Ferry, *Max Perutz and the Secret of Life* (New York, 2007).

prevalence of certain diseases was high. For his investigations of sickle-cell anemia hemoglobin, most prevalent in the United States among African Americans, Linus Pauling secured a blood supply from a clinician in New Orleans to use in his laboratory at Caltech.[116] In Cambridge, England, Vernon Ingram relied on sickle-cell anemia blood brought by Anthony C. Allison from Kenya.[117] Later, Ingram explored the molecular differences in hemoglobin from patients with many different pathological conditions. In this case, the blood samples were taken from the blood collection that the clinician Hermann Lehmann had established in Cambridge from his trips in several African countries.[118]

The supply of biological material from wild animals posed a greater challenge to laboratory workers. As for earlier naturalist collections, local environments played a defining role. Indeed, protein sequences from deer were determined in a laboratory in Stockholm, those from camels in Udaipur, and those from rattlesnakes in Los Angeles.[119] Unlike museum naturalists, most laboratory biochemists had no prior experience of field collecting, and they gathered material from the immediate surroundings of their laboratories. The chemists Margareta and Birger Blombäck at the Karolinska Institute in Stockholm, for example, were leading researchers on the molecular basis of blood coagulation in the 1960s, but their only use for the outdoors had been recreational. For their studies on the mechanisms of coagulation, however, they needed and secured blood samples from a wide range of organisms, beginning with domestic animals, and moving later to wild ones. Their interest in samples from the latter turned them into field-workers. Together with a visitor from the United States, the biochemist Russell F. Doolittle, they flew to northern Sweden for the annual reindeer hunt, where "a Laplander and his lasso" captured a few specimens from which blood was drawn.[120] They also traveled Sweden's northern islands to hunt seals whose blood was then investigated in Stockholm.[121] The problem of storing biological samples at a freezing temperature, which had stymied so many field-collecting expeditions for blood in Africa and Central America, was easily solved in Lapland, with "nature providing excellent refrigeration."[122]

Like early field collectors of natural history, the Blombäcks were interested in expanding their collection whenever the opportunity arose to do so. In 1963, they had moved temporarily to Australia and seized this chance to gather blood from different species of kangaroos and sharks that were readily accessible in this new environment. That same year, they extended their interests in fibrinopeptide variation to human populations, again because a new diversity of types was available in their new surroundings. Margareta Blombäck wrote enthusiastically that they had gathered "blood

[116] Lily E. Kay, *The Molecular Vision of Life: Caltech, the Rockefeller Foundation and the Rise of the New Biology* (New York, 1993).

[117] Ingram, "Sickle-Cell Anemia Hemoglobin: The Molecular Biology of the First 'Molecular Disease'—the Crucial Importance of Serendipity," *Genetics* 167 (2004): 1–7.

[118] De Chadarevian, "Following Molecules" (cit. n. 98).

[119] Margaret O. Dayhoff, *Atlas of Protein Sequence and Structure* (Silver Spring, Md., 1972).

[120] John F. Henahan, "Dr. Doolittle—Making Big Changes in Small Steps," *Chemical and Engineering News*, February 9, 1970, 22–32, on 23.

[121] Margareta Blombäck, personal communication to the author, 19 May 2010; Blombäck, "Thrombosis and Haemostasis Research: Stimulating, Hard Work and Fun," *Thrombosis and Haemostasis* 98 (2007): 8–15.

[122] Henahan, "Dr. Doolittle" (cit. n. 120), 23.

from different [human] races, as pure as they possibly can be, such as Maoris (New Zealand), New Guinea natives, East Africans and Australian Negros" and that they had started "a new field of biochemical anthropology."[123]

In addition to field collecting, the Blombäcks, like many other biochemists and naturalists, relied on gifts from individual colleagues around the world who had access to local species. The method had its limitations, mainly because the regions hosting the most exotic species also had the least number of laboratories. When he was unable to obtain blood from a rare species of monkey living on just a few Southeast Asian islands for his hemoglobin studies of primates, the anthropologist John Buettner-Janusch, at Yale University, complained, "We have not yet been able to beg, borrow, or steal a sample of Tarsius hemoglobin."[124] Most researchers adopted the same strategy as Doolittle, who worked in San Diego and relied extensively on the exceptionally rich animal collection present in its public zoo. Marine stations, such as the Marine Biological Laboratory in Woods Hole, Massachusetts, and the marine station of the Collège de France in Concarneau, Brittany, were used as sources for aquatic animals.[125]

Examined from the perspective of where and how specimens were obtained, laboratory research on the molecular basis of protein function begins to resemble the collecting endeavors usually associated with natural history. The same logic of place prevailed: collectors first assembled local species, and then more distant ones, in a quest to have the broadest number of species represented. The same logic of assembling and using a collection prevailed too: these researchers brought the different protein sequences into a common format to make them comparable, performed systematic comparisons, and drew general conclusions about the structure, function, and history of these proteins. This story of collecting and comparing begins to differ from the story that attributes the successes of the molecular life sciences to experimental virtuosity and single model organisms.

THE COLLECTORS' STANDING IN SCIENCE

The vast majority of collectors of experimental data were not naturalists, although a few naturalists collected experimental data in the twentieth century, such as Alan Boyden, in his Serological Museum at Rutgers University, or Charles Sibley, in his collection of bird DNA at Yale University. Most were trained experimental scientists, and many had doctoral degrees, in fields such as physical chemistry or crystallography. Their experiences, however, in some ways paralleled those of naturalist collectors of an earlier period. Perhaps most obviously, data collectors, such as Dayhoff (of the *Atlas*), Berman (of the PDB), and Olga Kennard (of the Cambridge Crystallographic Data Centre), were often women and they relied on extensive female staffs, not unlike many earlier collecting enterprises. Certainly, several men devoted their (late) career to collecting,[126] such as Walter Goad with GenBank, but women

[123] M. Blombäck to her parents, 19 September 1963, Margareta Blombäck personal archives.

[124] Buettner-Janusch and R. L. Hill, "Molecules and Monkeys," *Science* 147 (1965): 836–42.

[125] Doolittle, "Characterization of Lamprey Fibrinopeptides," *Biochemical Journal* 94 (1965): 742–50; R. Acher, J. Chauvet, and M. T. Chauvet, "Phylogeny of the Neurohypophysial Hormones," *Nature* 216 (1967): 1037–8.

[126] In another field, see Michael D. Gordin, "Beilstein Unbound: The Pedagogical Unraveling of a Man and His *Handbuch*," in *Pedagogy and the Practice of Science: Historical and Contemporary Perspectives*, ed. David Kaiser (Cambridge, Mass., 2005), 11–39.

made up an unusually high proportion of collectors, especially in comparison to their marginality in the fields from which the data were being collected. In addition, their professional identity within the experimentalist community was very unstable. Although they were trained in the experimental sciences and worked with experimental data, their collecting work was generally considered not to be of a scientific nature. Experimentalists writing to Dayhoff, for example, addressed her as a "compiler," an "editor," or a "librarian," none of which was a very enviable status to aspire to for a scientist.[127]

As I have discussed elsewhere, viewing Dayhoff's work as part of the collecting tradition, resting on different epistemic, social, and cultural norms than the experimental sciences, helps us understand the difficulties in the development of her professional career.[128] She was denied membership in the American Society of Experimental Biologists because, according to one of its members, the "compilation of the Atlas of Protein Sequence and Structure" could not be considered her "own research."[129] In other words, the problem with Dayhoff's collection-based work was not only that it was not experimental, but also that it was collective, and thus did not fit into the highly individual reward ethos of the experimental sciences. Similarly, the NIH remained reluctant until the 1980s to fund data collections, because they did not fit a grant system geared toward individual experimental research. In 1981, after the NIH had turned down one of her grant requests for the *Atlas*, Dayhoff lamented, as she had at other times, "Databases do not inspire excitement."[130]

Most experimentalists considered the work of data collectors mundane, clerical, or even trivial. They overlooked the data collectors' wide range of expertise. First, the experimental data gathered by the collectors were often plagued with errors, many due to simple transcription mistakes by the authors and publishers of the data. It took a precise understanding of the nature of proteins and of the methods (biochemical or crystallographic) that had been used to produce the data to spot the possible errors and resolve them with the authors. Since the data collections were intended to be not mere repositories, but tools for the production of knowledge, the collectors organized the data in ways that would be most productive epistemically. This task required collectors to understand how the data could be used in research. For example, the protein sequences contained in the *Atlas* were aligned in order to highlight their similarities and differences. This was no trivial task. Since sequences were not identical, "gaps" were inserted in them to optimize the extent of the alignment between a given two, making an implicit assumption about their evolution. Collectors also used the taxonomies they created to organize their data collections. After 1974, for example, the *Atlas* was structured around "superfamilies," a concept introduced by Dayhoff, after an extensive study of all sequences present in the *Atlas*.[131]

Just as the long history of collecting pointed to the importance of mobilizing large networks of collectors in the making of collections, so too does the history of recent collecting call attention to the importance of numerous "field" collectors and to the

[127] E.g., Richard Synge to "Compilers," 7 April 1966, NBRF Archives.

[128] Bruno J. Strasser, "Collecting and Experimenting: The Moral Economies of Biological Research, 1960s–1980s," *Preprints of the Max Planck Institute for the History of Science* 310 (2006): 105–23; Strasser, "Collecting, Comparing, and Computing" and "Experimenter's Museum " (Both cit. n. 100).

[129] John T. Edsall to Dayhoff, 4 November 1969, NBRF Archives.

[130] Dayhoff to D. M. Moore, 24 September 1981, NBRF Archives.

[131] Dayhoff, "Computer Analysis of Protein Sequences," *Federation Proceedings* 33 (1974): 2314–6.

moral economies on which these coordinated collecting enterprises were based. In 1965, the first edition of the *Atlas* included contributions from just over 150 researchers; in 2011 more than 20,000 scientists submitted DNA sequences to GenBank. How were these individuals brought to participate in the collection of data?

Dayhoff encountered great difficulties in obtaining sequence data from researchers. In an earlier paper, I have shown how the failure of Dayhoff's efforts at securing the participation of individual experimentalists was a result of divergent moral economies.[132] Experimentalists had a strong sense of ownership in the data they produced and were unwilling to share them openly for others to use. The fact that Dayhoff copyrighted the data she received and used them for her own research was deemed unacceptable to many experimentalists who had spent months or even years producing these data. Many wanted to retain a symbolic form of ownership over them, in addition to being able to exploit them further.

But Dayhoff's difficulties in obtaining sequence data from researchers paled in comparison to the resistance encountered by collectors of crystallographic data.[133] Those who set up the PDB to collect all known protein structures were often unable, despite repeated calls and pleas, to secure crystallographic data from individual researchers. There was a common agreement that data supporting published interpretations should be publicly available. But the very nature of what constituted "data" was hotly debated. Researchers were most reluctant to share "raw" data as opposed to "processed" data, or "results," arguing that raw data belonged to the inner workings of a laboratory. Others argued that the failure of many crystallographers to make their data public, either in print or electronically through the PDB, undermined the very notion of a "publication." As one crystallographer put it, "Results without data are unproven, and interpretations without results are hearsay." After noting that in three-quarters of the cases raw data were unavailable in publications of certain molecular structures, he concluded that they were "not really published at all, in the literal sense of making the information public." In macromolecular crystallography, he noted, "a custom of non-publication" had been "allowed to grow from an idiosyncrasy, to an inconvenience, to an outright scandal."[134]

By the end of the 1980s, after years of intense negotiations, several crystallographers succeeded in convincing scientific journals to enforce a mandatory submission policy.[135] Only those papers for which the data had been deposited in the PDB would be cleared for publication. At the same time, the managers of GenBank, the collection of nucleic acid sequences, arrived at similar arrangements with journal editors, effectively solving the problem of data collection. What appeared from the 1990s to be a spontaneous communal effort to produce and share data was, in fact, the result of a hard-fought battle that succeeded in balancing the risks and benefits of sharing scientific data.

The struggles encountered by collection managers in securing data from individual researchers invite comparisons with naturalists' collections, such as those in museums of natural history. Naturalists relied and continue to rely on large numbers of

---

[132] Strasser, "Collecting and Experimenting" (cit. n. 128); see also Strasser, "Experimenter's Museum" (cit. n. 100).

[133] Marcia Barinaga, "The Missing Crystallography Data," *Science* 245 (1989): 1179–81.

[134] Richard E. Dickerson to president of the American Crystallographic Association, July 1987, Protein Data Bank Archives, Rutgers University, New Brunswick, N.J.

[135] J. L. Sussman, "Protein Data Bank Deposits," *Science* 282 (1998): 1993.

amateurs (including many women in botany) to gather data and specimens. These field collectors were often content to give their findings to a local natural history museum, and they felt honored to be mentioned in a scientific publication. Their exclusion from authorship, either in the naming of species or in the publication of taxonomic descriptions, for example, was made easier by the difference in status, and often the gender divide, between professionals and amateurs. These convenient arrangements, however, were not available to modern data collectors: the data they gathered had been produced by scores of professional experimenters who intended to be fully credited for any interpretive work that was based on their data. Furthermore, experimenters believed that the first interpretation of the data belonged to them. For example, in 2002, a genomic researcher from the Marine Biological Laboratory in Woods Hole complained that he had been "scooped with his own data," in an episode that a *Nature* writer called the "latest in a string of clashes between those who collect and those who interpret data."[136]

There was increased acknowledgment that data producers and data analyzers filled different professional niches (as did field collectors and museum taxonomists), but what was the proper place of the data curators and those who assembled data collections? Curators were legitimate figures in natural history (Mayr and Simpson were curators), but had no comparable position in the experimental sciences. If the data in their collections were to be made public and they were denied privileged access to these data (through which they could make scientific contributions), their role would be reduced to that of infrastructure managers, not scientists. In the nineteenth century, Augustin Pyramus de Candolle considered the ownership of an herbarium to be a prerequisite to being a botanist;[137] in twentieth-century molecular sciences, managing a data collection almost prevented one from being a scientist. Unsurprisingly, data collectors (the majority, PhD-carrying scientists) have been dissatisfied by this lack of professional recognition. Some have been able to derive their professional legitimacy from publishing original methods of data analysis; for example, David Lipman, head of GenBank since 1989, has contributed to the development of BLAST (the Basic Local Alignment Search Tool), the most widely used algorithm to compare sequences.[138] However, the professionalization of the role of database curator and manager, aligned with the development of similar professional roles in the digital information and library sciences,[139] has produced an ambiguous legitimacy for researchers in the natural sciences who work with databases of experimental knowledge. As a recent paper put it, database curators "dread the immortal cocktail party question 'So, what do you do?'"[140]

[136] The Woods Hole group had determined the sequence of a bacterium and made the data available online, only to see another group publish an evolutionary interpretation of these data before they were able to propose one themselves. E. Marshall, "Data Sharing—DNA Sequencer Protests Being Scooped with His Own Data," *Science* 295 (2002): 1206–7, on 1206.

[137] Peter F. Stevens, *The Development of Biological Systematics: Antoine-Laurent de Jussieu, Nature, and the Natural System* (New York, 1994).

[138] S. F. Altschul et al., "Basic Local Alignment Search Tool," *Journal of Molecular Biology* 215 (1990): 403–10.

[139] E.g., *Database: The Journal of Biological Databases and Curation* was launched in 2009, two years after the *International Journal of Digital Curation*.

[140] Kyle Burkhardt, Bohdan Schneider, and Jeramia Ory, "A Biocurator Perspective: Annotation at the Research Collaboratory for Structural Bioinformatics Protein Data Bank," *PLoS Computational Biology* 2 (2006): 1186–9, on 1186.

HOW ARE DATABASES USED?

Since their inception in the 1960s, electronic databases, like earlier collections, have been used for generating knowledge on a variety of topics, but always through comparison. Comparison has been the key epistemic practice in producing knowledge about the relationship between form and function, the history of organisms and their parts, and the systematic relationships between organisms. Like Vicq d'Azyr, Cuvier, and other comparative anatomists a century earlier, biochemists have assembled collections of structures that have then served as the basis for their comparative studies.[141] The American biochemist Christian B. Anfinsen, in his 1959 book, *The Molecular Basis of Evolution*, did much to popularize the comparative approach among protein researchers, as well as the idea that similarities in sequence would indicate "the minimum structure which is essential for biological function."[142] These and other biochemists have relied on the diversity of nature—as accessed through their collections—to gain insights into the relationship between the structure and the function of proteins.

The reconstruction of the history of life has long relied on the collection of existing and extinct specimens. Unsurprisingly, phylogenetic research became one of the key uses of molecular databases, such as the *Atlas* and GenBank, following the development of methods in molecular evolution. Dayhoff, for example, pioneered methods to infer phylogenetic distances from numbers of differences between protein sequences.[143]

Databases have also been widely used to elaborate taxonomies of their elements, whether protein structures or DNA sequences. As mentioned previously, Dayhoff proposed the concept of "protein superfamilies," a clear analogy to the taxonomy of species, to designate groups of proteins that shared a similar structure and that had evolved from a unique protein. She derived this concept from the comparison of the data present in her collection and then used it to reorganize the collection according to these categories, much in the same way that natural history collections were (and are) structured by their taxonomic rank. Similarly, the PDB has been used to classify proteins according to their three-dimensional shape. The comparison of shapes, unlike that of sequences, does not lend itself so easily to numerical approaches. Thus, those who have attempted to develop taxonomies of protein structures have resorted to strategies very similar to those used by naturalists in comparing specimens.

Among the many researchers who have adopted the comparative approach in classifying protein structures, the case of Jane S. Richardson is particularly illuminating, as an example not only of this approach but also of the recognition among some scientists of the alignment of their practices with those of natural history collecting and comparing. Without a graduate degree in science (she had a master's in philosophy and had taken some courses in plant taxonomy and evolution at Harvard), she joined a chemistry laboratory at MIT as a technician.[144] She became interested in pro-

---

[141] On d'Azyr, see Stéphane Schmitt, "From Physiology to Classification: Comparative Anatomy and Vicq D'Azyr's Plan of Reform for Life Sciences and Medicine (1774–1794)," *Sci. Context* 22 (2009): 145–93.

[142] Anfinsen, *The Molecular Basis of Evolution* (New York, 1959), 143.

[143] Joseph Felsenstein, *Inferring Phylogenies* (Sunderland, Mass., 2004), chap. 10.

[144] S. Bahar, "Ribbon Diagrams and Protein Taxonomy: A Profile of Jane S. Richardson," *Biological Physicist* 4, no. 3 (2004): 5–8.

tein structures and elucidated several of them, before focusing on their classification. In the mid-1970s, she started to systematically survey all known protein structures, visually identifying different patterns. She used these patterns, which she compared to geometric motifs common on Greek and American Indian weaving and pottery, as a basis for her classification, which made the cover of *Nature* in 1977.[145] Her work culminated a few years later in an almost two-hundred-page review of the "anatomy and taxonomy of protein structure," which made extensive use of the data contained in the PDB.[146] She grouped all known proteins into classes according to their structures and provided simplified representations of each that would make their common features more apparent. For the same reason, she conceived a new representation of a structural pattern (the beta-sheet) that soon became a standard in protein science.

Richardson explicitly acknowledged how much her comparative approach derived from natural history:

> The vast accumulation of information about protein structures provides a fresh opportunity to do descriptive natural history, as though we had been presented with the tropical jungles of a totally new planet. It is in the spirit of this new natural history that we will attempt to investigate the anatomy and taxonomy of protein structures.[147]

Richardson confessed her "love of complex primary data and what is essentially a new kind of natural history."[148] The objects that Richardson classified might have been the product of experimental virtuosity, but the ways in which she approached them were clearly in line with the natural history tradition. Furthermore, her approach to taxonomy, like that of traditional naturalists, relied not only on the visual inspection of structure, but also on an intimate, personal, and even intuitive grasp of similarities. She later explained that she believed in the importance of

> exhaustively *looking*, in detail, at each beautifully quirky and illuminating piece of data with a receptive mind and eye, as opposed to the more masculine strategy of framing an initial hypothesis, writing a computer program to scan the reams of data, and obtaining an objective and quantitative answer to that one question while missing the more significant answers which are suggested only by entirely unexpected patterns in those endless details.[149]

In this quote, Richardson draws a gender division between "hypothesis-driven" science (done computationally) and a more intuitive and visual approach, reflecting the traditional gendering of experimentation as a male activity and natural history as female.[150]

Richardson and other protein taxonomists, in their various comparative approaches, experienced the same kinds of epistemic tensions as those who classified organisms in more typical natural history activities. From the 1930s to the present, proponents of different forms of "experimental taxonomy" have clashed with proponents of

---

[145] Richardson, "Beta-Sheet Topology and the Relatedness of Proteins," *Nature* 268 (1977): 495–500.

[146] Richardson, "The Anatomy and Taxonomy of Protein Structure," *Advances in Protein Chemistry* 34 (1981): 167–339.

[147] Ibid., 170.

[148] Cited in Bahar, "Ribbon Diagrams" (cit. n. 144), 5.

[149] Ibid., 6; emphasis in the original.

[150] Evelyn Fox Keller, *Reflections on Gender and Science* (New Haven, Conn., 1985).

morphological taxonomies over the issue of the objectivity of classifications.[151] Among the latter, Mayr, a leading systematist, valued subjectivity most, writing that the "good doctor and the good taxonomist make their diagnoses by a skillful evaluation of symptoms in the one case and of taxonomic characters in the other."[152] Simpson, a paleontologist, likewise argued that the identification of species depended "on the personal judgment of each practitioner of the art of classification." To this, he added that classification could not be objective: "To insist on an absolute objective criterion would be to deny the facts of life, especially the inescapable fact of evolution."[153] The experimental taxonomists, such as the molecular evolutionists, disagreed strongly with these assessments and insisted that classification and phylogeny should be objective and quantitative. They argued that reliance on molecular data, not morphology, was necessary to reach these goals.[154]

Similarly, in protein science, a number of researchers remained somewhat skeptical about the validity of visual methods to classify proteins. They developed alternative methods that they claimed would "analyze automatically and objectively" the coordinates of proteins to identify protein domains. These researchers criticized those who relied on the visual inspection of three-dimensional models stored in the PDB.[155] Automated approaches, they argued, also conducted with data from the PDB, were far superior because they were objective. Similar concerns with the objectivity of visual comparison were widespread.

These classifications of proteins, whether produced visually or automatically, borrowed (sometimes consciously, sometimes not) from standard natural historical practices. By the beginning of the twenty-first century, some protein scientists were ready to acknowledge the similarity between their work and that of naturalists. In a 2002 review titled "The Natural History of Protein Domains," protein researchers drew these parallels explicitly:

> For over a century, zoologists have classified organisms using the Linnaean system in order to provide insights into their natural history. Biologists are beginning to appreciate the benefits of hierarchical domain classification systems based on sequence, structure, and evolution. The numerous parallels between these systems suggest that domain classifications will prove to be key to our further understanding of the natural history of domain families.[156]

This is not to say that current practices in protein classification have returned biological research to its natural historical origins. Rather, collecting and comparing practices have been essential to both natural historical and experimental research. As seen in protein taxonomies, bringing modern databases into the larger framework of collections highlights problems shared equally by naturalist collectors and the database users, such as the place of subjectivity in comparing biological shapes, the role

[151] Strasser, "Laboratories, Museums" (cit. n. 95).
[152] Mayr, Linsley, and Usinger, *Methods and Principles* (cit. n. 75), 106–7.
[153] George Gaylord Simpson, *Principles of Animal Taxonomy* (New York, 1961).
[154] On this debate, see Strasser, "Laboratories, Museums" (cit. n. 95).
[155] M. Levitt and J. Greer, "Automatic Identification of Secondary Structure in Globular Proteins," *Journal of Molecular Biology* 114 (1977): 181–239.
[156] C. P. Ponting and R. R. Russell, "The Natural History of Protein Domains," *Annual Review of Biophysics and Biomolecular Structure* 31 (2002): 45–71.

of visual and numerical approaches, and the function of taxonomies in organizing collections.

In his recent *Styles of Knowing: A New History of Science from Ancient Times to the Present*, Kwa claims that "a comparative method is a means of building a taxonomy, nothing more, nothing less."[157] Actually, it is a bit less and a lot more. The production of taxonomies has perhaps been the most visible use of the comparative method as applied to collections, with examples ranging from Linnaeus's herbarium to Mayr's vertebrates at the American Museum of Natural History. But it has not been the only one. The comparative method has also served to identify specimens, relying on the fact that collections are embodiments of knowledge systems. Most obviously, type specimens (holotypes) stored in natural history museums serve as the ultimate referent for a species.[158] Naturalists compare specimens of unknown status to a type specimen in order to assess whether they belong to the same species or to another one. Modern databases have served an identical role. The most precious help that computerized databases of DNA sequences can provide to experimentalists and that a dispersed set of printed sequences cannot is in identifying the function, and thus the identity, of genes. DNA sequences are often determined before the function of a gene is known. In the case of a gene coding for a protein, instead of attempting to find experimentally every possible biochemical reaction in which that protein might be involved, researchers compare their new DNA sequences with all other sequences available in a database, using algorithms such as BLAST. And if they find a "match," that is, a sequence that is sufficiently similar, and whose function is known, they can infer that the two sequences produce proteins of similar structure and function.[159] This function can then be explored further experimentally. Databases offer a unique shortcut for experimental investigations: they suggest possible roles for proteins that scientists had never even thought of.[160] Although journals first accepted the evidence of sequence comparisons as sufficient to warrant a publication, they soon required that the results be confirmed experimentally. Yet sequence comparison remains a crucial step in the process of producing knowledge experimentally.

CONCLUSIONS

By the beginning of the twenty-first century, public collections of molecular data numbered in the thousands; so many, in fact, that databases of databases were established to help researchers keep track of these resources.[161] One significant question remains, at least from the perspective of a generalist vision: Although these twenty-first-century databases certainly resemble earlier natural history collections, do they bear any historical connections to them?

---

[157] Kwa, *Styles of Knowing* (cit. n. 24), 167.

[158] On the epistemology of type specimens, see Lorraine Daston, "Type Specimens and Scientific Memory," *Crit. Inq.* 31 (2004): 153–82.

[159] Michel Morange, *A History of Molecular Biology* (Cambridge, 2000), chap. 17.

[160] On the beginnings of this approach, see Russell F. Doolittle, "Some Reflections on the Early Days of Sequence Searching," *Journal of Molecular Medicine* 75 (1997): 239–41.

[161] M. Y. Galperin and G. R. Cochrane, "The 2011 *Nucleic Acids Research* Database Issue and the Online Molecular Biology Database Collection," *Nucleic Acids Research* 39, supplement (2011): D1–D6.

To take a metaphor from evolutionary theory, the databases that grew in the twentieth century are certainly *analogous* to earlier collections, because they share a similar
structure and have a similar function.[162] Databases, like earlier natural history collections, are organized assemblages of standardized objects. The physical (or virtual)
proximity of these objects, their mobility within the collection, the temporary order
in which they are arranged, and the uniform format in which they are kept make it
possible for the investigator to approach these objects comparatively. This comparative perspective is perhaps the most distinctive epistemic practice associated with all
kinds of collections. It has been most important for natural history, especially studies
of systematics and evolution. It has been much less relevant in most of the experimental approaches to life, which have relied on a different perspective centered on
exemplary phenomena, usually produced in model organisms (more on this later).
One interesting exception is comparative embryology in the nineteenth and early
twentieth centuries, which was at the same time experimental and comparative (obviously). Comparative embryologists experimented on developing embryos, but (unlike physiologists, e.g.) they also relied extensively on collections of objects such as
wax embryos, preserved tissues, and microscopic slides.[163] Here again, the collecting
and comparative approaches were closely associated, in the same way as they have
been associated around the experimental data and data collections examined in this
article. The most striking similarity between modern databases and earlier collections in natural history is not simply that they are all collections of some sort, but that
they have been constituted through similar collecting practices and have been put to
use in similar ways for the production of knowledge.

One might object that databases and biological collections cannot be subsumed
under the same heading because the former deal with information and the latter with
material things. However, as noted above, it would be erroneous to equate natural
history collections solely with the cataloging of whole or parts of organisms. Take
Candolle's herbarium: it contained whole plants, although they were dried between
sheets of paper, but for large plants, such as trees, it contained only parts of the plants,
usually flowers and some leaves. It also contained, in the same format of large sheets,
drawings of plants, or even verbal descriptions. This particular collection thus contained an entire ontological range of collected objects, from material things to abstract ideas. The difference between things and data is very real, but it is more a matter of degree than a matter of kind.[164]

Obviously, the contemporary biomedical sciences have not simply returned to
the old methods of natural history. Current electronic databases differ in many ways
from former natural history collections. But they have reincorporated collecting and
comparing approaches into the experimental tradition. What is most distinctive about
current biomedical research is its hybrid character that produces knowledge through
both experimentation and collection.[165] Establishing this successful hybrid culture
has not been simple, and incorporating collecting practices has had deep consequences for the entire research enterprise. It has forced a reexamination of notions of

---

[162] I thank Robert Kohler for suggesting this useful analogy.

[163] Nick Hopwood and Friedrich Ziegler, *Embryos in Wax: Models from the Ziegler Studio* (Cambridge, 2002).

[164] See Latour's wonderful essay on the ontological range of scientific objects in Latour, *Pandora's
Hope—Essays on the Reality of Science Studies* (Cambridge, 1999), chap. 2.

[165] Strasser, "Laboratories, Museums" (cit. n. 95).

authorship, which can no longer be understood solely in individual terms as it was in the experimental sciences, and has brought about serious changes in attitudes toward data ownership and data sharing, challenging the very meaning of "publication."

If one accepts that these various collections are indeed analogous, there remains a question of whether modern databases are *homologous* to earlier natural history collections. In addition to sharing a common structure and function, do they share a common descent? Is there some historical connection, whether social, intellectual, or cultural, between collections and databases? This question is far more difficult to answer, although the most probable answer is *no*. Most data collectors of the twentieth century were not naturalists trained in the arts of collecting specimens in the field, preparing them for herbaria or museum collections, and using a broad, systematic comparative approach to produce knowledge; they were trained as experimentalists (or theoreticians of some sort). When they began collecting and comparing, they became alienated from their experimentalist colleagues, many of whom no longer considered them scientists. And naturalists did not have much more sympathy for these new collectors, many of whom had never been in the field.[166] The data collectors of the twentieth century went through the difficult process of creating a new professional identity for themselves. Only in the 1980s did science funding agencies fully recognize the importance of data collectors for the progress of the experimental sciences. At the same time, the term *bioinformatician* was gaining wider acceptance (the journal *Bioinformatics* was created in 1985); this term seems to point solely to the use of computers in biology, but in fact designates a professional group committed to producing knowledge through the collection and comparison of data.

If we accept, for heuristic purposes, that modern and ancient collections are at least analogous, though not homologous, how does this contribute to a generalist vision of the life sciences from the early modern period to the present? First, the rise of experimentalism did not mean an irreversible movement away from all other practices of biological investigation. Other ways of knowing (e.g., those based on collecting and comparing) that were centrally important in the early modern period continue to be central for naturalist research, but have also, as this article has argued, grown in importance in experimental research as of the beginning of the twenty-first century. The twentieth century can still be safely qualified as the "experimental century," but the twenty-first might well be a "collecting century." If one were to take an even broader view, the twentieth century might no longer be considered the culmination of methodological progress leading to experimentalism, but rather a brief, albeit significant, interlude in the history of the life sciences.

To be sure, this article does not have the pretension to replace the standard narrative crafted by Coleman and Allen with another narrative centered on collecting practices. Nor does it claim to have identified an actor's concept that historians have ignored. It claims that *collecting*, as an analytical category of practices, can be productively deployed in writing a history of the life sciences, not only in the field and the museum, where these practices have been most closely examined, but in the laboratory, where the focus has almost exclusively been on experimentation. It also suggests that this approach might be successful in connecting recent developments in the biomedical sciences, such as the proliferation of databases and data-driven

---

[166] See, e.g., field naturalists' attitudes toward the experimental taxonomist Alan Boyden, described ibid.

methods, diachronically with the earlier history of the life sciences and synchronic-
ally with sciences such as systematics or ecology, which have developed mainly outside the laboratory. Thus, instead of seeing these recent developments in data-driven
science as yet another revolution (or worse, a "fourth paradigm") and isolating them
from other changes taking place in other disciplines, one may succeed in producing
a narrative that brings them into historical perspective and offers critical insights into
these transformations.[167]

Seeing the persistence of collecting practices over time leads to another crucial
question. Why did collecting approaches develop so prominently in the life sciences?
Was it a historical accident, or does something about the objects studied by life scientists lend itself to comparative perspectives? One key reason why comparative perspectives and, thus, collections have been so important in the study of life is that
natural selection operates on functions but is blind to structures. Evolution has resulted in a variety of structures performing similar functions, making it particularly
difficult for researchers, whether the biological components they consider are molecules or morphological traits, to infer functions from single structures. Physicists
and chemists do not have that problem: all the entities of one kind that they explore
are believed to be structurally identical. Collecting and comparing, then, is intimately
linked to the historicity of the objects investigated. This is borne out in considering the nature of other historical disciplines: geology and cosmology, like the life
sciences, have both been heavily dependent on collecting practices.

Bringing collecting to the fore thus leads to new questions about the boundaries between the sciences. Rather than following historically contingent disciplinary boundaries, it might be more productive to think about the deep commonalities
between the sciences that are historical (biology, geology, cosmology) and those that
are not (physics, chemistry, mathematics). It can also help us question the divisions
between the natural and the social sciences. What would happen if we began to think
about the aims of biology and history on one side, and those of physics and sociology
on the other? Writing history from the vantage point of specific practices, such as collecting, makes such connections visible, in a way that disciplinary histories do not.

Whether scholars follow up on this latter suggestion or not, I hope that this article
has at least made clear that it can be productive to think about current databases as
collections that follow a long tradition of collecting in the life sciences. Other readings are possible, of course, and worth exploring. One might, for example, bring databases into the context of the encyclopedist movement and the development of library
sciences. But this article has offered a first attempt to ask some of the same kind of
questions about modern databases that historians of natural history have asked about
wonder cabinets, herbaria, and museums. This historical contextualization draws attention to the variety of collecting strategies, to the challenges of managing a heterogeneous network of collectors, and to the epistemic challenges of working comparatively, especially in an experimentalist age. The analytic focus on collecting practices
(in the field *and* the laboratory) and the contrasts between experimental and relational
systems, between exemplary and comparative perspectives, and between centrifugal
and centripetal places make it possible to overcome the distinction between natural
history and experimentation, the museum and the laboratory, and hypothesis-driven

[167] Bruno J. Strasser, "Data-Driven Sciences: From Wonder Cabinets to Electronic Databases,"
Stud. Hist. Phil. Biol. Biomed. Sci. 43 (2012): 85–7.

and data-driven science. It also helps us bring current claims about the uniqueness of contemporary science into perspective.

Indeed, broadening the perspective, one might question the role of the laboratory in defining modern science. In concluding *Reinventing Knowledge* (2008), a pointed overview of the six major institutions of knowledge developed during the last two and a half millennia, Ian F. McNeely and Lisa Wolverton claim that "by the mid-twentieth century, the laboratory had ascended to an almost impossibly dominant status as an institution of knowledge" and that "laboratory science and its accomplishments now act as the chief means by which Western knowledge systems manifest their superiority to the rest of the world" (Herbert Butterfield would have approved of the style and content).[168] The laboratory remains obviously indispensable to and powerful for the production of knowledge about the natural world, but the proliferation of data collections and comparative approaches seems to indicate that it no longer enjoys this dominant position alone. As McNeely and Wolverton show, the fortunes of different institutions of knowledge have changed over time. Museums, for example, which once "performed indispensable functions in legitimating knowledge," now thrive in different roles, such as "education, entertainment, and outreach."[169] But the epistemic qualities that made museums so vital to the production of knowledge—their role as stable referents of the natural world and the possibility of applying comparative approaches to their collections—are now also present elsewhere, in digital databases that might be thought of as "data museums."

Reframing the history of recent science in this perspective also illuminates the recent politics of knowledge. The increasing use of databases for the production of knowledge has only made the question of access more acute, a question that has been addressed extensively by naturalists in the case of natural history collections—for example, by defining rules about the borrowing of museum specimens.[170] Because databases were the product of broad collective efforts, many argued that they should be freely accessible and open to all. This position also facilitated the collection of data and maximized the potential use of databases. What represented a pragmatic decision for the managers of databases, such as GenBank, also lent support to broader initiatives to make knowledge more accessible. The success of GenBank's open-access policy was used as an argument to promote PubMed Central, an open repository of published scientific literature, and eventually the NIH's open-access policy (all publications resulting from NIH-funded research must be deposited on PubMed Central within a year).[171] The greater availability of scientific knowledge in a format that lends itself to the further production of knowledge has made possible a broader participation in science, including anyone from secondary school teachers and their students in the classroom to computer-game amateurs competing to solve protein-folding problems.[172] The availability of data collections and their increasing legitimacy for the production of knowledge has fueled the growth of "citizen science" and

[168] McNeely and Wolverton, *Reinventing Knowledge* (cit. n. 28), 251, 271; Butterfield, *The Origins of Modern Science, 1300–1800* (London, 1949).

[169] McNeely and Wolverton, *Reinventing Knowledge* (cit. n. 28), 256.

[170] On the norms about the borrowing of specimens, see Mayr, Linsley, and Usinger, *Methods and Principles* (cit. n. 75).

[171] R. J. Roberts et al., "Building a 'Genbank' of the Published Literature," *Science* 291 (2001): 2318–9.

[172] S. Cooper et al., "Predicting Protein Structures with a Multiplayer Online Game," *Nature* 466 (2010): 756–60.

made its expansion more plausible than ever before, especially for the experimental sciences.

As I have suggested in this article, looking at collections beyond their alleged decline in the late nineteenth century offers promising venues to contextualize some of the deep transformations currently taking place in science. For one, data-driven science now seems more familiar and less a product of our "information age." As Robert Darnton has reminded us, "Every age was an age of information, each in its own way,"[173] and the age when natural history was most flourishing was no exception. The extensive use of collections by naturalists and the existing scholarship on natural history collecting provides the historian of recent science with rich material to ask fresh questions about the use of current databases in science. The insights of earlier naturalists about the epistemic, social, and cultural challenges of working with collections help us understand some of the current difficulties faced by the participants in data-driven science. Indeed, when Simpson referred to "a science that is most explicitly and exclusively devoted to the ordering of complex data," he was not referring to current data-driven science but to animal taxonomy.[174] Same questions, different times.

[173] Darnton, "An Early Information Society: News and the Media in Eighteenth-Century Paris," *Amer. Hist. Rev.* 105 (2000): 1–35.
[174] Simpson, *Principles of Animal Taxonomy* (cit. n. 153), 10.

Notes on Contributors

Thomas Broman teaches History of Science and History of Medicine at the University of Wisconsin–Madison. He is currently writing a survey of science in the Enlightenment.

Harold J. Cook is John F. Nickoll Professor of History at Brown University, having previously served as Director of the Wellcome Trust Centre for the History of Medicine at University College London. He has written extensively on medicine and science in early modern Europe, in recent years with a particular interest in seeing changes through the perspective of the Low Countries and their connections to the rest of the world. His *Matters of Exchange: Commerce, Medicine, and Science in the Dutch Golden Age* (New Haven, Conn., 2007) was awarded the Pfizer Prize by the HSS in 2009.

Lorraine Daston is Director at the Max Planck Institute for the History of Science, Berlin, and Visiting Professor in the Committee on Social Thought at the University of Chicago. Her books include *Classical Probability and the Enlightenment* (Princeton, N.J., 1988), *Wonders and the Order of Nature, 1150–1750* (with Katharine Park; Cambridge, Mass., 1998), *Things That Talk: Object Lessons from Art and Science* (Cambridge, Mass., 2004), and *Objectivity* (with Peter Galison; Cambridge, Mass., 2007). Her current research concerns the history of rules, algorithms, and the mechanization of rationality.

Peter Dear is Professor of History and of Science and Technology Studies at Cornell University. He is the author of *Revolutionizing the Sciences: European Knowledge and Its Ambitions, 1500–1700* (Princeton, N.J., 2008 [2nd ed.]) and is currently preparing a study titled *The Roots of Modern Reason*.

Fa-ti Fan is the author of *British Naturalists in Qing China: Science, Empire, and Cultural Encounter* (Cambridge, Mass., 2004) and many articles on science and empire, science in modern East Asia, and other related topics. He is Associate Professor of History at the State University of New York at Binghamton. He is currently writing two books, one on science and mass politics in communist China and the other on the formation of modern science in Republican China.

Paul Forman ended his fourth decade as Curator of the Modern Physics Collection at the Smithsonian Institution and is now retired. Over this last decade his work has been increasingly devoted to tracing out the main lines of postmodernity—that is, of our present historical epoch, in which science must adapt to the absence of the cultural supports that made it central to modernity.

Jan Golinski is Professor of History and Humanities at the University of New Hampshire, where he currently serves as Chair of the Department of History. He is the author of *Science as Public Culture: Chemistry and Enlightenment in Britain, 1760–1820* (Cambridge, 1992), *Making Natural Knowledge: Constructivism and the History of Science* (Chicago, 2005), and *British Weather and the Climate of Enlightenment* (Chicago, 2007).

Edward Grant is Distinguished Professor Emeritus of History and Philosophy of Science at Indiana University, Bloomington. His special area of research is medieval natural philosophy, with emphasis on cosmology and the interrelations of science and religion. He is the author and editor of twelve books, including *A Source Book in Medieval Science* (Cambridge, Mass., 1974), *Much Ado about Nothing: Theories of Space and Vacuum from the Middle Ages to the Scientific Revolution* (Cambridge, 1981), *Planets, Stars and Orbs: The Medieval Cosmos, 1200–1687* (Cambridge, 1994), *God and Reason in the Middle Ages* (Cambridge, 2001), and *A History of Natural Philosophy from the Ancient World to the Nineteenth Century* (Cambridge, 2007).

David Kaiser is Professor of the History of Science in the Program in Science, Technology, and Society at the Massachusetts Institute of Technology and Senior Lecturer in MIT's Department of Physics. He is author of the award-winning book *Drawing Theories Apart: The Dispersion of Feynman Diagrams in Postwar Physics* (Chicago, 2005) and of *How the Hippies Saved Physics: Science, Counterculture, and the Quantum Revival* (New York, 2011). He has also edited several books, including *Pedagogy and the Practice of Science: Historical and Contemporary Perspectives* (Cambridge, Mass., 2005) and *Becoming MIT: Moments of Decision* (Cambridge, Mass., 2010). His current work concerns American physics during the Cold War.

Robert E. Kohler is Emeritus Professor of the History and Sociology of Science at the University of Pennsylvania. He writes on the institutions and practices of modern science, especially the field sciences. His books include *Lords of the Fly: Drosophila Genetics and the Experimental Life*

(Chicago, 1994), *Landscapes and Labscapes: Exploring the Lab-Field Border in Biology* (Chicago, 2002), and *All Creatures: Naturalists, Collectors and Biodiversity, 1850–1950* (Princeton, N.J., 2006).

Lynn K. Nyhart is Professor of the History of Science at the University of Wisconsin–Madison. She is the author of *Biology Takes Form: Animal Morphology and the German Universities, 1800–1900* (Chicago, 1995) and *Modern Nature: The Rise of the Biological Perspective in Germany* (Chicago, 2009), which won the press's 2009 Susan E. Abrams Award. She is currently working on a collaborative project with Scott Lidgard of the Field Museum, on the history of concepts of biological individuality.

Kathryn M. Olesko is Associate Professor of History at Georgetown University. She has served as Editor of twelve volumes of *Osiris*, of which the current volume is the last. She has written extensively on science in Germany, the history of scientific and technical education, and the history of precision measuring practices. The author of *Why Prussians Measured* (forthcoming) and *Physics as a Calling* (Ithaca, N.Y., 1991), she is now working on water management in the North German Plain and a popular book on measuring nature from antiquity to the present.

Theodore M. Porter is Professor of History at UCLA. His most recent books are *The Cambridge History of Science*, vol. 7: *Modern Social Sciences* (coedited with Dorothy Ross; Cambridge, 2003) and *Karl Pearson: The Scientific Life in a Statistical Age* (Princeton, N.J., 2004). In an earlier era he wrote *The Rise of Statistical Thinking* (Princeton, N.J., 1986) and *Trust in Numbers* (Princeton, N.J., 1995). Just now he is exploring the uses of statistical recording practices and fieldwork in Europe and North America from about 1820 to 1920 to investigate heredity at insane asylums and schools for the "feebleminded." On the side, he has some recent papers on issues of science and public reason, such as "How Science Became Technical," *Isis* 100 (2009): 292–309.

Bruno J. Strasser is Assistant Professor in the Section of the History of Medicine and the Department of History at Yale University. He is the author of a book on the history of molecular biology in postwar Europe, *La fabrique d'une nouvelle science: La biologie moléculaire à l'âge atomique, 1945–1964* (Florence, 2006), and has published on the history of international scientific cooperation during the Cold War, the interactions between experimental science and clinical medicine, the transformations of the pharmaceutical industry, the relations between museums and laboratories, and the roles of collective memory. He is currently finishing a book on the history of biomedical collections and databases, *Collecting Experiments: The New Production of Biomedical Knowledge* (Chicago, forthcoming).

Index

SUGGESTIONS FOR CONTRIBUTORS TO OSIRIS

OSIRIS is devoted to thematic issues, conceived and compiled by guest editors who submit volume proposals for review by the OSIRIS Editorial Board in advance of the annual meeting of the History of Science Society in November. For information on proposal submission, please write to the Editor at osiris@etal.uri.edu.

1. Manuscripts should be submitted electronically in Rich Text Format using Times New Roman font, 12 point, and double-spaced throughout, including quotations and notes. Notes should be in the form of footnotes, also in 12 point and double-spaced. The manuscript style should follow *The Chicago Manual of Style*, 16th ed.

2. Bibliographic information should be given in the footnotes (not parenthetically in the text), numbered using Arabic numerals. The footnote number should appear as superscript. "Pp." and "p." are not used for page references.

    a. References to books should include the author's full name; complete title of book in *italics*; place of publication; date of publication, including the original date when a reprint is being cited; and, if required, number of the particular page cited (if a direct quote is used, the word "on" should precede the page number). *Example*:

      [1] Mary Lindemann, *Medicine and Society in Early Modern Europe* (Cambridge, 1999), 119.

    b. References to articles in periodicals or edited volumes should include the author's name; title of article in quotes; title of periodical or volume in *italics*; volume number in Arabic numerals; year in parentheses; page numbers of article; and, if required, number of the particular page cited. Journal titles are spelled out in full on the first citation and abbreviated subsequently according to the journal abbreviations listed in *Isis Current Bibliography*. *Example*:

      [2] Lynn K. Nyhart, "Civic and Economic Zoology in Nineteenth-Century Germany: The 'Living Communities' of Karl Möbius," *Isis* 89 (1999): 605–30, on 611.

    c. All citations are given in full in the first reference. For succeeding citations, use an abbreviated version of the title with the author's last name. *Example*:

      [3] Nyhart, "Civic and Economic Zoology" (cit. n. 2), 612.

3. Special characters and mathematical and scientific symbols should be entered electronically.

4. A small number of illustrations, including graphs and tables, may be used in each volume. Hard copies should accompany electronic images. Images must meet the specifications of The University of Chicago Press "Artwork General Guidelines" available from the Editor.

5. Manuscripts are submitted to OSIRIS with the understanding that upon publication copyright will be transferred to the History of Science Society. That understanding precludes consideration of material that has been previously published or submitted or accepted for publication elsewhere, in whole or in part. OSIRIS is a journal of first publication.

OSIRIS (ISSN 0369-7827) is published once a year.

Single copies are $33.00.

Address subscriptions, single issue orders, claims for missing issues, and advertising inquiries to *Osiris*, The University of Chicago Press, Journals Division, PO Box 37005, Chicago, IL 60637.

Postmaster: Send address changes to *Osiris*, The University of Chicago Press, Journals Division, PO Box 37005, Chicago, IL 60637.

OSIRIS is indexed in major scientific and historical indexing services, including *Biological Abstracts*, *Current Contexts*, *Historical Abstracts*, and *America: History and Life*.

Paperback edition, ISBN 978-0-226-02939-9

 A RESEARCH JOURNAL DEVOTED
TO THE HISTORY OF SCIENCE
AND ITS CULTURAL INFLUENCES

A PUBLICATION OF THE
HISTORY OF SCIENCE SOCIETY

EDITORIAL OFFICE
DEPARTMENT OF HISTORY
80 UPPER COLLEGE ROAD, SUITE 3
UNIVERSITY OF RHODE ISLAND
KINGSTON, RI 02881 USA
osiris@etal.uri.edu